MANUFACTURING ENGINEERING PROCESSES

MANUFACTURING ENGINEERING AND MATERIALS PROCESSING

A Series of Reference Books and Textbooks

FOUNDING EDITORS

Geoffrey Boothroyd
University of Rhode Island
Kingston, Rhode Island

George E. Dieter
University of Maryland
College Park, Maryland

SERIES EDITOR

John P. Tanner
John P. Tanner & Associates
Orlando, Florida

ADVISORY EDITORS

Gary Benedict
Allied-Signal

E. A. Elsayed
Rutgers University

Fred W. Kear
Motorola

Michel Roboam
Aerospatiale

Jack Walker
McDonnell Douglas

1. Computers in Manufacturing, *U. Rembold, M. Seth and J. S. Weinstein*
2. Cold Rolling of Steel, *William L. Roberts*
3. Strengthening of Ceramics: Treatments, Tests, and Design Applications, *Harry P. Kirchner*
4. Metal Forming: The Application of Limit Analysis, *Betzalel Avitzur*
5. Improving Productivity by Classification, Coding, and Data Base Standardization: The Key to Maximizing CAD/CAM and Group Technology, *William F. Hyde*
6. Automatic Assembly, *Geoffrey Boothroyd, Corrado Poli, and Laurence E. Murch*
7. Manufacturing Engineering Processes, *Leo Alting*
8. Modern Ceramic Engineering: Properties, Processing, and Use in Design, *David W. Richerson*
9. Interface Technology for Computer-Controlled Manufacturing Processes, *Ulrich Rembold, Karl Armbruster, and Wolfgang Ülzmann*
10. Hot Rolling of Steel, *William L. Roberts*
11. Adhesives in Manufacturing, *edited by Gerald L. Schneberger*
12. Understanding the Manufacturing Process: Key to Successful CAD/CAM Implementation, *Joseph Harrington, Jr.*

13. Industrial Materials Science and Engineering, *edited by Lawrence E. Murr*
14. Lubricants and Lubrication in Metalworking Operations, *Elliot S. Nachtman and Serope Kalpakjian*
15. Manufacturing Engineering: An Introduction to the Basic Functions, *John P. Tanner*
16. Computer-Integrated Manufacturing Technology and Systems, *Ulrich Rembold, Christian Blume, and Ruediger Dillman*
17. Connections in Electronic Assemblies, *Anthony J. Bilotta*
18. Automation for Press Feed Operations: Applications and Economics, *Edward Walker*
19. Nontraditional Manufacturing Processes, *Gary F. Benedict*
20. Programmable Controllers for Factory Automation, *David G. Johnson*
21. Printed Circuit Assembly Manufacturing, *Fred W. Kear*
22. Manufacturing High Technology Handbook, *edited by Donatas Tijunelis and Keith E. McKee*
23. Factory Information Systems: Design and Implementation for CIM Management and Control, *John Gaylord*
24. Flat Processing of Steel, *William L Roberts*
25. Soldering for Electronic Assemblies, *Leo P. Lambert*
26. Flexible Manufacturing Systems in Practice: Applications, Design, and Simulation, *Joseph Talavage and Roger G. Hannam*
27. Flexible Manufacturing Systems: Benefits for the Low Inventory Factory, *John E. Lenz*
28. Fundamentals of Machining and Machine Tools: Second Edition, *Geoffrey Boothroyd and Winston A. Knight*
29. Computer-Automated Process Planning for World-Class Manufacturing, *James Nolen*
30. Steel-Rolling Technology: Theory and Practice, *Vladimir B. Ginzburg*
31. Computer Integrated Electronics Manufacturing and Testing, *Jack Arabian*
32. In-Process Measurement and Control, *Stephan D. Murphy*
33. Assembly Line Design: Methodology and Applications, *We-Min Chow*
34. Robot Technology and Applications, *edited by Ulrich Rembold*
35. Mechanical Deburring and Surface Finishing Technology, *Alfred F. Scheider*
36. Manufacturing Engineering: An Introduction to the Basic Functions, Second Edition, Revised and Expanded, *John P. Tanner*
37. Assembly Automation and Product Design, *Geoffrey Boothroyd*
38. Hybrid Assemblies and Multichip Modules, *Fred W. Kear*
39. High-Quality Steel Rolling: Theory and Practice, *Vladimir B. Ginzburg*
40. Manufacturing Engineering Processes: Second Edition, Revised and Expanded, *Leo Alting*

Additional Volumes in Preparation

MANUFACTURING ENGINEERING PROCESSES

SECOND EDITION, REVISED AND EXPANDED

LEO ALTING
The Technical University of Denmark
Lyngby, Denmark

English Version Edited by
Geoffrey Boothroyd
University of Massachusetts
Amherst, Massachusetts

MARCEL DEKKER, INC. NEW YORK • BASEL

Library of Congress Cataloging-in-Publication Data

Alting, Leo
[Grundlaeggende mekanisk teknologi. English]
Manufacturing engineering processes / Leo Alting; English version edited by Geoffrey Boothroyd. -- 2nd ed., rev. and expanded.
p. cm. -- (Manufacturing engineering and materials processing; 40)
Translation of: Grundlaeggende mekanisk teknologi.
Includes bibliographical references and index.
ISBN 0-8247-9129-0 (alk. paper)
1. Manufacturing processes. I. Boothroyd, G. (Geoffrey). II. Title. III. Series.
TS183.A4713 1994
670.42--dc20 93-33384
CIP

This second edition contains figures and tables from *Grundlæggende Mekanisk Teknologi* by Leo Alting published by Akademisk Forlag © 1974 and new figures and tables prepared for Chapters 3, 8, 10, 11, and 12, unless otherwise stated.

The publisher offers discounts on this book when ordered in bulk quantities. For more information, write to Special Sales/Professional Marketing at the address below.

This book is printed on acid-free paper.

Copyright © 1994 by Marcel Dekker, Inc. All Rights Reserved.

Neither this book nor any part may be reproduced or transmitted in any form or by any means, electronic or mechanical, including photocopying, microfilming, and recording, or by any information storage and retrieval system, without permission in writing from the publisher.

Marcel Dekker, Inc.
270 Madison Avenue, New York, New York 10016

Current printing (last digit):
10 9 8 7

Foreword

In this book the subject of manufacturing is discussed within the framework of a fundamental classification of processes. This should help the reader understand where a particular process fits within the overall manufacturing scheme and what processes might be suitable for the manufacture of a particular component. The treatment of the subject matter is adequately descriptive for those unfamiliar with the various processes and yet is sufficiently analytical for an introductory academic course in manufacturing. One particularly attractive feature of the book is the presentation of summaries of the various manufacturing processes in data sheet form.

There are many textbooks that attempt to deal with manufacturing processes at the introductory level: some are formed from a collection of individual chapters having no common theme or underlying structure; most are purely descriptive and of little interest to those wishing to introduce analyses of processes into their teaching; one or two are too analytical. Some textbooks concentrate only on the mechanics of processes or on the mechanical types of processes such as machining, metal forming, and so on, while neglecting the metallurgical or chemical types of processes such as welding, casting, and powder metallurgy. None of these criticisms can be leveled at this book.

The enhancements included in this second edition bring the textbook right up-to-date. The new chapters on nontraditional processes, manufacturing systems (including the Japanese philosophy), and the life cycle approach to manufacturing are valuable additions. Professor Alting is a well-known world authority on the life cycle approach, emphasizing the recycling of products, a subject which is rapidly becoming a top priority for manufacturing organizations throughout the world.

Geoffrey Boothroyd

Preface to the Second Edition

This edition includes enhancements and extensions of several of the chapters in the First Edition as well as three new chapters on topics of great importance to the manufacturing industry today.

The major revisions are as follows: Chapter 3, on engineering materials, has been rewritten and sections on ceramics and composites have been added. In Chapter 8, on joining, the section on welding arc formation and maintenance has been completely rewritten and expanded, and data sheets on the most important welding processes are given. In Chapter 10, on casting, data sheets on the most important casting processes have been added. In Chapter 11, on plastics, data sheets on the most important plastic processes have been added.

Chapters 12 to 14 present new material. Chapter 12, on nontraditional manufacturing processes, discusses such important processes as electrical discharge machining, electron beam machining, laser processing, abrasive jet machining, ultrasonic machining, electrochemical machining, and layer manufacturing (rapid prototyping).

Chapter 13, on manufacturing systems, includes fundamentals of manufacturing systems, advanced equipment, flexible manufacturing systems, CIM-computer integrated manufacturing, efficient manufacturing, production planning, scheduling and control, and the Japanese production philosophy.

Chapter 14, on the life cycle approach in manufacturing, provides a basic understanding of the life cycle approach in manufacturing with emphasis on environmental, occupational health, and resource consequences. This life cycle perspective is a necessity in developing a sustainable manufacturing industry.

This new edition thus provides an expanded and more comprehensive treatment of the manufacturing processes and places the processes in a broader context in relation to manufacturing systems and the life cycle perspective. The book is now a more complete text for academia as well as for practicing designers and manufacturing and industrial engineers.

The development of the new edition has mainly been carried out by Associate Professor J. R. Dissing, of the Technical University of Denmark, who uses the text for about 200 students every year. His experiences and needs have been important in the selection of the new material. Dr. K. Siggaard has contributed the chapter on manufacturing systems.

I want to express my gratitude to Professor Dissing for his valuable and significant contributions and Dr. Siggaard for providing his knowledge on manufacturing systems.

Leo Alting

Preface to the First Edition

Manufacturing engineering is an important discipline in any industrialized society. For many years manufacturing has not been granted the stature and significance in engineering curricula that is necessary in order to fulfill the demands of the industry and thus of society.

This situation is partly due to the fact that engineering methods and scientific approaches have not been fully introduced in the manufacturing field. This field has mainly been considered the purview of technicians and skilled craftsmen. At engineering colleges and universities, manufacturing has generally been taught in the traditional descriptive manner, which is not very challenging for either the student or the practicing engineer.

Rapid technical developments in the last decade, for example, in computer technology and its applications in design (computer-aided design, CAD) and manufacturing (computer-aided manufacturing, CAM), have stressed the need for a more systematic engineering approach in manufacturing oriented toward practical problem-solving.

This books represents the first fundamental step in the development of a more systematic approach in manufacturing engineering. The mission is accomplished through the following main features:

The book gives a systematic and coherent picture of the manufacturing field.

The book allows a quick survey to be made of the possibilities and limitations of the processing methods available for the production of specific components.

The book creates a basis for systematic process development, systematic tool and die design, and systematic design of production machinery and production systems.

Finally, the presentation is based on a scientific and systematic approach that stimulates the imagination and utilizes a general engineering background.

To achieve the preceding goals it has been necessary to consider manufacturing engineering from a new point of view. Traditionally, specific processes are treated individually, each requiring a special description. When the different processes are analyzed, it appears that they can all be described by a common process model built up from a few fundamental elements. A combination of these elements gives a process morphology for all known (and unknown) processes. This model is described in Chapter 1. In Chapters 5 to 11 specific process areas are described and structured according to the model presented in Chapter 1. It should be mentioned that Chapter 10, on casting, and Chapter 11, on plastics technology, are not fully developed according to the new model but the reader is encouraged to do this as a valuable exercise.

In order to give the reader the necessary background to understand the processes, Chapter 2 introduces material properties, Chapter 3, engineering materials, and Chapter 4, the fundamentals of metalworking (plasticity theory).

As a reminder that the application of manufacturing processes is not determined solely from technical and economic viewpoints, Chapter 12 introduces the subject of industrial safety.

A version of this material has been used as a textbook at The Technical University of Denmark for 7 years and the results are very encouraging. The duration of the course is one semester of about 42 class hours with 80 hours of homework. Along with the course, problems are presented, some of which are discussed in special problem classes. The results of the course improve drastically when workshop training is given in parallel.

It is my hope that many engineering colleges and universities will be able to use the book as their textbook for a fundamental introductory course in manufacturing engineering.

For valuable editorial comments and suggested improvements Dr. G. Boothroyd is thanked. Professor Dell K. Allen, Brigham Young University, has also read the manuscript and stimulated several improvements for which I am very grateful.

Leo Alting

Contents

1

A Morphological Process Model

1.1 INTRODUCTION

In industrial production, many different processes or manufacturing methods are used. To be able to select the technically and economically best manufacturing sequence for a given product, it is necessary to have a broad, fundamental knowledge of the possibilities and limitations of the various manufacturing processes, including the work materials used and the geometries, surface finishes, and tolerances required.

In this first chapter the individual processes are not considered in detail, but a coherent picture of the common structure on which all processes are based is introduced. By defining and considering the elements in this structure, a systematic understanding of materials processing is obtained, which is based on a general engineering background and allows an evaluation of the possibilities and limitations of the different processes. This approach has a broad, general application since it reflects invariant relations, methods, or principles, but in the context of this book it will be related only to those processes characteristic of the manufacturing industry.

1.2 BASIC STRUCTURE OF MANUFACTURING PROCESSES

The term *process* can in general be defined as a change in the properties of an object, including geometry, hardness, state, information content (form data), and so on. To produce any change in property, three essential agents must be available: (1) material, (2) energy, and (3) information. Depending on the main purpose of the process, it is either a material process, an energy process, or an information process. In the following sections, only material processes will be considered, especially those producing geometrical changes or changes in material properties, or both (1). This does not, however, imply a limitation of the general principles.

1.2.1 General Process Model

The general process model can be illustrated as shown in Fig. 1.1. The model shows that a material process can be described by the associated flow system: material flow, energy flow, and information flow.

Material flow can be divided into three main types, as shown in Fig. 1.2:

1. Through flow, corresponding to *mass-conserving processes*
2. Diverging flow, corresponding to *mass-reducing processes*
3. Converging flow, corresponding to *assembly or joining processes*

Mass-conserving processes ($dM = 0$) can be characterized as follows:

- The mass of the initial work materials is equal to (or nearly equal to) the mass of the final work material, which means, when referring to geometrical changes, that the material is manipulated to change its shape.

Mass-reducing processes ($dM < 0$) can be characterized as follows:

- The geometry of the final component can be circumscribed by the initial material geometry, which means that a shape change is brought about by the removal of material.

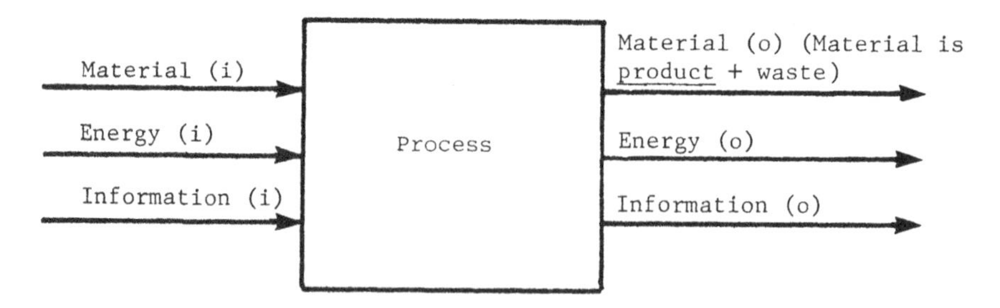

FIGURE 1.1 The general process model. Here *i* designates inputs and *o* outputs.

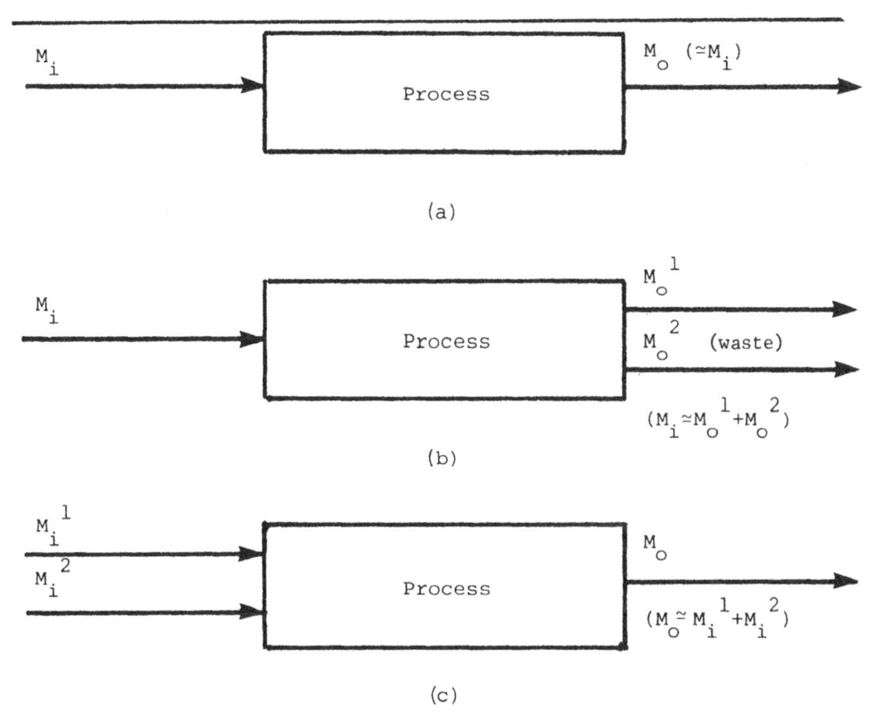

FIGURE 1.2 The three main types of material flow: (a) mass-conserving processes ($dM = 0$); (b) mass-reducing processes ($dM < 0$); (c) assembly or mass-increasing processes ($dM > 0$). Here M means mass of material, i input, and o output. The numbers 1 and 2 refer to the number of material elements.

Assembly or joining processes (sometimes expressed as $dM > 0$) can be characterized as follows:

- The final geometry is obtained by assembling or joining components so that the mass of the final geometry is approximately equal to the sum of the masses of the components which are manufactured by one or both of the previous methods.

These three types of material flow have been related to the work material but, depending on the process, auxiliary flow of material may be necessary, such as lubricants, cooling fluids, and filler material. Most processes aiming at a change in material properties without a change in geometry are mass-conserving processes.

The energy flow associated with the process can be characterized as energy supply, energy transmission to the workpiece, and removal or loss of energy.

Information flow includes what might be termed shape and property information. A certain geometry for a certain material can be characterized as the shape information for the material. In a geometry-changing process, shape-change information is impressed on the material so that the final shape information is equal to the sum of the initial shape information and the shape-change information impressed by the process. The shape-change information is created by an interaction between a tool or die (with a certain contour content) and a pattern of movement for the work material and the tool or die. This means that a geometry-changing process is characterized by a material flow on which, by means of an energy flow, the shape-change information corresponding to the information flow is impressed.

Impressing a change in geometry on a material can be carried out in one or more steps, which means that

$$I_o = I_i + \Delta I_{p1} + \Delta I_{p2} + \cdots + \Delta I_{pn} \tag{1.1}$$

where I_o is the desired geometry, I_i the initial shape information of the material, and I_{pn} the shape-change information for a single process. The number of processes necessary is determined partly for technical and partly for economical reasons.

Similarly, the property information flow, for example, hardness, strength, and so on, involves the sum of the properties of the initial material and the changes in properties produced by the various processes.

The proper interaction between these three fundamental flow systems, yielding the desired component, is governed by the control information, which includes knowledge of the forces, power, friction and lubrication, cutting data, and so on. This control information, which is partly analytical and partly empirical, is discussed later.

Based on the three flow systems described, a complete model of a manufacturing process is shown in Fig. 1.3. In this context, the various kinds of material flow, energy flow, and shape and property information flow associated with manufacturing processes will be considered.

1.2.2 Morphological Structure of the Processes

When analyzing manufacturing processes it appears that they can all be described by a general morphological model built up from a few fundamental elements related to the three flow systems. By combining these elements, a process morphology is obtained from which all manufacturing processes can be deduced.

The fundamental elements in this morphological model are:

Material flow
- State of material
- Basic process
- Type of flow (process type)

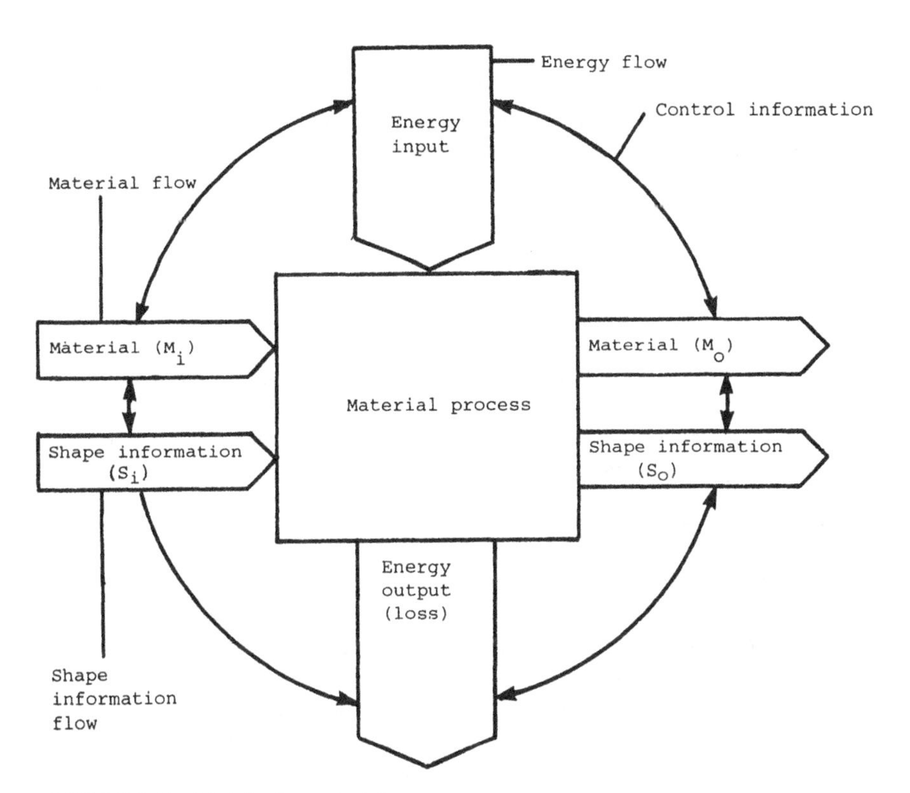

FIGURE 1.3 Model of a material process.

Energy flow
  Tool/die
    Energy supply
    Transfer medium
  Equipment
    Energy characteristics
    Type of energy
Information flow
  Surface creation (principles)
  Pattern of movement for
    Material
    Tool/die

Each of these elements can have different "values," as shown in Fig. 1.4. By choosing a value from each column the fundamental basis for a material process is obtained. Some of the combinations are physically impossible but, in general, the model contains all the basic ingredients necessary to establish a process. This morphological model gives a systematic and coherent picture of the process field, enabling a quick survey of the possibilities and limitations of the various processes; it can also be used to generate new process ideas.

To be able to use the model, a knowledge of the various single elements must be obtained; consequently, the three flow systems and their elements are discussed next.

1.3 MATERIAL FLOW SYSTEM

As shown in Fig. 1.4, the material flow deals with the state of the material for which the geometry and/or the properties are changed, the basic processes that can be used to create the desired change in geometry and/or properties, and the type of flow system characterizing the process.

1.3.1 State of Material

The various states in which the material can be processed are, as shown in Fig. 1.4, solid, fluid, granular, and gaseous. When processing composite materials, different states can appear at the same time. The granular state can be considered as a subdivision of the solid state, since solids can be divided into coherent solid or incoherent solid (granular) materials. Considering the technological differences in the processing sequences, the usual division into solid and granular materials is maintained. The different states of the materials will, as shown later, result in quite different process structures. In addition to the state of the material, its composition is also important. Here a division into homogeneous and heterogeneous materials may be helpful, partly to obtain new ideas for materials and partly to evaluate the forming properties in relation to the basic processes.

Homogeneous materials include homogeneous mixtures and pure materials in the form of chemical compounds and elements.

Heterogeneous materials include mechanical mixtures.

Materials can further be characterized by their thermal, chemical, mechanical, and manufacturing properties, depending on the purpose of the analysis to be conducted. Clearly, in a study of manufacturing processes, a broad knowledge and understanding of materials and their properties is important.

1.3.2 Basic Processes

Basic processes are defined as those processes that create changes in the geometry and/or the properties of the materials. The basic processes are characterized by the nature of their interaction with the material. A manufacturing process

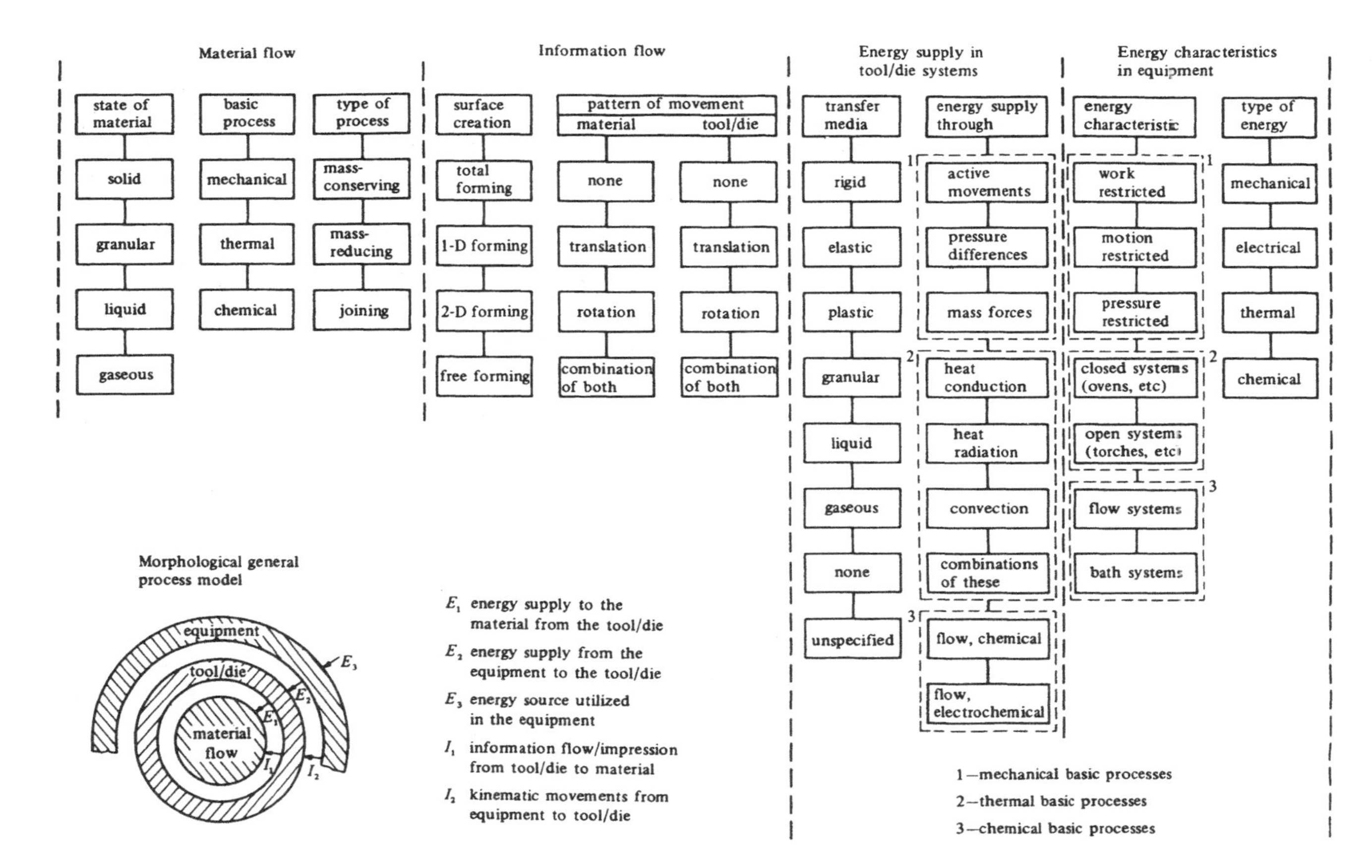

FIGURE 1.4 The morphological structure of material processes.

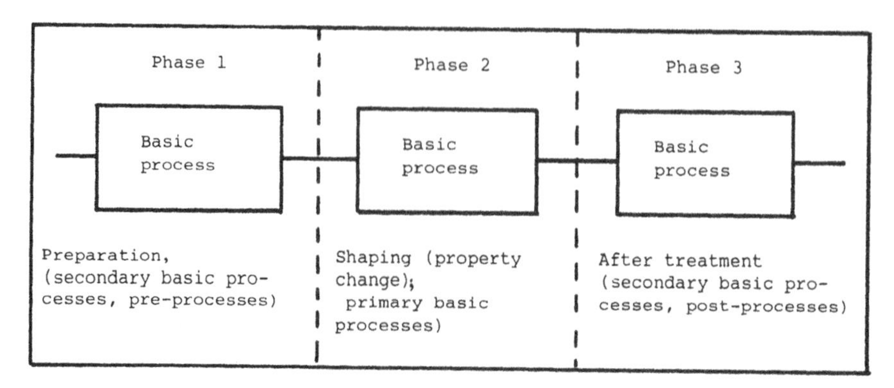

FIGURE 1.5 Division of a manufacturing process into three phases.

normally consists of a series of basic processes, which constitute the structure of the material flow. Any series of basic processes can be divided into three typical phases:

Phase 1, which consists of the basic processes that bring the material into a suitable state—geometry and/or properties (heating, melting, sawing, cropping, etc.)—for the primary change in geometry and/or properties

Phase 2, which consists of the basic processes that create the desired geometry and/or change in properties

Phase 3, which consists of the basic processes that bring the component into the specified end state (solidification, cooling, deburring, etc.)

This division is illustrated in Fig. 1.5, where the basic processes associated with phase 2 are called the primary basic processes (according to the primary goal). The basic processes associated with phases 1 and 3 are called secondary basic processes. The structure shown in Fig. 1.5 is very useful when analyzing and designing manufacturing processes.

The basic processes can be divided into three main categories, as shown in Table 1.1. Each of these categories is characterized by the nature of the interaction with the work material. The various single basic processes are described later.

When the main objective for a process has been established, relevant series of primary and secondary basic processes can be found. Here the actual type of material has a significant influence, since the materials react differently when subjected to mechanical, thermal, or chemical actions. If only processes aiming at geometrical changes are considered, the number of possible primary basic processes (phase 2 in Fig. 1.5) is reduced to those shown in Table 1.2.

TABLE 1.1 The Three Main Categories of Basic Processes

Mechanical	Thermal	Chemical
Elastic deformation	Heating	Solution/dissolution
Plastic deformation	Cooling	Combustion
Brittle fracture	Melting	Hardening
Ductile fracture	Solidification	Precipitation
Flow	Evaporation	Phase transformation
Mixing	Condensation etc.	Diffusion etc.
Separation		
Placing		
Transport etc.		

TABLE 1.2 Primary Basic Processes Used in Material Processes that Change Geometry

Category of basic process	Basic processes
Mechanical	Plastic deformation
	Fracture (brittle and ductile)
	Elastic deformation
	Flow (filling, placing, etc.)
Thermal	Melting
	Evaporation
Chemical	Solution-dissolution (electrolytical and chemical)
	Deposition (electrolytical and chemical)
	Combustion

It is the primary basic process and the way in which it is established that determines the types and number of secondary basic processes required. A close coupling with the information flow exists here.

1.3.3 Flow Type (Type of Processes)

The material flow system can be graphically illustrated in various ways. Figure 1.6 shows one example. To give more detailed examples it would be necessary to distinguish between the three flow types as shown in Fig. 1.2.

The manufacturing processes can be characterized according to the material flow system as shown in Table 1.3. Examples of processes are listed in the last column for metallic materials.

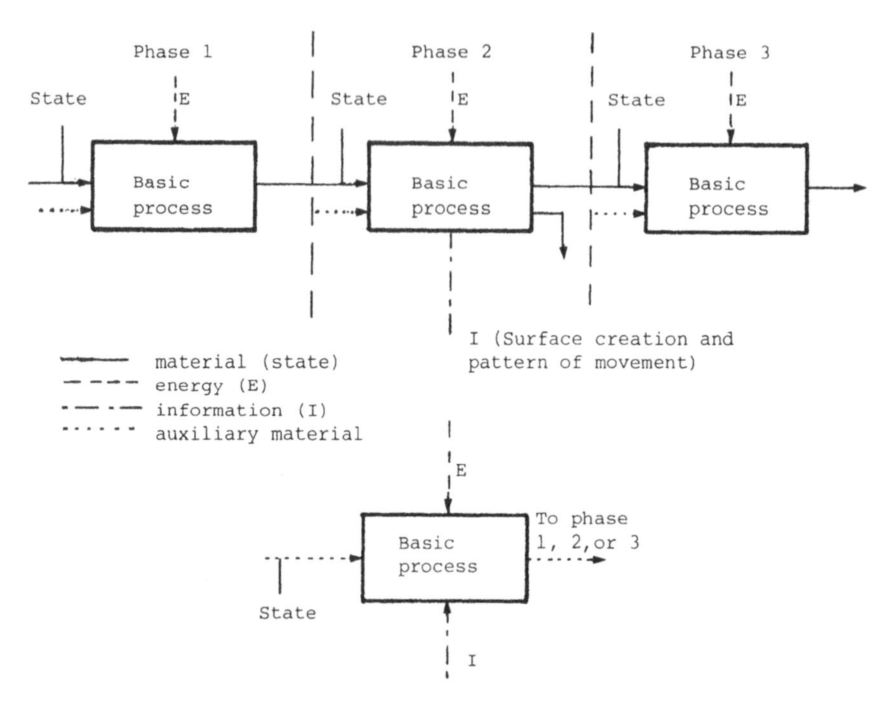

FIGURE 1.6 Schematic illustration of the material flow system. More detailed illustrations can be drawn within the main flow types.

As mentioned in Section 1.2.1, it is possible to distinguish among three types of flow: mass-conserving processes, mass-reducing processes, and assembly or joining processes. Selection of flow or process type depends on the requirements of material, geometry, surface, tolerance, number, price, and other factors.

1.4 EXAMPLES OF MANUFACTURING PROCESSES

In this section a short description of the process examples mentioned in Table 1.3 is given, partly to illustrate the foregoing discussion and partly to give a background for the following sections. Assembly and joining processes are not described here. The examples will be described in accordance with the structure of Table 1.3, and the individual processes are discussed in more detail in later chapters (1,2,4).

1.4.1 Forging

Forging can be characterized as: mass conserving, solid state of work material (metal), mechanical primary basic process—plastic deformation. A wide vari-

TABLE 1.3 Classification of Technological Material Processes Used in Shaping Materials[a]

Process or flow type	State of material	Category of basic process	Primary basic process	Process examples
Mass-conserving processes ($dM = 0$)	Solid	Mechanical	Plastic deformation	Forging and rolling
	Granular	Mechanical	Flow and plastic deformation	Powder compaction
	Fluid	Mechanical	Flow	Casting
Mass-reducing processes ($dM < 0$)	Solid	Mechanical	Ductile fracture and brittle fracture	Turning, milling, and drilling
		Thermal	Melting and evaporation	Electrical discharge machining (EDM) and cutting
		Chemical	Dissolution	Electrochemical machining (ECM)
			Combustion	Cutting
Joining processes				
Atomic bonding	Solid	Mechanical	Plastic deformation	Friction welding
	Fluid (vicinity of the joint)	Mechanical	Flow	Welding (fusion)
Adhesion	Solid (fluid filler material)	Mechanical	Flow	Brazing

[a]Only typical process examples are mentioned.

ety of forging processes are used, and Fig. 1.7a shows the most common of these: drop forging. The metal is heated to a suitable working temperature and placed in the lower die cavity. The upper die is then lowered so that the metal is forced to fill the cavity. Excess material is squeezed out between the die faces at the periphery as flash, which is removed in a later trimming process. When the term *forging* is used, it usually means hot forging. Cold forging has several specialized names. The material loss in forging processes is usually quite small.

Normally, forged components require some subsequent machining, since the tolerances and surfaces obtainable are not usually satisfactory for a finished product. Forging machines include drop hammers and forging presses with mechanical or hydraulic drives. These machines involve simple translatory motions.

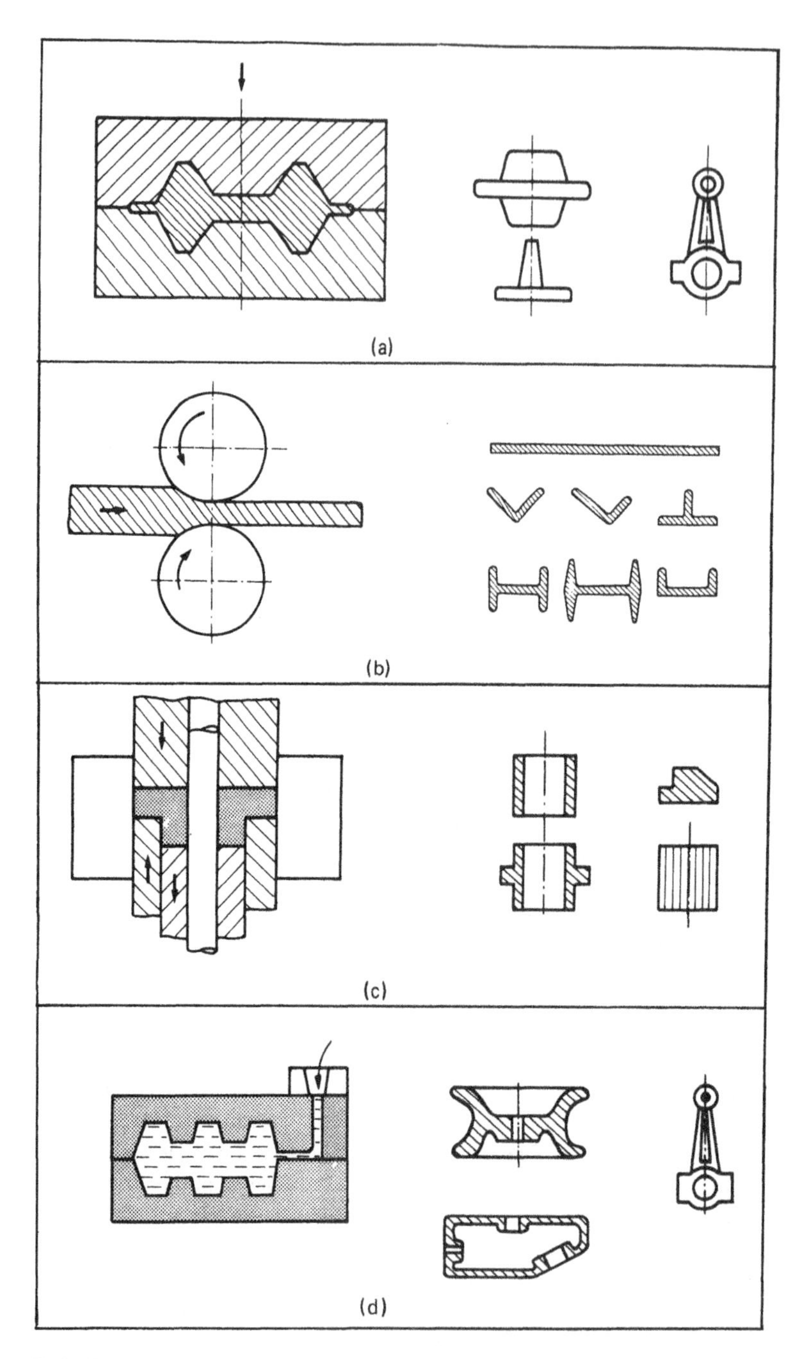

FIGURE 1.7 Mass-conserving processes in the solid state of the work material: (a) forging; (b) rolling, granular state of work material; (c) powder compaction and fluid state of work material; (d) casting.

1.4.2 Rolling

Rolling can be characterized as: mass conserving, solid state of material, mechanical primary basic process—plastic deformation. Rolling is extensively used in the manufacturing of plates, sheets, structural beams, and so on. Figure 1.7b shows the rolling of plates or sheets. An ingot is produced in casting and in several stages it is reduced in thickness, usually while hot. Since the width of the work material is kept constant, its length is increased according to the reductions. After the last hot-rolling stage, a final stage is carried out cold to improve surface quality and tolerances and to increase strength. In rolling, the profiles of the rolls are designed to produce the desired geometry.

1.4.3 Powder Compaction

Powder compaction can be characterized as: mass conserving, granular state of material, mechanical basic process—flow and plastic deformation. In this context, only compaction of metal powders is mentioned, but generally compaction of molding sand, ceramic materials, and so on, also belong in this category.

In the compaction of metal powders (see Fig. 1.7c) the die cavity is filled with a measured volume of powder and compacted at pressures typically around 500 N/mm^2. During this pressing phase, the particles are packed together and plastically deformed. Typical densities after compaction are 80% of the density of the solid material. Because of the plastic deformation, the particles are "welded" together, giving sufficient strength to withstand handling. After compaction, the components are heat-treated—sintered—normally at 70–80% of the melting temperature of the material. The atmosphere for sintering must be controlled to prevent oxidation. The duration of the sintering process varies between 30 min and 2 h. The strength of the components after sintering can, depending on the material and the process parameters, closely approach the strength of the corresponding solid material.

The die cavity, in the closed position, corresponds to the desired geometry. Compaction machinery includes both mechanical and hydraulic presses. The production rates vary between 6 and 100 components per minute. Powder compaction is described in more detail in Chapter 9.

1.4.4 Casting

Casting can be characterized as: mass conserving, fluid state of material, mechanical basic process—filling of the die cavity. Casting is one of the oldest manufacturing methods and one of the best known processes. The material is melted and poured into a die cavity corresponding to the desired geometry (see Fig. 1.7d). The fluid material takes the shape of the die cavity and this geometry is finally stabilized by the solidification of the material.

The stages or steps in a casting process are the making of a suitable mold, the melting of the material, the filling or pouring of the material into the cavity, and the solidification. Depending on the mold material, different properties and dimensional accuracies are obtained. Equipment used in a casting process includes furnaces, mold-making machinery, and casting machines.

1.4.5 Turning

Turning can be characterized as: mass reducing, solid state of work material, mechanical primary basic process—fracture. The turning process, which is the best known and most widely used mass-reducing process, is employed to manufacture all types of cylindrical shapes by removing material in the form of chips from the work material with a cutting tool (see Fig. 1.8a). The work material rotates and the cutting tool is fed longitudinally. The cutting tool is much harder and more wear resistant than the work material. A variety of types of lathes are employed, some of which are automatic in operation. The lathes are usually powered by electric motors which, through various gears, supply the necessary torque to the work material and provide the feed motion to the tool.

A wide variety of matching operations or processes based on the same metal-cutting principle are available; among the most common are milling and drilling carried out on various machine tools. By varying the tool shape and the pattern of relative work-tool motions, many different shapes can be produced (see Fig. 1.8b and c). A detailed description of machining processes is given in Chapter 7.

1.4.6 Electrical Discharge Machining

Electrical discharge machining (EDM) can be characterized as: mass reducing, solid state of work material, thermal primary basic process—melting and evaporation (see Fig. 1.8d). In EDM, material is removed by the erosive action of numerous small electrical discharges (sparks) between the work material and the tool (electrode), the latter having the inverse shape of the desired geometry. Each discharge occurs when the potential difference between the work material and the tool is large enough to cause a breakdown in the fluid medium, fed into the gap between the tool and workpiece under pressure, producing a conductive spark channel. The fluid medium, which is normally mineral oil or kerosene, has several functions. It serves as a dielectric fluid and coolant, maintains a uniform resistance to the flow of current, and removes the eroded material. The sparking, which occurs at rate of thousands of times per second, always occurs at the point where the gap between the tool and workpiece is smallest and develops so much heat that a small amount of material is evaporated and dispersed into the fluid. The material surface has a characteristic appearance composed of numerous small craters.

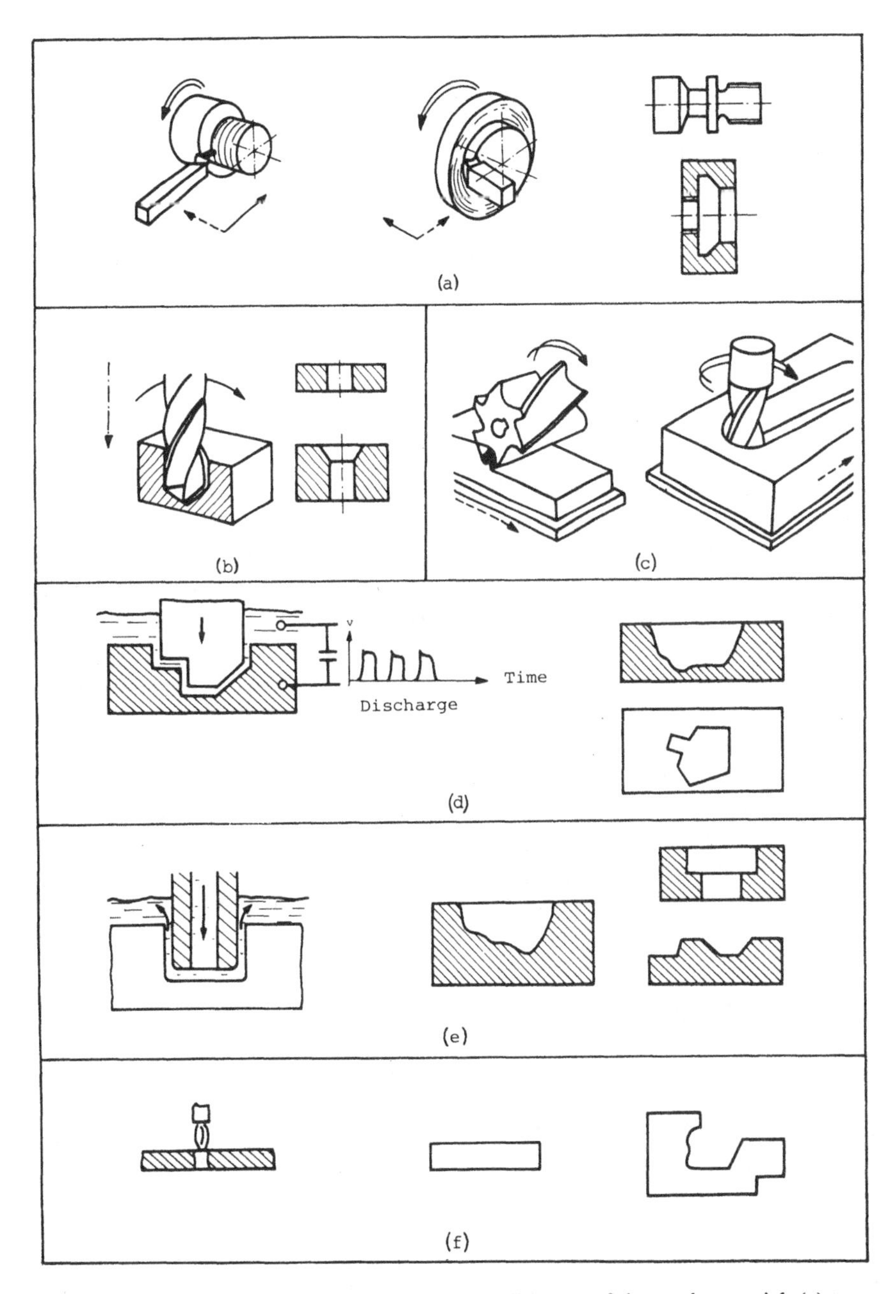

FIGURE 1.8 Mass-reducing processes in the solid state of the work material: (a) turning; (b) drilling; (c) milling; (d) electrical discharge machining (EDM); (e) electrochemical machining (ECM); (f) torch cutting.

1.4.7 Electrochemical Machining

Electrochemical machining (ECM) can be characterized as: mass reducing, solid state of work material, chemical primary basic process—electrolytic dissolution (Fig. 1.8e). Electrolytic dissolution of the workpiece is established through an electric circuit, where the work material is made the anode, and the tool, which is approximately the inverse shape of the desired geometry, is made the cathode. The electrolytes normally used are water-based saline solutions (sodium chloride and sodium nitrate in 10–30% solutions). The voltage, which usually is in the range 5–20 V, maintains high current densities, 0.5–2 A/mm^2, giving a relatively high removal rate, 0.5–6 cm^3/min · A 1000, depending on the work material.

1.4.8 Flame Cutting

Flame cutting can be characterized as: mass reducing, solid state of work material, chemical primary basic process—combustion (Fig. 1.8f). In flame cutting, the material (a ferrous metal) is heated to a temperature where combustion by the oxygen supply can start. Theoretically, the heat liberated should be sufficient to maintain the reaction once started, but because of heat losses to the atmosphere and the material, a certain amount of heat must be supplied continuously. A torch is designed to provide heat both for starting and maintaining the reaction. Most widely used is the oxyacetylene cutting torch, where heat is created by the combustion of acetylene and oxygen. The oxygen for cutting is normally supplied through a center hole in the tip of the torch.

The flame cutting process can only be used for easily combustible materials. For other materials, cutting processes based on the thermal basic process—melting—have been developed (arc cutting, arc plasma cutting, etc.). This is the reason cutting is listed in Table 1.3 under both thermal and chemical basic processes.

1.5 ENERGY FLOW SYSTEM

According to Fig. 1.4 the next system to consider in the morphological process model is the energy flow. To carry out the basic processes described above, energy must be provided to the work material through a transmission medium. The energy flow for mechanical, thermal, and chemical basic processes is discussed next.

The energy flow system (see Fig. 1.4) can be divided into two subsystems: the tool/die system and the equipment system. The tool/die system describes how the energy is supplied to the material and the transfer media used. The equipment system describes the characteristics of the energy supplied from the equipment and the type of energy used to generate this. In the following sections the possible energy supply principles, the various transmission media, and the energy sources are described without separation of the two systems, but it should be kept in mind by the reader.

1.5.1 Energy Flow for Mechanical Basic Processes

The primary mechanical basic processes are (Table 1.2) plastic and elastic deformation, brittle or ductile fracture, and flow. Energy to carry out a mechanical basic process (Fig. 1.9) can be provided through:

Relative motions between a transfer medium and the work material
Pressure differences across the work material
Mass forces generated in the work material

If the energy is supplied through active motions, the state of the transfer medium can, depending on the actual process, be rigid, granular, or fluid. When pressure differences are used to supply the energy, the state of the transfer medium can be plastic, elastic, granular, or gaseous (including vacuum). In the work material itself, mass forces are generated primarily through gravity, accelerations, or magnetic fields, which means that the medium situated between the field generator and the work material is unimportant as long as it does not interfere with the energy transmission.

How relative motions, pressure differences, and mass forces are generated and established and which energy sources are available will now be discussed. Figure 1.10 shows the energy supply schematically, and it can be seen that the necessary mechanical energy can be generated:

As mass forces directly in the work material itself (contour 1 on Fig. 1.10)
As relative motions and pressure differences outside the work material (indirectly) and transmitted through a suitable medium (contour 2 on Fig. 1.10)

Further, the energy can be provided either throughout the work material (total energy supply) or to portions of the material at different times (partial or local energy supply); in the latter case the energy source must be moved relative to the material.

The energy sources that can be used to create the relative motions, the pressure differences, or the mass forces necessary to carry out the mechanical basic processes are mechanical, electrical (including magnetic), and thermal or chemical (see Table 1.4). (The use of the same energy sources to carry out thermal or chemical primary basic processes is discussed later.)

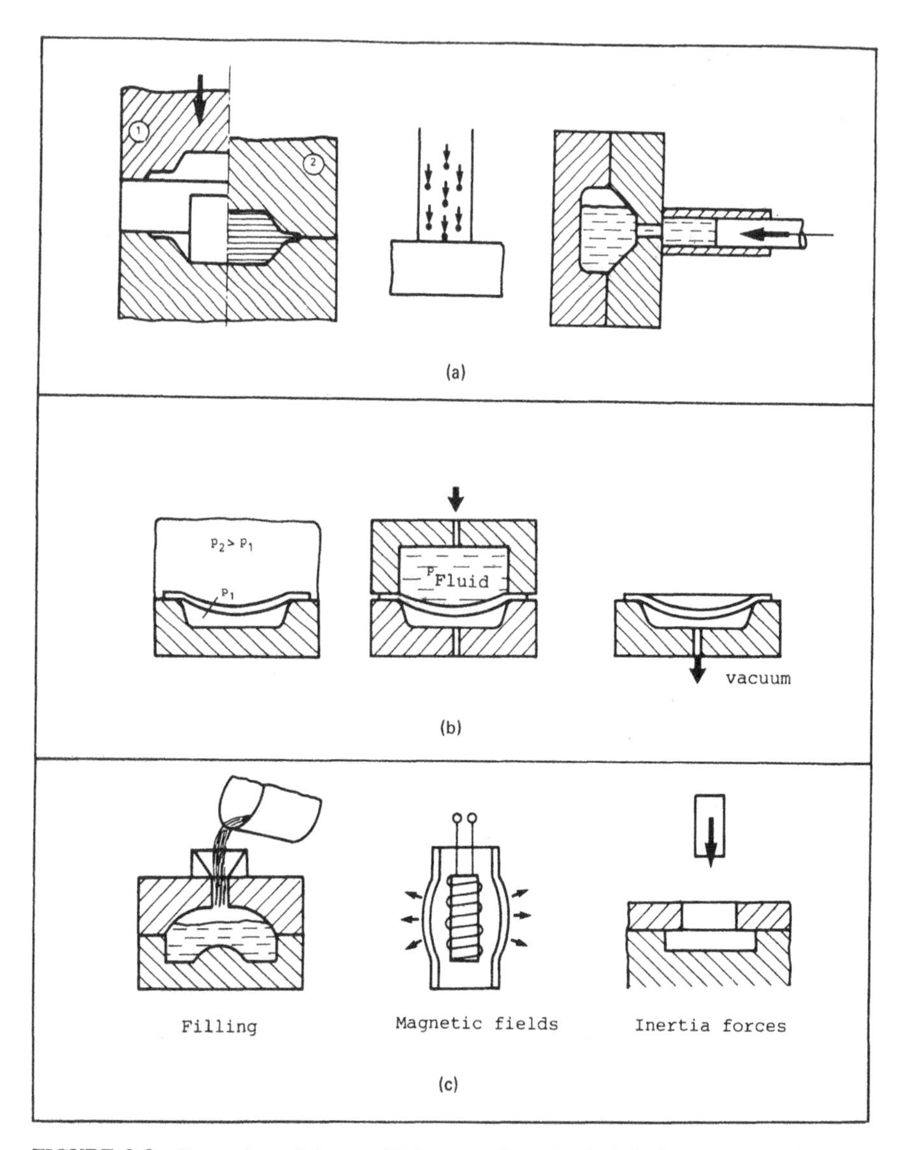

FIGURE 1.9 Examples of the establishment of mechanical basic processes: (a) mechanical basic processes established by relative motions; (b) mechanical basic processes established by pressure differences; (c) mechanical basic processes established by mass forces.

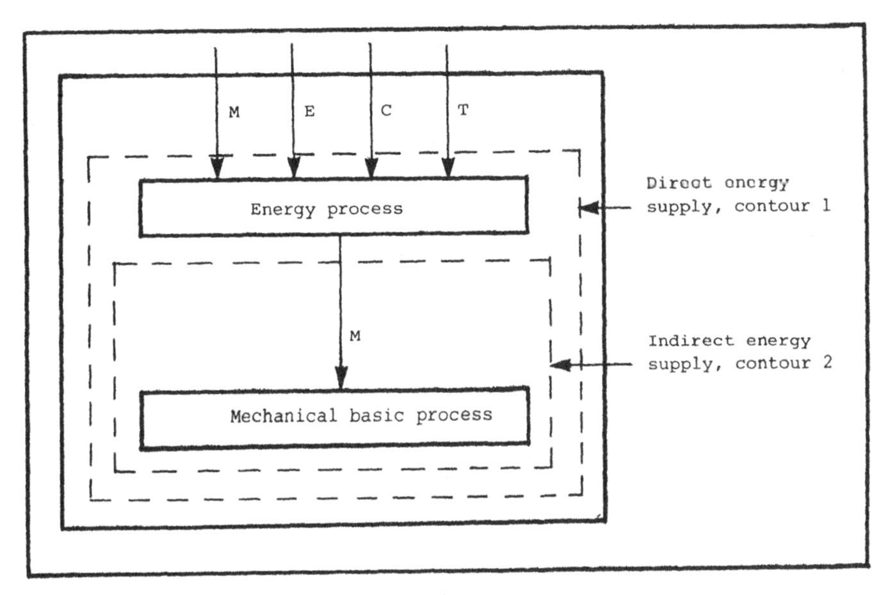

FIGURE 1.10 Energy supply for mechanical basic processes (interaction among basic process, work material, and energy source). M stands for mechanical energy, E for electrical, C for chemical, and T for thermal.

Mechanical Energy Sources

The available mechanical energy sources are:

Kinetic energy
- Translation
- Rotation
- Combinations

Potential energy
- Gravitational
- Elastic

Pressure in a medium (kinetic energy in the molecules)

Vacuum

These energy sources are used to create relative motions, pressure differences, or mass forces through the tool/die system. The media of transfer must be selected to fulfill the requirements involved.

Electrical Energy Sources

Electrical energy can be used directly or indirectly to create mechanical energy to be delivered through relative motions, pressure differences, or mass forces.

TABLE 1.4 Morphological Diagram for Energy Flow (Supply) Systems in Primary Mechanical Basic Processes

Energy	Mechanical	Electrical	Chemical	Thermal
Principles (generating mechanical energy)	Kinetic energy Potential energy Pressure in a medium Vacuum Gravitation and elastic energy	Discharge Electromagnetic fields Magnetostriction Piezoelectric effects	Combustion Explosion (detonation) Other reactions	Thermal expansion
Transfer media (state)		Rigid Plastic Elastic Granular Gaseous Liquid Vacuum Unspecified		
Transmission (principles of utilization in mechanical basic processes)		Relative (active) motions Pressure differences Mass forces		

Discharge Between Two Electrodes. A discharge of electrical energy stored in condensers between two electrodes in a fluid medium (usually water) will, because of the sudden evaporation of the fluid in the discharge channel, create a shock wave, which can be applied directly to the work material through pressure differences or indirectly in the form of kinetic energy through a suitable medium (see Fig. 1.11).

Electromagnetic Fields. The discharge of electrical energy through a coil can create sufficiently transient magnetic fields that can be utilized directly or indirectly. In the direct utilization of magnetic fields, the work material is placed inside or outside a coil, so that a magnetic field is induced in the material, attracting or repelling the coil field. These forces can be of sufficient strength to create plastic flow in metallic materials (see Fig. 1.12a). In the indirect utilization of the magnetic fields, these fields are used to create rotations or translations in a solid material in the form of an iron core. This produces mechanical energy which can be adapted for the process through gears, for ex-

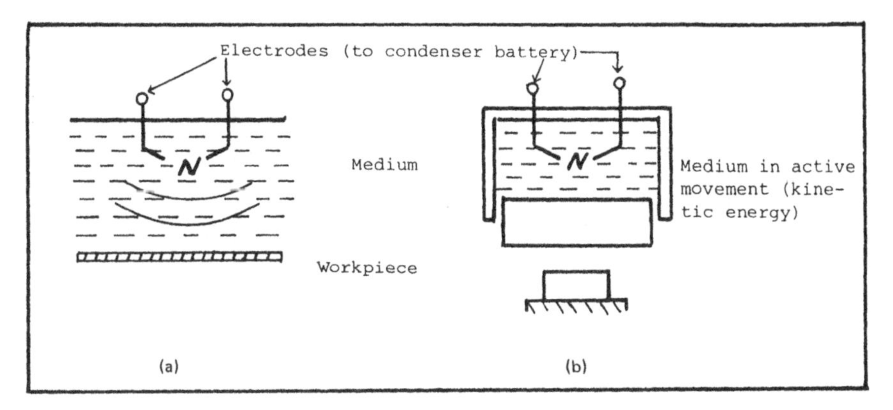

FIGURE 1.11 Discharge of electrical energy: (a) utilized directly (pressure differences); (b) utilized indirectly (relative motions).

ample (see Fig. 1.12b and c). The most common example here is the electric motor. As shown in Fig. 1.12c, in specifying the energy system, the requirements for the basic process are first determined, then the different principles to fulfill these requirements are investigated and specified.

Magnetostrictive Effect. Some materials, particularly ferromagnetic materials, change dimensions when they are subjected to a magnetic field. Nickel contracts, whereas ferrous and aluminum alloys expand. If these materials are subjected to a field fluctuating at a high frequency, an oscillator (20 kHz) which can be used in ultrasonic machining is obtained. The amplitude can be varied by suitably shaping the solid material connected to the oscillator (see Fig. 1.13).

Piezoelectric Effect. The piezoelectric effect is exhibited by some crystalline materials in which there is a reversible interaction between an elastic strain and an electric field. This means that when such a material is strained by the application of a stress, it becomes dielectrically polarized (i.e., a certain potential difference arises). Conversely, when the crystal is subjected to a potential difference, it will change dimensions corresponding to the elastic strain. This principle is used in various pressure or force transducers.

Chemical Energy

Chemical energy can be converted to mechanical energy in different ways, depending on the energy source (explosives, combustible gases, etc.), resulting in an increase in pressure in the medium. The resulting high pressure can be utilized directly in the form of pressure differences across the work material or indirectly by introducing relative motions between the transfer medium and the work material.

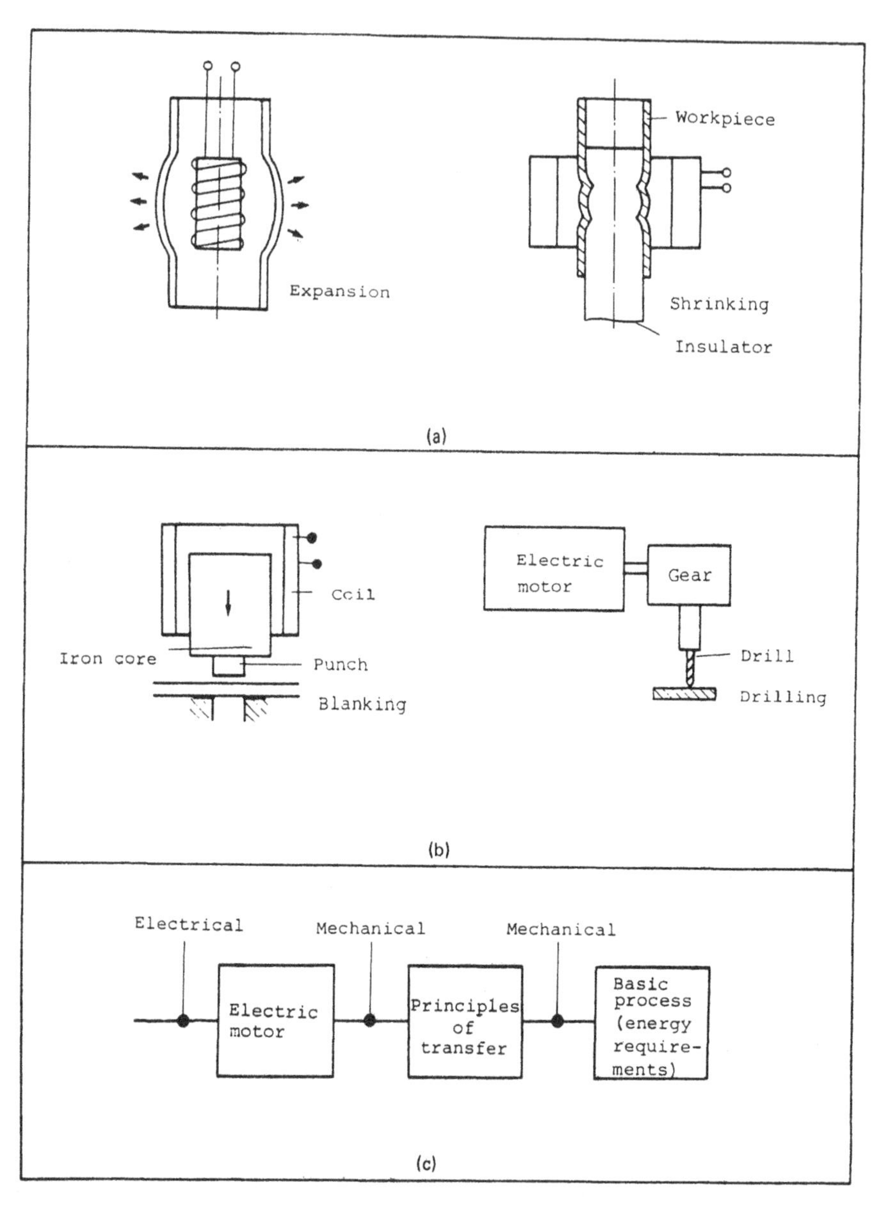

FIGURE 1.12 Indirect utilization of electrical energy through magnetic fields. The magnetic field can again be utilized (a) directly or (b) indirectly. (c) Typical steps in designing the energy system based on indirect utilization of electrical (magnetic) energy are shown.

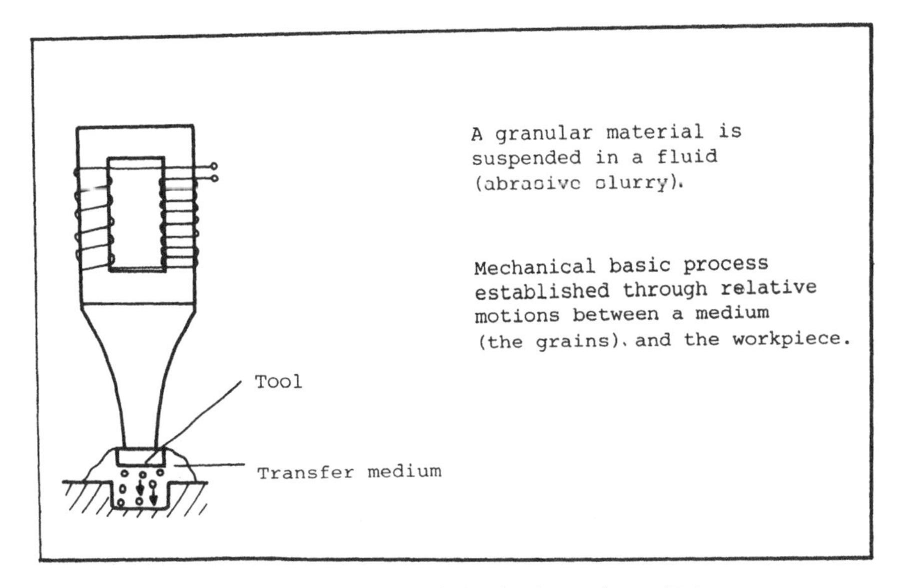

FIGURE 1.13 Utilization of magnetostriction in ultrasonic machining.

Figure 1.14 shows schematically a combustion process, involving gasoline, oil, or other materials, and the detonation of an explosive. These principles can be utilized in many different ways (Fig. 1.14). Combustion can be used in forming and forging machinery; detonation can be used in explosive forming, explosive welding, and compaction, for example.

Thermal Energy

Thermal energy or heat can be converted into mechanical energy through utilizing the thermal expansion of materials to provide relative motions or to generate pressures in a medium (Fig. 1.15).

The preceding descriptions of the energy flow for mechanical basic processes can be summarized in the morphological diagram shown in Table 1.4. This diagram is used to generate possible energy flow (supply) systems.

1.5.2 Energy Flow for Thermal Basic Processes

Here only the primary thermal basic processes that require heat (melting, evaporation) are discussed (Table 1.2). For a thermal basic process, a heat source must be available. The heat source may consist of an energy process where electrical, chemical, or mechanical energy is converted into heat, or of a heat reservoir. Figure 1.16 shows the relations among the heat source, the work material, and the basic process. The heat can be generated inside the work ma-

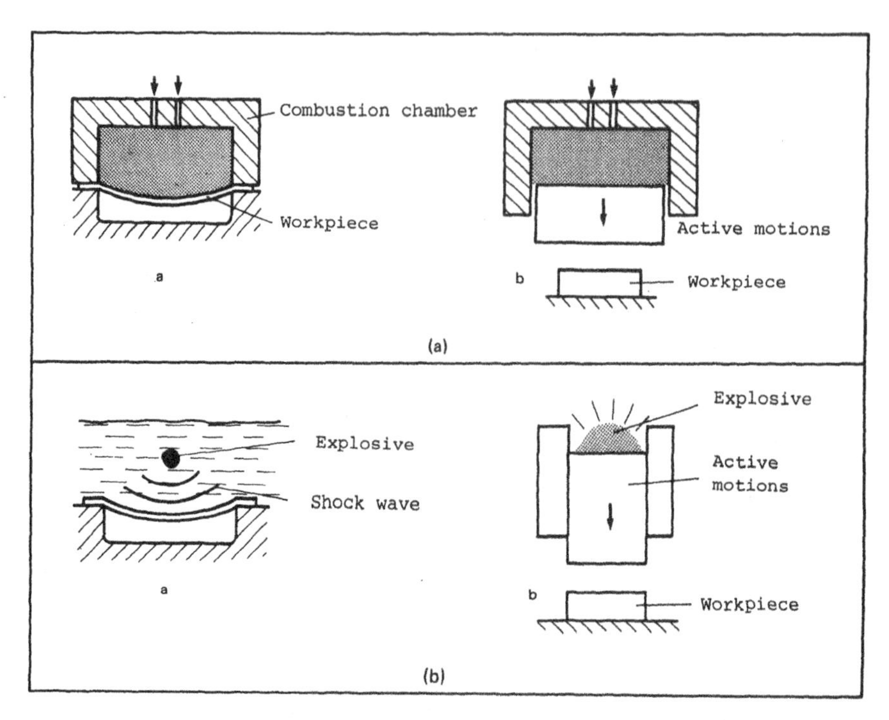

FIGURE 1.14 Utilization of chemical energy in (a) combustion and (b) detonation.

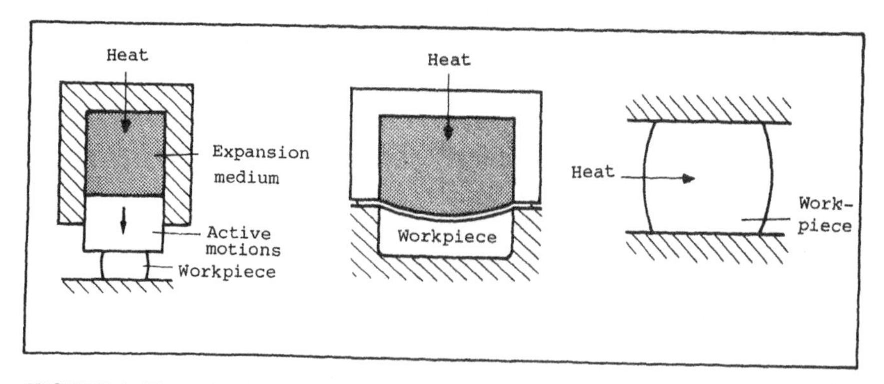

FIGURE 1.15 Thermal energy (heat reservoir) utilized in mechanical basic processes (1 and 2 directly and 3 indirectly).

terial itself (direct heat supply, contour 1) or outside the work material (indirect heat supply) and then transferred through a medium to the material. In establishing a thermal basic process, heat transfer and heat sources thus play an important role and, consequently, these will be discussed further.

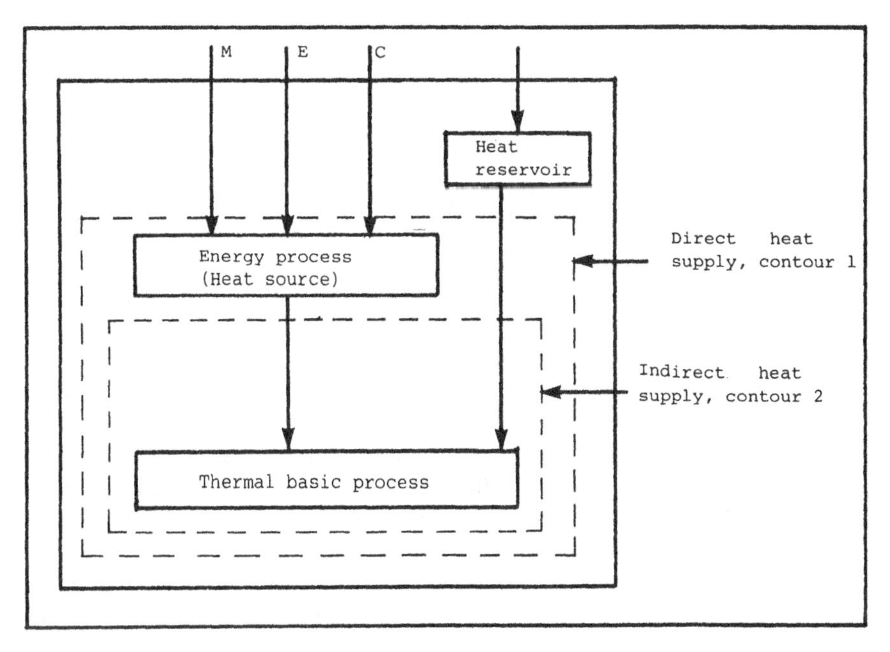

FIGURE 1.16 Relations among work material (contour), heat source, and thermal basic process.

Heat Transfer

Heat can be transferred by conduction (thermal), radiation, convection, and mass transportation. Conduction can take place in a rigid, granular, fluid, or gaseous medium. Radiation, which consists of electromagnetic waves, requires a medium that is transparent for the electromagnetic waves. Heat transfer by convection can normally take place in fluid or gaseous media. Mass transportation can be used as a link in heat transfer. A medium with a certain amount of heat is transported to the work material, where the heat is transferred through conduction, radiation, or convection.

Heat Sources

Heat Sources Based on Electrical Energy. The principles that can be used to generate heat from electrical energy are:

Electrical conduction (resistance)
Induction
Dielectric loss
Arcing (discharge between electrodes)
Sparking
Electron beams
Lasers

Generation of heat by conduction is based on the dissipation of heat due to the resistance of the conducting material. If the conducting material is the work material itself, a direct heat supply is the result. If a special conducting material (heat element with high resistance) is used to generate the heat, this heat must be transferred to the work material through a suitable medium by conduction, radiation, or convection, which means that an indirect heat supply is being utilized. The conversion of electrical energy into heat based on conduction can be used both in the process itself and in the process machinery.

Induction heating can be established in two different ways, corresponding to the eddy current principle and the transformer principle. In the *eddy current principle*, the conductive work material is placed in the field of an induction coil through which passes a high frequency (5 kHz to 5 MHz) alternating current. Heat is generated directly in the object by means of the eddy currents induced in it. In the *transformer principle*, the work material itself may act as the secondary coil, which means that a low-voltage, high-amperage current is induced in it; this forms a direct heat supply. If the secondary coil is made the heating element, we have generation of heat by conduction.

In all these cases the heating is basically due to the electrical resistance of the material. However, in ferromagnetic materials, heat is also generated by the alternating magnetic fields.

Dielectric heating refers to the heating of nonmetals, such as plastics and plywood, by placing them in the electric field of a capacitor to which a high-frequency voltage is applied. Heat is generated directly in the object by virtue of dielectric losses within the material when it is placed between the plates of a capacitor, and hence subjected to an alternating electric field. Unlike induction heating, dielectric heating is distributed uniformly throughout the material. The heat generated increases as the frequency is increased. If the work material is conductive, dielectric heating can only be used indirectly, as a heating element, for example.

The quantity of heat generated by electrical discharge between two conductive materials depends on the conditions under which the discharge is taking place. During the discharge, an ionized channel is established having a cross section that depends on the discharge time. If the discharge time is longer than 100 μs, the whole area between the two materials is ionized and continuous arcing is the result. If the discharge time is shorter, a narrow ionized channel is established, and sparking results.

Arcing creates heat at three different places: at the surfaces of the two electrodes and in the arcing column. From the arc, the heat is transferred to the material by conduction, radiation, and/or convection, depending on the utilization principle (see Fig. 1.17). If the medium in which the arcing takes place is air, the temperature in the arc itself can be approximately 6000°C. The surfaces of the electrodes will, depending on the material and the dimensions, have tempera-

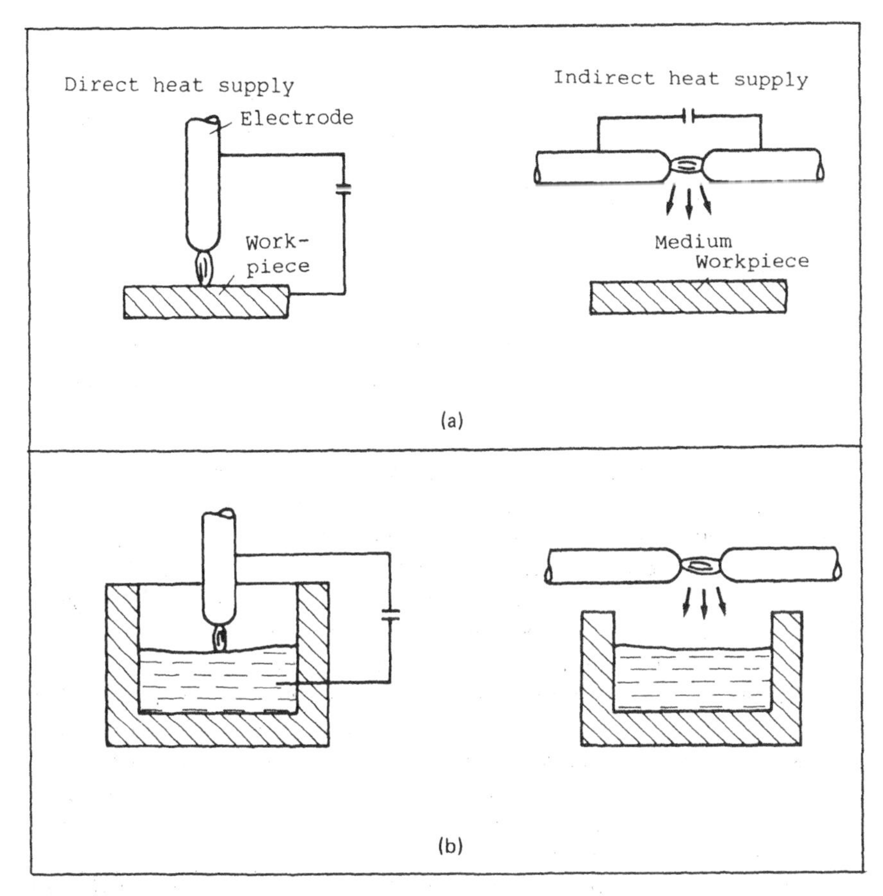

FIGURE 1.17 Heating by arcing: (a) local/partial heat supply; (b) total heat supply.

tures ranging from 1500 to 3000°C. Both arcing and sparking require a gaseous or fluid discharge medium.

In sparking (Fig. 1.18) the local discharge areas are very small and achieve temperatures on the order of 25,000°C. If the discharge time t_1 (Fig. 1.18) is small, the anode will reach the highest temperature, and if t_1 is large, the cathode will reach the highest temperature. During disruption of the sparks, cooling can occur, so that both the anode and the cathode can be kept at a low average temperature. This principle is utilized in electrodischarge machining (EDM), which is a common process. It should be mentioned that two consecutive sparks normally take place at different positions on the electrodes (where the gap, and therefore the resistance, are smallest), giving a typical surface appearance of numerous small craters.

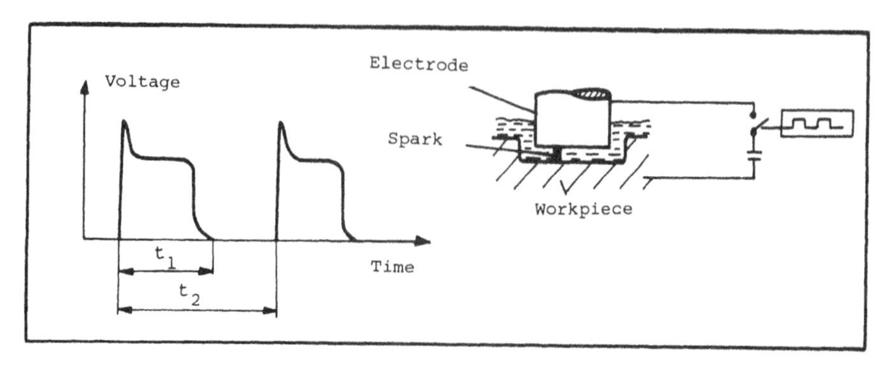

FIGURE 1.18 Heating by sparking utilized in electrical discharge machining (EDM).

In electron beam machining, an electron beam is created in a triode, consisting of a tungsten cathode, a negative grid, and an anode. The grid and the anode have shapes which ensure that the electrons emitted from the cathode are transferred directly to the work material, placed in a vacuum chamber. The energy density ($\sim 10^7$ W/cm^2) in the electron beam, which can be focused by magnetic lenses, is sufficiently high to melt and vaporize the work material.

Another heat source based on electrical energy is the laser beam. *Laser* is an abbreviation for "*l*ight *a*mplification by *s*timulated *e*mission of *r*adiation." Different materials (solid, fluid, or gaseous) can be used as the medium for the light emission, resulting in laser beams with different properties. When the laser beam reaches the work material, some of the energy is reflected, and the rest is converted into heat in the material. The heat generated can melt or vaporize most materials. The energy density in the laser beam is high ($\sim 10^2 - 10^8$ W/cm^2), since the beam is focused to very small diameters on the order of 10–100 μm. Tables 1.5 and 1.6 present a review of the different principles of heat generation from electrical energy.

Heat Sources Based on Chemical Energy. These energy processes are combustion or other exothermic chemical reactions. Combustion can be obtained from solid, granular, fluid, or gaseous fuels and the heat supply to the work material is indirect. Table 1.7 shows the various possibilities. The elements in the diagram can be described by a number, so that a given combination will appear as a four-digit number.

In addition to combustion, the exothermic reaction between Fe_3O_4 and Al should be mentioned as well as the association and dissociation of argon, hydrogen, or helium which is utilized in special arc-plasma torches.

Heat Sources Based on Mechanical Energy. Here the heat can be generated by friction or internal hysteresis losses. Generation of heat by friction is utilized

TABLE 1.5 Principles of Heat Generation from Electrical Energy

Heat generated from:	Illustration		Material requirement
1. Electrical conduction	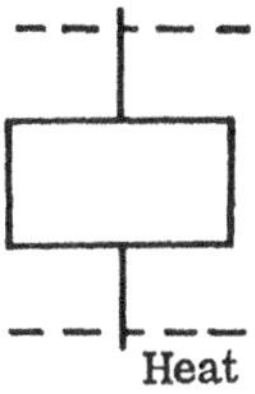 Heat	Electrical current	Electrical conductor
2. Stopping of free electrical particles	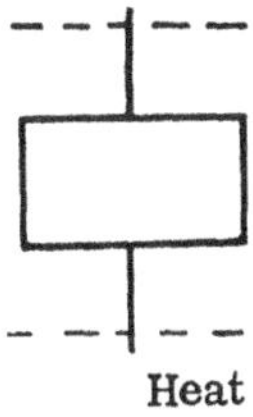 Heat	Free electrical particles in movement	
3. Placing a material in an electrical field	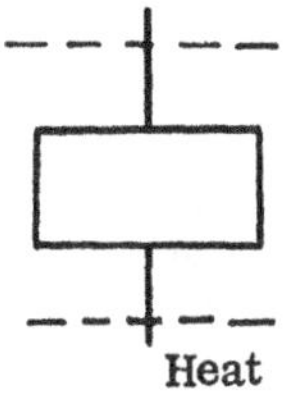 Heat	Electrical field	Dielectric
4. Placing a material in a magnetic field	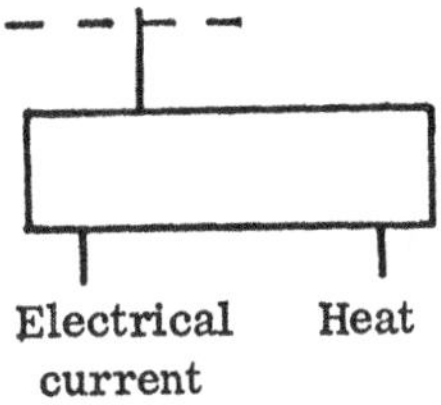 Electrical current; Heat	Magnetic field	Electric conductor Ferromagnetic
5. Placing a material in electromagnetic beams (light waves)	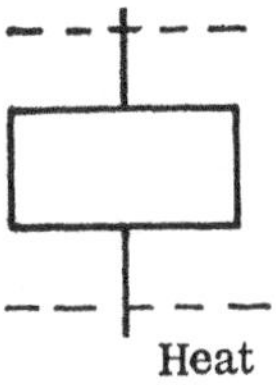 Heat	Electromagnetic beam	

TABLE 1.6 Relation Between Heating Process and Principle

Principle (Table 1.5)	Example	Material requirement
1	Conduction	Electrical conductor
2	Arcing	Electrical conductor
2	Sparking	Electrical conductor
2	Electron beam	—
3	Dielectric heating	Dielectric
4	Induction	Electrical conductor
5	Laser	—

TABLE 1.7 Morphological Relationships for the Heat Generated by Combustion

Heat supply		State of fuel		Transfer mechanism		Medium of transfer	
Local	(1)	Solid	(1)	Conduction	(1)	Rigid	(1)
		Granular	(2)	Radiation	(2)	Granular	(2)
		Fluid	(3)	Convection	(3)	Fluid	(3)
Total	(2)	Gaseous	(4)	Mass transportation	(4)	Gaseous	(4)
						Vacuum	(5)

TABLE 1.8 Morphological Relationships Showing the Possible Transfer Principles When a Thermal Heat Reservoir Is Available

State of heat reservoir medium	Transfer mechanism	State of transfer medium
Rigid (solid)	Conduction	Rigid (solid)
Granular	Radiation	Granular
Fluid	Convection	Fluid
Gaseous	Mass transportation	Gaseous
	Combinations	Vacuum

in friction welding, and generation of heat by internal loss is utilized in ultrasonic welding (which normally also includes heating by friction).

Heat Sources Based on Thermal Energy. If a heat reservoir is available, it can be utilized in the various ways shown in Table 1.8. The heat can be provided locally or throughout the work material.

Table 1.9 shows the morphological diagram for all the possible ways that thermal energy for thermal basic processes can be generated. The last row of this diagram shows the way in which heat is provided and corresponds to Table 1.4. Some of the combinations in Table 1.9 are not valid, but the diagram presents a systematic approach, which supports the efforts to find all the relevant possibilities.

TABLE 1.9 Morphological Relationships Showing the Possibilities of Generating Heat for the Primary Thermal Basic Processes

Energy	Electrical	Chemical	Mechanical	Thermal (reservoir)
Principles (generating thermal energy from)	Conduction Induction Dielectric Arcing Sparking Electron beam Laser beam	Combustion Dissociation/ association Exothermic reactions (others)	Friction Internal loss	(Heat in solid, granular, fluid, or gaseous media)
Transfer media (state)	Rigid (solid) Granular Fluid Gaseous Vacuum			
Transmission (principles of utilization in thermal basic processes)	Indirectly Heat conduction Heat radiation Convection Mass transportation Directly Heat generated in the work material			

1.5.3 Energy Flow for Chemical Basic Processes

The chemical basic processes (solution/dissolution, deposition, diffusion, phase transformation, etc.; Table 1.2) will not be treated fully here because they require a close study of the energy conditions in chemical reactions. In this section, therefore, only general principles are presented. Generally, chemical reactions can be influenced by either thermal (heating/cooling) or electrical energy, which are described by chemical thermodynamics and electrochemistry, respectively.

Chemical solution (dissolution) of metals is utilized, for example, in etching and polishing. The etching process is attracting increasing attention, and many new applications have been found in recent years. Electrolytic solution of electrically conductive materials is utilized in electrochemical machining, which is basically electroplating (deposition) in reverse. The plating processes, both electrolytic and chemical, are used widely for surface protection.

Phase transformation and diffusion play an important role in the heat treatment of metals. These processes require heating (or cooling), and a number of specialized furnaces have been developed for this purpose.

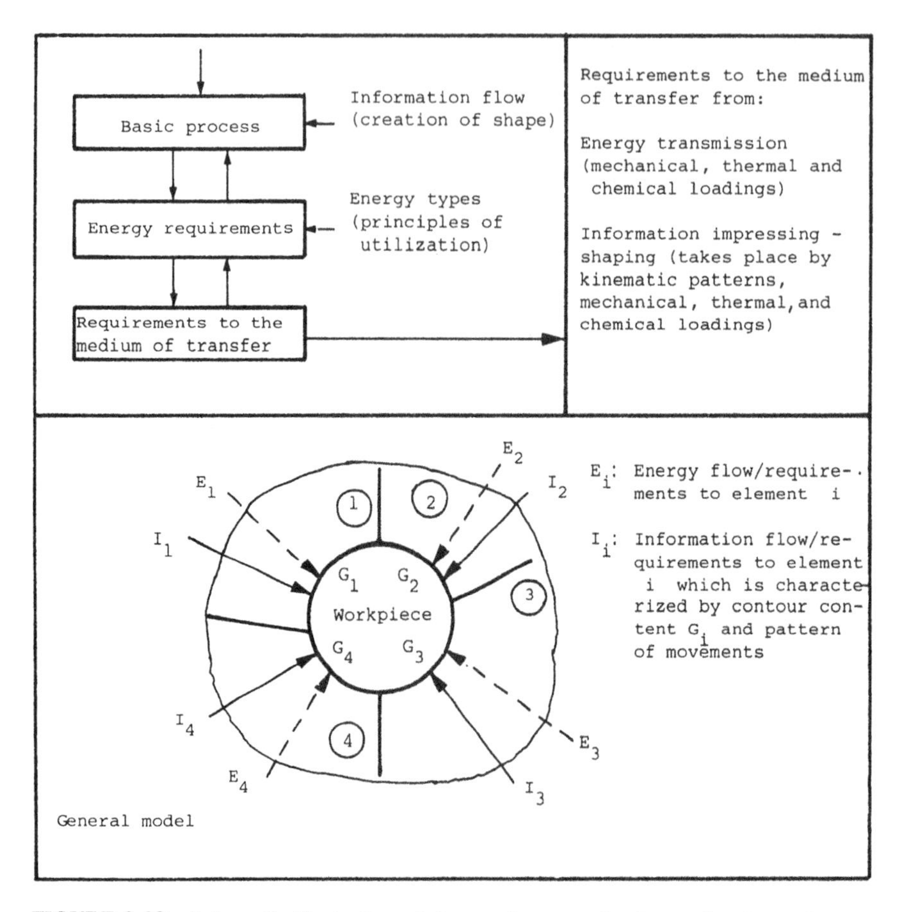

FIGURE 1.19 Schematic illustration of the requirements for the media of transfer.

The chemical reactions that occur in the hardening of plastics are exothermic, which means that cooling is necessary in many cases. Generally, for chemical basic processes it can be stated that they require either no external energy, electrical energy, or thermal energy.

1.5.4 Transfer Media

The media of transfer through which the energy is transmitted have, in the present context, only been described by their states. In a more detailed study of the structures of the material processes, this would be insufficient. It is necessary to define the actual requirements of the transfer medium by considering (Fig. 1.19) the energy transmission and the information impressing (shaping).

The determination of the basic process and the information flow (structure of geometry generation) allows an evaluation of both the total energy requirements and the energy to be distributed within single geometrical elements ($E_i \sim G_i$ in Fig. 1.19). The supply of external energy normally takes place through one or two elements. For each element it is now possible to determine the energy transmission requirements and the geometry creation requirements, which together specify the requirements of the media of transfer. Depending on the actual process, the external energy transmission and the geometry generation can be integrated (as in forging) or separated (as in hydraulic forming and casting).

Analyses of the type described above lead to detailed specifications for the transmission and geometry-creating media.

1.6 INFORMATION FLOW SYSTEM

The term *information flow* covers, as described earlier, the impressing of shape information on the work material. The principles on which information impressing can be based can be analyzed in relation to the type of process (material flow), the state of the material, and the basic process. By coupling the geometrical possibilities derived from the information flow for the specific material with the energy system, the various possible physical processes can be defined.

Figure 1.20 shows a general, schematic model for shape impressing. The creation of the desired geometry takes place for a given basic process by the interaction between the medium of transfer together with the contour of the desired geometry G_i and the pattern of motions for the medium of transfer ($G_1, \ldots, G_i$) and the work material (A). It should be mentioned that there can be more than one element in the medium or transfer, whereas there can be zero or a finite number of elements in the creation of geometry. In actual situations, the functional requirements of the transfer medium, including energy transmission (E_i) and information impressing (I_i), must be determined. Figure 1.21 shows a simplified version of Fig. 1.19. Figure 1.21a and b define the elements. For the two transfer elements, the energy transmission requirements (Fig. 1.21c) and the shape-impressing requirements (Fig. 1.21d) must be defined. These requirements necessitate a determination of the extension of the media in relation to the desired geometry and the pattern of motions for the media and the work material. Figure 1.21c shows how the extension of the energy supply can necessitate two, one, or no relative motions (scanning motions) between the energy source and the work material. Here it should be emphasized that the media (O_1 and O_2) can, at the same time, contribute to geometry creation because they contain the desired geometry.

Figure 1.21d shows the basic principles in surface creation. Here the medium O_2 is used as a reference, but it should be understood as the sum of the O_1's and O_2's contour contents. The four possibilities that arise are:

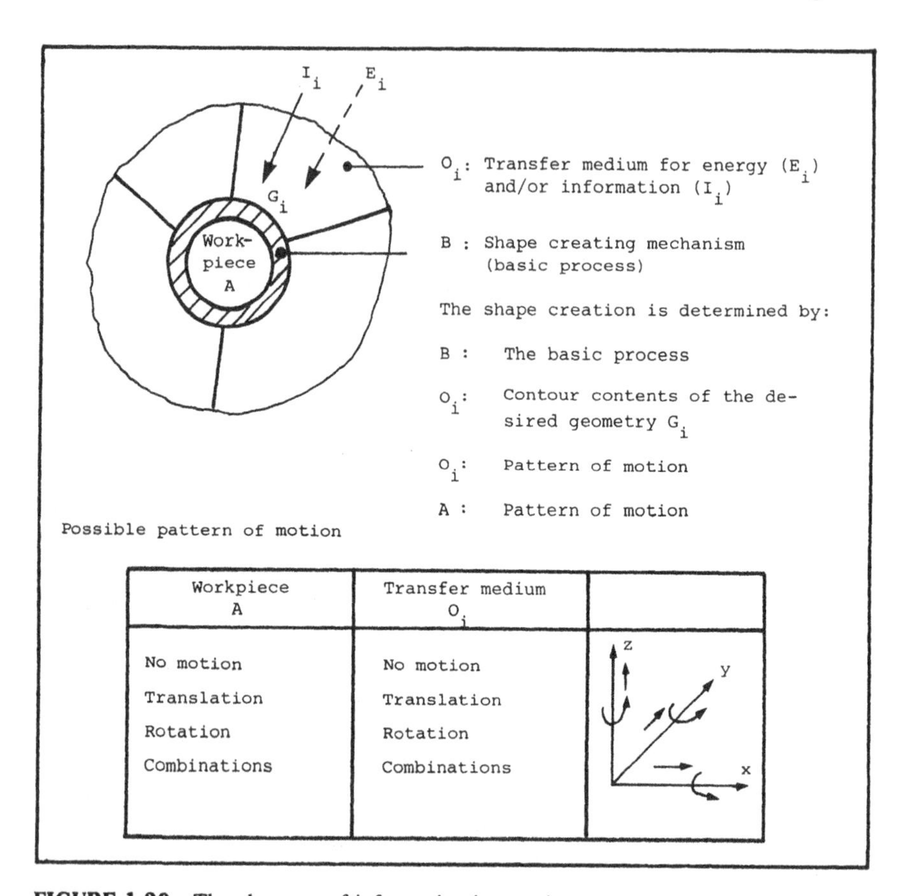

FIGURE 1.20 The elements of information impressing.

Free forming. Here the medium of transfer does not contain the desired geometry (i.e., the surface/geometry is created by stress fields).

Two-dimensional forming. Here the medium of transfer contains a point or a surface element of the desired geometry, which means that two relative motions are required to produce the surface.

One-dimensional forming. Here the medium of transfer contains a producer (a line or a surface area along the line) of the desired surface, which means that one relative motion is required to produce the surface.

Total forming. Here the medium of transfer contains (in one or more parts) the whole surface of the desired geometry, which means that no relative motion is necessary.

These four basic principles of surface creation can be established with a partial or total geometry-creating mechanism (Fig. 1.21c). For example, forging is total

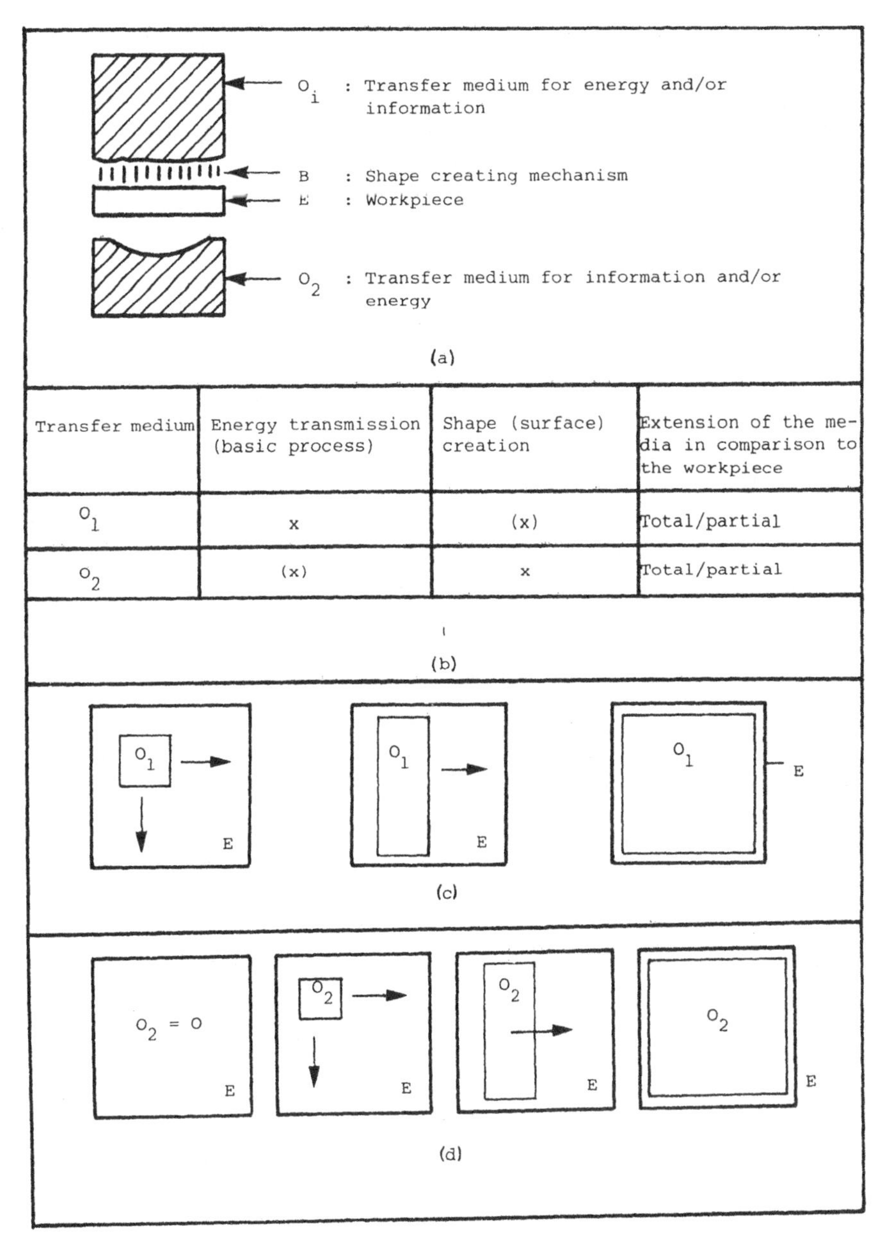

Transfer medium	Energy transmission (basic process)	Shape (surface) creation	Extension of the media in comparison to the workpiece
O_1	x	(x)	Total/partial
O_2	(x)	x	Total/partial

FIGURE 1.21 Information impressing: (a), (b) characterizing the function and size of the elements; (c) The extension of the energy transmission (basic processes + evt contour contents) in comparison to the workpiece, can necessitate two, one, or no relative motions (scanning motions); (d) the surface creation principles derived from the size of the contour contents of the lower medium of transfer and the desired geometry.

forming, rolling is one-dimensional forming, turning is two-dimensional forming, and torsion is free forming.

Table 1.10 shows, in a general way, the possibilities of shape impressing corresponding to the combination of Fig. 1.21c and d. The states of the media are classified simply as rigid (compared to the work material) or nonrigid, which cover the states elastic, plastic, granular, fluid, gaseous, and vacuum. It should be noted that one of the media may degenerate to a fixture or clamping device, as in turning, for example.

The following description of the types of information flow is divided according to the type of process and the state of the work material.

1.6.1 Information Flow for Mass-Conserving Processes ($dM = 0$)

Solid Material

The basic processes here are plastic and elastic deformation, and the information is impressed on the work material through the media of transfer having contour contents and pattern of motions corresponding to Table 1.10.

To decide if a given process can be carried out on a certain material, stresses in the material, strains, strain rates, and temperatures must be considered; this also allows the determination of energy and force requirements.

When analyzing the pattern of motions and the contour contents of the media of transfer (tools/dies), it is sometimes helpful to distinguish between open and closed dies.

Examples of processing corresponding to the four possibilities mentioned above are shown in Fig. 1.22, but they are described in more detail in Chapter 6.

Granular Material

Granular materials are shaped in a flow process followed by a stabilization. The term *flow* also includes filling and placing. Stabilization can, depending on the powder material, be carried out as a plastic deformation and/or a hardening process.

Figure 1.23 shows the series of basic processes involved in information impressing when forming granular materials (corresponding to phase 2 in Fig. 1.5). The production of green sand molds for casting includes filling and deformation (compaction). In the production of dry sand molds, the shape is further stabilized by hardening (baking); this is also the case in the production of cores. The surface creation is usually total, with a partial or total supply of stabilizing energy.

Compaction (stabilization by deformation) of metal powders can be isostatic or axial. Using *isostatic compaction*, only rough geometries which normally

TABLE 1.10 Principles of Information Impressing Corresponding to Fig. 1.21

Principles of surface creation	Energy transmission based on: Two relative motions	One relative motion	No motions (total energy supply)
Total forming (TF)	 O_1: Rigid / Not rigid	 O_1: Rigid / Not rigid	 O_2: Rigid / Not rigid
	 O_2: Rigid / Not rigid	 O_2: Rigid / Not rigid	 O_1: Rigid / Not rigid
One-dimensional forming (ODF)	 O_1: Rigid / Not rigid	 O_1: Rigid / Not rigid	
	 O_2: Rigid / Not rigid	 O_2: Rigid / Not rigid	
Two-dimensional forming (TDF)	 O_1: Rigid / Not rigid		
	 O_2: Rigid / Not rigid		
Free forming (FF)	 O_1: Not rigid	 O_1: Not rigid	 O_1: Not rigid
	 O_2: Not rigid	O_2: Not rigid	O_2: Not rigid

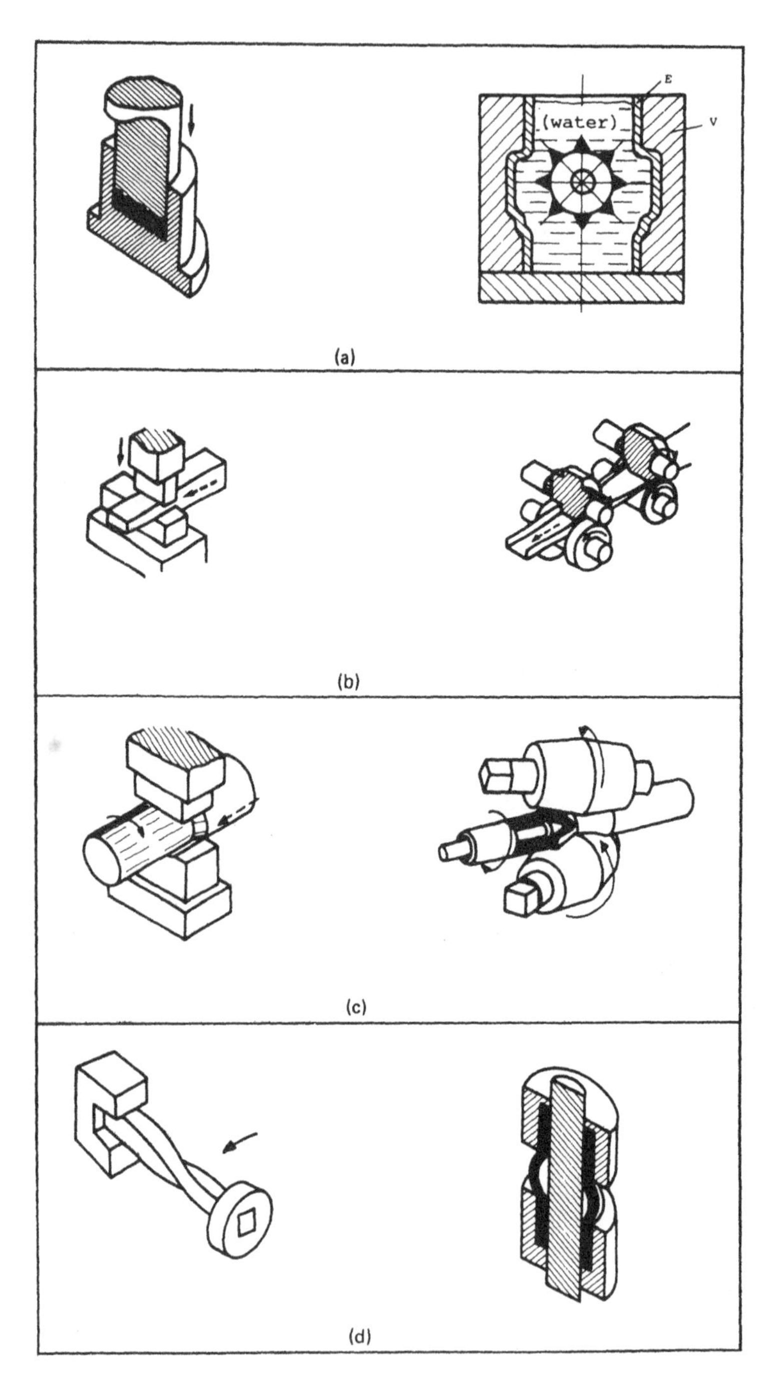

FIGURE 1.22 Examples of information impressing by mass-conserving processes with solid materials: (a) total forming; (b) one-dimensional forming; (c) two-dimensional forming; (d) free forming.

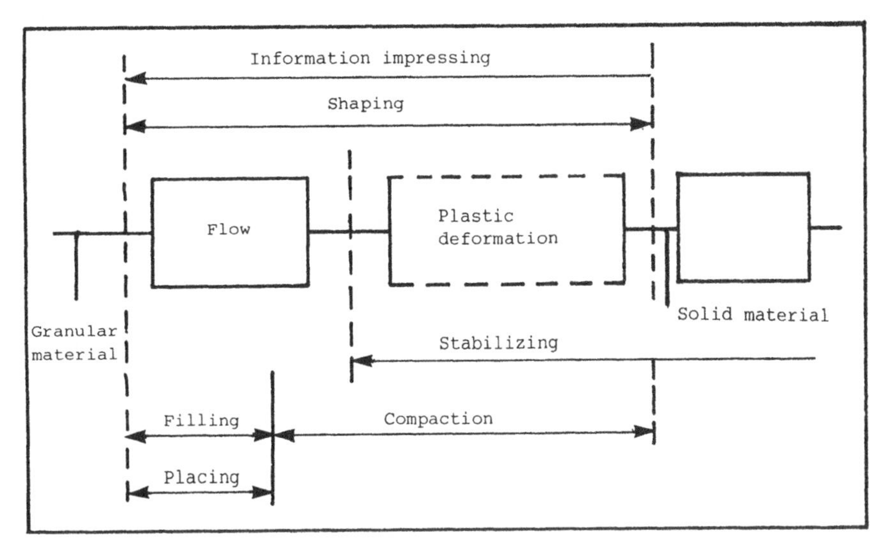

FIGURE 1.23 The series of basic processes (phase 2, Fig. 1.5) involved in shaping granular materials.

require finishing operations can be produced. In *axial compaction* of metal powders, geometries with several levels in height usually require several plungers, which complicates the total pattern of motions (see Chapter 9).

Liquid Materials

In Fig. 1.24 a series of basic processes, corresponding to phases 2 and 3 in Fig. 1.5, are shown for liquid materials. The information impressing can be carried out by flow alone followed by a separate stabilization (Fig. 1.25a) or by flow integrated with stabilization (Fig. 1.25b). Considering flow alone as shape impressing, this can take place in open or closed dies. Depending on the requirements of the product, the die can be permanent or temporary (used only once). In a closed die, the whole surface of the desired geometry is contained in the die geometry. In an open die, the shape information is impressed by the geometry in the open die and a field (gravity, acceleration, surface stress) (see Fig. 1.25a). In other words, flow and stabilization are now integrated.

Considering the processes where flow and stabilization are integrated (Fig. 1.25b, marked II in Fig. 1.24), the input material is liquid and the output solid. The geometry is created by a field, a die surface, and a stabilization mechanism. Actually, a continuous transition exists between Fig. 1.25a and b. More detailed descriptions are given in Chapter 10.

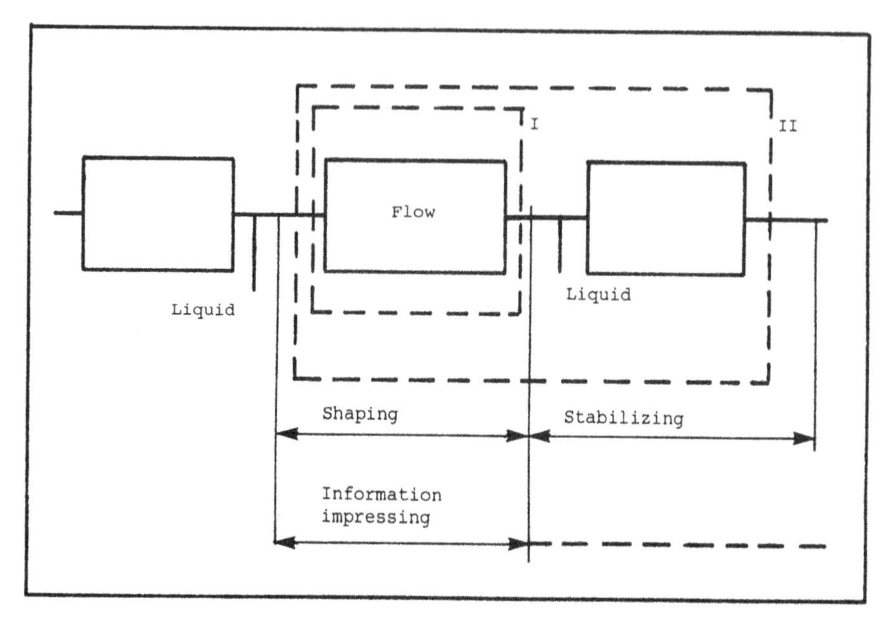

FIGURE 1.24 The series of basic processes (phases 2 and 3, Fig. 1.5) involved in shaping liquid materials.

Summarizing, the shape impressing for mass-conserving processes can thus be characterized by (1) the principles of surface creation (Table 1.10), and (2) the pattern of motions of media of transfer and work material.

1.6.2 Information Flow for Mass-Reducing Processes ($dM < 0$)

For mass-reducing processes that deal only with solid materials, the information impressing is based on mechanical, thermal, or chemical basic processes. These basic processes can be applied in four fundamental material-removal mechanisms or methods (see Fig. 1.26). Considering removal methods I, II, and III, it can be seen by comparison with Fig. 1.21 that the medium of transfer O_2 here is degenerated to a fixture or clamping device. This means that the medium of transfer O_1 has both an energy transmission function and a geometry creation function. For the removal method IV, which includes the blanking processes, both media of transfer are needed as active elements. This method is not in itself mass reducing, but since the applications produce scrap or waste, it can be considered as mass reducing.

In the following sections, these fundamental removal methods are briefly described.

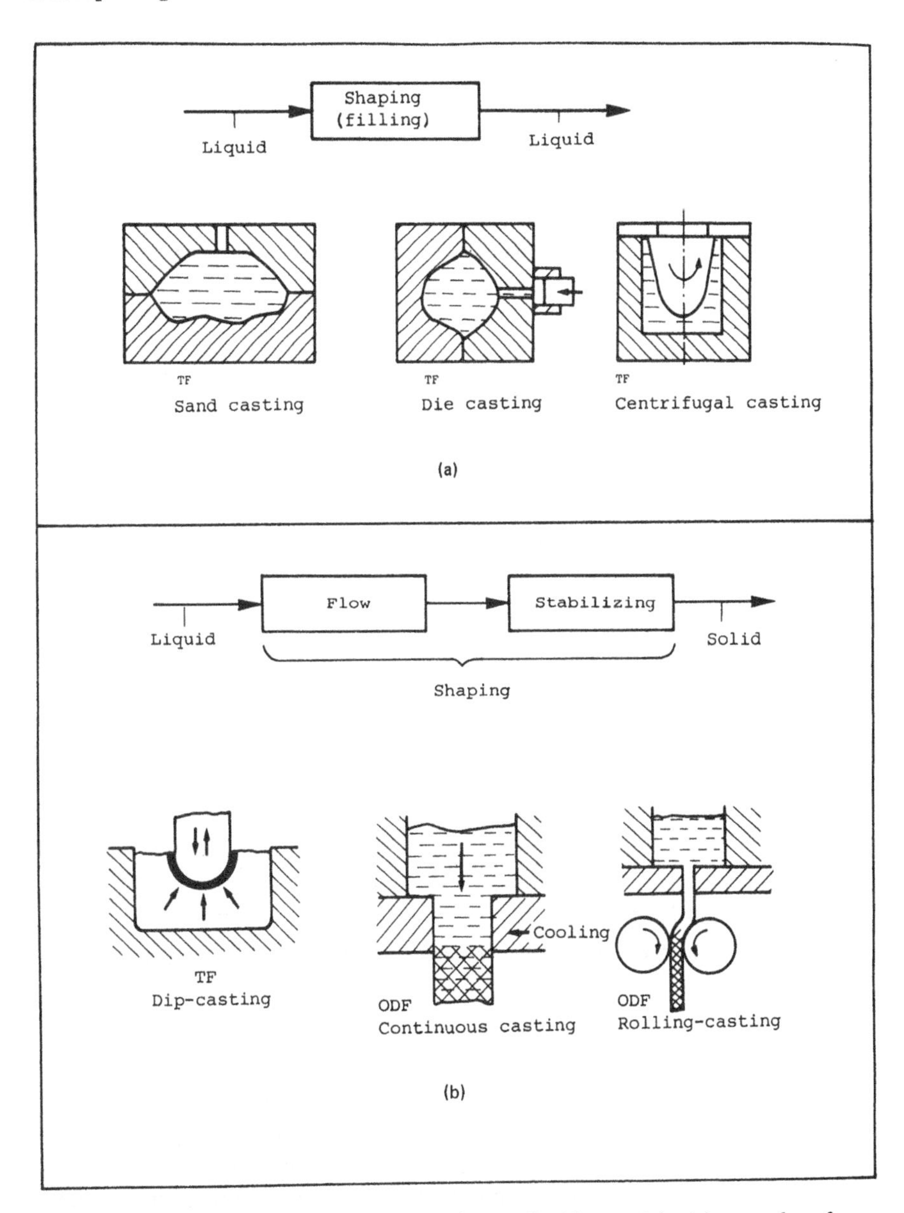

FIGURE 1.25 Examples of shape impressing on liquid materials: (a) examples of processes where shaping and stabilizing are separate; (b) examples of processes where shaping and stabilizing are integrated (TF, total forming; ODF, one-dimensional forming).

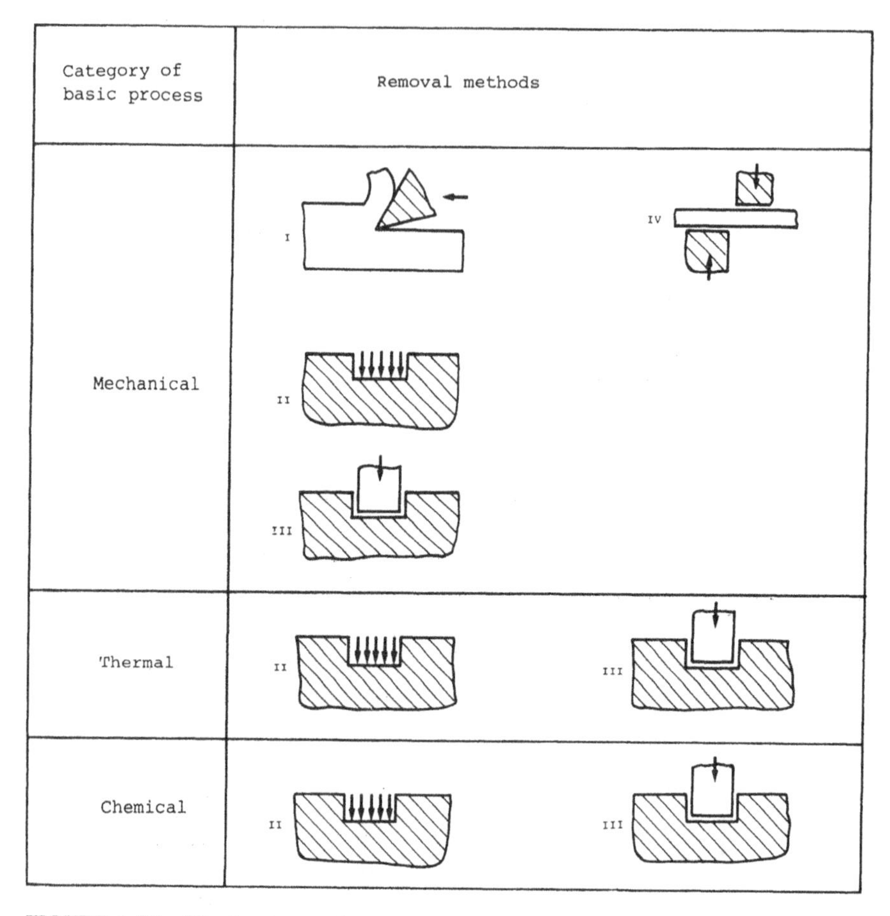

FIGURE 1.26 The fundamental removal methods in mass-reducing processes.

Fundamental Removal Method I, Cutting Processes

This method contains all the common cutting processes (see Fig. 1.26). The primary basic process is fracture, which is created by relative motions between the work material and a rigid medium of transfer (the tool). The motions can be classified as cutting, feeding, or positioning motions. The motions that create the desired surface are the cutting and the feeding motions.

The transfer media or tools can, depending on the number of cutting edges, be divided into:

Single-point tools (well-defined edge geometry)

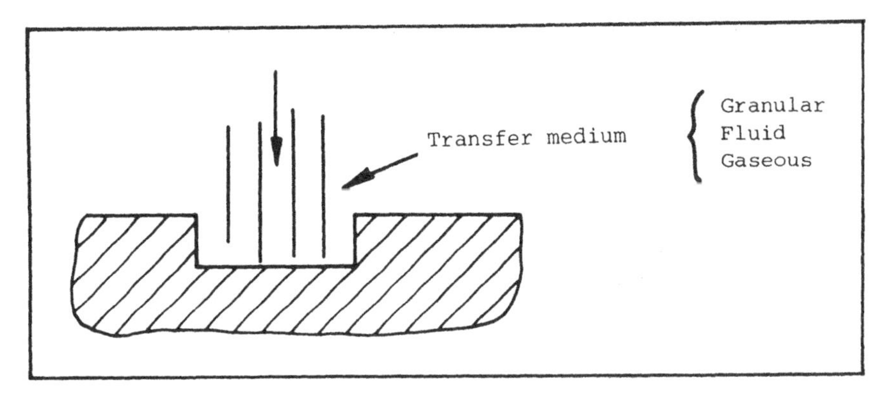

FIGURE 1.27 The fundamental removal method (see Fig. 1.26).

Multipoint tools
Well-defined edge geometry (milling, broaching, etc.)
Undefined edge geometry (grinding, etc.)

By analyzing

The structure of the tool
Single-point tools
Multipoint tools (inclusive of the geometrical arrangement of the cutting edges)
The pattern of motions for the tool
The pattern of motions for the material

a large number of possibilities for forming different geometries are obtained.

Generally speaking, since the contour contents of the tools are small, the pattern of motions play the main role in the creation of the desired surface.

In this field, a huge number of production machines with different patterns of motions (for tools and materials) are available. By choosing between these and using the principles of surface creation, an economic application situation can be obtained. This is discussed further in Chapter 7.

Fundamental Removal Method II

The basic processes are mechanical, thermal, or chemical. The necessary energy (and the impressing of information) is transmitted through granular, liquid, or gaseous media. The energy supply covers a certain surface area, which may be the whole surface, a producer, or a point (see Fig. 1.27).

By moving the energy source and the material, all different patterns in the material can be described. A principal factor here is the geometry of the cross section created by the energy source depending on the process parameters (see Fig. 1.28).

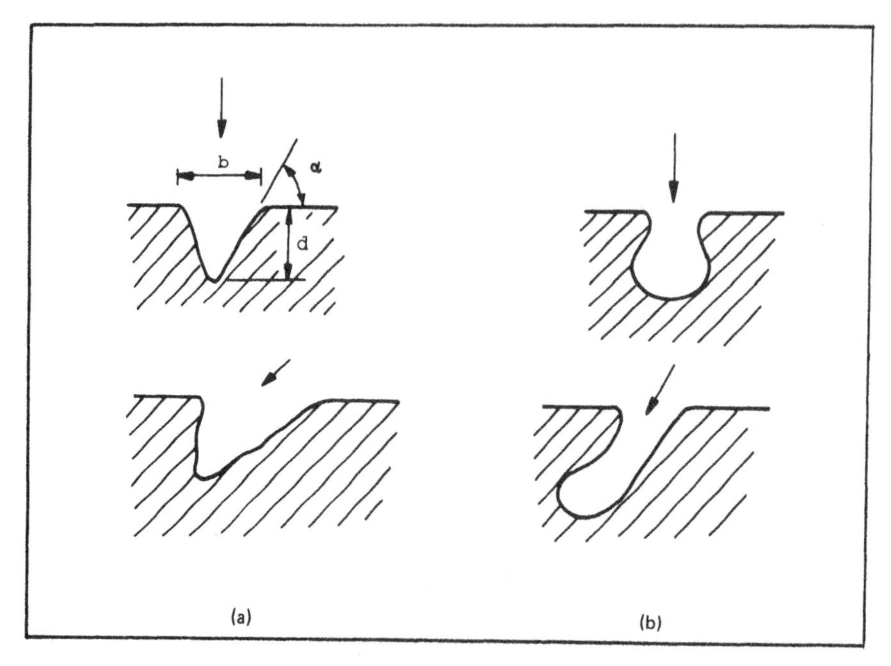

FIGURE 1.28 (a) Open and (b) semiopen cross sections.

The semiopen cross sections, which can, for example, be created by chemical etching, are rarely used. The patterns of motions can be combined or substituted by coating the surface in a given pattern with a medium that prevents the basic process (energy source) from attacking the work material. This principle is widely used in chemical etching.

Examples of manufacturing processes based on this fundamental method (see Fig. 1.26) are abrasive cutting, fluid-jet cutting, electron-beam cutting, thermal cutting, and laser cutting.

Fundamental Removal Method III

The basic processes here are mechanical, thermal, or chemical, and the impressing of information is carried out through a rigid transfer medium (see Fig. 1.29). The surface creation is then a result of the material removal mechanism, the geometry of the rigid medium of transfer, and the pattern of motions (see Fig. 1.29b). The rigid medium of transfer (the tool) is not in direct contact with the work material, since a fluid medium (with solid particles), necessary to establish the basic process, is placed between the rigid tool and the work material

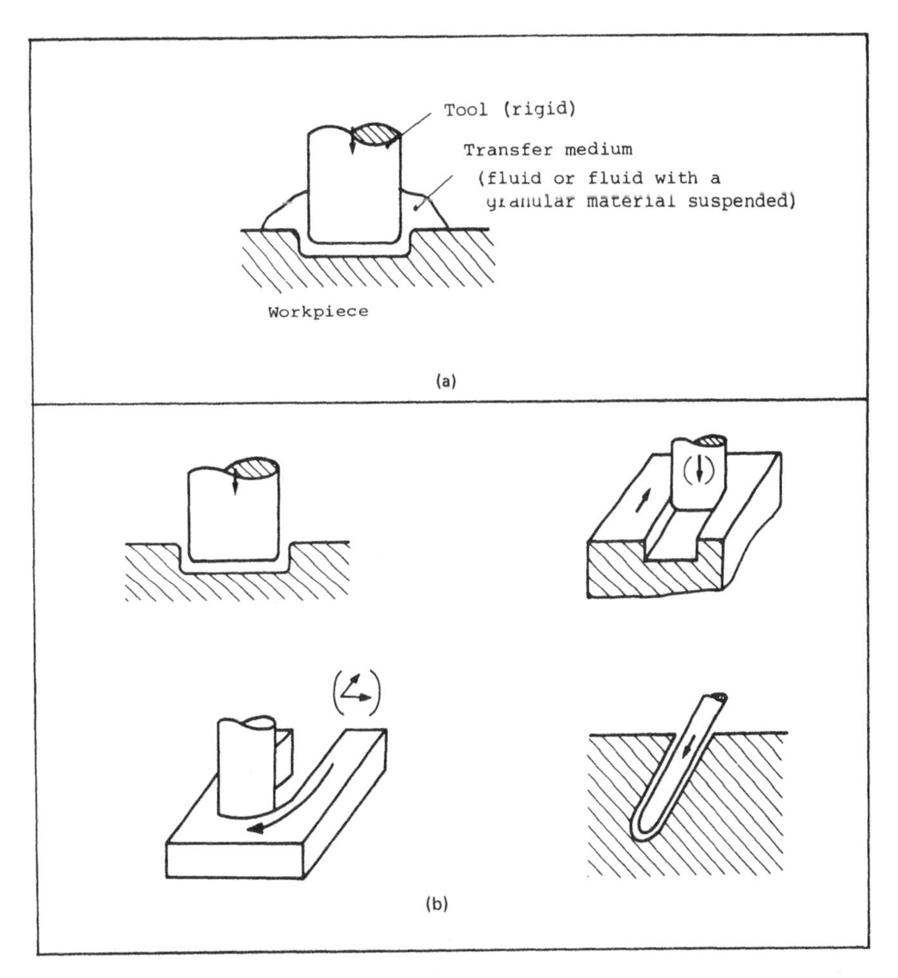

FIGURE 1.29 The fundamental removal method III (see Fig. 1.26): (a) method III (principle); (b) examples of pattern of motions for method III.

(see Fig. 1.29a). This fluid medium normally fills only the small gap between the tool and the work material.

Examples of manufacturing processes based on this fundamental method (see Fig. 1.26) are ultrasonic machining, electrodischarge machining, and electrochemical machining.

Fundamental Removal Method IV

The basic process here is mechanical (fracture) and the medium of transfer is rigid. By varying the geometry of the medium of transfer (the tools/dies) and the

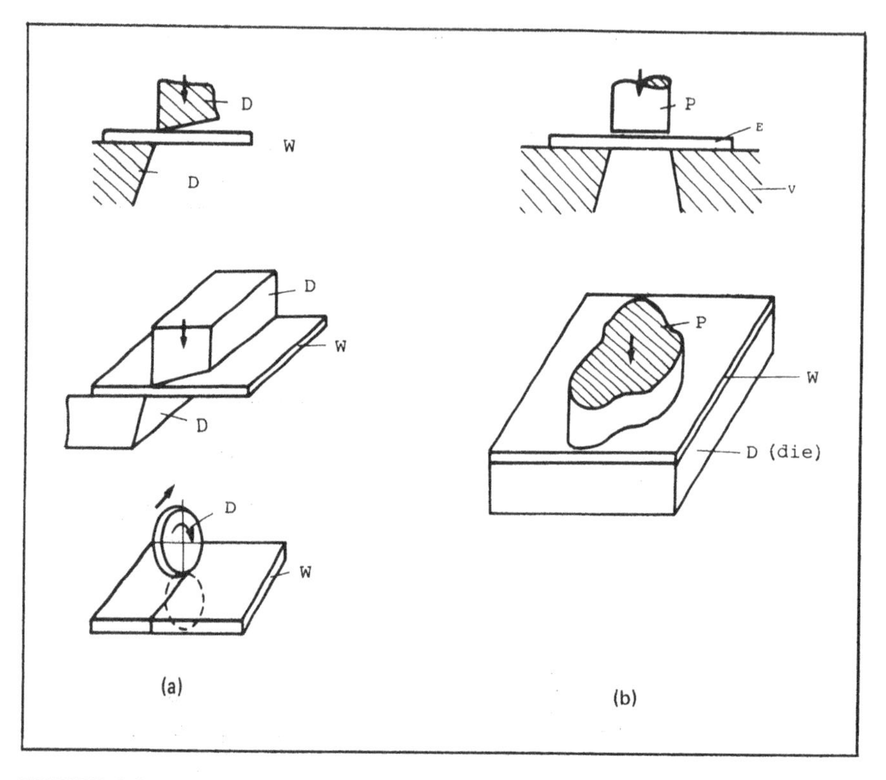

FIGURE 1.30 Fundamental removal method IV (see Fig. 1.26), examples: (a) shearing; (b) blanking (D, die; P, punch; W, workpiece).

pattern of motions, a number of different processes are obtained (see Fig. 1.30). These include blanking, punching, and shearing.

1.6.3 Information Impressing for Assembly and Joining Processes

Assembly and joining processes are not themselves information impressing. The geometry is obtained by positioning and locking together components produced by the former methods. The assembly process can be based on atomic bonding, adhesion, or mechanical locking (with or without separate locking elements). Depending on how the assembly is arranged and the chosen assembly mechanism, the pattern of motions (for the work material and/or the assembly mechanism) will contain none, one, or two relative motions.

Assembly and joining processes are described in Chapter 8.

1.7 SUMMARY

Based on the general, morphological process model (Fig. 1.4) of engineering manufacturing processes, the characteristic flow systems for material, energy, and information have been discussed. Their structure and the possibilities for their realization have been described.

This approach will, based on a general engineering knowledge of materials, physics, chemistry, electrical energy and energy conversion, and so on, give a coherent and systematic understanding of the materials processing field, enabling generic and imaginative applications. This approach is applicable to a wide variety of processes.

It may be difficult at first reading to understand fully the importance of Chapter 1. It is therefore strongly recommended that the reader return to this chapter often while reading the remainder of the book.

2

Properties of Engineering Materials

2.1 INTRODUCTION

For engineers to be able to select a suitable material fulfilling the functional requirements of a component (determined from the performance and economical production requirements), they must have a broad knowledge of:

- The properties of the materials available
- The manufacturing possibilities, including:
  - The suitability of the materials for the various processes (i.e., the important properties affecting processing)
  - The effects of the various processes on the material properties
  - The economics of the various material-process combinations

Only by considering these relationships coherently can a satisfactory selection be made. This means that it is not sufficient to choose the cheapest material that would satisfactorily perform the functions desired, because it may be very expensive to process. The functional properties of a material can be evaluated only when the manufacturing processes have been selected, as the processes normally change the properties. Certain mechanical, physical, and metallurgical changes are generally brought about; some beneficial and others detrimental. This change in properties may sometimes allow the choice of a cheaper material than if the functional requirements alone were considered.

2.2 MATERIAL PROPERTIES

The properties of materials may be divided into the following four groups:

1. Physical properties
2. Chemical properties
3. Mechanical properties
4. Manufacturing properties

Physical properties include color, density, melting point, freezing point, specific heat, heat of fusion, thermal conductivity, thermal expansion, electrical conductivity, magnetic properties, and so on.

Of the *chemical properties*, corrosion resistance plays an important role in the choice of materials and generally includes resistance to chemical or electrochemical attack. Corrosion resistance can also be important during the manufacturing processes because it can influence the formation of surface films, affecting friction and lubrication, and thermal and electrical conductivity.

Mechanical properties generally include the reactions of a material to mechanical loadings. In the majority of cases, it is the mechanical properties with which the engineer is principally concerned in material selection, because to evaluate their performance in terms of the desired functions he or she needs to know how materials would react to the design loadings.

The technological or *manufacturing properties* of a material, which describe the suitability of the material for a particular process, are very complex and can generally not be assessed by a single number. To evaluate these properties, various testing methods have been developed designed to describe the "machinability," "formability," "drawability," "castability," and so on, of a material. These testing methods will not be described in detail, but some of them are mentioned in later chapters. In this chapter, only the mechanical properties of materials and their determination are considered, since they are important to both the design and the manufacturing engineer.

In Chapter 3 the manufacturing properties of the materials and the effects of the various processes on material properties are discussed before a description of the more important materials is given.

2.3 MECHANICAL PROPERTIES OF MATERIALS

To determine the mechanical properties of materials, various standardized testing methods have been developed. The materials are subjected to these laboratory tests under controlled conditions so that their reactions to changes in the conditions may be determined. When using data obtained from such tests, the engineer must be careful to apply the data only to conditions similar to the testing conditions. With caution, and employing a general knowledge of ma-

FIGURE 2.1 Tensile test specimens: (a) cylindrical and (b) flat for sheet metal.

terials, it is sometimes possible to extend the results of tests to other conditions as a first approximation.

In the following sections the tensile test, hardness tests, impact test (Charpy V), fatigue test, and creep test are described [3,4].

2.3.1 Tensile Test (Stress–Strain Diagrams)

From the tensile test, considerable information about the properties of a material can be obtained. A specimen with a standardized geometry is gripped between two sets of jaws and pulled in tension until fracture occurs. To avoid fracture in the gripped sections and to produce uniaxial tension in the central part, the test specimen has enlarged ends. The transition between the enlarged ends and the reduced central portion is gradual (see Fig. 2.1) [3]. Usually, the specimen is cylindrical for bulk materials and rectangular for sheet materials with a relatively large width/thickness ratio.

Many different testing machines are available with a wide range of loading possibilities, and they are often equipped with some type of strain-measuring device, which may be attached to the specimen, permitting accurate measurement of the elongation over a certain gage length during loading. Corresponding values of loading and elongation are recorded automatically and presented as a graph of force against elongation. Of course, all the conditions of the test must be in accordance with the appropriate national standards [3].

The results of a tensile test can be converted into a stress-strain diagram. But before discussing these diagrams, stress and strain will be defined.

Figure 2.2 shows a bar with a uniform cross section in unloaded and loaded conditions. In the unloaded condition, the length of the bar is l_1 and the cross section A_1. When loaded with the force P the length becomes l_2, which means an elongation of $\Delta l = l_2 - l_1$. The elongation per unit length is called the *linear* or *engineering strain* or sometimes the *nominal strain* and is designated by e. Thus

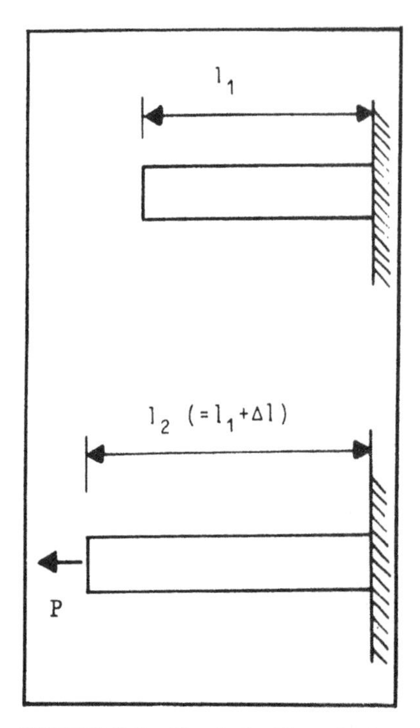

FIGURE 2.2 Tensile loading of a bar.

$$e = \frac{l_2 - l_1}{l_1} = \frac{\Delta l}{l_1}\left(= \frac{l_2}{l_1} - 1\right) \tag{2.1}$$

Nominal strain e is often quoted as a percentage.

The force P, distributed uniformly across the original cross section, results in the nominal or engineering stress, which will be designated by σ_{nom} and is given by

$$\sigma_{nom} = \frac{P}{A_1} \tag{2.2}$$

The stress is measured in N/m^2, lb/in.^{2}, and so on. When the applied load tends to elongate the specimen, the stress is called a *tensile stress*, and when the load tends to compress the specimen, the stress is called a *compressive stress*. Correspondingly, the strains are called *tensile* or *compressive strains*. Tensile stresses and strains are normally defined as positive and compressive stresses and strains as negative, but in some cases it is more convenient to use the reverse conventions.

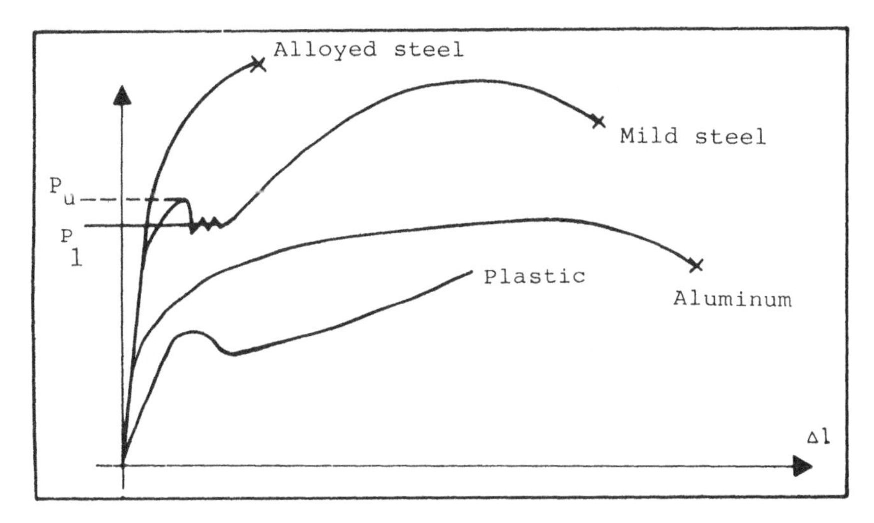

FIGURE 2.3 Typical tensile test diagrams obtained at room temperature and slow speed.

Figure 2.3 shows some typical force–elongation diagrams obtained at room temperature and slow elongation speed (static conditions) for different materials. It is necessary to specify both temperature and speed, as these conditions can influence the shape of the diagrams drastically, effects that are described later. The "static" diagrams play the major role in the choice of materials, since most loading situations can be considered to be static.

The four examples shown in Fig. 2.3 illustrate how the relation between load and elongation varies with the material. The curve for mild steel is unusual for a metallic material, in that after departure from elastic behavior at a load of P_u a nonuniform yielding occurs at a lower load P_l. Most metals exhibit a relation similar to aluminum, where no reduction in load occurs at yield. For plastic materials, the relation shown in Fig. 2.3 is typical. The force–elongation diagram will now be analyzed further with the aid of Fig. 2.4.

In accordance with Eqs. (2.1) and (2.2), the force–elongation diagram can be transformed into a stress-strain diagram by changing the units on the axis to $\sigma_{nom} = P/A_1$ and $e = \Delta l/l_1$. Consequently, the diagram in Fig. 2.4 can be used as both a $P - \Delta l$ and a $\sigma_{nom} - e$ diagram.

When the specimen is loaded, the deformation will be elastic up to point B, which means that for this region, if the load is removed, the specimen will return to its original length l_1. Up to point A the stress is proportional to the strain, and the material obeys Hooke's law, which can be expressed

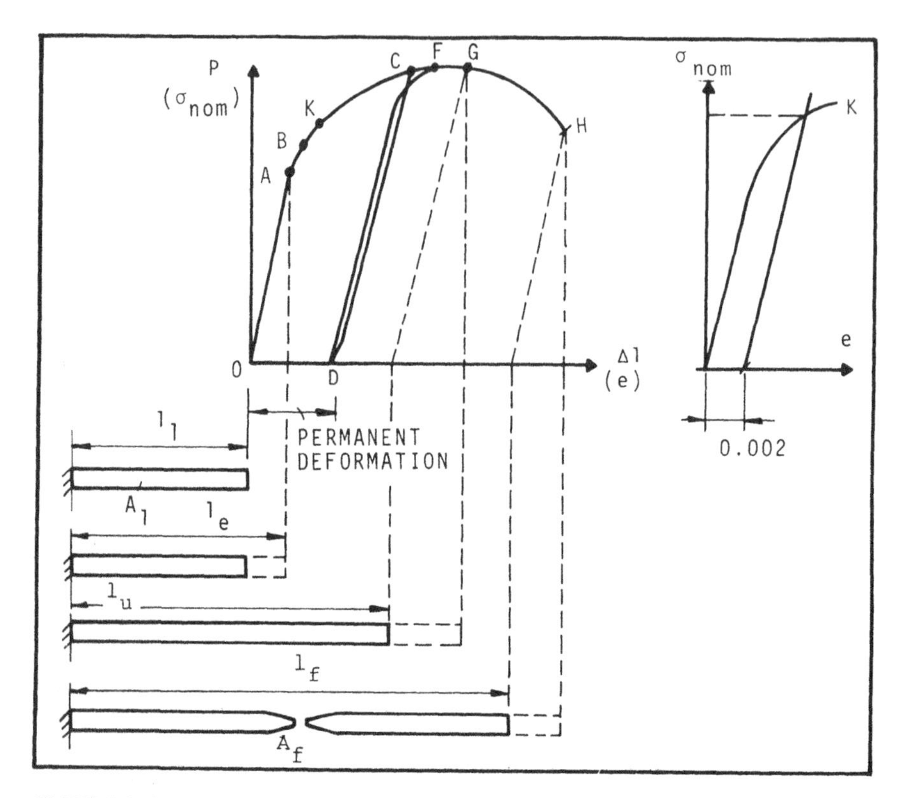

FIGURE 2.4 The force–elongation/stress–strain diagram from a tensile test.

$$\sigma_{nom} = Ee \tag{2.3}$$

where E is known as the *modulus of elasticity* or *Young's modulus*. Young's modulus is a characteristic number for the material or the material group.

If the deformation is continued from A to B, the deformation is still elastic, but the proportionality described by Eq. (2.3) is no longer valid. Point A is called the *proportional limit* and B the *elastic limit*. For most materials the elastic limit is only slightly higher than the proportional limit.

For deformation beyond B, the specimen will not return to its original length because it has now been deformed plastically (i.e., the length has been increased permanently).

The elastic limit B is defined in practice as the stress where the permanent deformation has reached a standardized value, normally between 0.001 and 0.03% of the gage length.

If the deformation is continued to point C, the permanent or plastic deformation after unloading will be OD. If the specimen is loaded again, the curve

DF is followed, and after F the curve that would have been described by uninterrupted loading will be followed. The line DF has the same slope as OA. On a stress–strain diagram, this slope is equal to Young's modulus E. Normally, a small amount of energy is lost in the unloading and reloading cycle as indicated by the exaggerated hysteresis loop between D and C. As shown in the figure, the specimen first starts to deform plastically again at point F, which means that the material has become harder and less ductile than in the original condition. This load increase is due to what is called *work hardening*; that is, as the load producing plastic deformation is increased, a greater load will be required to produce further deformation.

If the deformation is continued beyond point F, the load increases until point G, where it achieves a maximum value, after which it decreases until point H, where fracture occurs.

For metals it has been shown experimentally that no volumetric change occurs during plastic deformation, which means that plastic elongation must be accompanied by a corresponding decrease in cross section (or a contraction in the lateral dimension). For deformations less than that at point G (Fig. 2.4) the cross section decreases continuously, while the load required to continue the deformation increases due to work hardening. At point G, the increase in load due to work hardening is exactly balanced by the decrease in load due to the reduction in the cross-sectional area (i.e., the effects of reduction in area and work hardening balance each other). For deformations greater than that at point G, the contraction in area dominates, the deformation becomes unstable, and a localized necking—reduction in cross section—occurs. The necking takes place in the weakest part of the specimen, and subsequent elongation of the gaging length is entirely due to local elongation in the neck (Fig. 2.4).

Considering the stress in the specimen it is clear that the actual or true stress σ will be given by the load P divided by the current cross-sectional area A_c. Thus

$$\sigma = \frac{P}{A_c} \tag{2.4}$$

Hence, the true stress σ is greater than the nominal stress σ_{nom} (Fig. 2.2) since $A_c < A_1$.

For deformations greater than that at point G, the stress is, because of the necking, no longer uniaxial, and consequently Eq. (2.4) cannot be used without correction. This situation is discussed further in Chapter 4.

For materials that cannot, or can only to a slight degree, be deformed plastically, a diagram corresponding to that for alloyed steel shown in Fig. 2.3 is typical (i.e., fracture occurs before necking or instability arises). For many plastic materials instability does not give rise to necking (i.e., the specimen is deformed uniaxially until fracture occurs).

Referring again to Fig. 2.4, a few definitions should be given.

1. The stress occurring at point G is called the *ultimate strength* (or tensile strength) and is given by

$$\sigma_{nom,uts} = \frac{P_{max}}{A_1} \tag{2.5}$$

where P_{max} is the maximum load applied.
2. The stress corresponding to point H is called the *fracture strength* (or breaking or rupture strength) and is given by

$$\sigma_{nom,f} = \frac{P_f}{A_1} \tag{2.6}$$

where P_f is the load applied when fracture occurs.
3. The axial strain (nominal or engineering) corresponding to point G (before necking starts) is called the *uniformly distributed strain* (or uniform strain) and is given by

$$e_u = \frac{l_u - l_1}{l_1} \tag{2.7}$$

4. The total permanent strain (after fracture) is called *strain* or *percent elongation at fracture* and given by

$$e_f = \frac{l_f - l_1}{l_1} \tag{2.8}$$

5. The *reduction of area at fracture* is defined as

$$RA = \frac{A_1 - A_f}{A_1} \tag{2.9}$$

The last two quantities describe the ductility of the material: the higher the value, the better the ductility. The reduction of area RA is normally preferred as a measure of ductility.

An important property in materials processing is the *yield stress*, which is a stress value, above which the deformation is permanent or plastic (Fig. 2.4). The yield stress is given by

$$\sigma_{nom,0} = \frac{P_K}{A_1} \tag{2.10}$$

where P_K is the load at the actual plastic strain.

Most materials do not have a well-defined yield point, as do mild steel and some plastics (Fig. 2.3). Consequently, it has become standard practice to measure the yield stress at a point on the stress–strain curve corresponding to a permanent strain of 0.002. This yield stress is sometimes called *0.2% proof stress* (see Fig. 2.4) and is given by

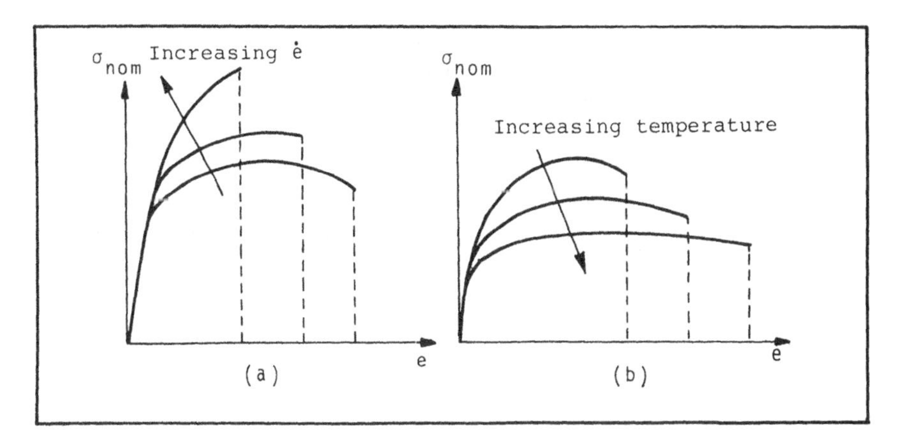

FIGURE 2.5 The influence of (a) strain rate and (b) temperature on the shape of the stress-strain curve.

$$\sigma_{\mathrm{nom},0.2} = \frac{P_K(= P_{0.2})}{A_1} \tag{2.11}$$

For the special case of mild steel (Fig. 2.3) both an upper ($\sigma_{\mathrm{nom},0u} = P_u/A_1$) and a lower ($\sigma_{\mathrm{nom},0l} = P_l/A_1$) yield stress can be defined.

The yield stress and the ultimate strength are each a measure of the strength of the material, whereas the elongation and reduction of area are measures of its ductility. The relative values depend on the work-hardening characteristics of the material.

The shape of the stress–strain curve is affected by both the temperature and the strain rate as shown in Fig. 2.5. The strain rate is defined by

$$\dot{e} = \frac{de}{dt} = \frac{\dfrac{x(l_2 - l_1)}{l_1}}{dt} = \frac{(l/l_1)xl_2}{dt} = \frac{v}{l_1} \tag{2.12}$$

where v is the test velocity. This means that the nominal or engineering strain rate is the testing velocity divided by the original length of the specimen. Figure 2.5a shows that to obtain the same strain for increasing strain rate, an increasing stress is required. Many materials have a higher strain rate sensitivity at elevated temperatures. At room temperature the strain rate sensitivity is generally small. Figure 2.5b shows how the temperature influences the shape of the stress-strain curve. At increasing temperatures the strength decreases and the ductility increases (i.e., the opposite of an increasing strain rate). At a certain temperature the yield stress becomes independent of the strain.

Since many manufacturing processes occur under compressive loads, it is often convenient to determine the properties of the material in a compression test so that the test situation is closer to the actual situation. The compression test is also particularly suitable for materials with low ductility (brittle materials such as concrete, glass, wood, and cast iron), for which the tensile test produces fracture at very low strains.

With ductile materials, the cylindrical compression test specimen develops a barrel shape, due to friction between the ends of the specimen and the compression plates. Thus, to determine the true material properties, the influence of friction must be minimized or allowed for. The compression test will not be described in detail.

2.3.2 Hardness Tests

The *hardness* of a material is an important property for many applications and can be defined as the resistance to indentation of a material or its resistance to scratching or wear. These definitions do not describe the same properties and, therefore, hardness has to be related to the testing method employed. There is no exact correlation between the test results of the various tests since they measure different phenomena. Hardness tests are therefore comparative, as it is difficult to relate the behavior of the material in the test to its behavior in other situations.

In the present context, hardness defined as the resistance to indentation is most important. The indenter may be a ball, a pyramid, or a cone with higher hardness than the material to be tested. In the following sections the most common standard hardness tests are described.

Brinell Hardness Test

In the Brinell hardness test, a hardened steel ball with a diameter D is pressed into a smooth surface of the material with a load P, and the mean diameter d of the resulting impression is measured with a low-powered microscope fitted with a suitable scale. The Brinell hardness number HB is defined as the load divided by the surface area of the indentation. Hence,

$$\mathrm{HB} = \frac{0.102P}{(\pi D/2)[D - (D^2 - d^2)^{1/2}]} \tag{2.13}$$

where P is the applied load measured in newtons, D the ball diameter in millimeters, and d the mean indentation diameter in millimeters. The hardness number is always quoted without units, although in reality it has the units of pressure. To maintain the hardness numbers from the previous system of units, it is necessary to include the factor 0.102 in Eq. (2.13).

The applied loads are standardized and correspond to masses of 500, 1000, and 3000 kg, depending on the material being tested. Usually, a 3000-kg mass is used and applied for a time of 10–15 s for steel and cast iron. Recommendations concerning the test conditions for the various metals are given in the appropriate standards. If standard conditions are not used, it should be indicated as

HB $D/P/t$. In practice the Brinell hardness number is determined from tables which give the hardness number for a given diameter of the indentation. The Brinell test can be compared roughly with the compression test. The following approximate relationship exists between the Brinell number of the ultimate strength for non-strain-hardening or heavily cold worked materials:

$$\sigma_{nom,uts} \approx 3.3 \text{ HB N/mm}^2 \tag{2.14}$$

Vickers Hardness Test

In the Vickers hardness test a diamond pyramid (angle 136°) with a square base is used as an indenter, and the hardness is defined as the load divided by the contact area:

$$\text{HV} = \frac{0.102P}{d^2/2 \sin 68°} = 0.189 \frac{P}{d^2} \tag{2.15}$$

where P is the load in newtons and d is the mean length of the diagonals of the indentation in millimeters. The hardness number is again given without units.

The mass of the applied load can vary between 2 and 120 kg (in standardized steps), depending on the material. The Vickers hardness number is independent of the load, and the Vickers test can be used for very hard materials, whereas the Brinell test depends on the load and can be used satisfactorily only for materials having a Brinell hardness of less than 500. The applied load in a Vickers test is indicated after the letters HV: for example, HV30.

Rockwell Hardness Test

In the Rockwell test the indenter is either a diamond cone (C) with an included angle of 120° or a steel ball (B) with a diameter of 1/16 in. The diamond cone is usually used and the Rockwell hardness is then identified by the letters HRC.

A preload is first applied to the indenter using a mass of 10 kg to seat it in the material and the indicator is then set at zero. The major load corresponding to a total of 150 kg is then applied, and the indentation depth e is measured after the load is removed. The HRC number is determined from the expression

$$\text{HRC} = 100 - e \tag{2.16}$$

where e is measured in multiples of 0.002 mm. The Rockwell test is suited to rapid and reliable routine inspection.

A summary of the three hardness tests is presented in Table 2.1. The Vickers test is the most commonly used hardness test.

Other Hardness Tests

To determine the hardness of very soft materials, for example, rubber and plastics, the durometer test is suitable. In this test the resistance to elastic deformation is used as a hardness number.

TABLE 2.1 Summary of the Brinell, Vickers, and Rockwell C Hardness Tests

Testing method	Indenter	Shape of indentation From side From above	Load: mass (kg)	Hardness
Brinell	10-mm ball	D, d; d	500, 1000, 3000	$HB = \frac{0.102P}{(\pi D/2)[D - (D^2 - d^2)^{1/2}]}$ (P measured in N, D and d in mm)
Vickers	Diamond pyramid	136°; d_1, d_2	2–120	$HV = 0.189 \frac{P}{d^2}$; $d = \frac{d_1 + d_2}{2}$ (P measured in N, d in mm)
Rockwell C	Diamond cone	120°, e	150 = 10 + 140	$HRC = 100 - e$ (e measured in multiples of 0.002 mm)

The hardness tests described above are based on one or another form of deformation, but as mentioned earlier, hardness can also be defined as resistance to being scratched. Here the Mohs hardness scale, which is based on an arrangement of 10 minerals in order of ascending hardness (talc 1, gypsum 2, calcite 3, fluorite 4, apatite 5, orthoclase 6, quartz 7, topaz 8, corundum 9, and diamond 10) is used. According to this scale, a given material should be able to scratch any material with a lower Mohs number. For example, glass has a hardness of 5.5; hardened steel, 6.5.

2.3.3 Dynamic Tests

In many applications, components are subjected to dynamic loads with a wide spectrum of characteristics, for example, very rapid loading, repeated variations in loads and stresses, sometimes changing from tension to compression, and so on. Most dynamic tests do not give results that can be used in design work, but they are very useful in the classification of materials relative to each other in terms of their behavior when subjected to certain loads. In the following sections some of the most commonly used dynamic tests are described briefly.

Impact Test (Charpy V-Notch)

The most commonly used impact test is the Charpy V-notch test, wherein the energy required to break a standard specimen using an impulse load is measured. The specimen is 10 mm square and 55 mm in length and centrally notched on one side with a 2-mm-deep, 45° included angle notch with a 0.25-mm bottom radius. The specimen is arranged as a simply supported beam with 40 mm between the supporting points, and a pendulum (variable height and mass) strikes it on the side opposite the notch. From the mass of the pendulum and its height before and after the impact, the energy absorbed can be calculated. The amount of energy absorbed is a measure of the brittleness of the material.

A brittle fracture is characterized by a small amount of absorbed energy. Since the tendency to fracture changes with temperature, the impact test is often used to determine the *transition temperature*, below which the material exhibits brittle behavior (notch brittle) and above which it exhibits ductile behavior. If a component is to be used at low temperatures under dynamic loadings, to avoid failure it is important to know the transition temperature of the proposed material.

Fatigue Test

It is a known phenomenon that metals, in general, cannot withstand cyclic variation of stress at high stress levels for a long time. The type of failure occurring under these circumstances is called *fatigue failure*. The stress situation can be characterized by R, the amplitude of the stress variations, and the mean stress M. If M is zero, it is found that the value of R that will cause failure if repeatedly applied is much smaller than the stress to cause failure in a single pull.

From fatigue tests, the greatest stress (*R*, *M*) at which failure does not take place after a certain number (10^6–10^8) of cycles of loading is determined and defined as the fatigue strength of the material. It is found that a limiting stress level exists below which the material will not fail regardless of the number of cycles. This limiting stress is called the *endurance limit*. The fatigue strength and the endurance limit vary over a wide range for different materials.

The usual form of a fatigue test is where a cylindrical specimen gripped at one end is simultaneously rotated about its axis and loaded as a cantilever beam. The specimen is thereby subjected to alternating bending stresses, that is, a sinusoidal variation of stress with (depending on the axial loads) different mean stresses. Different types of test equipment have been developed, but these will not be described here. In all cases testing is carried out in accordance with the national standards, which can also help the selection of testing equipment.

Dynamic Tensile and Compression Tests

In the tensile and compression tests described earlier, the loading rate is so slow that the stress–strain diagram represents, in reality, a continuous series of equilibrium states. If the loading rate is increased, the shape of the stress–strain curve ($\sigma_{nom} - e$) will change. The amount of change depends on the material, the loading rate (stress rate or strain rate), and the temperature. In general, the $\sigma_{nom} - e$ curve is raised as the loading rate is increased, which means that a higher stress is necessary to give the same strain (see Fig. 2.5). Changing the temperature has the opposite effect. Many plastic materials are particularly sensitive to the loading rate. As mentioned previously, most metals are not very sensitive to loading rate at room temperature, but at higher temperatures the sensitivity can become significant.

2.3.4 Creep Test

If a material is subjected to a constant load and the load is applied for a long period of time, the material will deform permanently with time (i.e., it creeps). All metals creep under load at sufficiently high temperatures. At temperatures below 40% of the absolute melting point, creep is normally not a problem; consequently, creep is generally of concern with materials subjected to elevated temperatures and is a long-term effect.

In the creep of a tensile test specimen under constant stress, three different stages exist (see Fig. 2.6). In the primary state, the strain increases rapidly toward a minimum constant rate, in the secondary stage the strain continues at a constant rate, and in the tertiary stage the strain rate increases until fracture.

For plastic materials, the creep is often a problem at room temperature or at slightly elevated temperatures. For metals creep usually becomes a problem at

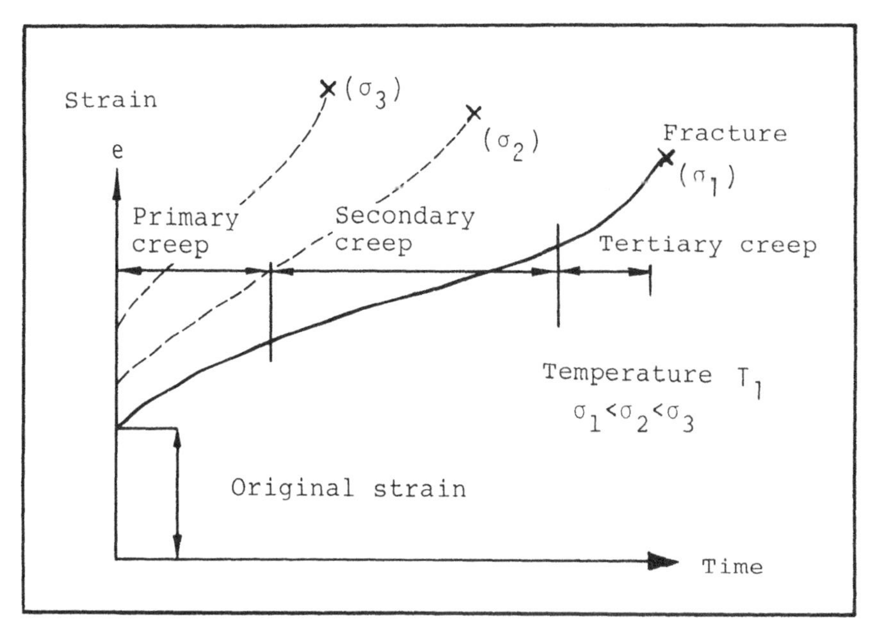

FIGURE 2.6 Typical creep curves for a tensile test illustrating the three stages in creep.

relatively high service temperatures. Many high-temperature, creep-resistant alloys have been developed for uses in steam and gas turbines, high-temperature pressure vessels, power plants in general, and so on.

More detailed descriptions of the material properties and their determination (testing methods) can be found in the literature and in the appropriate standards.

3
Engineering Materials

3.1 INTRODUCTION

In the last few decades, very rapid development of engineering materials has taken place, resulting in a huge number of commercially available materials with a wide spectrum of properties. Therefore, only a general, simplified introduction to engineering materials is given in this chapter. The structure of materials, the dependence of the properties on the structure, and so on, are not described in detail but are mentioned where necessary for a basic understanding [5,6,7].

As mentioned in Chapter 2, the engineer's choice of material is based not only on the physical, chemical, and mechanical properties but also on the technological properties, which describe the suitability of a material for a particular manufacturing process [8–12].

In this chapter a short discussion of the important material properties in manufacturing is given as an introduction to the description of the different material groups. The effects of the processes on the materials to which they are applied are mentioned only briefly, as a more detailed treatment requires a greater background knowledge of processes and materials.

3.2 IMPORTANT MATERIAL PROPERTIES IN MANUFACTURING

As mentioned in Chapter 2, it is very difficult to state exactly which properties or, more correctly, which combination of properties a material intended for a given process must possess. But it is often possible to identify certain dominating properties or characteristics which any material must have for it to be processed by a given process or process group. To evaluate these technological properties, many specialized test methods have been developed which describe in one way or another the suitability of a material for the particular process or group of processes. The testing methods can normally be applied only over a limited range, and the result should be judged with caution. A description of these testing methods can be found in the literature [12].

3.2.1 Forming from the Liquid Material State

Forming from the liquid state of a material includes the following phases (Chapter 1):

Phase 1: melting
Phase 2: forming (creation of shape)
Phase 3: solidification (stabilization of shape)

In practice, phases 2 and 3 can be more or less integrated.

Forming from the liquid state requires primarily that the material can be melted, and that furnace equipment to do this is available. This depends on the level of the range of melting points or temperatures and the requirements of the furnace equipment in producing a complete melt. These requirements depend on the chemical composition of the material, its affinity to the surroundings, its gas absorptions, and other factors. If the melt can be produced, the next question is the availability of a suitable mold or die material for an appropriate solidification.

The melting temperatures for some common pure metals are listed in Table 3.1a. The alloyed metals, which have the greatest industrial importance, do not have a melting point but, rather, a melting-temperature range defined by the *solidus temperature*, below which the material is solid, and the *liquidus temperature*, above which the material is liquid. Between the solidus and liquidus temperatures, a mixture of liquid and solid material exists. The melting-temperature range plays an important role in the solidification process, and is discussed later. The melting-temperature ranges for some common industrial alloys are given in Table 3.1b.

During solidification, the change in volume associated with the transition from the liquid to the solid state plays a very important role, as this determines how much molten metal it is necessary to supply during solidification. All met-

TABLE 3.1 Examples of Melting-Point Temperatures and Melting-Temperature Ranges

a. Pure metals	(°C)		(°C)
Iron	1535	Lead	327
Copper	1083	Tin	232
Aluminum	660	Magnesium	650
Nickel	1455	Chromium	1850
Zinc	419		

b. Alloys	(°C)
Stainless steel (18% Cr, 9% Ni)	1400–1420
Brass (35% Zn, 65% Cu)	905–930
Bronze (90% Cu, 10% Sn)	1020–1040
Aluminum-bronze	1050–1060
Aluminum (1% Si, 0.2% Cu)	643–657

Source: From Ref. 12.

als, except bismuth, antimony, and silicon, contract during solidification, which means that material will be missing in the central region of the component, as solidification starts at the outside of the component. Compensation for this is established by placing risers (reservoirs of molten metal) on the component. These risers must be arranged so that they are the last to solidify (see Chapter 10).

The volume of solidification contraction is, for example, for cast iron about 2%, for cast steel about 3%, and for aluminum alloys about 3.5–8.5%. The magnitude of the solidification contraction has a primary influence on the required size of the risers.

After solidification, the component is cooled down to room temperature, resulting in a uniform solid-state contraction determined by the difference between the melting temperature and the room temperature multiplied by the average thermal expansion. This solid-state contraction or shrinkage must be compensated for by having a slightly longer pattern or mold so that the cooled component will have the right dimensions (see Chapter 10).

As mentioned, the magnitude of the melting-temperature range plays an important role in the solidification of the material. Increasing solidification range increases the risks of internal porosity, hot tearing, and segregations. *Internal porosity* is created when partly solidified material stops adequate feeding of molten material from the risers. *Hot tears* occur due to high temperatures in the mold, where contraction is physically prevented resulting in large tensile strains. *Segregation*, nonuniform distribution of the material constituents, is generally produced by a large freezing range, where the composition of the remaining molten material gradually changes as the temperature reduces.

Other properties of importance in forming from the liquid state include the specific heat, thermal conductivity, and viscosity of the material.

To minimize some of the problems described, many different casting alloys have been developed. It should also be mentioned that continuous research and development into casting processes is being carried out to increase the range of applicable materials.

3.2.2 Forming from the Solid Material State

Forming from the solid material state can be carried out by mass-conserving processes, mass-reducing processes, or joining processes.

Mass-Conserving Processes

In the forming of metals, the primary basic process is mechanical plastic deformation (see Chapter 1). The suitability of a material to undergo plastic deformation is determined primarily by its ductility (measured by the reduction of area in the tensile test). The amount of plastic deformation necessary to produce the desired component depends on the chosen surface creation principle and the intended increase in shape information. In other words, the ductility of a given material decides the surface creation principle and the information increase obtainable without fracture.

Stress-strain curves are the most important information source when evaluating the suitability of a material to undergo plastic deformation. The strain at instability, the percent elongation, and the reduction of area are the most important characteristics. For most forming processes, there is a good correlation between the reduction of area and the "formability" of the material. The stress–strain curves also reveal the stresses necessary to produce the desired deformation. The stresses and strains and the resulting forces, work, and energy are important for tool or die design and for the choice of process machinery.

As mentioned previously, the conditions under which a given process is carried out can influence "formability" to a great extent. The important parameters are state of stress, strain rate, and temperature. Concerning the state of stress, it can be stated that forming under compressive stresses is generally easier than under tensile stresses, since the tendencies toward instability and tensile fracture are suppressed. Furthermore, a super-imposed hydrostatic pressure increases formability (ductility) and is utilized in certain processes. In most processes the state of stress varies throughout the deformation zone; therefore, it can sometimes be difficult to identify the limiting state of stress. These problems are discussed further in Chapters 4 and 6.

As seen in Fig. 2.5, the strain rate also influences the ductility of a metal. Increased strain rate leads to decreased ductility and an increase in the stresses required to produce a certain deformation. The most commonly utilized industrial processes are carried out at room temperature; consequently, the strain rate does not create problems. However, for those processes that are carried out at elevated temperatures, the effects of strain rate must be taken into consideration (see Fig. 2.5). High temperatures can result in a material with a constant flow stress (yield stress) which is independent of the strain. In this state the material is able to undergo very large deformations, as the temperature is above the recrystallization temperature, where new strain-free grains are produced continuously and almost instantly. These "hot working processes" do not create serious problems in the deformation phase when the strain rate is controlled.

The discussion above is valid for most metals, with some exceptions: for example, cartridge brass, which exhibits a tendency to brittleness at temperatures above the recrystallization temperature.

Mass-Reducing Processes

The basic processes of the mass-reducing type are mechanical, fracture (ductile or brittle); chemical, dissolution and combustion; or thermal, melting.

Industrially, mass-reducing processes based on fracture are the most important, as they include all the cutting processes. The suitability of a material for cutting processes is often called its machinability. *Machinability*, which depends on many different material properties, is a measure of how well the interaction between the cutting tool and the material takes place. The parameters covered by a machinability index can be tool wear, surface quality, cutting forces, or chip shape. Tool wear is often considered to be the main criterion, and standardized testing procedures have been developed (see Chapter 7).

Machinability depends primarily on:

1. The mechanical properties of a material (ductility and hardness)
2. Its chemical composition
3. Its heat treatment (structure)

Concerning mechanical properties, it can be stated that low ductility, low strain hardening, and low hardness give good machinability. Correspondingly, this means that materials with high ductility and high strain hardening are difficult to machine. For many materials (e.g., cast iron) hardness is a reasonably good indication of the ease with which the material can be machined.

The composition of a material has a great influence on its machinability. By adding small amounts of lead, manganese, sulfur, selenium, or tellurium, machinability can be increased considerably without changing the mechanical properties.

As to the structure of a material, it must be as homogeneous as possible without abrasive particles and hard inclusions, as these increase tool wear and result in poor surfaces.

For those mass-conserving processes that use chemical basic processes, the mechanical properties play a minor or no role at all, with the chemical and electrochemical properties playing a major role. This means that a hardened material is as easy to process as a nonhardened material. For example, in the electrochemical machining process, the material removal rate is determined solely by Faraday's laws.

Combustion, which is utilized in torch cutting, for example, requires that it is possible to burn the material using a supply of oxygen. It is possible to cut steel and cast iron (<2.5% C); however, stainless steel cannot be cut by this process.

The processes based on the thermal basic process of melting (cutting and electrodischarge machining, for example) require that the material can be melted by an appropriate energy source. After melting, the material must be removed from the machining zone. These processes are largely influenced by the thermal properties of the material (i.e., its thermal conductivity, heat capacity, specific heat, etc.). Low thermal conductivity as well as low heat capacity decreases the energy requirements and minimizes the heat-affected zone.

Joining Processes

Only the main type of joining process, fusion welding, is discussed here. The *weldability* of a material is, like the other technological properties, difficult to define. Many factors, such as those mentioned under forming from the liquid material state, influence the welding properties of a material. Chemical composition and the constituent's affinity to the surroundings have a great influence, as contaminations, gas absorptions, structure, and so on, depend on these factors. In addition, cooling conditions influence the resulting internal stresses and the final hardness of the material.

3.2.3 Forming from the Granular Material State

For this process area, it is rather difficult to define the material properties that determine the suitability of the granular material for compaction and sintering.

All materials that can be produced in the granular state can be compacted and sintered but, depending on the particular material, it may be difficult to develop suitable compaction and sintering processes. In general, the functional requirements, not the process itself, dictate the material to be used.

3.3 EFFECT OF THE PROCESSES ON THE MATERIAL PROPERTIES

The original properties of the material, the actual basic process, and the conditions under which it is carried out determine the final complex of material properties. By varying the governing parameters of the process, it is possible to vary the final properties, sometimes within rather wide limits.

Depending on the process and the material, the properties that are affected will fall within one or more of the following groups: physical properties (corrosion resistance, metallurgical changes); mechanical properties (strength, hardness, ductility); and technological properties ("formability," "machinability," "weldability"). It must be remembered that some of the changes that take place are beneficial and some detrimental. In the actual situation, normally only a few of the affected properties are important for the performance of the desired functions.

During processing, various defects (micro- or macrofractures, porosities, nonuniform distribution of properties, etc.) are often introduced into the material and may influence the performance of the component drastically. The type and character of these defects must be carefully analyzed.

In forming from the liquid material state, the final material properties depend mainly on the composition (including solidification temperature range), the thermal and mechanical properties of the molding or die material, and the solidification conditions (direction, rate, etc.). In forming from the solid material state by plastic deformation, the amount of deformation, the temperature, and the rate of deformation primarily determine the final properties. Cold deformation increases the strength and decreases the ductility of the material. Hot deformation gives poor surface quality and reasonably good mechanical properties. Solid-state forming by machining (mass-reducing processes) primarily influences the surface properties (roughness, hardness, internal stresses, etc.).

The examples mentioned only serve to illustrate the complexity of the evaluation of the final material properties of a component. These problems are discussed in more detail in some of the later chapters.

3.4 CLASSIFICATION OF MATERIALS

As mentioned previously, it is very difficult to provide broad information regarding all the important engineering materials in this context. Consequently, only a general survey will be given to allow a rough evaluation of the suitability of the different material groups for various processes. From this survey and the process descriptions in the later chapters, a reasonable background for the evaluation of the final properties of the materials will be available.

Engineering materials can be divided into groups showing important relationships. In this context the traditional classification shown in Fig. 3.1 will be followed.

The main groups are metallic materials, nonmetallic materials, and composite materials. Composite materials are built up from two or more materials, so that new and special properties are obtained. Metallic materials are subdivided into ferrous and nonferrous metals. The nonmetallic materials are subdivided into polymers, ceramics, and glasses, but the group covers many other

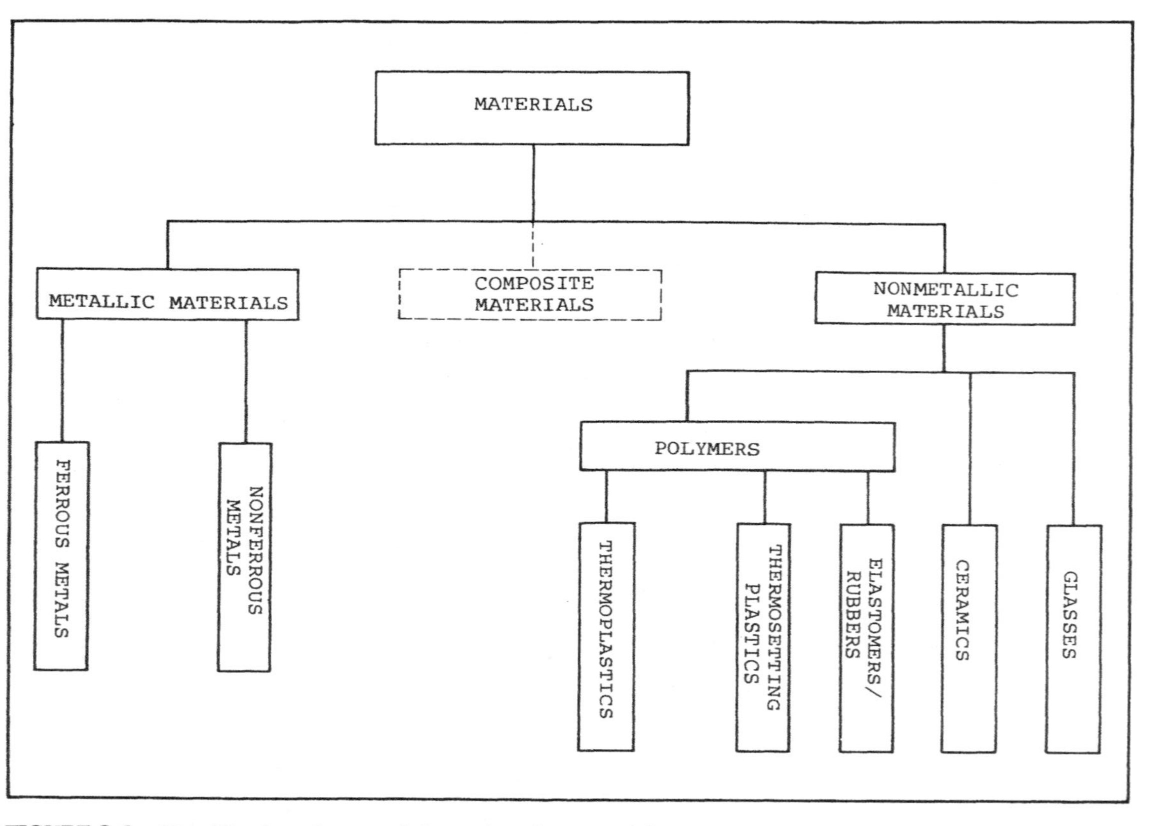

FIGURE 3.1 Classification of some of the engineering materials.

materials (wood, concrete, bricks, etc.) that are not important for the present discussion.

3.5 METALLIC MATERIALS

3.5.1 Bonding and Structure

Metals are characterized by the *metallic bonding*, where the metal ions are held together by an "electron cloud." This type of bonding has a high mobility of the free (valence) electrons and accounts in general for the high strength level, the ductility (ability to be deformed without fracture), and the relatively high melting temperature of metals. These general tendencies can be influenced by many factors; consequently, exceptions are common.

Metals have a crystalline structure with predominantly body-centered cubic, face-centered cubic, or close-packed hexagonal lattice structures. Crystalline materials normally consist of thousands of small individual crystals or grains, depending on the production method. During solidification, many individual lattices begin to form at various points within the melt. As solidification proceeds these crystals or grains, which have random orientation, grow, meet, and form the *grain boundaries* (Fig. 3.2), where a high degree of disorder in the atomic arrangement exists [1,5,6].

The grain boundaries have a dominating influence on the properties of the metal. The number and size of the grains are functions of the rate of nucleation of grains and the rate of grain growth. Once a metal has solidified, the number and size of the grains can be changed by deformation or heat treatment, which will allow its mechanical properties to be varied within rather wide limits. The following equation illustrates, for iron, the influence of the grain size on the yield stress:

$$\sigma_0 = k_1 + \frac{k_2}{\sqrt{D}} \tag{3.1}$$

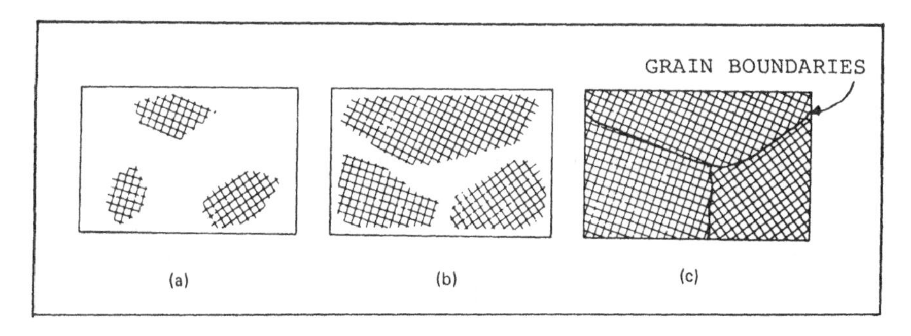

FIGURE 3.2 The formation of grain boundaries [1].

where σ_0 is the yield stress, k_1 and k_2 are material constants, and D is the average grain size. This means that a small grain size gives a high yield stress.

Depending on the composition of the metal and the solidification conditions, different characteristic grain structures or patterns will be formed. The grain structure, which also has a significant influence on the properties of the metal, can be changed by deformation and heat treatment.

The individual grains or crystals have various faults or defects in their lattice structure; these faults can influence the material properties strongly. The size and number of defects are dependent on the history of the material, including cooling conditions after solidification and deformation, as, for example, in rolling, forging, or extrusion.

In normal polycrystalline materials the grain boundaries can also be considered as defects in the lattice structure. The type, size, number, and distribution of all the defects largely determine the mechanical properties of the material. Also, as mentioned, heat treatment and deformation can influence the defects and thus the properties.

The most important properties affected by the defects are yield stress, ductility, ultimate stress, hardness, and electrical conductivity. Properties such as melting point, heat capacity, thermal expansion, and elastic constants are not influenced by the defects.

3.5.2 Strength-Increasing Mechanisms

The mechanical properties of metals are generally the most important for the engineer; consequently, great effort has been directed toward increasing the strength of metals leading to more favorable strength/weight ratios. The strength of metallic materials can, in general, be increased by:

1. Phase transformations in the solid state, that is, hardening from:
   a. Martensitic (diffusionless) transformations
   b. Precipitation and solution (diffusion)
2. Strain hardening
3. Dispersion hardening (see Section 3.11.1)

Hardening by Solid-State Phase Transformations

The main purpose of a manufacturing process is to produce a component with a desired geometry and desired properties. During processing the material properties may be changed in a beneficial way, but very often it is necessary to increase the strength properties of the component to obtain the intended functional performance. Most metallic materials allow phase transformations in the solid state after the shaping process without changing the general geometry, which makes it possible to control the structure and thus the properties within rather wide limits.

Phase transformations are usually achieved in several stages: (1) heating to and holding at an elevated temperature for such time that phase equilibrium is closely approached, (2) cooling (i.e., controlled lowering of the temperature), so that the existing phases are no longer in equilibrium, causing a phase change. The most desirable structures are normally obtained by interrupting progress toward a new equilibrium structure before it is reached. Practical technical operations, which are associated with such phase transformations, are called heat treatment.

Depending on the type of phase transformation, two main categories can be identified: (1) diffusionless or martensitic transformations and (2) diffusion transformations. Many different diffusion-based transformations exist, but only a few will be discussed.

Martensitic transformations are closely associated with steel, as the most important applications are with steels; however, martensitic transformation can also occur in other metallic materials. For steel the temperature in the first stage is kept at 800–900°C until the face-centered cubic equilibrium structure (austenite) has been obtained. By quenching, a body-centered tetragonal structure is formed due to the fact that the body-centered cubic structure (of ferrites) becomes supersaturated with carbon atoms; the result is a deformed and stressed tetragonal structure called martensite. The martensitic structure is hard and brittle; the degrees of hardness and brittleness depend on the carbon content. For example, for a steel with 0.2%C the martensitic hardness is about HV $\simeq$ 400 while the original hardness is HV $\simeq$ 100; for steel with 0.4%C the figures are HV $\simeq$ 650 against HV $\simeq$ 130; for steel with 0.6%C the figures are HV $\simeq$ 830 against HV $\simeq$ 160. This shows that the *hardenability* increases strongly with the carbon content, owing to the fact that it is the carbon which creates the tetragonal structure.

For most applications, martensite is too hard and brittle; consequently, either slower cooling conditions giving less hard and less brittle structures must be chosen, or a tempering of the martensite (heating to 200–500°C, allowing diffusion to create some stress relieving) must be carried out. Various alloying elements can modify the transformations without changing the fundamental characteristics of the material.

More detailed descriptions of the diffusionless solid-state transformations can be found in the literature.

Precipitation and *age hardening* are strength-increasing methods based on decreasing the solid-state solubility of one element in another as the temperature is decreased. This means that the metal is heated at first to bring all the elements into solid solution and then cooled rapidly to retain the elevated temperature structure. From this supersaturated structure, the element for which the solubility has been exceeded precipitates out of the matrix. If this precipitation takes place at room temperature, it is called *aging* or *age hardening*; if it takes place at an elevated temperature, it is called *artificial aging* or *precipitation hardening*.

It should be mentioned that after quenching, alloys are soft, in general, and many alloys are even softer after quenching than in the annealed state, so that forming, machining, and so on, can easily be carried out before precipitation, which can be suppressed by holding the material at low temperatures.

This can be illustrated by most aluminum rivets used in aircraft. After quenching, the rivets are kept in a refrigerator, which means that they are soft and can be driven easily. They attain full strength and hardness at room temperature.

The precipitated atoms coalesce into particles which act as obstacles toward deformation. As these particles grow in size, the hardness of the material is increased. If a certain critical particle size is exceeded, hardness starts to decrease; this is known as *overaging*. The precipitated particles are hard and brittle and lie in a soft matrix.

An important example is that of aluminum with 4% copper (Duralumin). At 550°C, 5.7% Cu is soluble in aluminum; at 20°C, only 0.5% Cu is soluble. The solution treatment is carried out at about 490°C. Aging (or precipitation) at room temperature takes about 4 days, but higher strength can be obtained by artificial aging at about 200°C.

For some alloys, deformation can increase the rate of precipitation. Many casting alloys will, after some time, undergo age hardening directly from the cast state. Ordinary solution treatment, quenching, and artificial aging normally increase strength and hardness further.

As mentioned previously, metals are soft after quenching. In this state the atoms of the alloying elements are uniformly distributed in the matrix. If the alloying elements differ substantially in atomic size (substitutional atoms or interstitial atoms) from those of the parent matrix, the resulting lattice is distorted, giving higher strength and hardness. This mechanism, which occurs in single-phase solid solutions, is called *solution hardening*.

Figure 3.3 shows the solution hardening of copper alloyed with different elements. The hardening effect is measured by the yield stress at 1% strain, $\sigma_{1\%}$.

For many applications only the surface of a component needs to have high strength and hardness; consequently, many different methods have been developed to carry out surface hardening. These may be based on martensitic transformations or solution hardening.

For steel, the carbon content determines the hardness obtainable. In low-carbon steels, carbon can be added to the surface by carbonizing from solid carbonaceous compounds or from gaseous atmospheres having an excess of CO at elevated temperatures. The component is then quenched, resulting in considerable hardness depending on the carbon content. Steels containing elements that will form nitrides (which are very hard) can be heated to approximately 500°C

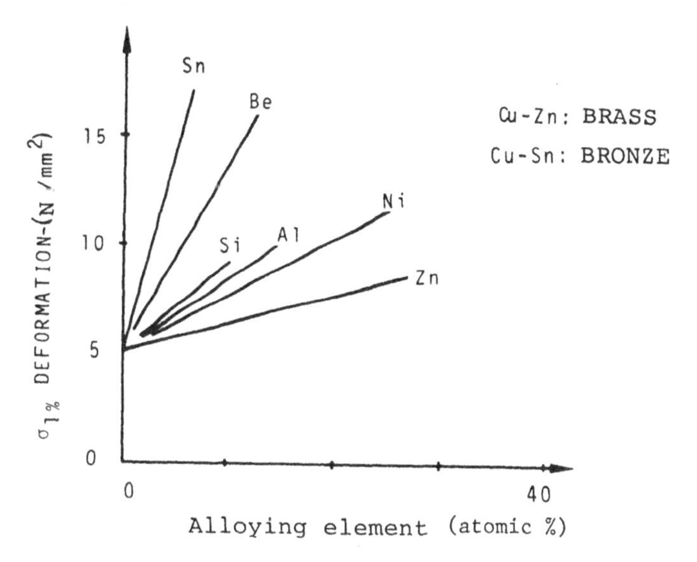

FIGURE 3.3 Solution hardening of copper with various alloying elements [5].

in an ammonia gas, allowing nitrides to be formed. This process is called *nitriding* and gives a harder surface than can be obtained with other methods. In the carbonitriding process, both carbon and nitrogen are added to the surface but at lower temperatures than those used for carburizing.

More detailed descriptions of these and other methods can be found in the literature.

Strain Hardening

As described in Chapter 2 where the stress–strain curve was discussed, strain hardening occurs in metals when they are cold-worked. Figure 3.4 shows in general how the yield stress, ultimate stress, and percent elongation change with increasing cold working, which could be brought about, for example, by forging between two parallel plates.

After deformation, the grains become elongated in certain directions and contracted in others, which results in *anisotropy*; that is, the material has different properties in different directions. By heat treatment—recrystallization—it is possible to change the distorted grains into new stress-free grains (see Fig. 3.5), a procedure eventually accompanied by a growth in grain size. Small deformations, high temperatures, or long periods at elevated temperature favor grain growth.

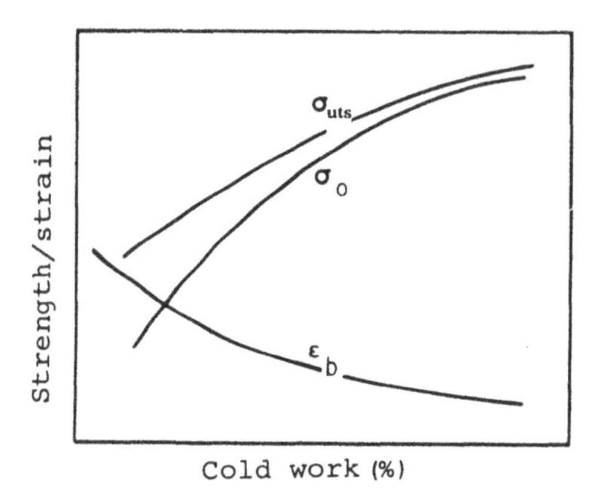

FIGURE 3.4 The change of yield stress (σ_0), ultimate stress (σ_{uts}), and elongation (ϵ_b) by increasing amounts of cold working.

The recrystallization temperature can be estimated as roughly 0.4 times the melting temperature on the absolute scale (Kelvin). Table 3.2 gives approximately the lowest possible recrystallization temperatures for different alloys.

The term *cold working* refers to deformations carried out at temperatures below the recrystallization temperature, and the term *hot working* refers to deformations carried out at temperatures above the recrystallization temperature. Some metals recrystallize at room temperature (lead, tin, and zinc), which means that these metals are normally hot-worked. Strain hardening induced only in the surface layers of the component will often increase its fatigue strength as well as its hardness.

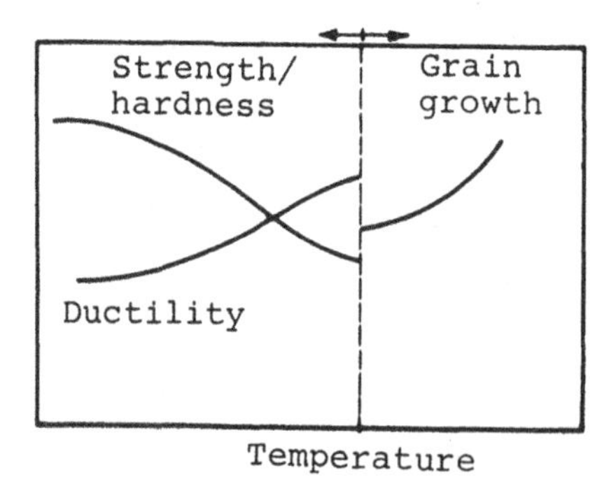

FIGURE 3.5 The change in properties by recrystallization [1].

TABLE 3.2 Lowest Possible Recrystallization Temperature, Melting Point, and Upper Limit for Hot Working for Four Metals

Metal	Lowest recrystallization temperature (°C)	Melting point (°C)	Upper limit for hot working (°C)
Mild steel	600	1520	1350
Copper	150	1083	1000
Brass (60/40)	300	900	850
Aluminum	100	660	600

3.6 FERROUS METALS

In a description of metals, different primary characteristics can be chosen as structuring parameters. In this context, composition and application areas will be chosen as principal characteristics, as they represent useful and practical guides for the engineer engaged in material selection. It must be emphasized that, because of limited space, the description here will consist only of the more general guidelines.

In practical situations, more detailed literature and material suppliers' catalogs must be studied carefully when considering functional and manufacturing requirements. In the following, only constructional steels, tool steels, and cast iron will be discussed.

3.6.1 Composition and Alloying Possibilities

With ferrous materials the base matrix is iron (Fe, the parent metal) having varying amounts of carbon (C). Steels are generally alloys of Fe and C with less than 2%C; cast irons contain 2–4%C. Depending on the cooling conditions and additional alloying elements, the carbon may be present mainly in the combined form of iron carbide, Fe_3C, called cementite, as for example in steel, or in the form of graphite, as in gray cast iron. Cementite itself is very hard and brittle.

Pure carbon steels (alloys of Fe and C) represent only a very small fraction of the steels used today. Most steels are alloyed with a variety of elements to obtain:

Greater strength
Better hardenability
Improved high- and low-temperature properties
Better corrosion resistance
Better technological (manufacturing) properties

The alloying elements in steels can generally be used in two different ways, depending on the purpose:

1. In small amounts, less than 5%, to increase strength and hardenability
2. In larger amounts, in the range 5–30%, to produce special properties: for example, high corrosion resistance or high-temperature properties

In Table 3.3 a survey is given of the effects of some of the important alloying elements in steel [4].

Steels can, depending on the amount of alloying elements, be roughly classified into four groups:

1. Carbon steels with less than 1–2% alloying elements (microalloyed). This group contains the widely used plain carbon steels.
2. Low-alloy steels with more than 1–2% and less than 5% alloying elements, but having the same base matrix as carbon steels.
3. High-alloy steels with more than 5% alloying elements (5–30%). This group contains the stainless steels and the high-temperature steels.

TABLE 3.3 Main Effects of Some Important Alloying Elements in Steel

Alloying element	Amount (%)	Main effect
Manganese	0.25–0.40	Prevent brittleness when combined with sulfur
	>1	Increase hardenability
Sulfur	0.08–0.15	Increase machinability
Nickel	2–5	Increase ductility
	12–20	Increase/give corrosion resistance
Chromium	0.5–2	Increase hardenability
	4–18	Increase/give corrosion resistance
Molybdenum	0.2–5	Increase hardenability and form stable carbides (inhibits grain growth)
Vanadium	0.15	Form stable carbides, give small grain size, increase strength at retained ductility
Boron	0.001–0.003	Increase hardenability considerably
Tungsten		Increase/give hardness at high temperature
Silicon	0.2–0.7	Increase strength
	2	Increase hardness and strength (spring steels)
	>2	Improve magnetic properties
Copper	0.1–0.4	Increase corrosion resistance
Aluminum	Small	Increase hardenability by nitriding

Reprinted with permission of Macmillan Publishing Co., Inc., as adapted from *Materials and Processes in Manufacturing*, 3rd ed., by E. Paul DeGarmo. Copyright © by E. Paul DeGarmo.

4. Alloys with such large amounts of alloying elements that the parent metal is no longer iron. This group contains the special alloys, such as Nichrome and Superalloys.

The development of the many available steels has been carried out to obtain their functional and applicational properties as well as their manufacturing properties. This is reflected in the following description, which refers to both applications and manufacturing methods.

It can generally be stated that if the engineer requires only "common" mechanical strength, with no special corrosion resistance and so on, the carbon steels normally represent the best and cheapest solution. With the slightly more expensive low-alloy steels, better hardenability and better strength at higher temperatures are obtained. The high-alloy steels—which are rather expensive—are used only where their special properties (stainlessness, heat resistability, etc.) can be utilized. The same is true for the special materials in group 4 above.

The following section presents a classification of steels by applications; later, cast iron is described.

3.6.2 Classification of Steels by Applications

The first rough classification of steels results in two main groups: constructional steels (<0.9%C) and tool steels (0.5–2%C).

Constructional Steels

Constructional steels may be subdivided into many groups, of which only a few will be mentioned below. It should be remembered that much help and guidance can be found in the various standards (ANSI, AISI, SAE, etc.) and handbooks (e.g., ASM *Metals Handbook*). In the following, the numbers in parentheses refer to the previous four-group steel-alloy classification.

General structural steels (*1 and 2*) are mainly used for bridges, buildings, machine structures, vessels, trucks, trains, moderately loaded machine components, and so on. These steels are cheap, weldable, and have average good strength and manufacturing properties. Some of the most important groups are the plain low-carbon steels (<0.35%C) and the low-alloy high-strength structural steels, which are playing an increasing role in industry.

General engineering steels suitable for machining (*1 and 2*), sometimes called constructional steels, are used for a large variety of machine components that require high strength and medium to high hardness. Among the most important are the plain medium-carbon steels (0.35–0.55%C), the low-alloy engineering steels (>0.3%C), the quenched and tempered steels, carburizing and nitriding steels, and so on.

Corrosion-resistant (or stainless) steels (3) are generally used where corrosive media (either internal or external) are encountered. In many applications high strength is also required. Most of the corrosion-resistant steels are difficult to machine. For applications in the chemical and dairy industries the good clean appearance is also important. These steels normally contain large amounts of Cr and Ni.

Heat-resistant steels (1, 2, 3, and 4) are especially used where high creep resistance at elevated temperatures, often combined with good strength characteristics and corrosion resistance, is required.

Free-machining steels (1 and 2) are used where the machining properties play an important role. These steels are mostly plain low-carbon and medium-carbon steels modified either with small amounts of sulfur (0.1–0.3%) and manganese (0.5–1.5%) or with sulfur (0.25–0.35%) and lead (0.15–0.35%) with small amounts of tellurium, selenium, or bismuth.

Plain carbon and low-alloy steels (1 and 2) are sometimes more or less refined or modified for fabrication of special sheets, tubes, wires, and so on. Examples are steels for electrical machinery (silicon steels with a low carbon content and 2–4% Si), deep drawing steels (~0.1%C), wire (~0.1%C) for nails, screws, and so on.

Many other alloys developed for special applications could be mentioned (steel for springs, valves, etc.), but for these the literature and the standards must be studied.

The preceding examples illustrate the large variety of steels available and emphasize that material selection must be carried out carefully, giving due consideration to the functional requirements and the manufacturing possibilities.

Tool Steels

The increasing requirements for higher productivity and the wider use of metal-forming processes, which often impose severe mechanical and thermal loads on tooling, have contributed to the rapid development of tool materials. Tool steels are among the most important of all the steels produced, and a wide variety is available. It is not possible to give a comprehensive survey in a limited space, so only a few important groups will be mentioned. The suppliers' catalogs and the literature must be consulted for more detailed information [8].

Tool steels can be roughly classified into the following major groups (the term *tool* includes dies):

Cold-working tool steels (1, 2, and 3) with 0.5–2.0%C. Application examples: cutting tools, press tools, forging dies, blanking tools.

Hot-working tool steels (3) with 0.3–0.6%C and various amounts of W, Cr, V, and Ni. Application examples: forging dies, casting dies, extrusion dies.

High-speed tool steels (3) with 0.7–1.3%C and various amounts of W, Cr, Co, Mo, and V. Application examples: cutting tools and, at an increasing rate, various press tools.

Cemented carbides (4) consist of hard particles (WC, TiC, TaC, NbC, etc.) in a softer matrix. For the softer matrix Co is generally used, but Ni and Mo are also used. Application examples: cutting tools (inserts or tips), forming dies, blanking dies.

Some of these tool materials are discussed in later chapters.

3.6.3 Cast Iron

Cast iron having 2–3.8%C and varying amounts of Si, Mn, P, and S and some other special elements is technically a very important material. This material has excellent castability, machinability, applicational properties, and a low price.

As mentioned previously, the carbon is present in most cases as graphite, depending on cooling conditions and Si content. The graphite structure (its shape) has a major influence on the mechanical properties; therefore, cast irons are classified in accordance with the shape of the graphite particles, their distribution, and their size. Depending on the shape of the graphite particles, cast irons can be grouped into gray cast iron with flake graphite and nodular cast iron with nodular or spheroidal graphite (called SG iron). Cast iron with flake graphite, normally called *gray cast iron* after the gray appearance of a fractured surface, is used most widely. It has a relatively low ultimate tensile strength, a good compression strength, and low ductility (brittle); its machinability is very good. The ultimate tensile strength of gray cast iron, which normally varies between 100 and 500 N/mm^2, depends on the size of the graphite flakes, their distribution, the structure of the material, the composition, and the cooling conditions.

For SG iron (nodular cast iron) strength varies between 350 and 750 N/mm^2, and elongation lies between 2 and 15%. This type of cast iron is being applied in industry at a rapidly increasing rate.

Depending on the rapidity of cooling and the Si content, the carbon can be precipitated in the combined form Fe_3C (cementite), giving a hard and brittle material called *white cast iron*, named after the appearance of a fractured surface. It cannot be machined and has a very high wear resistance. It is mainly used in combination with gray cast iron as a surface layer on a component where extremely high wear resistance is required, as in car wheels and as components in crushing machines. The necessary rapid cooling is produced by chilling the surfaces.

White cast iron is an intermediate product in the manufacturing of malleable cast iron. The white cast iron is given a prolonged annealing at about 800–900°C and cooled slowly. The resulting material has very fine graphite nodules in a nearly pure iron matrix (ferrite), has considerable shock resistance, good strength (~350N/mm^2), high ductility (10–20% elongation), and is used in the railroad, automobile, pipe-fitting, and agricultural industries.

3.7 NONFERROUS METALS

Only a few groups of the most important nonferrous metals are discussed in the following sections, and it must be emphasized that only general guidelines are given.

The industrial importance of the nonferrous metals is steadily increasing, and they provide several important properties that cannot be obtained in steels, for example:

High corrosion resistance
Ease of fabrication
High electrical and thermal conductivity
Low density
High strength/weight ratio
Attractive color

Not all nonferrous metals possess all these qualities, but nearly all have at least two without adding special alloying elements. In general, it is the combination of several of these properties that makes the nonferrous metals so attractive.

The strength of the nonferrous metals is, in general, lower than for steels, but because of the low density, the strength/weight ratio can be rather high. The modulus of elasticity is relatively low, which is a disadvantage where stiffness is required. Most nonferrous metals have a relatively low melting point and they are, in general, easy to cast in sand molds or permanent dies and can often be cold-worked to provide complicated shapes because of their high ductility and low yield stress.

The following groups are described: copper/copper alloys, aluminum/aluminum alloys, magnesium/magnesium alloys, and zinc/zinc alloys.

3.7.1 Copper/Copper Alloys

Pure copper, having a density of 8.96 g/cm^3 and a melting point of 1083°C, is widely utilized in the electrical industry for cables, wires, coils, contacts, and so on, due to its high electrical conductivity, its high corrosion resistance, and its good manufacturing properties. It is also used in coolers, heat exchangers, vessels, and so on, where its high thermal conductivity can be utilized.

Copper alloys have a wide applicational spectrum, and a wide variety of alloys are commercially available. The most important copper alloys are *brass*, which is copper alloyed with 10–40% zinc, and *bronze*, which is copper alloyed with tin, aluminum, or nickel and correspondingly called tin bronze, aluminum bronze, and nickel bronze.

Brass is the most important of all the copper alloys. A copper content of about 60% (~40% zinc) is typical, and this gives an alloy with good strength and hot-working properties. If the copper content is increased to 65–70% (zinc ~30–35%), alloys with high ductility and excellent cold-working properties are obtained. Besides the normal brasses, many special brasses are available, for example, those which are alloyed with aluminum, iron, or manganese. Some of these alloys are used for screws, nuts, and so on.

If 10–20% nickel is added to brass (60–70% Cu and 10–30% Zn), a nickel silver alloy is obtained, which is named for its color. This alloy is used for electrical contacts, springs, etc.

Bronzes are the second most important copper alloys. Here the tin bronzes play a major role, for example, for bearings, where 5–22% Sn is used. Often, a tin bronze is alloyed with lead. For casting purposes tin bronzes are often alloyed with zinc and lead in relatively small amounts. Typical applications include bearings, bushings, pipe fittings, and machine components.

Aluminum bronzes have high strength and high corrosion resistance.

In the utilization of copper alloys as well as for most other metals, it is important to remember that different alloys are used for casting, forming, and machining.

3.7.2 Aluminum/Aluminum Alloys

Pure aluminum, having a density of 2.7 g/cm^3 and a melting point below 660°C, has become one of the most important industrial materials in the last few years. This is due to its high strength/weight ratio, high corrosion resistance, and good electrical conductivity. It should be mentioned that the relative cost of aluminum has favored this situation. Applications include high-voltage cables; equipment for the chemical, dairy, and building industries; and kitchen appliances. Aluminum is often alloyed with silicon, magnesium, copper, manganese, and zinc in varying amounts and sometimes with nickel, iron, chromium, titanium, and beryllium in small amounts.

Aluminum alloyed with silicon is used extensively in the casting of machine components, parts for the automotive industry, and so on. Alloys with silicon and magnesium can be hardened by heat treatment.

If aluminum is alloyed with copper, very high strength can be obtained by heat treatment, but the corrosion properties are rather poor.

When using aluminum alloys, a careful selection between the commercially available alloys must be carried out to satisfy the functional and manufacturing requirements. Most aluminum alloys used for forming processes have good ductility and good cold and hot formability.

3.7.3 Magnesium/Magnesium Alloys

Magnesium, having a density of 1.7 g/cm^3 and a melting point of 650°C, is, in general, utilized in an alloyed condition, where high strength/weight ratios can be obtained. The most important alloying element is aluminum in the proportion of 6–8% for sand and die casting alloys and in the proportion of 3–8% for other alloys, which are eventually further alloyed with zinc and manganese. The highest-strength values can be obtained by the heat treatment of alloys containing zinc, zirconium, and thorium.

Magnesium alloys are used primarily where weight reduction is important: in the air- and spacecraft industries, the automotive industry, and certain fields of the general mechanical industry.

3.7.4 Zinc/Zinc Alloys

Pure zinc, having a density of 7.13 g/cm^3 and a melting point of 420°C, is produced in huge amounts, but only a small but increasing fraction is used as a construction material. Forty percent of the zinc produced is used for surface coatings, 25% is used as an alloying element in copper, and 10% is used for sheets for the building industry. Rather large amounts of zinc are used in the battery and printing industries.

As construction materials, fine zinc alloys with very low contents of tin, lead, and cadmium, to obtain high corrosion resistance, are predominantly used. The major alloying elements are aluminum (4–6%) and copper (1–2%).

Fine zinc alloys have good strength and ductility properties, and they are shaped mainly by casting. Applications include automobile accessories, components in kitchen appliances, office machinery, tools, and toys. Casting dies have a long production life because of the low melting point of zinc alloys. The tolerances obtained are usually so good that machining and other processing are unnecessary.

3.8 PLASTICS (HIGH POLYMERS)

Of the nonmetallic material group polymers, only the thermosetting plastics and the thermoplastics will be discussed. The term *plastics* covers a special group of polymer materials. Many natural polymers are known and widely used: for example, asphalt, cellulose, and wool. In general, the term "plastics" covers the enormous variety of synthetic polymers, which in the last couple of decades have become some of the most important industrial materials because of their

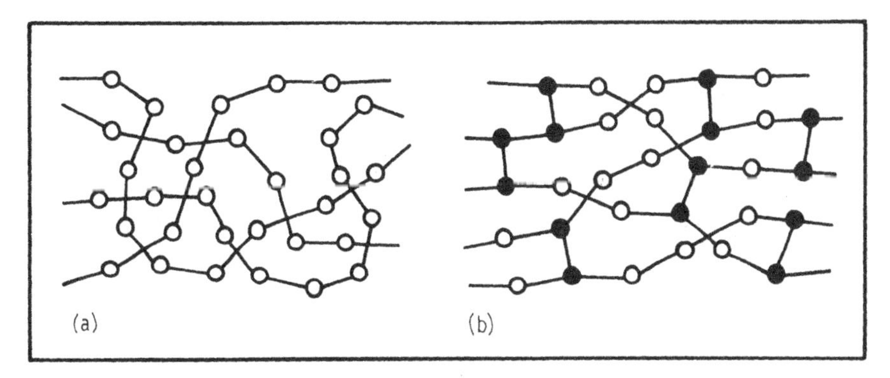

FIGURE 3.6 Schematic representation of (a) chain structures and (b) cross-linked (net) structures in plastics.

applicational and manufacturing properties. Plastics are a relatively new material group, as their development started after 1940.

3.8.1 Thermoplastic and Thermosetting Plastics

Both in applicational and manufacturing respects, the properties of plastics are dependent primarily on their molecular structure. On this basis the plastics can generally be classified into two main categories: thermoplastics and thermosetting plastics. *Thermoplastics* are characterized by the *chain structure* (see Fig. 3.6a), where large molecules (molecular chains, macromolecules) are bonded together by the weak secondary van der Waals forces. The type of bonding determines the mechanical and physical properties, depending on the actual type of plastic, the molecular weight, and the size and geometry of the chains (i.e., different types of thermoplastics have different properties).

Temperature has a major influence on the strength of the secondary bonds, which are weakened by increasing temperature, allowing the molecular chains to move more freely relative to each other. Decreasing temperature thus gives harder and stronger materials. This softening and hardening process can be repeated as often as desired, but the characteristics of the change are dependent on the structure (i.e., whether it is amorphous or crystalline). In *amorphous state* the chains are arranged completely at random, whereas in the *crystalline state* they are arranged in definite crystalline regions called *crystallites* embedded in an amorphous matrix. It should be mentioned that plastics are never completely crystalline. In the crystalline regions a closer packing of the chains is obtained, so that the secondary bonds act more strongly.

Thermoplastics can be machined in the solid state (at room temperature) and at increased/elevated temperatures, they can be formed either in a rubber-elastic or liquid state. If the temperature is too high or maintained for too long a time, the material will be destroyed.

Thermosetting plastics are characterized by their *cross-linked structure* (or net structure) (see Fig. 3.6b), where strong primary bonds exist between the chains. The cross-linked structure is created after or during the forming of the desired component, as it is an irreversible chemical process (hardening). The cross-linked structure gives a hard and strong material, which maintains its hardness at elevated temperatures. Furthermore, thermosetting materials are resistant to chemical attacks and creep. Once hardened, they cannot be softened and can only be shaped by machining.

The density and character of the cross-links have a major influence on the properties of thermosetting plastics, depending on the specific material.

3.8.2 Design of Plastic Materials

An extremely wide range of plastics are commercially available and, within each type, properties can be modified in different ways, so that the same type of plastic is available with a wide spectrum of properties. It is thus possible to ''tailor'' the material for specific applications.

In the production of high polymers (plastics) it is possible to vary the molecular structure, the molecular weight, size and geometry of the chains, and so on, which has a major influence on the mechanical and physical properties. But most plastics are not used in their ''pure'' state. Different additives are normally employed to modify the applicational and/or manufacturing properties. The most frequently used additives are:

Stabilizers (e.g., to provide protection against thermal and atmospheric actions, ultraviolet light, etc.)
Softeners (to increase ductility, e.g., at low temperatures)
Lubricants (e.g., wax, stearates, etc., to improve moldability and facilitate removal from the mold)
Fillers (to improve strength and reduce cost, e.g., wood flour, cloth fibers, glass fibers, asbestos fibers, etc.)
Coloring agents (dyes, colored pigments, etc.)

Inorganic fillers normally improve thermal stability and electrical properties. Fibrous fillers are generally used as reinforcing materials, giving considerable mechanical strength. The fillers comprise, in general, a considerable percentage of the total volume.

3.8.3 General Applicational Properties

The main reason for the rapidly increasing industrial importance of plastics is that the properties or combinations of properties obtainable are difficult or im-

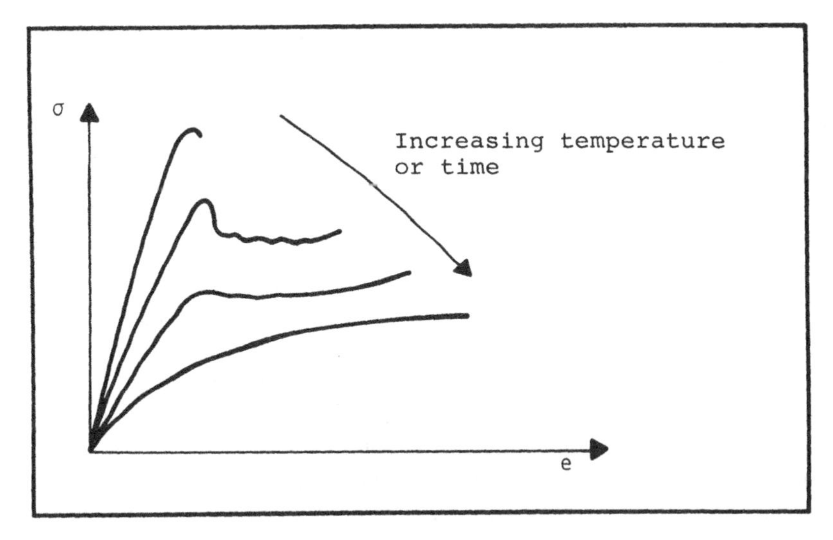

FIGURE 3.7 Schematic representation of the influence of temperature and time on the stress-strain curve for a thermoplastic material.

possible to obtain with other materials, combined with manufacturing methods lending themselves to mass production.

When selecting a plastic material for a specific purpose, the service conditions must be carefully analyzed, as the properties of plastic materials can vary much more under service than do those of metals. This necessitates a thorough testing of the plastic material to be used. Only when the testing conditions and the service conditions are very similar can the test data be used directly.

The service temperature and the duration of mechanical loading are especially important for thermoplastic materials. This is illustrated in Fig. 3.7, which shows schematically the influence of temperature and time (creep) on the stress–strain curve. Increasing temperature causes decreasing strength and decreasing rigidity (increasing ductility). Mechanical loads applied for long durations cause creep in the material, which is accelerated by increased temperature.

The strength of plastics is, in general, about one-tenth that of metals, but since the density is relatively low (0.9–2 g/cm^3), the strength/weight ratio is reasonably good. Especially with glass fiber-reinforced plastics, a high strength/weight ratio can be obtained which is comparable to that of metals.

Plastics have good electrical resistance, giving them extensive applications as insulating materials. Correspondingly, their low thermal conductivity makes them excellent heat insulators.

Resistance to chemical agents varies from plastic to plastic, but, in general, it will always be possible to find a material that can resist a given chemical agent. For this reason plastics are extensively used in the chemical industry and where the corrosion risks are high.

In this short presentation, it has not been possible to describe the enormous number of commercially available plastics; consequently, the literature [10,11] and the suppliers catalogs must be studied in specific cases.

3.9 CERAMICS

Ceramics are compounds of metallic and nonmetallic elements, such as metal oxides, borides, carbides, nitrides, or silicides. Ceramics have long been used in the electrical industry because of their high electrical resistance but have also assumed importance in a variety of other engineering applications such as exhaust-port liners, coated pistons, cylinder liners, cutting tools, grinding wheels and other components needing desirable properties such as strength at high temperatures, hardness, resistance to wear, thermal stability, creep resistance, and so on. Table 3.4 presents some general characteristics of a number of carbides and oxides.

Ceramics have crystal structures that are among the most complex of all materials, containing various elements of different sizes. The bonding between the atoms is generally covalent (electron sharing) and ionic (primary bonding between oppositely charged ions). The bonds are much stronger than metallic bonds, resulting in properties that differ significantly from those for metals. The

TABLE 3.4 Mechanical Properties of Some Oxides and Carbides[a]

Material	Density (g/cm^3)	Compressive ultimate strength (N/mm^2)	Softening or melting temperature (°C)	Hardness
Oxides				
Al_2O_3	3.76	2940	2040	2100
BeO	3.00	770	2730	—
MgO	—	1000	2800	500
ZrO_2	5.78	2030	2500	1150
ThO_2	11.08	1470	2760	1000
Carbides				
B_4C	—	—	2450	2750
SiC	—	2060	2200	2500
TiC	—	3500	3100	2450
ZrC	—	—	—	2100
WC	—	3500	2800	1900

[a]Approximate.

absence of free electrons makes the ceramic materials poor electrical conductors. Ceramics are available as single crystals or, more often, in polycrystalline form, consisting of many grains.

Generally, these materials are divided into two categories: traditional ceramics and industrial ceramics, also called engineering, high-tech, or fine ceramics. In this section some general characteristics and applications of engineering ceramics are presented, whereas the traditional ceramics used for pottery and bricks will not be discussed.

3.9.1 Types of Ceramics

Raw materials for ceramics found in nature are clay, flint (very fine grained SiO_2), and feldspar (a group of materials consisting of aluminum silicates, potassium, calcium, or sodium). These raw materials generally contain impurities which have to be removed before the materials are processed further into useful products. This removal is often difficult to achieve satisfactorily, and many ceramics are now produced almost exclusively from synthetic components so that their quality may be controlled to very strict specifications.

Alumina (Al_2O_3), also called corundum or emery, is the most widely used oxide ceramic, and is an example of a synthetic material. It is obtained by fusion of molten bauxite, iron filings, and coke. After fusion, it is crushed and graded by passing the particles through standard screens and by precipitation. Parts of aluminum oxide are cold pressed and sintered. Properties are improved by minor additions of other ceramics, such as titanium oxide and titanium carbide. Typical fields of application include electrical and thermal insulation, cutting tools, and abrasives.

Another important member of the family of oxide ceramics is zirconia (ZrO_2), which has good toughness, resistance to thermal shock, wear, and corrosion, low thermal conductivity and low friction coefficient. Zirconia and its derivatives are consequently very suitable for heat-engine components such as cylinder liners and valve bushings.

A characteristic feature of oxide ceramics is their anisotropy of thermal expansion, resulting in a variation in thermal expansion in different directions by as much as 50% for quartz. This behavior causes thermal stresses that can lead to cracking of the component.

Carbide ceramics, typical examples of which are tungsten carbide (WC) and titanium carbide (TiC), are extensively used as cutting tools and die materials. Table 3.5 gives the melting point and room temperature hardness of some of the carbides. All the values are very high compared with those of steel. The compounds undergo no major structural changes up to their melting points, and their properties are therefore stable and unaltered by induction of heat. These car-

Table 3.5 Melting Point and Room Temperature Hardness of Some Important Carbides

	Melting point (°C)	Hardness (HV)
TiC	3200	3200
V_4C_3	2800	2500
NbC	3500	2400
TaC	3900	1800
WC	2750 (decomposes)	2100

bides are strongly metallic in character, having good electrical and thermal conductivities and a metallic appearance. They have only slight ability to deform plastically without fracture at room temperature.

Figure 3.8 shows the hardness of four of the more important carbides measured at temperatures from 15°C to 1200°C. It is seen that the hardness drops rapidly with increasing temperatures, but they remain much harder than steel under almost all conditions. The very high hardness and the stability of properties when subjected to a wide range of thermal treatments are favorable to the use of carbides in cutting tools.

In cemented carbide alloys, the carbide particles normally constitute at least 80% by volume of the structure.

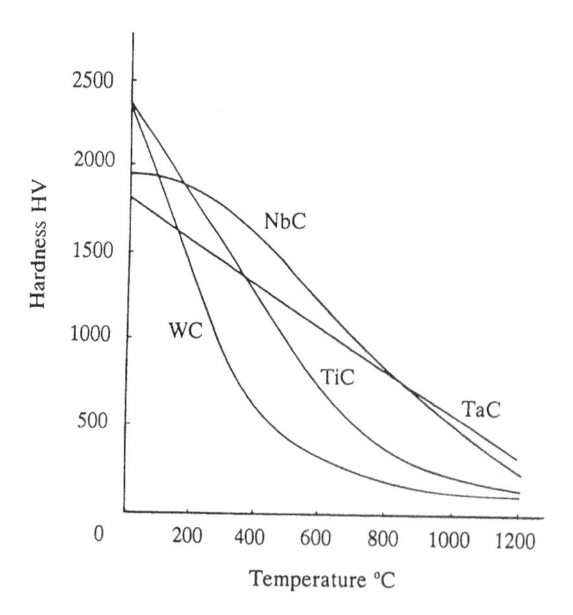

FIGURE 3.8 Hot hardness of some important carbides [42].

Cemented carbides are produced by the powder metallurgy process, where—in the case of WC—carbon and tungsten powders are mixed and then heated in a graphite crusible until a chemical combination occurs. The tungsten carbide particles are then milled together with cobalt in large drums containing steel balls so that the WC particles become coated with cobalt, which is to act as a binder. The powder is now pressed into forms and presintered at about 810°C. After presintering, the blanks have the consistency of chalk and can easily be machined to near final geometry after which they are sintered once again at 1540°C to obtain full hardness and strength. Other additives include titanium and tantalum. Titanium is added to increase the hot hardness of the tool. Tantalum is added to help prevent wear on the top of the tool surface, called cratering, by reducing the friction between the chip and the tool. The amount of binder has a major influence on the material's properties, as toughness increases with cobalt content, whereas hardness, strength, and wear resistance decrease. Titanium carbide has nickel and molybdenum as the binder and is not as tough as tungsten carbide.

Another class of ceramics is the nitrides, particularly cubic boron nitride (CBN), titanium nitride (TiN) and silicon nitride (Si_3N_4).

Cubic boron nitride, the second hardest known substance, after diamond, has special applications, such as abrasives in grinding wheels and as cutting tools. It does not exist in nature and is made synthetically with techniques similar to those used in making synthetic diamonds.

Titanium nitride is widely used to coat cutting tools. The coating is achieved by a process known as chemical vapor deposition, where the tool is placed in a sealed retort with an inert atmosphere until the coating temperature is reached in the range of 950–1050°C. Titanium tetrachloride ($TiCl_4$) and nitrogen/hydrogen gas are introduced into the reactor, resulting in the following reaction for the formation of TiN:

$$2TiCl_4 + 4H_2 + N_2 \rightarrow 2TiN + 8HCl$$

Coating thickness, normally about 2–4 μm is a function of reactant concentration, coating temperature and time at temperature.

Silicon nitride is stronger than steel, extremely hard, and as light as aluminum. It has high resistance to creep at elevated temperatures, low thermal expansion, and high resistance to thermal chocks. It is suitable for high-temperature structural applications, such as engine combustion chambers, bearings, turbocharger rotors, and so on. It is worth mentioning that the higher operating temperatures made possible by the use of ceramic components mean more efficient fuel burning and reduced emissions.

Cermets are combinations of metals and ceramics, usually oxides, carbides, or nitrides. They combine the high-temperature characteristics of ceramics and the toughness, thermal shock resistance, and ductility of metals. Cermets are used for nozzles for jet engines, aircraft brakes, and similar applications requiring strength and toughness at elevated temperatures. They can be regarded as composite materials (see Sect. 3.11).

3.9.2 Mechanical and Physical Properties

Individual characteristics of important ceramic materials have been given in Sect. 3.9.1. This section describes the general mechanical and physical properties of ceramics. These materials are very sensitive to cracks, impurities, and porosities, resulting in a strength in tension that is approximately one order of magnitude lower than the strength in compression, since the defects lead to the initiation and propagation of cracks under tensile stress. This deficiency may in part be overcome by prestressing the components, as in prestressed concrete, by subjecting them to compressive stresses. Methods used include heat treatment and chemical tempering, laser treatment of surfaces, coating with ceramics having different thermal expansion, or surface-finishing operations, in which compressive risidual stresses are induced on the surface.

Generally speaking, ceramics also lack impact toughness because of their inherent lack of ductility, so that a crack, once initiated, propagates rapidly. In addition to undergoing fatigue failure under cyclic loading, ceramics exhibit a phenomenon called static fatigue, so that they may suddenly fail after being subjected to a static tensile load over a period of time. Static fatigue, which occurs in environments where water vapor is present, but not in vacuum or dry air, has been attributed to a mechanism similar to stress corrosion in metals.

Hardness, wear resistance, and strength at elevated temperatures are common and attractive properties, as are light weight (specific gravities of 2.3–3.85 g/cm^3) dimensional stability, corrosion resistance, and chemical inertness. On the negative side, reliability is still rather low, and failure occurs with little prior warning as a result of brittle fracturing. The cost of ceramics is relatively high, and it is difficult to form reliable bonds to the various engineering materials. Machining is problematic, so products must be fabricated through the use of net-shape processes.

3.10 GLASSES

Glasses are generally based on silica (SiO_2) with additives that alter the structure or reduce the melting point. Presently there are some 750 different types of commercially available glasses, ranging from window glass, bottles, and cookware to types with special mechanical, electrical, high-temperature, chemical, or optical characteristics. All glasses contain at least 50% silica, known as a glass former. By adding oxides of aluminum, sodium, calcium, barium, and so on, properties, with the exception of strength, can be modified greatly.

For all practical purposes glasses are regarded as perfectly elastic and brittle. The modulus of elasticity is in the range 55–90 kN/mm^2, that is, 25–45% of the modulus of elasticity for steel. Hardness ranges from 5 to 7 on the Mohs scale. Strength is greatly influenced by the presence of small flaws and microcracks that may reduce the strength by two to three orders of magnitude, compared to the ideal, defect-free strength, which theoretically can be as high as 35 kN/mm^2. More realistically, glass fibers drawn from molten glass can reach a tensile strength of up to 7 kN/mm^2, with an average value of about 2 kN/mm^2. The fibers are thus stronger than steel and are used to reinforce plastics in applications such as boat hulls, automobile body parts, furniture, and sports equipment.

Glasses have low thermal conductivity, high electrical resistivity and dielectric strength. Their thermal expansion coefficient is lower than those for metals and plastics and may even approach zero. Optical properties, such as reflection, absorption, and refraction can be modified by varying the composition and treatment of the glass.

3.11 COMPOSITE MATERIALS

Composites are materials in which the desirable properties of separate materials are combined by mechanically or metallurgically binding them together. Each of the components retains its structure and characteristics, but the composite generally possesses better properties, such as stiffness, strength, or weight, than those of the constituent parts.

The idea of composites is not new, as examplified by the ancient use of straw in a clay matrix for mud huts and bricks. Another example is the reinforcing of concrete with iron rods to impart the necessary tensile strength to the composite, since concrete in itself is brittle with no useful tensile strength. Today, cutting tools, electrical components, golf clubs, prosthetic devices, military helmets and bulletproof vests, sailboats, aircraft, automobiles, and so on, all utilize advanced composite materials.

There are many types of composites and several methods of classifying them. One such method is based on the way the composite materials are built up:

1. Dispersion-hardening materials, consisting of a matrix filled with up to 15% particles (size <0.1 μm) of different materials.
2. Particle-reinforced materials, consisting of a matrix filled with more than 20% particles (size >1 μm) of different materials.
3. Fiber-reinforced materials, consisting of a matrix with up to 70% fibers of different materials.

3.11.1 Dispersion-Hardened Materials

Dispersion-hardened materials are normally produced by dispersing a small quantity of hard, brittle, fine particles in a softer, more ductile matrix. Pronounced strengthening and creep resistance can be induced, which decrease only gradually as temperature is increased. Examples include sintered aluminum or copper powder, consisting of an aluminum or copper matrix strengthened by particles of aluminum oxide. Applications include gas turbines and electrical components.

3.11.2 Particle-Reinforced Materials

This group contains large amounts of rather coarse particles and covers the previously mentioned combination of metals including alloys and ceramics, which give reasonably good ductility, high hardness, and good strength at elevated temperatures. Particle-reinforced materials are used primarily as cutting and forming tools as described in Section 3.9.1 for carbide ceramics. The hard, stiff carbide can withstand the high temperatures of cutting, but is extremely brittle. Toughness can be imparted by combining the carbide with cobalt, pressing the mixture into desired shape, and sintering the compacted material. Other areas of applications include electrical contacts designed for good conductivity and spark-erosion resistance by mixing tungsten powder with powdered silver and processing via powder metallurgy.

Table 3.6 lists some possible combinations of ceramics and metals, suitable for cutting tools and for grinding and cutting wheels, which are often formed from alumina (Al_2O_3), silicon carbide (SiC), cubic boron nitride (BN), or diamond bonded in a matrix of glass or polymeric material. As the hard particles wear, they fracture and are pulled out of the matrix, exposing new cutting edges. The potential speeds with alumina tools, which are three to four times higher than those normally used with carbide tools, represent an increase in metal removal rate as great as that achieved by high-speed steel and by cemented car-

TABLE 3.6 Some Possible Combinations of Ceramics and Metals

Ceramics	Metal
SiC	Ag, Co, Cr
TiC	Mo, Fe, Ni, Co
WC	Co
Al_2O_3	Al, Co, Fe, Cr
SiO_2	Cr

bides. Foundry molds and cores made from sand (particles) and a binder (matrix) are also made from particle-reinforced materials.

Because the reinforcing particles are distributed uniformly in the matrix, the properties of both dispersion-hardened and particle-reinforced materials may be expected to be isotropic, that is, uniform in all directions.

3.11.3 Fiber-Reinforced Materials

The most widely used types of composite materials are those based on a combination of fibers as the reinforcing material and a compatible binder as matrix. The fibers, which carry most of the force on the material and may be oriented in directions corresponding to the loading directions, consist of metals, carbon, ceramics, etc., and they are embedded in metals, polymers, or glasses that support and transmit loads to the fibers and provide ductility and toughness. The mean diameter of fibers is usually about 0.01 mm. The fibers are very strong and rigid because the molecules are oriented in the longitudinal direction, and their cross sections are so small that it is unlikely that any defects exist in the fiber. Glass fibers, for example, can have tensile strengths as high as 4.6 kN/mm^2, whereas the strength of glass in bulk form is much lower. Glass fibers are stronger than steel.

Matrix materials are usually epoxies or polyesters. Polyamides, which resist temperatures in excess of 300°C are being developed for use with graphite fibers. Above this temperature, metal matrices should be considered. Some thermoplastics are also being considered as possible matrix materials since they have a higher toughness than the thermosetting types.

According to the kind of constituents used, fiber composites are usually divided into two main groups, fiber-reinforced plastics and fiber-reinforced metals, to be described next.

Fiber-Reinforced Plastics

Glass-fiber-reinforced resins were the first of the modern fiber composites and were developed during World War II to fulfill the U.S. government's need for materials having electromagnetic transparency, impact resistance, corrosion and erosion resistance, lightness, and high strength. The glass fibers, made by drawing molten glass through small openings in a platinum die, were bonded in a variety of polymers, generally epoxy resins. The content of fibers vary between 30 and 65% by volume. Today, there are two principal types of glass fibers, the E-type, a borosilicate glass, which is the most common, and the S-type, a magnesia–alumina–silicate glass, which has higher strength and stiffness—and is more expensive.

Further efforts were made to improve strength and modulus of elasticity through the development of improved fibers. Boron fibers consist of boron deposited by chemical vapor deposition on tungsten core fibers. These fibers have high strength and stiffness in tension and compression (tensile strengths in excess of 3.5 kN/mm^2) and resistance to high temperatures (up to 250°C). They are expensive, and because of the use of tungsten, they have a rather high density. If it is desired, they can be incorporated into a cast metal matrix.

Graphite fibers, although more expensive than glass, have a very desirable combination of low density, high strength, and high stiffness. All graphite fibers are made from a textile-grade acrylic tow by pyrolyzing the acrylic fiber in such a way that an oriented graphitic structure is obtained in the final fiber. *Pyrolysis* is the term for inducing chemical changes by heat, such as burning a length of yarn, which becomes carbon and black in color. The temperatures for carbonizing range up to 3000°C. The basic product form is a tow containing about 40,000 filaments, a loose, untwisted bundle of parallel fibers of continuous length. The tow is easily impregnated with liquid resins, and the composites can be fabricated with many of the same techniques used in glass-fiber-reinforced plastics technology, including matched metal die, vacuum bag, hand layup, and other methods described in Chapter 11.

Kevlar, which is the trade name of an aramid fiber, has a tensile strength of 3kN/mm^2, making it the strongest of any organic fiber. Kevlar can undergo some plastic deformation before fracture and thus have a higher toughness than other types of fibers. Its density is about half that of aluminum, and it is flame retardant and transparant to radio signals, making it attractive for a number of military and aerospace applications. However, aramids absorb moisture.

Other fibers used in fiber-reinforced plastic materials are nylon, silicon carbide, silicon nitride, aluminum oxide, boron carbide, molybdenum, and so on. Whiskers are also used as reinforcing fibers. They are tiny, needlelike single crystals that grow to 1–10 μm in diameter and have length-to-diameter ratios from 100 to 15,000. Because of their small size they are virtually free from imperfections, so their strength approaches the theoretical strength of the material.

Fiber Orientation

Many variables must be taken into consideration when combining the fiber and matrix to form a composite. As has been mentioned, a fiber is essentially unidirectional in properties. This means that if the fibers are all laid in the same direction, the strength transverse to that direction will be only that of the matrix. Therefore, fiber placement is important. In addition, fiber length can also influence the performance of a composite, short fibers being less effective than long fibers, and their properties are strongly influenced by time and temperature. Long fibers transmit the load through the matrix better and are thus commonly used in critical applications, particularly at elevated temperatures.

One means to circumvent some of the problems associated with unidirectionally laid fibers is to use a fabric in which the fibers are woven into a pattern.

Such cloths are commercially available, for example, as graphite yarns woven together with kevlar, fiberglass, or other fibers. The fabrics can be prepregged with various matrix materials, allowing them to be designed for specific application requirements.

Some graphite and carbon fibers are available in the form of mats, composed of high-modulus filaments, mechanically bonded together to form a web. Graphite fibers have natural lubricity and a low coefficient of friction. Therefore, the mats can be used as reinforcements for bearings, seals and gears.

A critical factor in reinforced plastics is the strength of the bond between the fiber and the polymer matrix, since the load is transmitted through the fiber–matrix interface. Weak bonding results in fiber pullout and delamination. Bonding can be improved by special surface treatments, such as coatings and the use of coupling agents for better adhesion at the interface, but careful inspection and testing is essential in critical applications.

Applications

Composites with a matrix of epoxy were used for the first time in the 1930s, and beginning in the 1940s, boats were made with fiberglass, and reinforced plastics began to be seen in aircraft parts, electrical equipment, and sporting goods. Today, the advanced composites are used extensively in everyday as well as in advanced components such as automotive bodies, pipes, ladders, pressure vessels, boat hulls, and military and commercial aircraft and rocket components. The cargo bay doors of the space shuttle are made from graphite/epoxy, which is also used for the upper aft rudder of the McDonnell/Douglas DC10 passenger plane. This company also uses graphite/epoxy for what it calls the largest composite part ever built for an aircraft, the wing skin of the Harrier II fighter plane, which is 26% by weight composite material. The skin is laminated from 0.25 mm thick plies. It is 8.5 m from wingtip to wingtip and 11 m^2 in area. The completed wing weighs 20% less than a conventional wing and is an important factor in the increase in range and payload of the aircraft, to which may be added improved fatigue and corrosion resistance.

Fiber-Reinforced Metals

Metal matrix composites are used where operating temperature is high or extreme strength is desired. The ductile matrix material is usually aluminum, titanium, or nickel, although other metals are also being investigated, and the reinforcing fibers may be of graphite, boron, alumina or silicon carbide, with beryllium and tungsten as other possibilities. Compared to metals, these materials offer higher stiffness and strength, especially at elevated temperatures, and a lower coefficient of thermal expansion. Compared to the organic matrix material, they offer higher heat resistance, as well as improved electrical and thermal conductivity.

Boron fibers in an aluminum matrix are used in many aerospace structures, especially truss structures. This is because the boron fibers have high strength and stiffness and can therefore be used unidirectionally, and because the aluminum matrix in itself is strong enough to support the fibers without crossplying, as would have been necessary with resin–matrix composites. This material was selected for the structural tubular supports in the midfuselage section of the space shuttle. Table 3.7 shows some other applications of metal–matrix composites:

TABLE 3.7 Metal-Matrix Composite Materials and Applications

Fiber	Matrix	Applications
Graphite	1. Aluminum	1. Satellite, missile, helicopter structures
	2. Magnesium	2. Space and satellite structures
	3. Lead	3. Storage battery plates
	4. Copper	4. Electrical contacts and bearings
Boron	1. Aluminum	1. Compressor blades and structural supports
	2. Magnesium	2. Antenna structures
	3. Titanium	3. Jet-engine fan blades
Alumina	1. Aluminum	1. Superconductor restraints in fusion power reactors
	2. Lead	2. Storage-battery plates
	3. Magnesium	3. Helicopter transmission structures
Silicon carbide	1. Aluminum/titanium	1. High-temperature structures
	2. Superalloy (Co-base)	2. High-temperature engine components
Molybdenum, tungsten	1. Superalloy	1. High-temperature engine components

Source: From Ref. 44.

4

Basic Theory of Metalworking

4.1 INTRODUCTION

The purpose of metalworking theory is to provide a background from which reasonable evaluations may be made of the deformations obtainable without instability and fracture, and the forces and work necessary, which in turn determine the equipment and tooling required. Thus, the theory is important for both the design and production engineer.

In this chapter the fundamental principles, rules, and laws that can be used to describe plastic flow or deformation of solid materials when subjected to external loads are discussed. For this purpose, the materials are considered as homogeneous, continuous, and isotropic media, not as built up from thousands of small grains as described in Chapter 3. However, this assumption will not lead to any serious discrepancies at this level.

A short introduction to two- and three-dimensional systems of stress is followed by discussions of the stress–strain curve, true stress, logarithmic or natural strain, volume constancy, and instability. Finally, the yield criteria, which determine the stresses necessary to initiate and maintain plastic flow and the work necessary to carry out the deformation, are analyzed.

4.2 TWO- AND THREE-DIMENSIONAL SYSTEMS OF STRESS

The forces that act on a solid body may be classified as either volume or mass forces or surface forces. *Volume* or *mass forces*, which include acceleration forces, are not considered in this discussion because they are usually negligible in metalworking processes. *Surface forces* include forces acting on the surface of every volume element, transmitted either by the interaction of the surrounding material or by external forces.

When analyzing design situations for dimensioning purposes or forming processes for planning or design purposes, it is more appropriate to use the force per unit area as a measure of the load rather than the total force distributed over the area.

The force per unit area (described in Chapter 2) is called the *stress* and is usually designated by the symbol σ (sigma):

$$\sigma = \frac{\text{force}}{\text{area}} \tag{4.1}$$

Considering the tensile specimen shown in Fig. 4.1, the stress on a cross section (1) perpendicular to the longitudinal axis is defined by

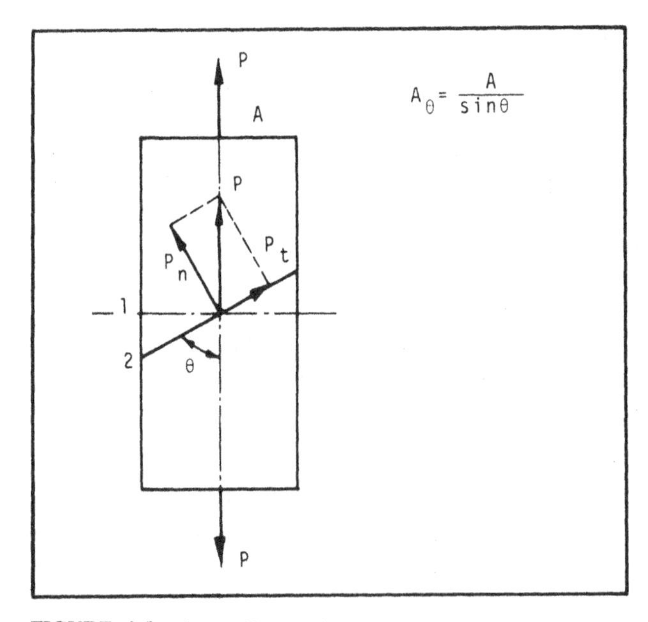

FIGURE 4.1 A tensile specimen with the cross-sectional area A subjected to the load P.

$$\sigma = \frac{P}{A} \tag{4.2}$$

where P is the force and A is the cross-sectional area.

If a cross section (2) that is inclined at an angle θ to the longitudinal axis is considered, the mean oblique stress σ_m is found to be

$$\sigma_m = \frac{P}{A_\theta} = \frac{P}{A} \sin \theta \tag{4.3}$$

where A_θ is the cross-sectional area of cross section 2. This stress [$<\sigma$, Eq. (4.2)] lies in the direction of the longitudinal axis of the specimen. The force P can be resolved or decomposed into two components P_n perpendicular to cross section (2) and P_t parallel or tangential to cross section (2), so that the state of stress can be described by:

The stress normal to cross section (2), called the *normal stress* (σ):

$$\sigma_\theta = \frac{P_n}{A_\theta} = \frac{P \sin \theta}{A/\sin \theta} = \frac{P}{A} \sin^2 \theta \tag{4.4}$$

The stress parallel to cross section (2), called the *shear stress* (τ):

$$\tau_\theta = \frac{P_t}{A_\theta} = \frac{P \cos \theta}{A/\sin \theta} = \frac{1}{2}\frac{P}{A} \sin 2\theta \tag{4.5}$$

The stresses σ_θ and τ_θ both vary with θ. The normal stress σ_θ is zero for $\theta = 0$ and maximum (P/A) for $\theta = \pi/2$. The shear stress τ_θ is zero for $\theta = \pi/2$ and maximum ($P/2A$) for $\theta = \pi/4$ and $\theta = 3\pi/4$.

It must be emphasized that the description of the state of stress depends on the chosen coordinate system, but that the most appropriate system is based on normal stresses and shear stresses (σ, τ). The normal stresses are most often defined as positive when they are tensile stresses, whereas the sign of the shear stresses is chosen as arbitrarily (see Fig. 4.2).

Figure 4.2a shows a two-dimensional load situation, that is, a sheet stressed in two mutually perpendicular directions. It is now our intention to describe the state of stress in terms of σ and τ on every plane passing through point P (Fig. 4.2a). A coordinate system (x, y) is chosen so that an appropriate and simple description is obtained, and based on the stresses in these directions, the stresses σ and τ on every plane through P can be calculated as shown in Fig. 4.2b and c.

From Fig. 4.2b it can be seen that the normal stress acting in the direction of the x-axis is called σ_x (acting on a plane parallel to the y-axis and in the direction of the x-axis) and that σ_y is acting on a plane parallel to the x-axis and in the direction of the y-axis. The subscript thus designates the normal to the plane on which the stress is acting (i.e., it identifies both plane and direction). To identify the shear stresses, two subscripts are necessary, the first identifying the plane (by its normal) on which the stress is acting, and the second identifying the direction. Thus τ_{xy} represents a shear stress acting on the plane normal to the x-axis and in the direction of the y-axis. Correspondingly, τ_{yx} represents a shear stress acting on the plane normal to the y-axis and in the direction of the x-axis.

On Fig. 4.2c the point P is shown surrounded by an infinitesimal element *ABCD*. The system of stress acting on this element must be in equilibrium if it is not to change its position (i.e., translate or rotate). From moment equilibrium it is found that the shear stresses τ_{xy} and τ_{yx} must be complementary and equal in size, which means that it is not necessary to distinguish between them. It is assumed that the element *ABCD* is so small that the stresses do not vary across it (i.e., it is in reality only a point). If the element is too large, the stress system will vary within the element (i.e., different points will have different states of stress).

Figure 4.3 shows a two-dimensional stress system represented by the stresses acting on a right-angled prism. It is assumed that the stresses on the planes normal to x and y are known, and that the stresses on the plane which is inclined at the angle θ to the y-axis (Direction of y') are to be determined. Since the system must be in equilibrium, the forces can be resolved and equated in any direction.

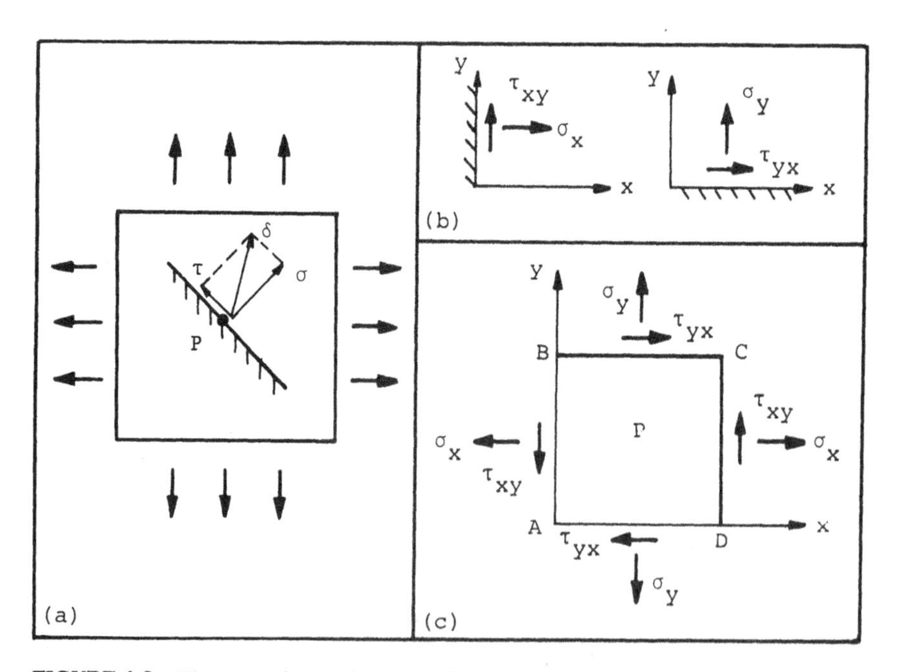

FIGURE 4.2 The state of stress in a biaxially loaded sheet (i.e., a two-dimensional system of stress).

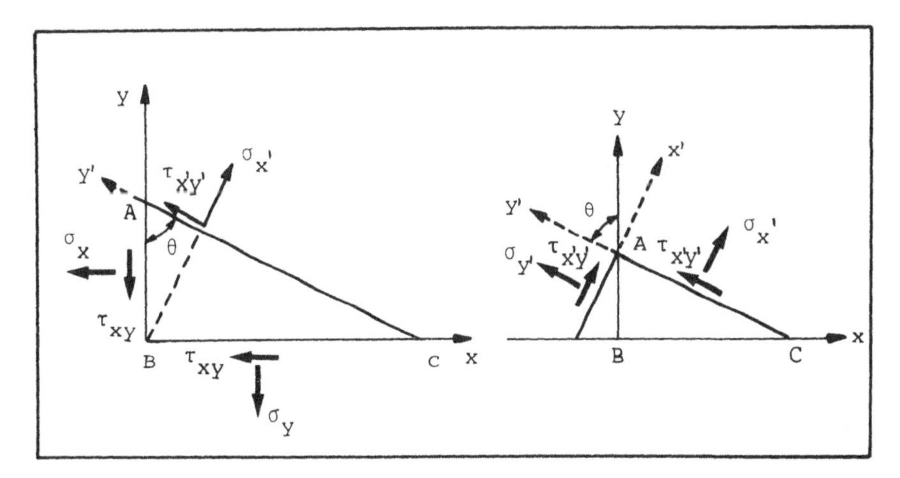

FIGURE 4.3 A two-dimensional stress system; determination of stresses acting on an arbitrary plane inclined at the angle θ to the *y*-axis.

Force equilibrium perpendicular to *AC* (in the direction *x'*) gives

$$\sigma_{x'}(AC) = \sigma_x(AB)\cos\theta + \sigma_y(BC)\sin\theta + \tau_{xy}(AB)\sin\theta + \tau_{xy}(BC)\cos\theta \tag{4.6}$$

Force parallel to the plane *AC* (direction y') gives

$$\tau_{x'y'}(AC) = -\sigma_x(AB)\sin\theta + \sigma_y(BC)\cos\theta + \tau_{xy}(AB)\cos\theta - \tau_{xy}(BC)\sin\theta \tag{4.7}$$

Since

$$AB = AC\cos\theta \qquad \text{and} \qquad BC = AC\sin\theta$$

Equations (4.6) and (4.7) can be modified to

$$\sigma_{x'} = \sigma_x\cos^2\theta + \sigma_y\sin^2\theta + 2\tau_{xy}\sin\theta\cos\theta$$
$$\tau_{x'y'} = -\sigma_x\sin\theta\cos\theta + \sigma_y\sin\theta\cos\theta + \tau_{xy}(\cos^2\theta - \sin^2\theta)$$

If the double angle 2θ is introduced, the following modified equations are obtained:

$$\sigma_{x'} = \tfrac{1}{2}(\sigma_x + \sigma_y) + \tfrac{1}{2}(\sigma_x - \sigma_y)\cos 2\theta + \tau_{xy}\sin 2\theta \tag{4.8}$$

$$\tau_{x'y'} = \tfrac{1}{2}(\sigma_y - \sigma_x)\sin 2\theta + \tau_{xy}\cos 2\theta \tag{4.9}$$

For the direction y' (see Fig. 4.3), the following equation for $\sigma_{y'}$ can correspondingly be found:

$$\sigma_{y'} = \tfrac{1}{2}(\sigma_x + \sigma_y) - \tfrac{1}{2}(\sigma_x - \sigma_y)\cos 2\theta - \tau_{xy}\sin 2\theta \tag{4.10}$$

Since $\tau_{y'x'}$ is equal to $\tau_{x'y'}$, these three equations constitute the complete description of the state of stress on a plane inclined at the angle θ to the y-axis (i.e., the x', y'-system is rotated through the angle θ counterclockwise relative to the x, y-system).

It must be noted that a change in sign occurs when θ is situated in the second and fourth quadrants, and consequently it is recommended that to avoid mistakes the system x', y' be chosen so that θ is situated in the first quadrant.

From Eq. (4.9) it can be seen that $\tau_{x'y'}$ vanishes; that is, no shear stress exists on the plane AC when θ is chosen so that

$$\tan 2\theta = \frac{\tau_{xy}}{\tfrac{1}{2}(\sigma_x - \sigma_y)} \tag{4.11}$$

Equation (4.11) defines two planes at angles θ and $\theta + \pi/2$, where the shear stress is zero. This means that for any two-dimensional stress system, there exist two mutually perpendicular planes on which the shear stress is zero. These planes are called the *principal planes*. The normal stresses acting on these planes are called the *principal stresses*. It can be shown by differentiating equation (4.8) that the principal stresses are the maximum and minimum normal stresses in the system. The principal stresses are designated as σ_1 and σ_2 and chosen so that $\sigma_1 > \sigma_2$. Thus the directions of the principal stresses are called 1 (maximum) and 2 (minimum).

Equations (4.8) and (4.10) determine the magnitudes of the principal stresses when the condition given by Eq. (4.11) is substituted. Thus

$$\left.\begin{matrix}\sigma_1\\ \\ \sigma_2\end{matrix}\right\} = \frac{1}{2}(\sigma_x + \sigma_y) \pm \left[\left(\frac{\sigma_x - \sigma_y}{2}\right)^2 + \tau_{xy}^2\right]^{1/2} \tag{4.12}$$

Often, the principal stresses are used to describe the state of stress because of the resulting simplifications in the calculations.

By differentiating Eq. (4.9) it can be shown that the maximum shear stress occurs on two mutual perpendicular planes inclined at 45° to the principal planes. The magnitude of the maximum shear stress is given by

$$\tau_{max} = \frac{\sigma_1 - \sigma_2}{2} \tag{4.13}$$

The normal stress on the planes of maximum shear can be shown to be equal to $(1/2)(\sigma_1 + \sigma_2)$.

EXAMPLE 1. A state of stress is described by $\sigma_x = \sigma_y = 0$ and $\tau_{xy} \neq 0$. Determine the principal stresses.

Equation (4.12) gives

$$\left.\begin{matrix}\sigma_1\\ \sigma_2\end{matrix}\right\} = \pm\tau_{xy} = \begin{cases}\tau_{xy}\\ -\tau_{xy}\end{cases}$$

The directions of the principal stresses can be found from Eq. (4.8), since

$$\sigma_{x'} = \sigma_1 = \tau_{xy} \Rightarrow$$

$$\tau_{xy} = \tau_{xy} \sin 2\theta \Rightarrow$$

$$\sin 2\theta = 1 \Rightarrow \theta = \frac{\pi}{4} \quad \text{(for direction 1)}$$

This implies that a state of pure shear is equivalent to a compressive stress and a tensile stress equal in magnitude and mutually perpendicular to each other acting on planes inclined at 45° to the x,y-system for the state of pure shear (see Fig. 4.4).

In Example 1 only two-dimensional systems of stress were discussed. For three-dimensional systems of stress the analysis is a little more complicated.

To describe a two-dimensional system of stress (biaxial stress system), three stresses were necessary: σ_x, σ_y, and τ_{xy}. To describe a three-dimensional system, six stresses are necessary: σ_x, σ_y, σ_z, τ_{xy}, τ_{yz}, τ_{xz}. If the principal stresses are used, only three are necessary: σ_1, σ_2, σ_3. The corresponding maximum shear stresses, occurring on planes inclined at 45° to the principal planes, are:

$$\begin{aligned}
1-2&: \tau_{1-2} = \frac{\sigma_1 - \sigma_2}{2} \ (= \tau_3)\\
1-3&: \tau_{1-3} = \frac{\sigma_1 - \sigma_3}{2} \ (= \tau_2) = \tau_{\max} \ \text{(absolute)}\\
2-3&: \tau_{2-3} = \frac{\sigma_2 - \sigma_3}{2} \ (= \tau_1)
\end{aligned} \tag{4.14}$$

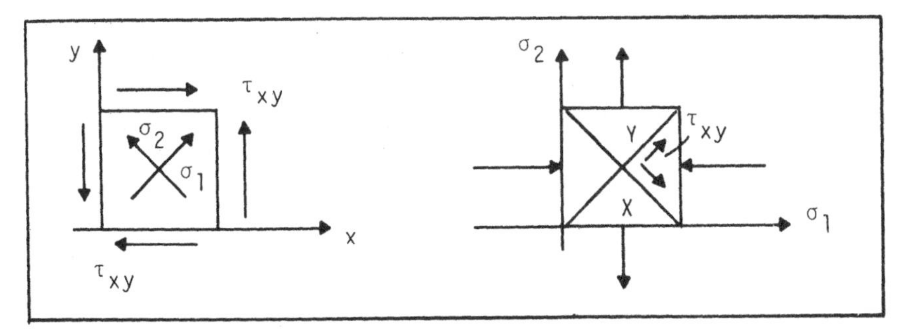

FIGURE 4.4 A state of pure shear.

The numbers 1, 2, and 3 refer to the principal directions, and the principal stresses are arranged so that $\sigma_1 > \sigma_2 > \sigma_3$. The shearing stresses in the parentheses are sometimes called the principal shearing stresses, and the suffixes designate the directions of the stresses. The absolute maximum shear stress is τ_{max} and is given by $(\sigma_1 - \sigma_3)/2$ $(= \tau_2)$. If the state of stress is given by σ_1, σ_2, $\sigma_3 = 0$, the maximum shear stress τ_{max} is given by $\sigma_1/2$. Thus in a three-dimensional state of stress, the greatest shear stress occurs on a plane bisecting the angle between the planes on which the greatest and the smallest principal stresses act.

4.3 TRUE STRESS–NATURAL STRAIN CURVES AND INSTABILITY

4.3.1 True Stress and Natural Strain

In Chapter 2 the stress–strain curves obtained from tensile tests were discussed. Nominal stresses were defined as

$$\sigma_{nom} = \frac{P}{A_{original}}$$

that is, the load was distributed uniformly on the original cross-sectional area. The nominal stress does not always describe the stress that the material experiences, since the cross-sectional area decreases as the specimen elongates plastically. To be able to carry out reasonable calculations concerning the deformation of materials, it is necessary to know the true stress (i.e., the stress that the material experiences).

The *true stress* is defined as

$$\sigma = \frac{P}{A_c} \tag{4.15}$$

where A_c is the current or instantaneous cross-sectional area. The true stress is thus obtained by dividing the instantaneous force by the instantaneous area (i.e., the force and the cross-sectional area must be measured simultaneously).

If the stress–strain curves are now plotted as true stress–nominal strain curves, the dashed curves in Fig. 4.5 are obtained. The solid curves are the nominal stress–nominal strain curves. It should be remembered that the curve for mild steel is an exception. The curve labeled "other metals" is typical.

From Fig. 4.5 it can be seen that the true stress–nominal strain curve does not show the maximum load/force point M as does the nominal stress curve. At M the strain hardening and the decrease in cross-sectional area exactly balance each other. Beyond M the decrease in area dominates, resulting in a necking leading to a three-dimensional system of stress here. Dividing the instantaneous

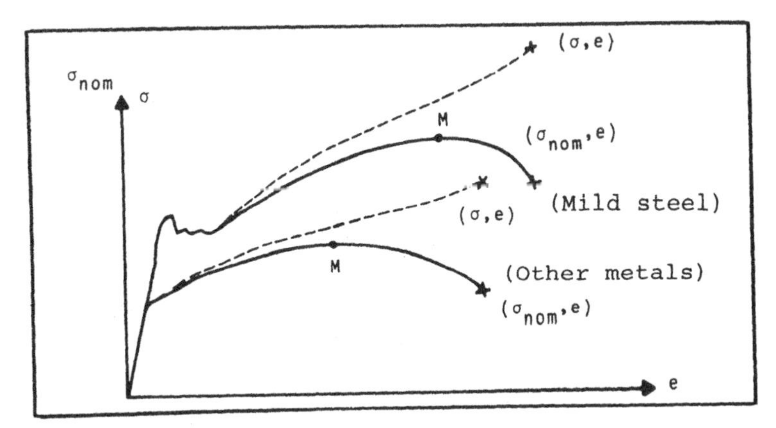

FIGURE 4.5 Stress–strain curves plotted as nominal stress-nominal strain curves (σ_{nom}, e) and true stress–nominal strain curves (σ, e).

force by the instantaneous smallest area an average "true" stress is obtained; that is, after the beginning of necking the real "true" stress can only be determined by correcting the average "true" stress for the three-dimensional stress system. In metalworking calculations, only deformations up to the beginning of necking where instability occurs are of interest; consequently, the correction of the "true" stress will not be discussed in this context.

The *nominal* or *engineering strain* was defined in Chapter 2 as

$$e = \frac{l_2 - l_1}{l_1} \times 100\%$$

where l_1 is the original length and l_2 the final length. The nominal strains—in the same way as the nominal stresses—are generally unsuitable for calculations involving plastic deformations, where large strains occur. Therefore, the natural, true, or logarithmic strain—sometimes called the *incremental strain*—has been defined.

A tensile specimen has been plastically elongated to the length l. It is now given a further elongation dl and this incremental increase of strain is defined by

$$d\epsilon = \frac{dl}{l}$$

If the specimen is deformed from the length l_1 to the length l_2, the total strain is obtained by integration:

$$\epsilon_{1-2} = \int_{l_1}^{l_2} \frac{dl}{l} = \ln \frac{l_2}{l_1} \tag{4.16}$$

where "ln" denotes the logarithm to the base e. This strain (4.16) is the natural strain and is designated by ϵ, whereas the nominal strain was designated by e.

Some of the advantages of the natural strain compared to the nominal strain are:

1. Natural strains are additive. A specimen is first deformed from l_1 to l_2 and then from l_2 to l_3 (case I). Another specimen is deformed directly from l_1 to l_3 (case II). The following table shows the results as natural and nominal strains. It can be seen that an interruption of the deformation does not influence the final natural strain, whereas it changes the nominal strain (i.e., they are not additive).

Case		ϵ	e
I	$l_1 \rightarrow l_2$	$\ln \frac{l_2}{l_1}$	$\frac{l_2 - l_1}{l_1}$
	$l_2 \rightarrow l_3$	$\ln \frac{l_3}{l_2}$	$\frac{l_3 - l_2}{l_2}$
	Adding	$\ln \frac{l_2}{l_1} + \ln \frac{l_3}{l_2}$	$\frac{l_2 - l_1}{l_1} + \frac{l_3 - l_2}{l_2}$
		$= \ln \frac{l_3}{l_1}$	$\neq \frac{l_3 - l_1}{l_1}$
II	$l_1 \rightarrow l_3$	$\ln \frac{l_3}{l_1}$	$\frac{l_3 - l_1}{l_1}$

2. The natural strain gives the same numerical values in compression and in tension, which is not the case for nominal strain. A specimen is deformed from l_1 to $l_2 = 2l_1$ (case I) and a specimen is compressed from h_1 to $h_2 = (1/2)h_1$ (case II). The results are shown in the following table.

Case		ϵ	e
I	$l_1 \rightarrow 2l_1$	$\ln \frac{2l_1}{l_1} = \ln 2$	$\frac{2l_1 - l_1}{l_1} = 1$
II	$h_1 \rightarrow \frac{1}{2}h_1$	$\ln \frac{(1/2)h_1}{h_1} = -\ln 2$	$\frac{(1/2)h_1 - h_1}{h_1} = -\frac{1}{2}$

The natural and nominal strains are related, below the maximum load, as follows:

$$e = \frac{l_2 - l_1}{l_1} = \frac{l_2}{l_1} - 1$$

that is,

$$\frac{l_2}{l_1} = 1 + e \qquad \text{or} \qquad \ln \frac{l_2}{l_1} = \ln (1 + e)$$

or, alternatively,

$$\epsilon = \ln (1 + e) \quad \text{(until instability)} \tag{4.17}$$

For small strains, e and ϵ give approximately the same result (i.e., $\epsilon \simeq e$).

Figure 4.6 shows the true stress/natural (true) strain curves (σ, ϵ) as well as the nominal stress/nominal strain curves (σ_{nom}, e). In the following work the true stress and the natural strain (σ, ϵ) will be used extensively since, as mentioned, they reflect the conditions that the material experiences.

4.3.2 Volume Constancy

Based on experimental evidence it is found that, for metals, the volume of the material is constant during plastic deformation. This is not true for elastic deformations, but since the elastic deformation is, in general, very small compared to the plastic deformation, it can be neglected without any measurable error.

Volume constancy can be expressed by

$$\frac{dV}{d\epsilon} = 0 \tag{4.18}$$

where V is the volume of the material undergoing plastic deformation and ϵ is the natural strain.

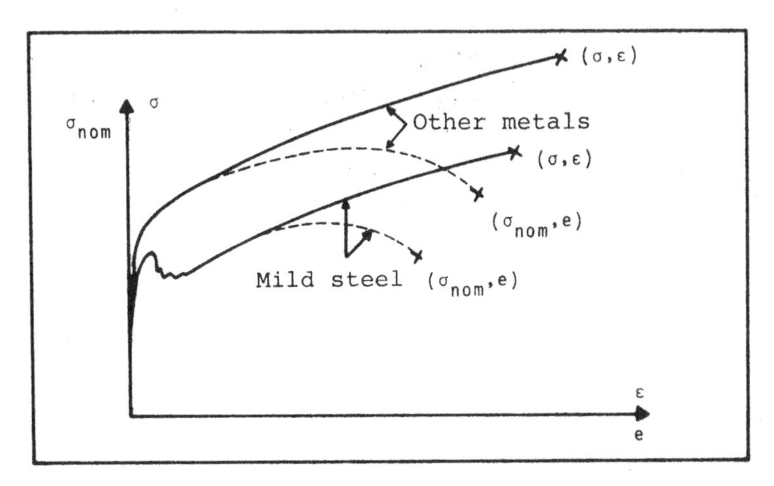

FIGURE 4.6 True and nominal stress–strain curves.

Consider a parallelepiped with the dimensions l_1, l_2, and l_3, which are deformed to $l_1 + \Delta l_1$, $l_2 + \Delta l_2$, and $l_3 + \Delta l_3$. Volume constancy gives

$$(l_1 + \Delta l_1)(l_2 + \Delta l_2)(l_3 + \Delta l_3) = l_1 l_2 l_3$$

This can be written as

$$\left(1 + \frac{\Delta l_1}{l_1}\right)\left(1 + \frac{\Delta l_2}{l_2}\right)\left(1 + \frac{\Delta l_3}{l_3}\right) = 1$$

or

$$(1 + e_1)(1 + e_2)(1 + e_3) = 1$$

where e is the nominal strain. Taking logarithms gives

$$\ln (1 + e_1) + \ln (1 + e_2) + \ln (1 + e_3) = 0$$

Using Eq. (4.17), this becomes

$$\epsilon_1 + \epsilon_2 + \epsilon_3 = 0 \qquad (4.19)$$

Volume constancy can also be expressed as $Al = A_1 l_1 = A_2 l_2$, where A is the cross-sectional area and l the length. This, combined with Eq. (4.16), results in

$$\epsilon_{1-2} = \ln \frac{l_2}{l_1} = \ln \frac{A_1}{A_2} \qquad (4.20)$$

For specimens with a circular cross section, this gives

$$\epsilon_{1-2} = \ln \frac{l_2}{l_1} = \ln \frac{A_1}{A_2} = \ln \frac{D_1^2}{D_2^2} = 2 \ln \frac{D_1}{D_2}$$

where D is the diameter of the cross section.

Equation (4.20) allows the natural strains to be calculated, even after necking has started, which makes the area strain very meaningful.

In addition to the relation between natural and nominal strain [Eq. (4.17)], a relation between nominal and true stress can be found.

$$\sigma = \frac{P}{A} = \frac{P}{A_1}\frac{A_1}{A} = \sigma_{nom} \frac{A_1}{A} = \sigma_{nom} \frac{l}{l_1}$$

where A is the cross-sectional area and the suffix 1 denotes the original state. Since $1 + e = l/l_1$, the preceding equation can be expressed as

$$\sigma = \sigma_{nom} (1 + e) \qquad (4.21)$$

This relation is valid only until necking starts.

4.3.3 Instability

As mentioned previously, the point where the strain hardening and the decrease in area during plastic deformation exactly counteract each other is called the *point of onset of instability* and corresponds to the ultimate load. Beyond this, necking occurs at a weak point of the specimen, and the deformation changes from being uniformly distributed along the gage length to local deformation in the necking region.

The point of the onset of instability is where the slope of the load-strain curve becomes zero (i.e., a strain increase takes place without any increase in load). This can be expressed as

$$\frac{dP}{d\epsilon} = 0 \tag{4.22}$$

Since $P = \sigma A$, Eq. (4.22) can be expressed as

$$A\frac{d\sigma}{d\epsilon} + \sigma\frac{dA}{d\epsilon} = 0$$

Volume constancy $dV/d\epsilon = 0$ gives

$$\frac{dV}{d\epsilon} = \frac{d(Al)}{d\epsilon} = A\frac{dl}{d\epsilon} + l\frac{dA}{d\epsilon} = 0$$

which is combined with the equation above to give

$$A\frac{d\sigma}{d\epsilon} + \sigma\left(-\frac{A}{l}\frac{dl}{d\epsilon}\right) = 0$$

Since $d\epsilon = dl/l$, the condition of instability becomes

$$\frac{d\sigma}{d\epsilon} = \sigma \tag{4.23}$$

This equation means that instability occurs when the slope of the stress–strain curve (rate of work hardening) equals the magnitude of the applied stress. Strain hardening continues beyond the point of instability, and hence there is no sudden change in the true stress–strain curve at this point.

Figure 4.7 shows how the point of instability is determined graphically on a stress–strain curve.

In many metalworking processes taking place under the action of tensile stresses, the maximum amount of deformation that a ductile material can withstand without failure is determined by the strain at instability. This is because failure by necking normally ruins the product. It is therefore important to be able to predict the stress and strain at the onset of instability, so that the process parameters to avoid failure can be safely specified. In Fig. 4.7 it was shown how the point of instability could be located on the true stress–natural strain curve

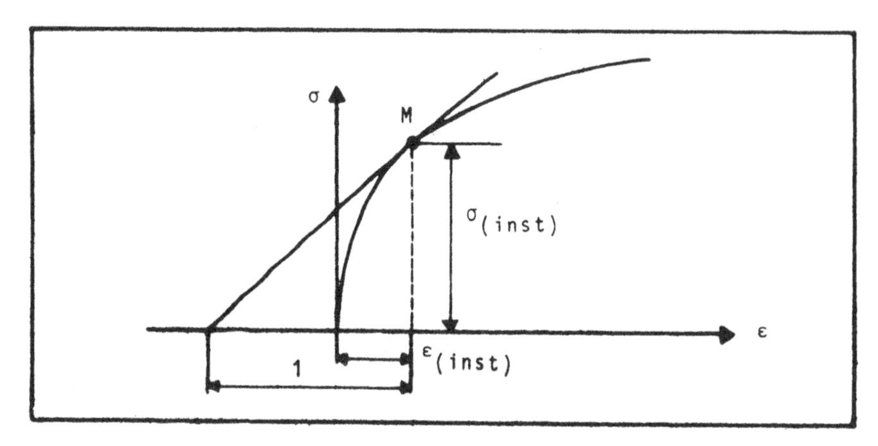

FIGURE 4.7 Graphic location of the point of instability on true stress–natural strain curve (uniaxial).

obtained by tensile test. It would be far more convenient, however, if the stress–strain curve could be expressed analytically, since this would permit an easy estimation of both strain and stress at instability. In the following section, various analytical models of the stress–strain curve are discussed.

4.3.4 Analytical Stress–Strain Curves

Different analytical models of the stress–strain curve can be employed depending on the actual material and the accuracy required. Here the most common model for strain-hardening materials (model 1) and a model for non-strain-hardening materials (model 2) are described. It should be mentioned that the availability of computers, which can easily handle the numerical test data, has decreased the need for analytical expressions, but these expressions are convenient in most calculations.

MODEL 1 (Fig. 4.8a).

$$\sigma = c\epsilon^n \tag{4.24}$$

This model represents with reasonable accuracy annealed metals with a cubic lattice structure. The symbols c and n represent material constants and n is called the strain-hardening exponent.

MODEL 2 (Fig. 4.8b).

$$\sigma = \sigma_0 \tag{4.25}$$

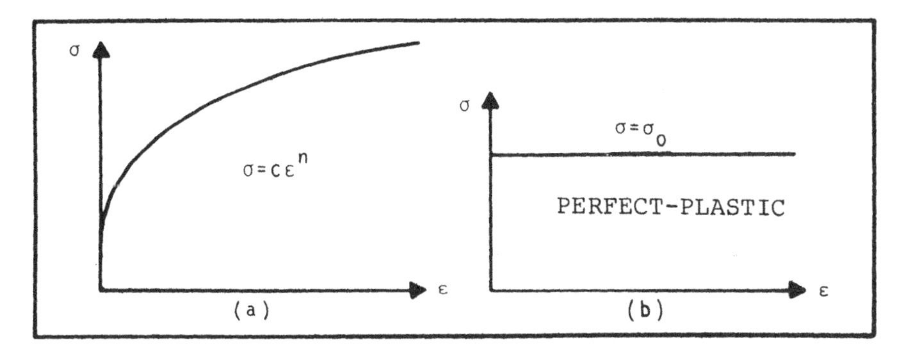

FIGURE 4.8 Approximate models for true stress–natural strain curve: (a) strain-hardening materials; (b) non-strain-hardening materials.

This model, which represents a perfect plastic, can be used with reasonable accuracy for materials with no, or slight, strain hardening ($n \simeq 0$). This simple model is often used in approximate calculations of average stresses and forces. The stress σ_0 is often defined as the mean yield stress $\sigma_{0m} = (\sigma_{01} + \sigma_{02})/2$, where the suffixes 1 and 2 indicate the yield stress before and after the deformation.

Since the elastic strains are very small compared to the plastic strains, they are generally neglected. Model 1 may include elastic strains if ϵ is considered as $\epsilon = \epsilon_{\text{elastic}} + \epsilon_{\text{plastic}}$, but model 2 cannot include elastic strains. It is possible to combine model 2 with Hooke's law $\sigma = E \cdot \epsilon$, so that an elastic perfect plastic material is described.

The values of the constants in the analytical models are usually chosen so that the best agreement between the models and the experimental curves is obtained. Considering materials that can be represented by model 1 (i.e., $\sigma = c\epsilon^n$), the point of instability can be determined from Eq. (4.23):

$$\frac{d\sigma}{d\epsilon} = nc\epsilon_{\text{inst}}^{n-1} = \sigma = c\epsilon_{\text{inst}}^{n}$$

$$nc\epsilon_{\text{inst}}^{n-1} = c\epsilon_{\text{inst}}^{n}$$

$$\epsilon_{\text{inst}} = n \tag{4.26}$$

This means that the strain at instability is equal to the strain-hardening exponent. This implies that n is a measure of the ability of the material to undergo plastic deformation without failure. Table 4.1 shows typical examples of the values c and n for model 1 for different materials. These values can be used as rough first approximation, but in the actual situation c and n must be determined from the experimental stress-strain curve, since rather large variations may occur for the same types of materials.

TABLE 4.1 Typical Values of c and n in Model 1 [Eq. (4.24)]

Metal	$c(\text{N/mm}^2)$	n
Mild steel	640	0.22
Stainless steel	1560	0.50
Aluminum, soft	156	0.25
Copper, soft	525	0.38
Brass	745	0.48

In most metal-forming processes, the material is subjected to a more complex system of stresses than a uniaxial stress system such as biaxial tension, for example. In such cases the same material may sustain only a fraction of or, at the other extreme, more than the strain of instability in uniaxial tension, depending on the actual stress system. Therefore, the instability value ($\epsilon_{inst} = n$) found in uniaxial tension should be used with caution in complex situations. In such situations the same procedure as described by Eq. (4.23) can be used when σ and ϵ are expressed as effective or equivalent values (see Section 4.5),$d\bar{\sigma}/d\bar{\epsilon} = \bar{\sigma}/z$ and where z is used to determine the instability by $\epsilon_{inst} = zn$. The constant n is the uniaxial strain of instability.

Under compressive systems of stress, instability such as necking does not occur, and here the limits of deformation are set by fracture.

EXAMPLE 2. For a given material the true stress/nominal strain curve is available. Determine the condition of instability. Equation (4.23) states that, for true stress–natural strain curves, instability occurs when

$$\frac{d\sigma}{d\epsilon} = \sigma$$

From Eq. (4.17), which expresses the relation between nominal and natural strain,

$$\epsilon = \ln (1 + e)$$

it is found that

$$d\epsilon = \frac{de}{1 + e}$$

This is substituted in the σ, ϵ-instability condition, giving the σ, e-instability condition

$$\frac{d\sigma}{de} = \frac{\sigma}{1 + e} \tag{4.27}$$

EXAMPLE 3. The stress–strain curve for a material is described by $\sigma = c\epsilon^n$. Determine the ultimate strength $\sigma_{nom,uts}$. The ultimate tensile strength was defined by [Eq. (2.5)].

$$\sigma_{nom,uts} = \frac{P_{max}}{A_1} = \frac{P_{inst}}{A_1}$$

where A_1 is the original cross-sectional area. Expressed in true stress, $P_{inst} = \sigma_{inst}A_{inst}$, that is,

$$\sigma_{nom,uts} = \frac{\sigma_{inst}A_{inst}}{A_1}$$

The instability equation (4.23) can be combined with $\sigma = c\epsilon^n$ and Eq. (4.26), giving

$$\frac{d\sigma}{d\epsilon} = \sigma \Rightarrow \epsilon_{inst} = n$$

This allows the calculation of σ_{inst}. Thus,

$$\sigma_{inst} = c\epsilon_{inst}^n = cn^n$$

Volume constancy gives

$$Al = A_1 l_1 = \text{constant}$$

$$\epsilon = \ln\frac{l}{l_1} = ln\frac{A_1}{A}$$

At instability this becomes

$$\epsilon_{inst} = ln\frac{A_1}{A_{inst}} = n$$

From this, A_{inst} can be determined:

$$A_{inst} = \frac{A_1}{\exp n}$$

The expressions for σ_{inst} and A_{inst} are now substituted in the equation for $\sigma_{nom,uts}$, giving

$$\sigma_{nom,uts} = \frac{\sigma_{inst}A_{inst}}{A_1} = \frac{cn^nA_1}{A_1 \exp n}$$

that is,

$$\sigma_{nom,uts} = \frac{cn^n}{\exp n} \tag{4.28}$$

4.4 YIELD CRITERIA

As discussed both in Chapter 2 and in this chapter, plastic deformation occurs in a material subjected to tensile stress along its axis (uniaxial stress), when the stress exceeds the yield stress σ_0. The yield stress can be determined from the stress–strain curve either as a characteristic value—the yield point—or as the yield strength, normally defined as the stress at 0.2% permanent strain. In the following, σ_0 will be used as the yield condition for uniaxial states of stress.

In most forming and cutting processes, deformation takes place or occurs under more complex states of stress, and it is therefore necessary to be able to predict the state of stress at which yielding is initiated and maintained. This means that a yield criterion, enabling consideration of all the combinations of stresses that will provide plastic flow, must be established. The establishment of a yield criterion is based on the following assumptions or empirical observations:

The metals are homogeneous, continuous, and isotropic (i.e., have the same properties in all directions).
The metals have the same yield stress in compression and tension.
A superimposed hydrostatic pressure does not influence the initiation of yielding.

4.4.1 Tresca's Yield Criterion

In 1864, Tresca put forward his criterion saying that plastic flow occurs when the maximum shear stress exceeds a critical value. Since the maximum shear stress, Eq. (4.14), is equal to half the difference between the greatest and the smallest principal stress, Tresca's criterion may be expressed as

$$\tau_{max} = \frac{1}{2}(\sigma_1 - \sigma_3) \geq \text{constant } k = \tau_{crit} \quad (4.29)$$

where $\sigma_1 > \sigma_2 > \sigma_3$. This criterion implies that yielding is independent of the intermediate principal stress, which is not strictly true. In spite of this, the difference in terms of stresses between Tresca's criterion and the more correct criterion of von Mises never exceeds 15%. By introducing a factor on the right-hand side of Eq. (4.29) (e.q., 1.075), the difference between these stresses can be reduced.

Since Eq. (4.29) is applicable to all systems of stress, the constant k (the critical value) can be found, for example, by considering the yield stress σ_0 in uniaxial tension. Here the state of stress at yielding is given by $\sigma_1 = \sigma_0$, $\sigma_2 = \sigma_3 = 0$. Equation (4.29 then gives)

$$\tau_{max} = \frac{\sigma_1 - \sigma_3}{2} = \frac{\sigma_0}{2} \geq k \quad (= \tau_{crit})$$

which means that the critical shear stress is related to the yield stress in simple tension by

$$k = \frac{\sigma_0}{2}$$

This is very important, since σ_0 is the most easily obtained material property.

Tresca's criterion can thus be expressed as

$$\sigma_1 - \sigma_3 \geq \sigma_0 \qquad (\sigma_1 > \sigma_2 > \sigma_3) \tag{4.30}$$

4.4.2 von Mises' Yield Criterion

In 1913, von Mises proposed a yield criterion, stating that yielding occurs when the work of deformation per unit volume provided by the system of stress exceeds a critical value for the particular material, which can be expressed mathematically as

$$(\sigma_1 - \sigma_2)^2 + (\sigma_2 - \sigma_3)^2 + (\sigma_3 - \sigma_1)^2 \geq \text{constant } C$$

Since the constant C is the same for all systems of stress, simple tension can be used to determine C. Uniaxial tension where $\sigma_1 = \sigma_0$, $\sigma_2 = \sigma_3 = 0$ gives

$$\sigma_1^2 + \sigma_1^2 = 2\sigma_0^2 = C$$

The von Mises' criterion can then be expressed as

$$(\sigma_1 - \sigma_2)^2 + (\sigma_2 - \sigma_3)^2 + (\sigma_3 - \sigma_1)^2 \geq 2\sigma_0^2 \tag{4.31}$$

For the state of pure shear, Tresca's and von Mises' criteria give different results. Pure shear is equivalent to (see Example 1): $\sigma_1 = k$, $\sigma_2 = 0$, $\sigma_3 = -k$:

$$\text{Tresca: } k + k = \sigma_0 \Rightarrow k = \frac{\sigma_0}{2}$$

$$\text{von Mises: } 6k^2 = 2\sigma_0^2 \Rightarrow k = \frac{2}{\sqrt{3}}\frac{\sigma_0}{2} \tag{4.32}$$

This means that the critical shear stress at yielding differs as mentioned previously by $2/\sqrt{3} = 1.15$; that is, von Mises' criterion requires a 15% higher critical shear stress value to initiate yielding than does Tresca's criterion. For ductile metals it has been shown experimentally that von Mises' criterion gives the best agreement, but due to its simplicity, Tresca's criterion is applied in many cases, particularly by design engineers.

EXAMPLE 4. In the following table the results of Tresca's and von Mises' yield criteria are shown for different systems of stress. The abbreviation PS means a plane stress system, where one of the principal stresses is zero and CS means a cylindrical stress system where two of the principal stresses are equal.

Principal stresses			Yield criteria	
σ_1	σ_2	σ_3	von Mises	Tresca
σ_1	σ_1	0	PS $\sigma_1 = \sigma_0$	$\sigma_1 = \sigma_0$
σ_1	σ_2	0	PS $\sigma_1^2 + \sigma_2^2 - \sigma_1\sigma_2 = \sigma_0^2$	$\sigma_1 = \sigma_0$
σ_1	0	$-\sigma_1$	PS $\sigma_1 = \frac{2}{\sqrt{3}}\frac{\sigma_0}{2}$	$\sigma_1 = \frac{\sigma_0}{2}$
σ_1	0	σ_3	PS $\sigma_1^2 + \sigma_3^2 - \sigma_1\sigma_3 = \sigma_0^2$	$\sigma_1 - \sigma_3 = \sigma_0$
0	σ_2	σ_2	PS $-\sigma = \sigma_0$	$-\sigma_2 = \sigma_0$
0	σ_2	σ_3	PS $\sigma_2^2 + \sigma_3^2 - \sigma_2\sigma_3 = \sigma_0^2$	$-\sigma_3 = \sigma_0$
σ_1	σ_3	σ_3	CS $\sigma_1 - \sigma_3 = \sigma_0$	$\sigma_1 - \sigma_3 = \sigma_0$

EXAMPLE 5. Many metalworking processes take place under a state of deformation called *plane strain*; that is, the strain in one principal direction is zero. This means that the flow everywhere is parallel to the plane (1, 3) and independent of the position along the normal (2) to this plane (see Fig. 4.9). Determine for deformations taking place under plane strain conditions (ϵ_1, $\epsilon_2 = 0$, ϵ_3), the initiation of yielding by Tresca's and von Mises' yield criteria.

When a material being plastically deformed has a tendency to flow in all directions, a plane strain condition implies that the flow in one direction is prevented either by the tooling or by the geometry of the component (sheet rolling, for example).

A plain strain condition (ϵ_1, $\epsilon_2 = 0$, ϵ_3) does not imply that the stress σ_2 is zero. It can be shown from the flow rules [13] that

$$\sigma_2 = \tfrac{1}{2}(\sigma_1 + \sigma_3) \tag{4.33}$$

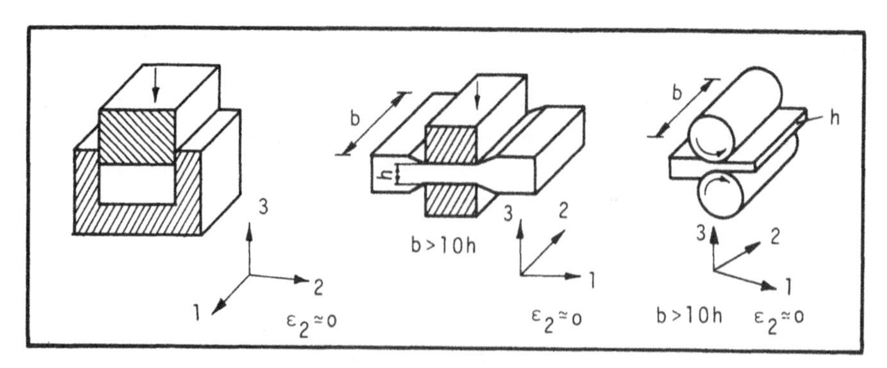

FIGURE 4.9 Examples of plane strain conditions.

Tresca's criterion implies then, for a plane strain condition [Eq. (4.30)]

$$\sigma_1 - \sigma_3 = \sigma_0$$

Von Mises' criterion implies [Eq. (4.31)] that

$$\sigma_1 - \sigma_3 = \frac{2}{\sqrt{3}} \sigma_0 \tag{4.34}$$

If the frictional forces are small and no external stresses are provided in the 1-direction (rolling and plane strain compression, for example), Eq. (4.34) is modified to become

$$-\sigma_3 = \frac{2}{\sqrt{3}} \sigma_0 \tag{4.35}$$

Finally, it can be mentioned that the plastic deformation or flow of metals can basically take place in three different ways: (1) stretching (elongation), (2) compression, and (3) shear (simple and pure). Figure 4.10 shows these three basic types of deformation. In most metalworking processes deformations occur as combinations of these types.

4.5 EFFECTIVE STRESS AND EFFECTIVE STRAIN

The purpose of introducing these terms is to obtain a convenient way of expressing the more complex systems of stress and strain acting on an element. This is done by defining effective or equivalent stresses and strains by which the complex systems are transformed to equivalent uniaxial situations. A major advantage here is that it is now possible to make use of the uniaxial stress–strain curve, giving, for example, the strain-hardening properties.

The definition of the effective or equivalent stress $\bar{\sigma}$ is based on von Mises' yield criterion and is given by

$$\bar{\sigma} = \{\tfrac{1}{2}[(\sigma_1 - \sigma_2)^2 + (\sigma_2 - \sigma_3)^2 + (\sigma_3 - \sigma_1)^2]\}^{1/2} \tag{4.36}$$

where the constant 1/2 is chosen so that $\bar{\sigma} = \sigma_0$ for uniaxial tension (σ_1, 0, 0). Basically, this can be explained as another way of expressing von Mises' yield criterion (i.e., flow occurs when $\bar{\sigma} \geq \sigma_0$).

Correspondingly, the effective or equivalent strain $\bar{\epsilon}$ is defined by

$$\bar{\epsilon} = \left[\frac{2}{3}(\epsilon_1{}^2 + \epsilon_2{}^2 + \epsilon_3{}^2)\right]^{1/2} \tag{4.37}$$

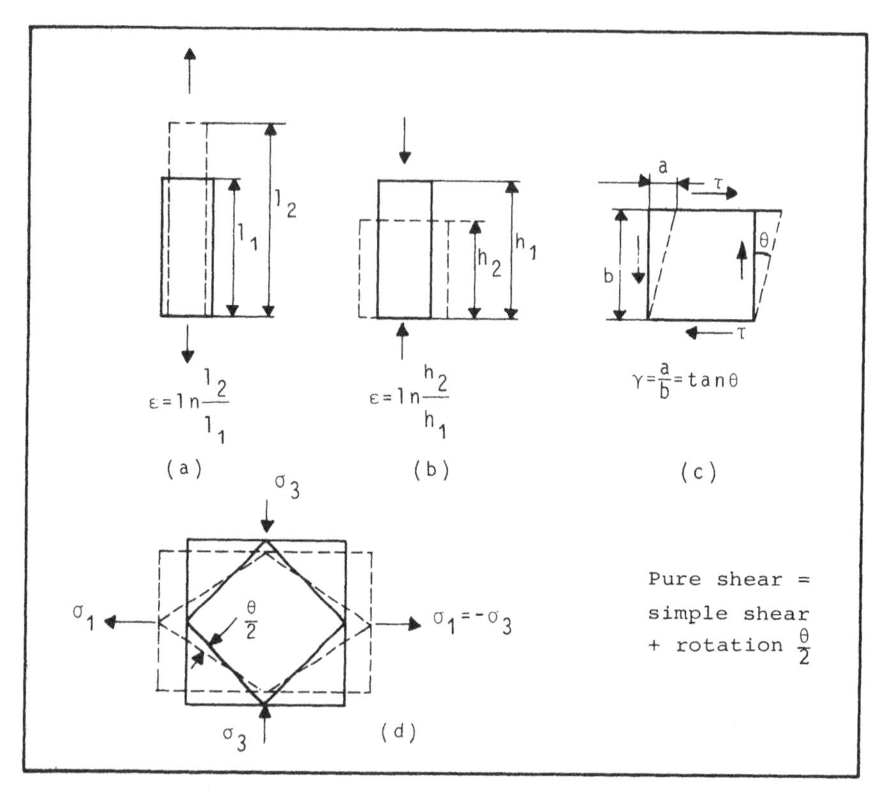

FIGURE 4.10 The basic types of deformation: (a) stretching; (b) compression; (c) simple shear; (d) pure shear.

where the constant 2/3 is chosen so that $\bar{\epsilon} = \epsilon_1$ for uniaxial tension (ϵ_1, $-\epsilon_1/2$, $-\epsilon_1/2$).

Equations (4.36) and (4.37) fulfill the intention of obtaining a simple method of describing complex systems of stress and strain. The stress–strain curve (σ, ϵ) obtained from a tensile test can be considered as a special case of an effective stress–strain curve ($\bar{\sigma}$, $\bar{\epsilon}$). Testing results ($\bar{\sigma}$, $\bar{\epsilon}$) obtained under complex situations can thus be directly compared to the results obtained in simple tension or compression tests. The reverse is true in normal situations; that is, the uniaxial stress–strain curves can be used directly in complex situations when they are expressed in terms of $\bar{\sigma}$ and $\bar{\epsilon}$.

EXAMPLE 6. As exercises, the effective stresses and strains are calculated for the different stress–strain systems shown in Table 4.2.

TABLE 4.2 Effective Stresses and Strains for Different Complex Systems of Stress and Strain

Process	ϵ_1	ϵ_2	ϵ_3	$\bar{\epsilon}$
	σ_1	σ_2	σ_3	$\bar{\sigma}$
Rolling 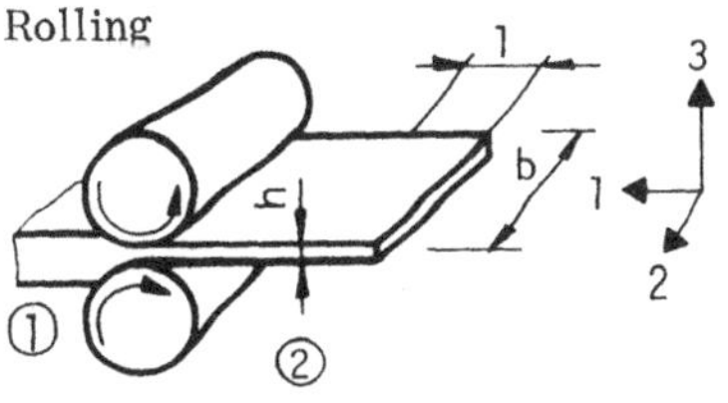	$\ln \frac{l_2}{l_1}$	$\ln \frac{b_2}{b_1} \simeq 0$	$\ln \frac{h_2}{h_1}$	$\frac{2}{\sqrt{3}} \epsilon_1 = -\frac{2}{\sqrt{3}} \epsilon_3$
	0	$\simeq \frac{\sigma_3}{2}$	$\simeq \sigma_3$	$-\frac{\sqrt{3}}{2} \sigma_3$
Forging (cylindrical workpiece) 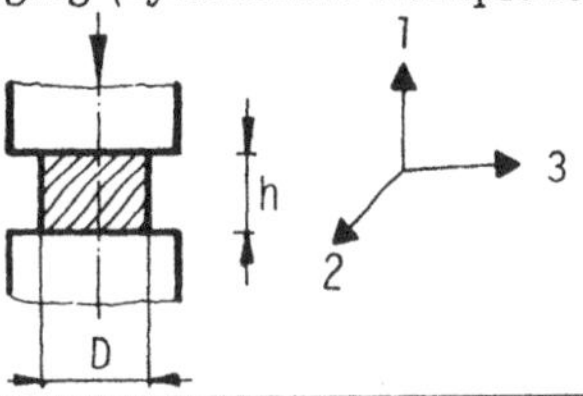	$\ln \frac{h_2}{h_1}$	$\ln \frac{D_2}{D_1}$	$\ln \frac{D_2}{D_1}$	$-\epsilon_1$ $(= 2\epsilon_2 = 2\epsilon_3)$
	σ_1	0	0	$-\sigma_1$
Extrusion 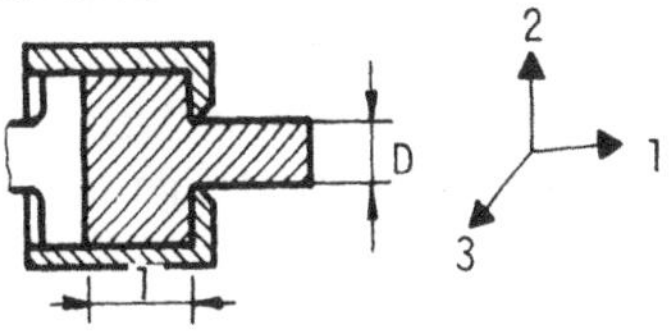	$\ln \frac{l_2}{l_1}$	$\ln \frac{D_2}{D_1}$	$\ln \frac{D_2}{D_1}$	ϵ_1 $(= -2\epsilon_2 = -2\epsilon_3)$
	σ_1	σ_3	σ_3	$-(\sigma_1 - \sigma_3)$
Bulging 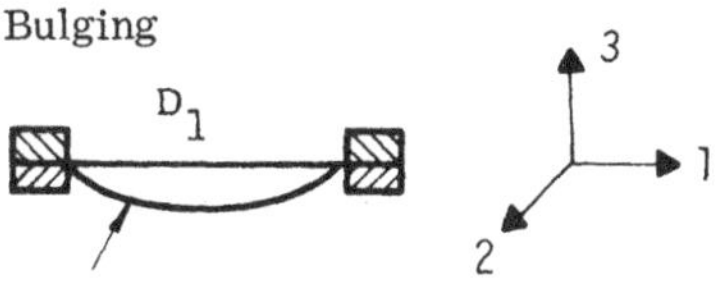 (Spherical segment)	$\ln \frac{D_2}{D_1}$	$\ln \frac{D_2}{D_1}$	$\ln \frac{t_2}{t_1}$	$-\epsilon_3$ $(= 2\epsilon_2 = 2\epsilon_1)$
	σ_1	σ_1	$\simeq 0$	σ_1
Tube expansion 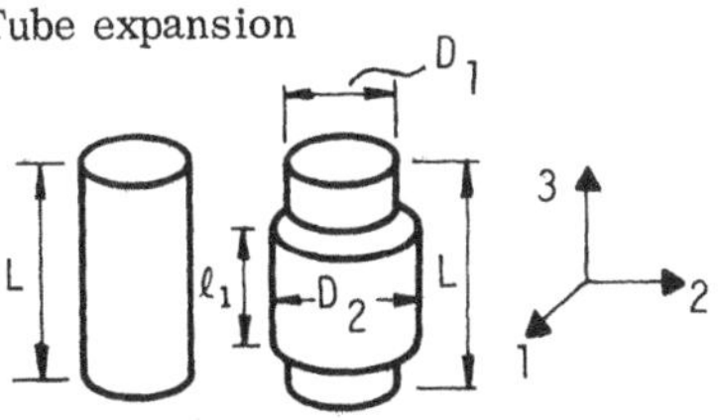 ($\ell_2 = \ell_1 + (D_2 - D_1)$, uniform deformation)	$\ln \frac{D_2}{D_1}$	$\ln \frac{t_2}{t_1}$	$\ln \frac{l_2}{l_1}$	$\bar{\epsilon}$
	σ_1	σ_2	$\simeq 0$	$\sqrt{\sigma_1^2 + \sigma_2^2 - \sigma_1\sigma_2}$

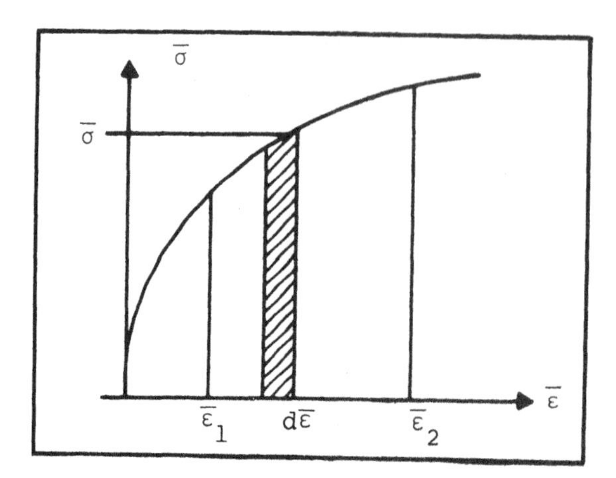

FIGURE 4.11 Determination of work of deformation.

4.6 WORK OF DEFORMATION

The deformation of a material requires a certain amount of work, depending on the conditions under which the deformation takes place. The deformation work is an important quantity, as it allows a determination of the energy necessary to carry out the deformation and allows a determination of the forces involved. Both parameters are necessary for the selection of machinery or the design of machinery.

From the stress–strain curve in Fig. 4.11, it can be seen that the work of deformation per unit volume to accomplish a strain increase of $d\bar{\epsilon}$ is

$$dw = \bar{\sigma}\, d\bar{\epsilon}$$

If the deformation is carried out from the strain $\bar{\epsilon}_1$ to the strain $\bar{\epsilon}_2$, the work per unit volume becomes

$$w = \int_{\bar{\epsilon}_1}^{\bar{\epsilon}_2} \bar{\sigma}\, d\bar{\epsilon} \tag{4.38}$$

The work necessary to deform the whole volume V then becomes

$$W = \int_V\!\!\int w\, dV = \int_V\!\int_{\bar{\epsilon}_1}^{\bar{\epsilon}_2} \bar{\sigma}\, d\bar{\epsilon}\, dV \tag{4.39}$$

If every element in the volume V is supplied with the same amount of work (homogeneous deformation), Eq. (4.39) can be written as

$$W = V \int_{\bar{\epsilon}_1}^{\bar{\epsilon}_2} \bar{\sigma}\, d\bar{\epsilon}$$

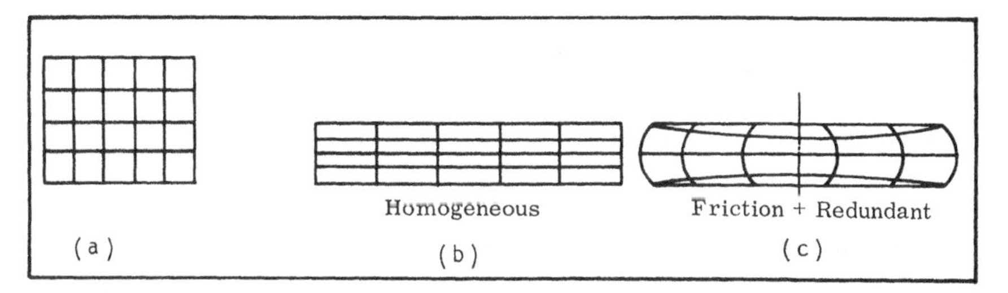

FIGURE 4.12 Work of deformation: (a) original workpiece; (b) homogeneous deformation of the workpiece (i.e., homogeneous work of deformation); (c) nonhomogeneous deformation of the workpiece (i.e., the work of deformation is equal to homogeneous work of deformation + frictional work + redundant work).

The work per unit volume can also be expressed approximately by the mean yield stress $\bar{\sigma}_m$ multiplied by the strain of deformation, giving

$$w = \bar{\sigma}_m(\bar{\epsilon}_2 - \bar{\epsilon}_1)$$

which combined with Eq. (4.38) gives

$$\bar{\sigma}_m = \frac{1}{\bar{\epsilon}_2 - \bar{\epsilon}_1} \int_{\bar{\epsilon}_1}^{\bar{\epsilon}_2} \bar{\sigma}\, d\bar{\epsilon} \tag{4.40}$$

If the stress–strain curve for the material can be represented by the model $\bar{\sigma} = c\bar{\epsilon}^n$, Eq. (4.39) becomes

$$W = \int_V \frac{c}{n+1} (\bar{\epsilon}_2^{n+1} - \bar{\epsilon}_1^{n+1})\, dV \tag{4.41}$$

If the initial strain $\bar{\epsilon}_1$ is zero and the deformation is homogeneous (i.e., all volume elements are supplied with the same amount of work), Eq. (4.41) becomes

$$W = V \frac{c}{n+1} \bar{\epsilon}^{n+1} \tag{4.42}$$

The work represented by Eqs. (4.39), (4.41), and (4.42) is the minimum work. To find the total work consumed, it is necessary to include the work of friction between the tool and the workpiece and the redundant work. The latter includes the work necessary to deform portions or all of the components, without changing the overall geometry. Considering one of the elements in Fig. 4.12, it can be seen that the deformation from (a) to (b) is homogeneous, but the deformation from (a) to (c) is no longer homogeneous, owing to friction and redundant deformation (the straight sides have become curved). The work to do this curving is the redundant work.

In this context the homogeneous work of deformation will predominantly be considered, and it must be remembered that this only constitutes the lower limit of the necessary work. When possible, the friction work will be taken into consideration, but redundant work will not be included.

EXAMPLE 7. Determine the work necessary to deform a tensile specimen from $\bar{\epsilon}_1 = 0$ to $\bar{\epsilon}_2 = \bar{\epsilon}$.

The material can be described by the stress–strain curve $\bar{\sigma} = c\bar{\epsilon}^n$. The original dimensions are D_1 and l_1, and the final dimensions are D_2 and l_2. Equation (4.20) gives

$$\bar{\epsilon}_2 = \bar{\epsilon} = 2 \ln \frac{D_1}{D_2}$$

With the assumption of homogeneous deformation, the work necessary can be calculated from Eq. (4.42):

$$W = \frac{\pi}{4} D_1^2 l_1 \frac{c}{n + 1} \left(2 \ln \frac{D_1}{D_2}\right)^{n+1} \tag{4.43}$$

5

Classification of the Manufacturing Processes

5.1 INTRODUCTION

In Chapter 1 a morphological model of the manufacturing processes was presented. This model was built up from a few fundamental elements arranged as a material flow, an energy flow, and an information flow. The description of the processes within the framework of the morphological model showed that the processes could appropriately be gathered in groups with certain common features. The features that distinguish these groups might be the state of the material, the process type, the basic process, and so on.

The more detailed descriptions presented in the following chapters are based on a classification of the processes into a few major groups with one or more basic common feature(s).

The aim of this text, in general, is to enable the engineer to distinguish among the various processes and to characterize them by means of their possibilities and limitations concerning material, geometry, tolerances, and surface finish. For the application of the processes, both in design and in production, it is important that the description of the processes rely on the basic principles covered by the morphological model, since this enables ingenious and imaginative utilizations of the existing processes and production equipment. This approach has been found to be much more challenging and fruitful than the traditional descriptive approach, which has the major disadvantage that only those processes remembered or known are considered. On the other hand, the systematic approach employed here ensures that all possibilities are included.

The classification discussed in the next section is based directly on the morphological model described in Chapter 1. As shown in Fig. 5.1, the material, the state of material, and the process type are used as characteristic grouping parameters, giving the headlines of the following chapters.

5.2 CLASSIFICATION OF THE PROCESSES

A classification of the technological manufacturing processes may be based on many different criteria, depending on its purpose. This classification will, as mentioned, be based on the morphological model discussed in Chapter 1 to obtain a structure oriented toward generation of possible manufacturing methods to produce specific components.

The structure of the classification is thus:

- Material flow
  - Type of material
  - State of material
  - Type of process
  - Basic process
- Energy flow
  - Type of energy
  - Medium of transfer
- Information flow
  - Surface creation principle
  - Pattern of motion

This structure is shown schematically in Fig. 5.1.

In this context, only processes aiming primarily at geometrical changes (obtaining specific geometries) will be discussed, but, as described in Chapter 2, the geometrical changes are normally accompanied by changes in other properties (mainly mechanical properties) depending on the process. These process-dependent changes in properties must be taken into consideration when selecting processes, as they may be decisive. An example of a process-dependent change would be the increase in mechanical strength of a metal due to strain hardening by deformation.

As discussed previously, many processes aiming primarily at changes in material properties without changing the geometry are available. A major group, and a very important one, constitutes the heat-treatment processes discussed in Chapter 3. Further information concerning these processes can be found in the literature.

Next the elements in the classification structure are discussed briefly.

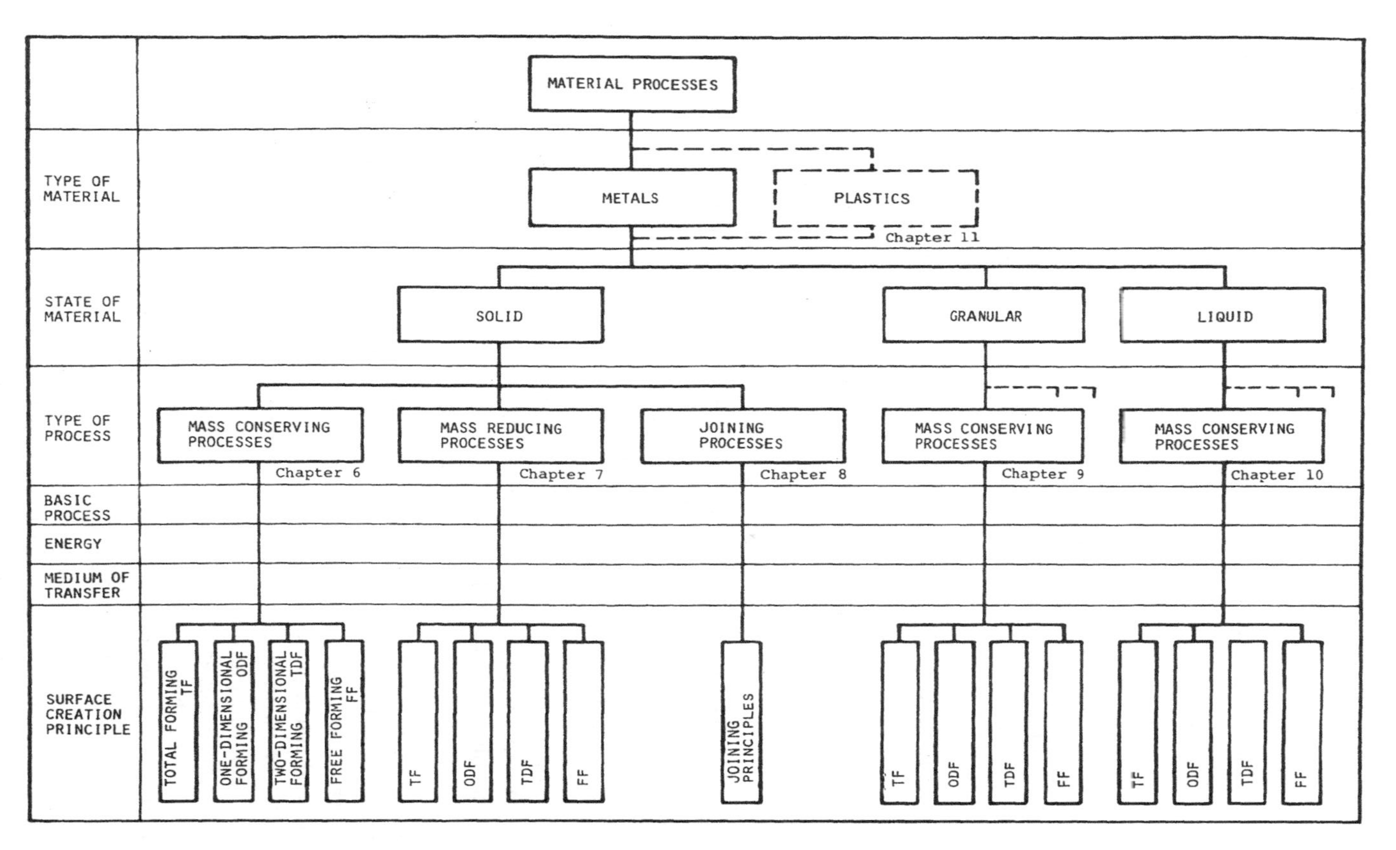

FIGURE 5.1 Classification of the technological manufacturing processes into groups having common features.

Type of Material. In Chapter 3 the materials were divided into metallic, nonmetallic, and composite materials. The classification structure (Fig. 5.1) should cover all the different materials, but only metals and plastics are shown and discussed further in this text. The major consideration will be the metallic materials—Chapters 6 through 10) and 12, 13, and 14—and consequently, Fig. 5.1 is primarily drawn for metals. The production of plastic components is discussed in Chapter 11, shown dashed in Fig. 5.1.

State of Material. A given type of material can be shaped in the solid, granular, or liquid state. The state of the material describes the situation in the shaping phase.

Type of Process. Considering materials in the solid state, shaping can be carried out by:

Mass-conserving processes ($dM = 0$): the mass of the component is equal to (or closely equal to) the mass of the original material. The basic process is plastic deformation.

Mass-reducing processes ($dM < 0$): the final component shape can be circumscribed by the shape of the original material, and the excess is removed by mechanical, thermal, or chemical basic processes.

Joining processes: the final geometry is obtained by joining the subgeometries. The subgeometries are produced by one or both of the abovementioned types of processes.

Concerning materials in the granular and liquid states, shaping is in general carried out only by mass-conserving processes.

In the blocks showing the type of process in Fig. 5.1, the chapters dealing with the specific types of processes for metallic materials are noted.

Basic Processes. Three types of basic processes exist: mechanical, thermal, and chemical. These basic processes are not specified in Fig. 5.1, since different basic processes may be utilized for each combination of material, state of material, and actual process.

It can be mentioned that, for solid materials, the basic processes within the mass-conserving processes are mechanical, the basic processes within the mass-reducing processes are predominantly mechanical but some are thermal or chemical, and the basic processes within joining processes are predominantly mechanical. For material in the granular and liquid states, the basic processes are predominantly mechanical.

Type of Energy. The main types of energy that can be utilized to create the specific type of energy necessary to carry out a given basic process are mechanical, electrical, thermal, and chemical. The type of energy is not specified in

Fig. 5.1, as more than one type can often be utilized for each combination of the previous parameters, depending on the conditions.

Medium of Transfer. The requirements of the media of transfer are determined by the type of basic process, the type of energy, and the way in which the surface creation is brought about. The media of transfer can be classified according to their state as follows: rigid, elastic, plastic, granular, gaseous, liquid, and none (unspecified). Since the requirements for the media of transfer can only be established when the previous parameters have been selected, it is not possible to specify the media in general; consequently, the spaces in Fig. 5.1 are left blank.

Surface Creation and Pattern of Motion. It was seen in Chapter 1 that a surface can be produced as a result of

Total forming (TF)
One-dimensional forming (ODF)
Two-dimensional forming (TDF)
Free forming (FF)

For each of these, the pattern of motions for the work material and the medium of transfer must now be selected, so that the desired component is obtained.

The surface creation and the pattern of motions (the information system) describe the geometrical possibilities of the processes. It is especially important that at this point, the systems are imaginatively utilized. The specification of the information system must be carried out as an iteration by detailing the information system and the energy system.

It should be mentioned that the joining processes are exceptions, as they do not themselves generate geometries.

Based on the morphological structure (or model) and the description in the following chapters, the engineer will be able to judge the material properties important in the production, the changes of properties by the processes, the geometrical possibilities, and the tolerances and surfaces obtainable.

6

Solid Materials: Mass-Conserving Processes

6.1 INTRODUCTION

In this major group of manufacturing processes—often called metal-forming processes—the desired geometry is produced by the mechanical basic process, plastic deformation.

Within the last couple of decades, this field has developed rapidly, resulting in an increasing number of applications. This is mainly due to the fact that mass-conserving processes used with solid materials provide good material utilization (low waste of expensive material) and excellent final material properties.

The mass-conserving processes can—according to their location in the series of processes necessary to produce a component—be classified as (1) primary processes and (2) secondary processes.

The purposes of the primary processes are twofold: first, to break down the initial structure of the materials in the form of ingots produced by casting in order to improve the material properties (in particular the mechanical properties) and second, to provide products (e.g., rods, bars, plates, sheets, tubes, etc.) that can be processed by secondary processes. The primary processes include rolling, forging, extrusion, and so on. The secondary processes aim at the production of semifinal or final components based on the products of the primary processes. The secondary processes include mass conserving processes, such as forging, sheet metal forming (including bending, deep drawing, stretching, spinning, etc.); mass-reducing processes, such as cutting, electrodischarge, and electrochemical machining; and assembly processes.

A classification of processes into primary and secondary categories is not completely suitable because, depending on certain parameters, a given process may be regarded as belonging to either group. However, the classification is useful because for each group, it is possible to establish some overriding characteristics, allowing general judgments to be made of the processes, their possibilities, and their limitations. Thus the primary processes are in general hot-working processes that are based on plastic deformations applied to materials heated above the recrystallization temperature. Under these conditions, the metals can be considered to be perfectly plastic, allowing large deformations without fracture in compression.

Hot-working processes normally have the following advantages:

The coarse (dendritic) crystal structure from the casting is broken down to form a refined structure with small and equiaxial grains.

Impurities are broken up and distributed more evenly throughout the material.

Pores or voids are closed up.

Mechanical properties are improved considerably (especially ductility and impact strength), because of the refined structure.

Forces and energy necessary to carry out the processes are relatively small, due to the lower yield strength of the material at elevated temperatures (note, however, that strain rate has the opposite influence; see Chapter 3 and Fig. 2.5).

Shape can be changed quite drastically (i.e., large deformations are obtainable in compression).

Some of the disadvantages associated with hot working are:

Rapid oxidation (i.e., formation of scales, resulting in rough surfaces).

Relatively wide tolerances (2–5%), due to the rough surfaces.

Hot-working machinery is expensive and requires considerable maintenance.

Basically, the same principles of processing are utilized in both the hot and cold working of metals. Since no distinction between these two categories is made in the following description, the main advantages and disadvantages for cold working are mentioned first.

In general, compared to hot working, cold working of metals will give:

Better surfaces and tolerances

Better mechanical properties (strength)

Better reproducibility

Anisotropy (i.e., directional properties of the material—this is only an advantage when it is possible to utilize the effect)

Some of the disadvantages of cold working compared to hot working are:

Increased force and energy requirements, due to strain hardening (i.e., heavier and more powerful equipment is required).
Less ductility in the work material.
Anisotropy is produced in the workpiece (an advantage in many sheet-forming processes).
Clean and scale-free surfaces are required on the original workpiece.

The distinction made here between primary and secondary processes must not be confused with the classification into primary and secondary basic processes described in Chapter 1.

6.2 CHARACTERISTICS OF MASS-CONSERVING PROCESSES

In this section some of the general characteristics of mass-conserving processes for solid materials are discussed.

As mentioned previously, a close relationship exists between the information system and the material system [i.e., among the geometry, the basic process, and the material (see Fig. 6.1)]. These systems cannot be selected independently. Here the process is mass conserving and the material is in the solid state. The conditions under which the process is carried out (i.e., the pressure, temperature, velocity, etc.) play an important role, since they can influence the possibilities and limitations of the process to a high degree.

6.2.1 Geometrical Possibilities

In Chapter 1 the information system was described by principles of surface creation and the pattern of motions for work material and tooling (media of transfer). In the following description the information system will not be related directly to specific materials, types of energy, and media of transfer, as only the overriding characteristics are considered. More details are given in later sections.

The principles of surface creation are:

1. Total forming (TF)
2. One-dimensional forming (ODF)
3. Two-dimensional forming (TDF)
4. Free forming (FF)

The pattern of motions must be one of the following: translations (T), rotations (R), combinations of translations and rotations (T/R), and no motion. In general, it can be stated that the more shape information that is built into the media

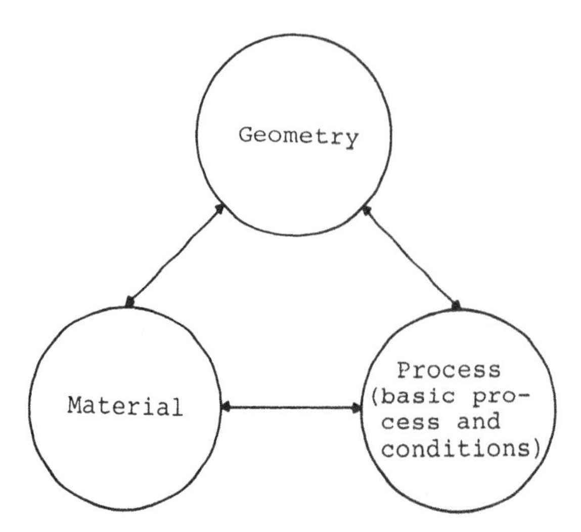

FIGURE 6.1 The relationship among geometry, process, and material.

of transfer (the tooling), the less freedom there is in the selection of the pattern of motions. Tables 6.1 through 6.4 show examples of processes within the four surface creation principles; the corresponding pattern of motions is listed for each process. These examples serve to illustrate the many possibilities for shape generation. Here it must be emphasized that an imaginative utilization of the surface creation principles and pattern of motions is very important in evaluating the geometrical possibilities. More detailed descriptions of important processes are given in Section 6.3.

In practice, the geometrical possibilities must be judged in relation to the specific material. For example, is it possible to obtain a desired geometry in a given material, and under which conditions? (See Fig. 6.1). This is discussed in Section 6.2.3.

6.2.2 Process Conditions

To be able to judge if a desired geometry and final material properties can be produced, the conditions under which the process is carried out must be known. The major influencing factors are the state of stress in the deformation zone, the temperature, and the velocity.

The State of Stress

The deformation zone in a process can be characterized by the magnitudes and state of the stresses. The size of the deformation zone for a fixed state of stress determines the forces and energy necessary to carry out the process. This

TABLE 6.1 Examples of Total Forming (TF)

Pattern of motions			
Workpiece	Tool	Total forming	Examples of processes
	T	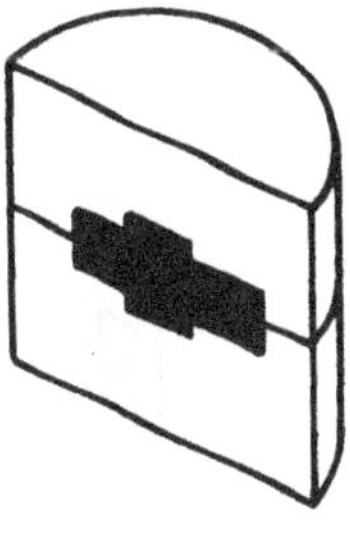 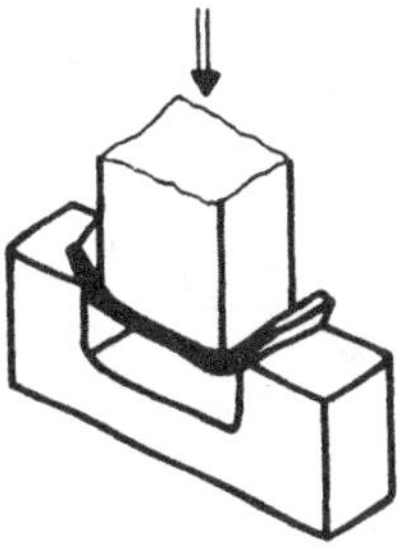	Forging Bending
		 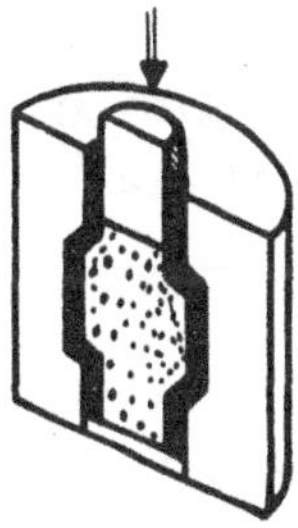	Impact forging Tube expansion
T			Upsetting

information is needed both for the design of the equipment and for the determination of the maximum size or yield strength of the components that can be processed using given equipment or machinery.

The size of the deformation zone is determined primarily by the contact area between the workpiece and the medium of transfer (the tool or die). Here a distinction must be made between total deformation and partial deformation. In total deformation, the contact area covers the whole or most of the desired surface (i.e., the deformation zone is simultaneously extended through the whole

TABLE 6.2 Examples of One-Dimensional Forming (ODF)

Pattern of motions		One-dimensional forming	Examples of processes
Workpiece	Tool		
T		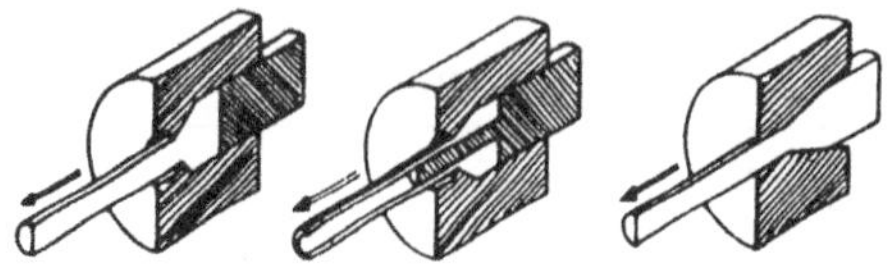	Direct extrusion Wire drawing
	T	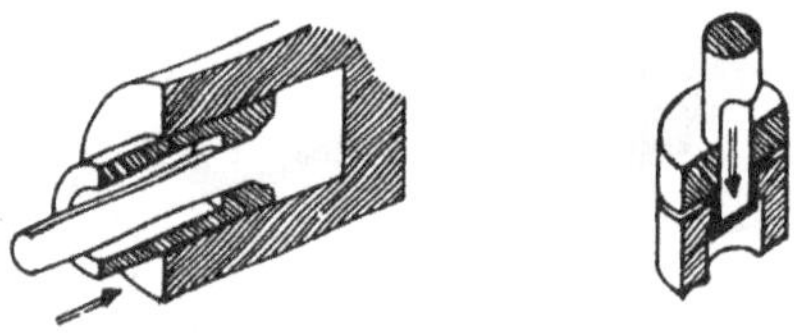	Indirect extrusion Deep drawing
	R	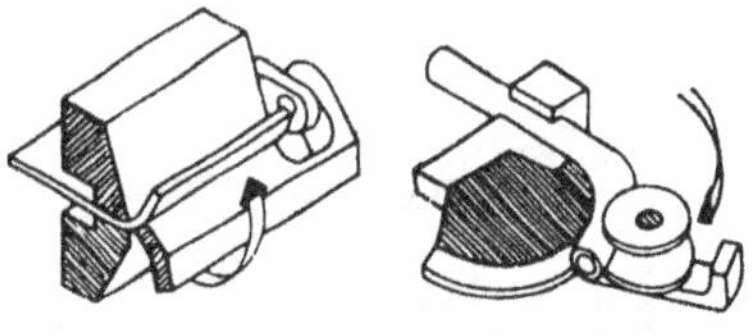	Sheet and tube bending
T	R	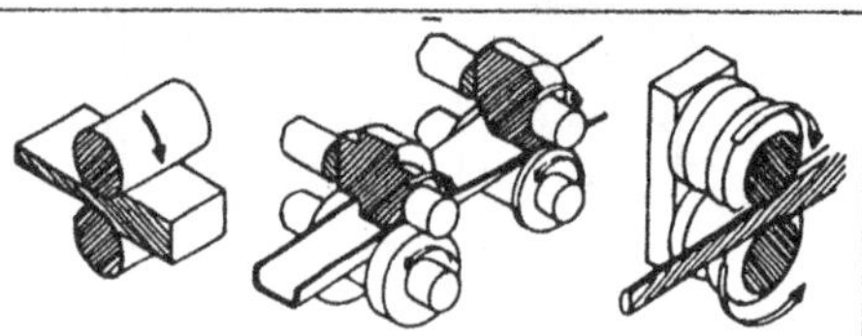	Rolling
T	T	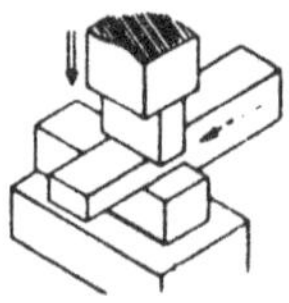	Bar forging
R	T	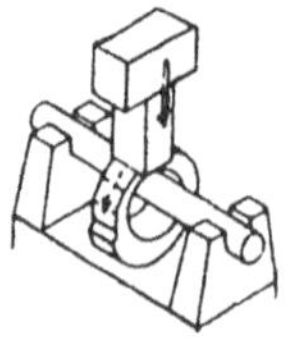	Ring forging
R	R	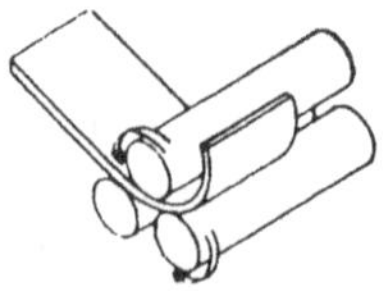	Roll bending

TABLE 6.3 Examples of Two-Dimensional Forming (TDF)

Pattern of motions		Two-dimensional forming	Examples of processes
Workpiece	Tool		
T/R	T	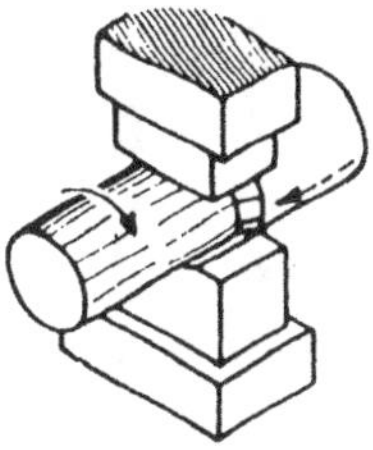 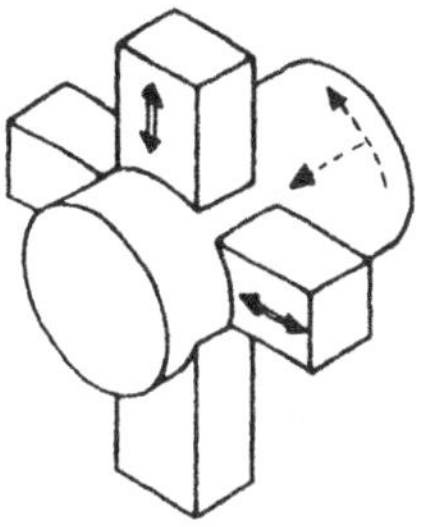	Bar forging Swaging
R/T	R	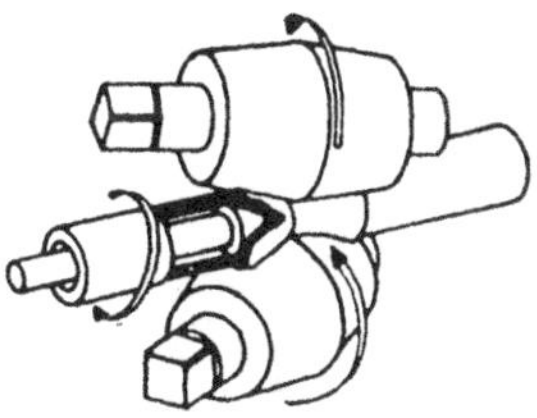	Tube rolling
R	T	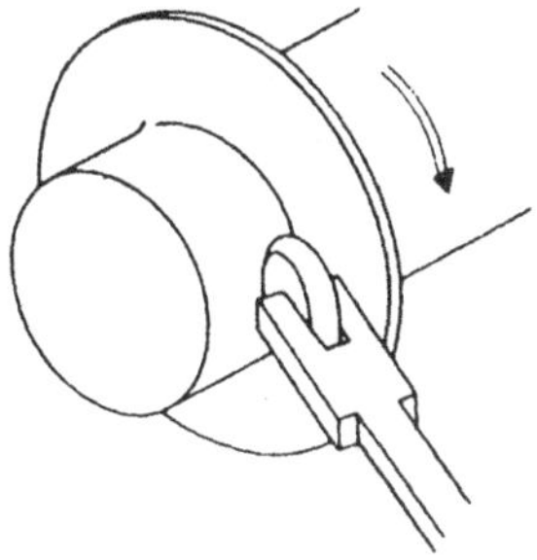	Spinning

TABLE 6.4 Examples of Free-Forming Processes

Pattern of motions		Free forming	Examples of processes
Workpiece	Tool		
	T	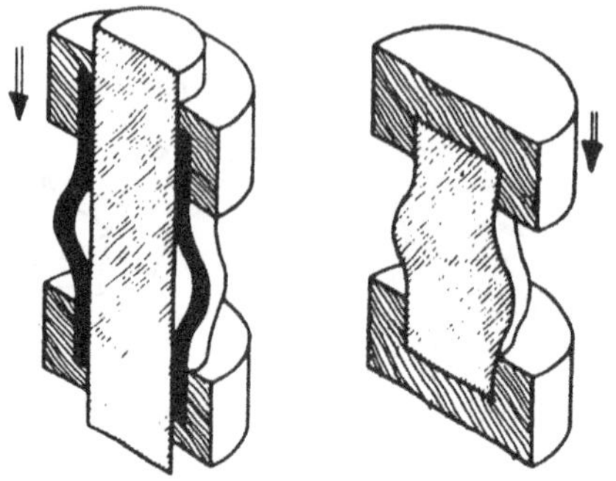	Upsetting
	R	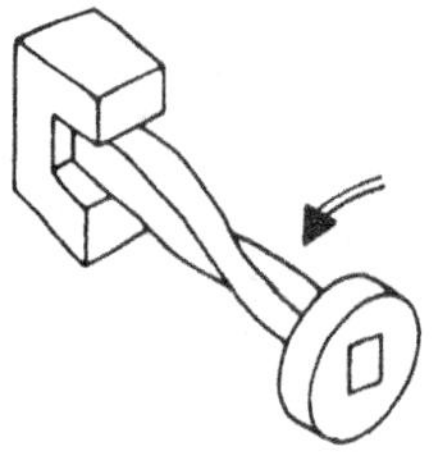	Torsion

component). In partial deformation, the contact area covers only a fraction of the surface; that is, at a given instant deformation is occurring within only a fraction of the total volume of the component, thus requiring a particular pattern of motions to describe the whole volume (see Fig. 6.2).

The principles shown in Fig. 6.2 will, when considering the same product or component, result in different force and energy requirements. If the same reduction is produced by forging and rolling (Fig. 6.2a and b), rolling will require much smaller forces but will take more time. Further, the geometry in rolling puts no limitations on the length of the component, whereas forging does.

The reduction of the contact area, and thus the deformation zone, as a method to obtain the same final deformation with smaller forces and energies has been known for centuries, and is the philosophy behind many processes today (including mass-reducing processes such as turning, grinding, etc.). Processes based on this principle are often called *incremental processes*, as opposed to *total processes*.

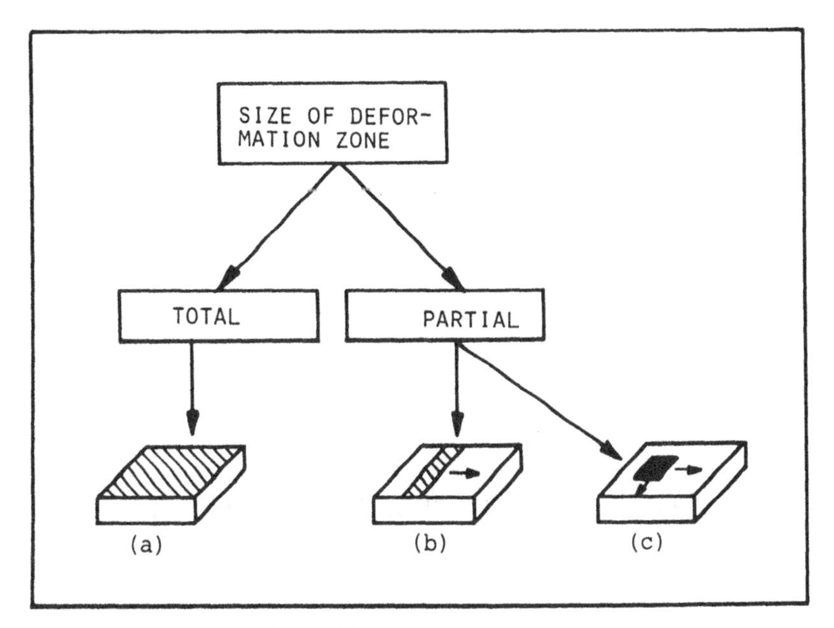

FIGURE 6.2 Total and partial deformation.

Figure 6.3 shows how an incremental process can be developed [1]. The purpose is to reduce the wall thickness of a tube without changing the internal diameter. In the basic process, the tube is pushed or pulled through a conical die, and the constancy of the internal diameter is maintained by an internal mandrel. The die is subjected to high radial forces, and large forces are required to pull or push the tube through the die. The deformation zone is circular, extending from the contact area between the die and the tube, through the tube, to the mandrel. The contact area can be reduced by replacing the die with a number of balls or conical rollers supported by an outer ring, which rotates during the deformation. A reduction of the forces to push or pull the tube is accomplished, but the ring is still subjected to high radial forces. Furthermore, both these processes require a special die system for each tube diameter. If the number of rollers is reduced to one, and the smaller forces necessary are supplied by the machine structure, a spinning process (Fig. 6.3c) is the result. This is a much more versatile process, since different wall thicknesses and tube diameters can be obtained by simply changing the position of the roller.

The principles of reducing the contact area can also be utilized in mass-reducing processes [1]. For example, the punching of circular holes (a total process) can, by reducing the contact area, be transformed into an incremental

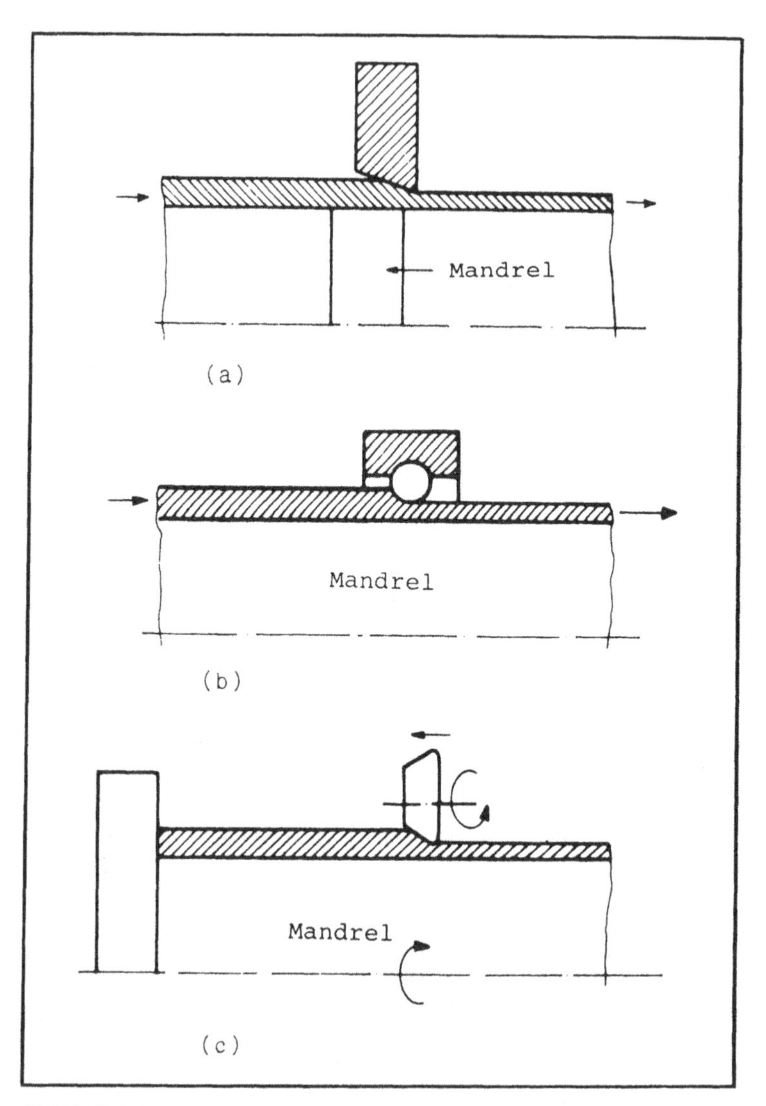

FIGURE 6.3 The development of incremental processes by reducing the contact area/deformation zone: (a) tube drawing/tube extrusion; (b) ball rolling; (c) spinning.

process by a pair of rollers or die elements or by a saw blade. In the same way, torch cutting can be considered an incremental (thermal) process.

The size of the deformation zone has now been discussed, and the next important question is: What state of stress exists in this zone? The state of stress is important, because it determines the deformation obtainable before instability and fracture occur and the forces required.

Most manufacturing processes take place under complex states of stress, and it is, in general, difficult to characterize a process by a single state of stress, since the state of stress varies throughout the deformation zone. The processes can be approximately classified into four groups, according to the dominant state of stress:

1. Tension (one-, two-, or three-dimensional)
2. Compression (one-, two-, or three-dimensional)
3. Shearing
4. Bending (nonhomogeneous)

Often, two or more of these states can be found in the same process. Considering the deep-drawing process (p. 155), the state of stress in the cylindrical wall is tension, and in the flange, compression. But as described previously (Chapters 2 and 4), the classification of the state of stress is very important in evaluating the maximum deformation that the material can sustain before instability occurs.

When planning a particular process or developing a new process concept, study of the deformation zone (size and state of stress) is fundamental. In Section 6.3, where some important processes are described, the states of stress are listed, allowing a basic evaluation of the deformation characteristics.

Temperature

The temperature in the deformation zone is an important parameter. Above the recrystallization temperature very large deformations in compression can be obtained without fracture. In tension instability occurs at very low strains. Below the recrystallization temperature strain hardening in tension increases the possible strains up to instability, and in compression it reduces the strains up to fracture.

Velocity

The velocity with which the process is carried out can influence the maximum deformation quite strongly, as shown in Fig. 2.5. Different materials react differently to the deformation velocity (strain rate). Some will exhibit increased ductility, and some decreased ductility. In actual situations, the strain rate must be estimated, and the influence on the properties of the particular material evaluated.

In general, it is found that the strain rate at room temperature has no significant influence on the stress–strain curve, but elevated temperatures normally

TABLE 6.5 Typical Deformation Velocities

Process	Tool/die velocity (deformation velocity) (m/s)
Tension test	10^{-6}–10^{-2}
Hydraulic press	2×10^{-2}–3×10^{-1}
Tube drawing	5×10^{-2}–5×10^{-1}
Sheet rolling	5×10^{-1}–25
Forging	2–10
Wire drawing	5–40
High-velocity forging	20–50
Explosive forming	30–200

Source: From Ref. 1.

increase the strain rate sensitivity. In hot-working processes it is therefore necessary to analyze carefully the strain rate situation and its consequences.

Table 6.5 shows typical values of deformation velocities (not strain rates) for different processes.

Other Important Factors

The geometry and surface of the tools, friction, lubrication, and the state of stress determine the final surface finish. If the desired surface quality is considered, good lubrication sometimes leads to poor surfaces, as the lubrication can be entrapped in the small cavities on the surface. A substitution of the tool/die material from a metal to an elastic material such as rubber can result in better surfaces with a given work material without destruction occurring.

The tolerances obtainable are difficult to describe in general, since they are dependent on the size of the deformation zone, the state of stress, the workpiece geometry, the tool/die system, and the equipment. If the deformations are small, elastic recovery must be considered. It may be difficult to obtain fine tolerances when the elastic deformation is of the same order of magnitude as the plastic deformation. In summary, it can be stated that the major factors affecting processing by mass-conserving processes are workpiece geometry, deformation zone (size and state of stress), temperature, deformation, velocity, lubrication, the properties of the workpiece material, and the tool/die material.

6.2.3 Important Material Properties

The amount of deformation that a material can sustain without instability or fracture depends, as described earlier, on the state of stress, the temperature, and the strain rate. In Chapter 4 it was found that instability occurs when a strain equal to the strain-hardening exponent was reached in a material following the relation $\bar{\sigma} = c\bar{\epsilon}^n$. If the same material is deformed under compressive stresses, considerably higher deformations are obtainable, limited only by fracture at locations where high tensile stresses are generated.

In a drawing operation, a material may be elongated 40%, whereas in rolling it can be elongated 400%. A major advantage of mass-conserving processes of the cold-working type is the strength improvement of the material, due to strain hardening. Compared to casting or hot working, the strength improvement is often so high that a cheaper work material can be selected.

The final properties of the material can be evaluated with reasonable accuracy from the stress–strain curve and a knowledge of the strains and the conditions under which the process is carried out.

Force and energy calculations are discussed in Section 6.4.

6.3 TYPICAL EXAMPLES OF MASS-CONSERVING PROCESSES

In the following pages short descriptions are given of a number of mass-conserving processes. The processes are classified according to the fundamental elements, basic process, energy to carry out the process, media of transfer, principle of surface creation, and the predominant state of stress. All the processes are mass conserving and solid materials are used.

The following abbreviations are used in the classification:

Basic process
- M, mechanical
- T, thermal
- C, chemical

Energy
- Me, mechanical
- El, electrical (including magnetic)
- Th, thermal
- Ch, chemical

Media of transfer
- Ri, rigid
- Ea, elastic
- Pl, plastic
- Ga, gaseous
- Gr, granular
- Fl, fluid

Principles of surface creation
- TF, total forming
- ODF, one-dimensional forming
- TDF, two-dimensional forming
- FF, free forming

State of stress
Te, tension
Co, compression
Sh, shearing
Be, bending

The description of the processes covers:

Name, classification code
General description
Applications/geometry (possibilities)
Material requirements
Tolerances/surfaces (possibilities)
Machinery/energy (in general)

In the figures, the deformation zone is characterized by the size of the zone (the whole workpiece or portions of it) and the state of deformation (steady, non-steady). A *steady state* of deformation occurs when the deformation picture is fixed in time and position (e.g., rolling, extrusion) during the process. A *non-steady state* of deformation occurs when the deformation picture changes continuously with time and position during the process.

The information sheets are to be considered only as an introduction to the processes; more details may be found in the literature.

Tables 6.1 through 6.4, the data sheets, and the morphological process model allow an evaluation of the possibilities and limitations of the mass-conserving processes for solid materials.

PROCESS 1: Rolling (M, Me, Ri, ODF, Co)

Description. The (sheet, plate, structural beam, etc.) rolling process is characterized by a solid material, one-dimensional forming, and a compressive state of stress. The workpiece (W) passes between two rolls (R) with a gap ($\sim h_2$), which is less than the initial thickness of the material (h_1). Since the width of the material is nearly constant during the deformation, the reduction in thickness results in a corresponding increase in length because of the constancy of volume.

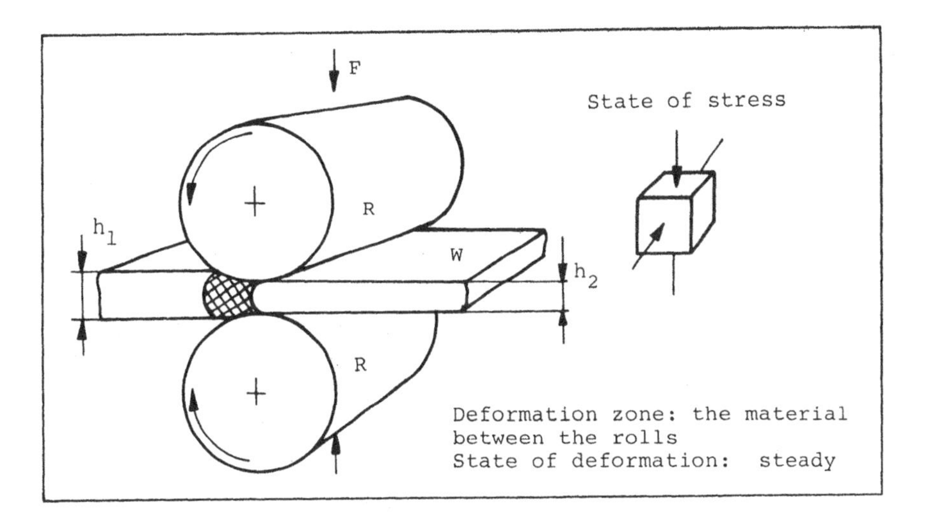

Applications. When considering production volume, hot rolling is the most important process. The most common products are plates, bars, rods, structural beams, and so on. Products that have smooth surfaces, accurate dimensions, and high strengths, like sheets, strips, bars, rods, and so on, are cold-rolled with varying reductions, depending on the purpose.

Material Requirements. The materials (ferrous and nonferrous metals) must possess sufficient ductility at the forming temperature.

Tolerances/Surfaces. Hot rolling produces slightly rough surfaces and tolerances in the range 2–5%. Cold rolling produces very smooth surfaces and tolerances in the range 0.5–1%.

Machinery/Energy. Very specialized and massive equipment is necessary, designed especially for either hot or cold rolling.

PROCESS 2: Forward extrusion (M, Me, Ri, ODF, Co)

Description. The forward extrusion process is, in general, characterized by a solid material, one-dimensional forming, and a compressive state of stress. The workpiece (W) is placed in the die/container (L), and the punch (P) squeezes the material through the die orifice in the direction of the applied force.

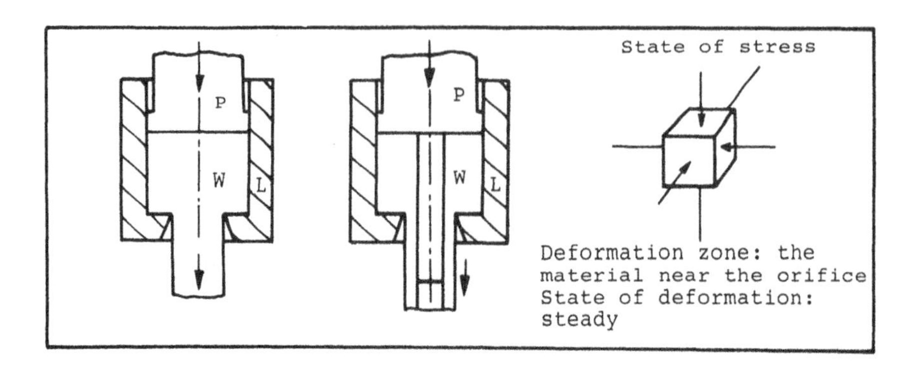

Applications. As a hot-working process, extrusion is used extensively for the production of a wide variety of regular and irregular structural profiles, such as window moldings, angle sections, I- and U-beams, and circular and noncircular tubing. As a cold working process, it is a variant of cold forging used alone or combined with cold heading, backward extrusion, and so on.

Material Requirements. In hot working, ferrous and nonferrous materials must possess sufficient ductility at elevated temperatures. In cold working, nonferrous metals and low-alloy steels are used, possessing sufficient ductility at room temperature.

Tolerances/Surfaces. Hot extrusion gives good tolerances and surfaces, and is best for nonferrous metals. Cold extrusion gives excellent tolerances (0.1–1%) and surfaces.

Machinery/Energy. For hot extrusion, special hydraulic presses are used; cold extrusion is carried out on general-purpose mechanical and hydraulic presses.

PROCESS 3: Hot drawing (M, Me, Ri, ODF, Te)

Description. The hot-drawing process is, in general, characterized by a solid material, one-dimensional forming, and a tension state of stress. The workpiece (W) is placed on a die (L), and the punch (P) pushes the metal through the die, forming a cup. The cup can be drawn through several dies with a single punch.

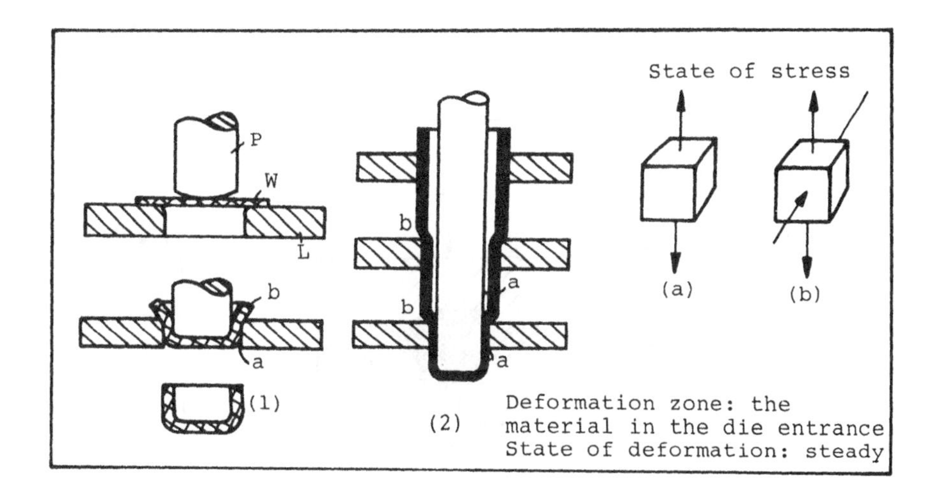

Applications. The hot-drawing process is generally used to produce relatively thick-walled cylindrical parts, such as oxygen tanks, artillery shells, tank heads, and short tubes.

Material Requirements. High ductility (low yield strength) at elevated temperature. Both ferrous and nonferrous metals are hot drawn.

Tolerances/Surfaces. Reasonably good tolerances are generally obtained (often below 0.5% of the diameter). The surface quality is good.

Machinery/Energy. Hydraulic presses (draw benches) for single or multiple draws are available.

PROCESS 4: Drop forging (M, Me, Ri, TF, Co)

Description. The drop-forging (or closed die forging/impression die forging) process is, in general, characterized by a solid work material, total forming, and a compressive state of stress. The workpiece (W) is placed in the lower die (L), and during the closing movement (one or more blows) of the upper die (U), the workpiece becomes plastic, and the die cavity is filled. Excess material is squeezed out between the die faces (peripherally) as a flash (F). The workpiece is removed from the die, eventually by an ejector (E), and the flash trimmed off in a special trimming die.

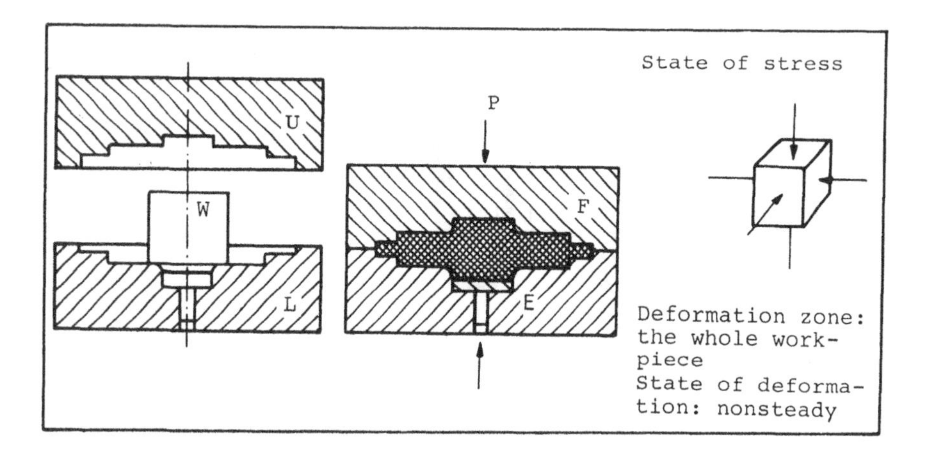

Applications. Production of a wide variety of part shapes limited only by process/die requirements (draft, radii, parting line, ribs, etc.) and the necessary forces. The process is, in general, a hot-working process, and the metal flow can be controlled by the original workpiece geometry and the die cavities, resulting in a favorable fiber structure. Examples include connecting rods, gear blanks, levers, and handles.

Material Requirements. All metals with a high ductility at the actual elevated temperatures (various steel alloys, nonferrous metals, etc.).

Tolerances/Surfaces. Components of about 1 kg mass give thickness tolerances in the range +0.6 to +0.2 mm. The surface quality is reasonably good, but in general, some further processing (turning, milling, etc.) is necessary to obtain the final surfaces.

Machinery/Energy. The energy necessary is mechanical, and many different machines are available (steam hammers, board hammers, hydraulic and mechanical presses, etc.). The selection of machinery must be based on the production volume and the actual process.

PROCESS 5: Upset forging (Upsetting) (M, Me, Ri, TF, Co)

Description. Upset forging is, in general, characterized by a solid work material, total forming, and a compressive state of stress. The workpiece (W) is placed into a stationary die (L) supported by a knockout pin or clamped in a split die. The heading tool (P) then moves longitudinally, upsetting the workpiece into the die cavity.

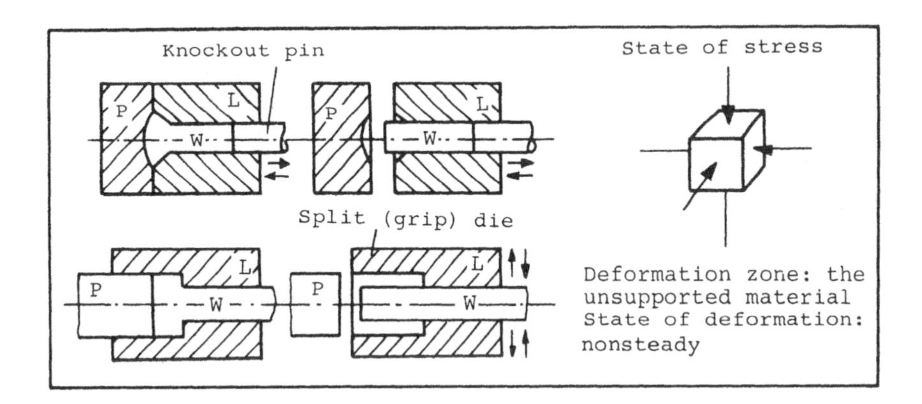

Applications. Upsetting as a hot-working process is used to produce heads on bolts, valves, flanges and shoulders on shafts, and so on. As a cold-working process, upsetting (called cold heading) is used to produce nails, rivets, small bolts, and so on.

Material Requirements. The work material must possess sufficient ductility at room temperature for cold heading or at elevated temperatures for hot upsetting (various steels and nonferrous metals).

Tolerances/Surfaces. In hot upsetting, the tolerances and surface quality are determined mainly by the amount of scale, lubricants, and so on. Cold heading gives good tolerances ($\sim \pm 0.2$ mm) and surfaces.

Machinery/Energy. A wide variety of mechanical upsetting machines, often horizontal, are available, allowing production rates of about 400 parts per minute.

PROCESS 6: Cold forging (M, Me, Ri, TF, Co)

Description. The cold-forging process is, in general, characterized by a solid material, total forming, and a compressive state of stress. The workpiece (W) is placed in the die (L), and the punch (P) moves down and squeezes the material to fill the die cavity.

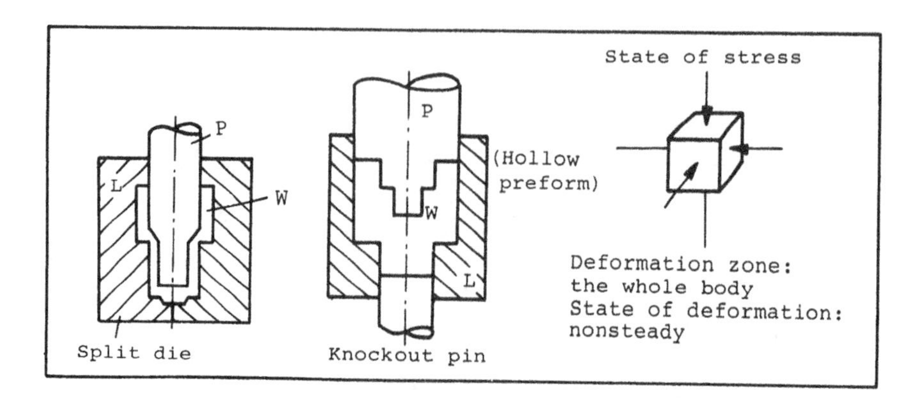

Applications. Cold forging allows production of relatively complex parts, using various types of closed dies and depending on the applied deformation pattern, different specialized names may be used (i.e., extrusion, impact extrusion, etc.). The process has a high material utilization and gives excellent material properties. The number of applications is increasing rapidly. The mass of the components may vary between a few grams and several kilograms.

Material Requirements. Sufficient ductility to sustain the deformations without fracture. Most nonferrous metals and low-alloy steel can be cold forged.

Tolerances/Surfaces. Tolerances in the range $\sim\pm 0.05$ mm on diameters and $\sim\pm 0.2$ mm on lengths are common. The surface quality is high, and machining can be nearly eliminated.

Machinery/Energy. A wide range of mechanical and hydraulic presses are available.

PROCESS 7: Back extrusion (cold) (M, Me, Ri, TF, Co)

Description. The back extrusion process, which is a variant within cold forging is, in general, characterized by a solid material, total forming, and a compressive state of stress. The workpiece (W) is placed in the die (L), and the punch (P) moves down to extrude the material around or into the punch.

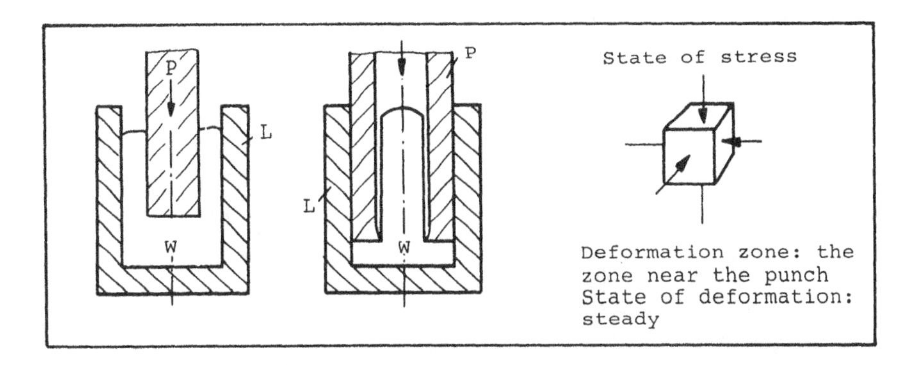

Applications. Back extrusion, as a cold-forging process, can be used to produce regular or irregular structural parts, tubes or tubular components, and so on. A special variant is can extrusion (see later). The process lends itself to large series or mass production, as the tooling is rather expensive.

Material Requirements. Sufficient ductility to sustain the deformations. Most nonferrous metals and low-alloy steel can be backward extruded.

Tolerances/Surfaces. As a cold-working process, good tolerances (± 0.05 to ± 0.2 mm) can be acquired, and the surface quality is correspondingly high.

Machinery/Energy. A wide variety of mechanical and hydraulic presses are available.

PROCESS 8: Can extrusion (impact extrusion) (M, Me, Ri, ODF, Co)

Description. The can extrusion process, which can be regarded as a variant of cold forging, is characterized by a solid material (sometimes even a metal powder), one-dimensional forming, and a compressive state of stress. The workpiece (W) is placed in the die (L), and as the punch (P) moves down, the material is squeezed or extruded up around it. The die may be provided with a conical bottom face and punch nose, as for example in toothpaste tubes.

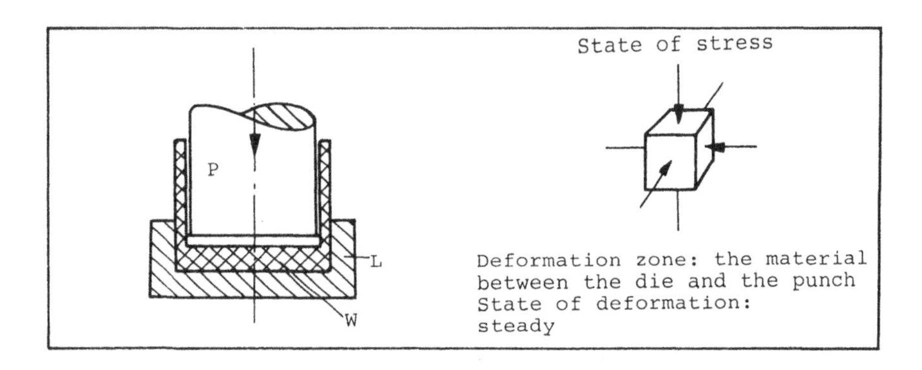

Applications. Can extrusion, which is a special back-extrusion process with a thickness/diameter ratio as low as 0.005, is used to produce collapsible tubes for toothpaste, cosmetics, and so on; cans for food and beverages; and shielding in electrical apparatus, batteries, etc.

Material Requirements. High ductility and low strength. Extensively used materials are aluminum and zinc alloys.

Tolerances/Surfaces. Tolerances in the range of ±0.05 to ±0.1 mm are obtainable, accompanied by a very high surface quality.

Machinery/Energy. A wide variety of mechanical and hydraulic presses are available. Because of production volumes, specialized machinery is usually developed.

PROCESS 9: Deep drawing (M, Me, Ri, ODF, Te)

Description. The deep-drawing process is, in general, characterized by a solid material, one-dimensional forming, and a tension state of stress. The workpiece (blank, W) is placed on the die (L), clamped by the pressure plate (C) to prevent wrinkling, and pushed through the die by the punch (P) to form a deep cup.

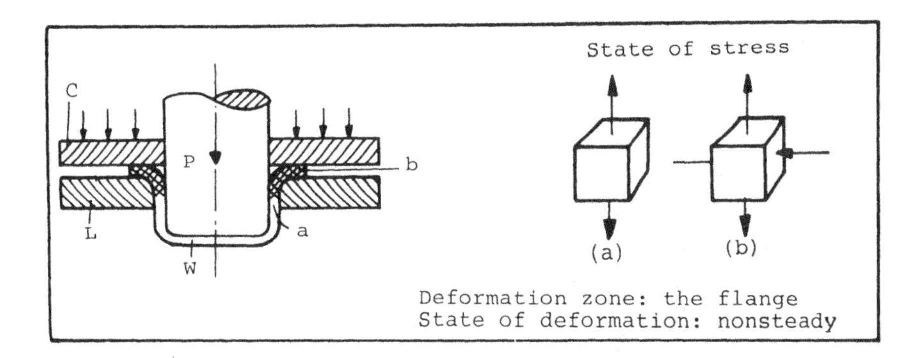

Applications. The deep-drawing process and variants of it are used extensively to produce a wide variety of shells, cylindrical or prismatic cups, and so on. Examples include bottle caps, automobile panels, tanks, appliance covers and bodies, and cans for food. Deep drawing is a cold-working process.

Material Requirements. Nonferrous and ferrous metals with sufficient ductility to sustain the actual strains (i.e., metals with anisotropy and high instability strains). In the deformations are large, intermediate annealing may be necessary.

Tolerances/Surfaces. Good tolerances are in general obtainable (± 0.2 mm for small diameters and increasing for increasing diameters). The surface quality corresponds closely to the original sheet.

Machinery/Energy. Double-acting hydraulic presses are used extensively, but other types (mechanical and hydraulic) are available, depending on the purpose.

PROCESS 10: Rubber forming (M, Me, Ea, TF, Be/Co)

Description. The rubber-forming process is, in general, characterized by a solid material, total forming, and a state of stress, including bending and some compression. The workpiece (blank, W) is placed on a rigid male die (L), and the punch (P) with a rubber pad (R) moves down, shaping the blank according to the shape of the male die.

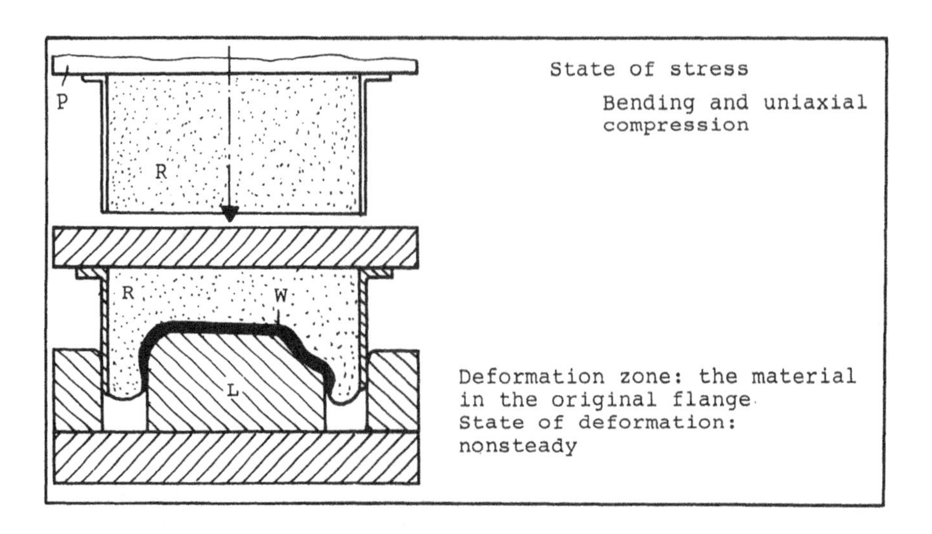

Applications. The rubber-forming process is used to produce relatively simple geometries in thin sheets (aluminum up to 3 mm, stainless steel and mild steel up to 1.5 mm thickness). The same punch can be used for various male dies (geometries). It is low-cost tooling, allowing easy and flexible fabrication. Examples include cans for food and parts for the electronic industry. The rubber-forming process can include piercing and blanking.

Material Requirements. High ductility and a relatively low yield strength. Nonferrous metals, mild steel, and stainless steels are typical materials.

Tolerances/Surfaces. The tolerances vary with the degree of deformation (elastic recovery). The surfaces are equal to the original sheet; even painted surfaces are not ruined.

Machinery/Energy. Most types of mechanical or hydraulic presses can be used.

PROCESS 11: Tube expansion (bulging) (M, Me, Ea/Ri, TF, Te)

Description. Tube expansion is, in general, characterized by a solid material, total forming, and a tension state of stress. The workpiece (W) is placed in a die (L), and by introducing a high pressure in a medium M (Fl, Ea), mechanical energy is transmitted to the workpiece. The high pressure may be obtained from high explosives (detonated in water) or a press ram (P).

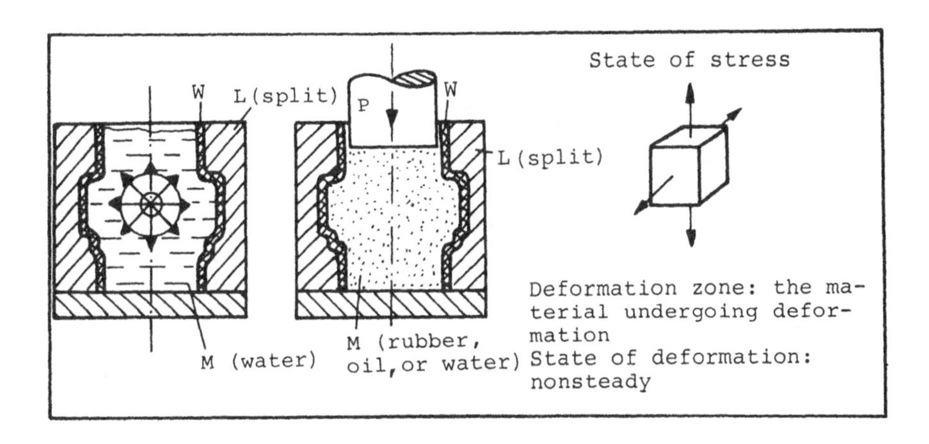

Applications. The tube expansion or bulging process is used to produce a wide range of sectional expanded or shaped (irregular) tubular components or shells for the aircraft industry, the chemical industry, and the mechanical industry.

Material Requirements. Sufficiently high instability strain to withstand the deformation without fracture.

Tolerance/Surfaces. Very fine tolerances are obtainable, depending on the dimensions (±0.05% of the diameter). The surface quality is determined mainly by the original material.

Machinery/Energy. Mechanical and hydraulic presses, high explosives, electrical discharges, and so on.

PROCESS 12: Spinning (M, Me, Ri, TF, Sh)

Description. The spinning process is, in general, characterized by a solid material, total forming, and shear state of stress. The workpiece (W) is placed on a rotating die or mandrel (L), and the movable roller (R) progressively forces the blank against the die.

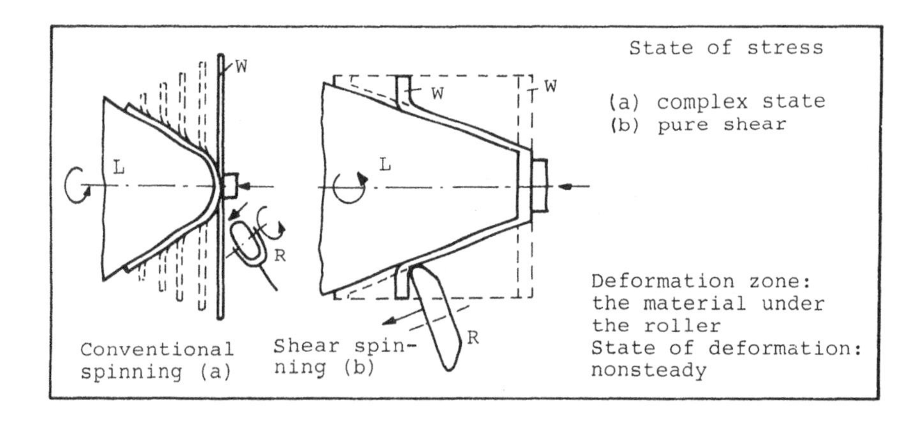

Applications. Both conventional spinning (no significant reduction in thickness) and shear spinning (reduction in thickness depending on the enclosed angle of the mandrel) are extensively used alone or combined in industry to produce bowls, lamps, reflectors, cooking utensils, bells, tubes (including changing diameters), and shafts (gas turbines, etc.).

Material Requirements. Sufficient ductility to withstand the actual strains without fracture.

Tolerances/Surfaces. Tolerances are good, for example 0.1–0.2% of the diameter. The surface quality depends on the process parameters (feed, speed, geometries of the rollers, etc.).

Machinery/Energy. A wide variety of spinning machines (manual or numerically controlled) are available.

PROCESS 13: Bending (braking) (M, Me, Ri, TF, Be)

Description. The bending process, carried out on press brakes, is characterized by a solid material, total forming, and a bending state of stress. The workpiece (blank, W) is placed on the die (L), and the punch (P) moves down and bends the sheet according to the die/punch geometry.

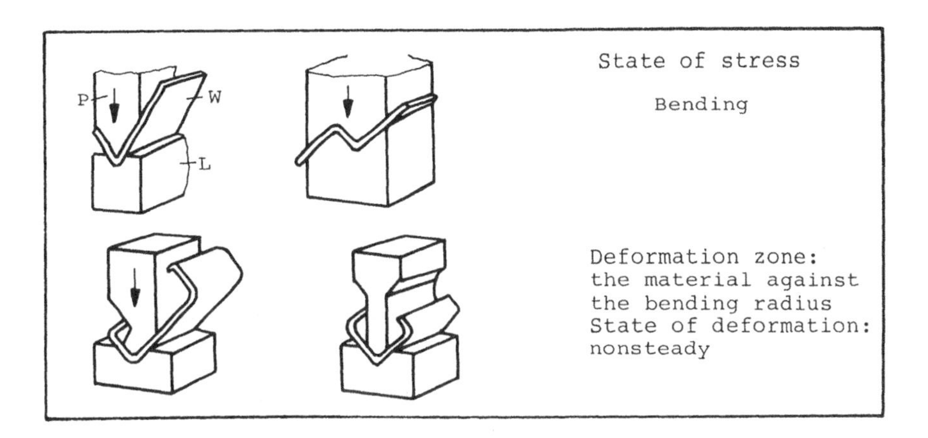

Applications. Press brake bending is used for the production of many structural shapes (i.e., angles, channels, etc.). A few examples are shown in the figure. It is used extensively in the aircraft industry, the automobile industry, and the lighter mechanical and electrical industries. A wide variety of dies gives a nearly unlimited number of shape possibilities. For small workshops, manual bar folders are used to carry out the bending, but the geometrical possibilities are much smaller.

Material Requirements. The strain at fracture must not be exceeded on the outside of the bend.

Tolerances/Surfaces. The tolerances depend on the sheet thickness and the bending geometry. Angular tolerances of $\pm 0.5°$ are common. The surface quality corresponds to the sheet.

Machinery/Energy. Press brakes of different sizes and types both mechanical and hydraulic are available.

PROCESS 14: Stretch forming (M, Me, Ri, TF, Te)

Description. The stretch-forming process is, in general, characterized by a solid material (sheet), total forming, and a tension state of stress. The workpiece (W) is clamped in the grips (G) and stretched and bent over the male die (L) to the final shape.

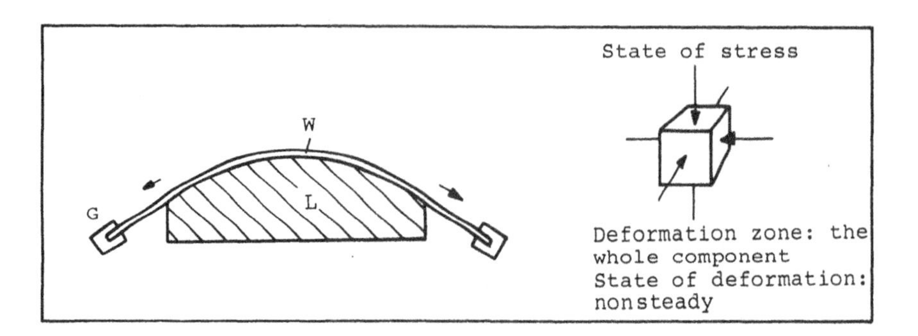

Applications. Stretch forming of sheet metals is a relatively new process, utilized extensively in the aircraft and automobile industries to produce large panels with varying curvatures, skin panels, engine cowlings, door frames, window frames, and so on.

Material Requirements. The process is limited by the strain at instability (i.e., the materials can be evaluated by their uniform elongation in tensile tests). Both nonferrous and ferrous metals are stretch formed.

Tolerances/Surfaces. The surface quality corresponds roughly to the original sheet, but the tolerances vary with the stress level (elastic recovery), the geometry, and the material.

Machinery/Energy. A wide variety of stretch-forming machines (types and capacities) are available.

PROCESS 15: Roll bending (M, Me, Ri, ODF, Be)

Description. The roll-bending process is characterized by a solid material, one-dimensional forming, and a bending state of stress. The workpiece (W) is fed between an adjustable upper roll (R) and two fixed lower rolls (R), which induces a bending state of stress in the plate, depending on the position of the adjustable roll in relation to the fixed rolls.

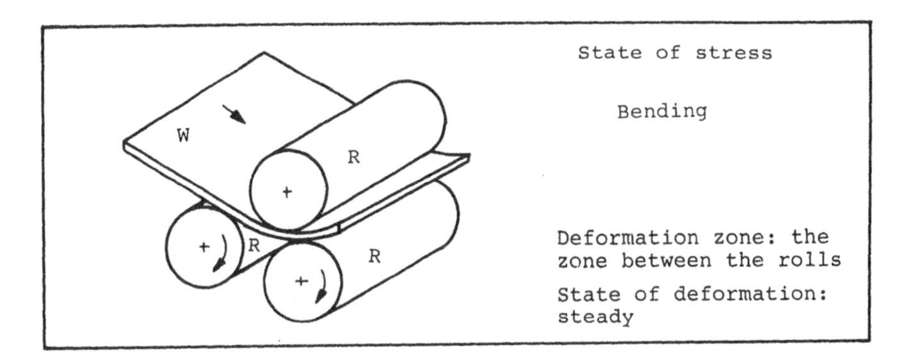

Applications. The roll-bending process is used to produce rings, vessels, and so on. By changing the position of the upper roll, the curvature of the plate changes. Roll-bending machines may be numerically controlled (NC), allowing easy production of regular and irregular shapes.

Material Requirements. Sufficient ductility, so that the strain at fracture is not exceeded at the outside of the bent plate. Both ferrous and nonferrous metals can be formed by this process.

Tolerances/Surfaces. The tolerances are mostly within 0.1–0.2% of the diameter. The surface quality is equal to the surface of the original plates/sheets.

Machinery/Energy. Bending rolls are available in a wide range of sizes (e.g., bending of plates up to 150–200 mm thickness).

6.4 DETERMINATION OF FORCES AND ENERGIES

The manufacturing processes involve, as discussed in Chapter 1, material flow, information flow (shape impressing), and energy flow. In previous sections, the major elements in the material and information flow have been elucidated. Based on this and the elementary plasticity theory, the main elements in the energy system can be determined. The energy system supplies the necessary forces and energies to carry out the desired deformations as determined by the planned information and material flow.

It should be remembered that the final specifications for each of the three systems are generally brought about by an iterative process.

6.4.1 Basic Principles in Force and Energy Determination

In this context only approximate methods to estimate the forces and energies necessary are discussed. More accurate and advanced methods are available, but they normally require an advanced theoretical knowledge of plasticity theory and solid mechanics.

The simplest possible method to estimate the forces necessary is based on yield in homogeneous deformation. This method can be applied only when the load or force is acting directly on the whole deforming cross section of the workpiece. For example, the maximum force necessary to compress a cylindrical workpiece in the direction of its axis is given by $P = \sigma_0 A_{max}$, where σ_0 is the yield strength at maximum strain and A_{max} the maximum cross-sectional area.

For most processes this approach is not applicable, as the force is not acting on the whole deforming cross section of the workpiece. A more general approach is to consider the work necessary to deform an element of the workpiece and integrate this over the whole deforming region. This total work of deformation is then related to the work carried out by the external force, allowing a determination of the latter.

In Chapter 4 it was shown that work W necessary to carry out a deformation is given by

$$W = \int_V \int_{\bar{\epsilon}_1}^{\bar{\epsilon}_2} \bar{\sigma}\, d\bar{\epsilon}\, dV \qquad (6.1)$$

where W is the work of deformation, V the volume of the deforming region, $\bar{\epsilon}_1$ and $\bar{\epsilon}_2$ the effective strains before and after the deformation, and $\bar{\sigma}$ the effective stress (which can be expressed as a function of $\bar{\epsilon}$).

If all elements in the workpiece volume (the deforming region) are supplied with the same amount of work, Eq. (6.1) can be written as

$$W = V \int_{\bar{\epsilon}_1}^{\bar{\epsilon}_2} \bar{\sigma}\, d\bar{\epsilon} \qquad (6.2)$$

If the stress–strain curve for the material is given by $\bar{\sigma} = c\bar{\epsilon}^n$, Eq. (6.2) becomes

$$W = V \frac{c}{n + 1} (\bar{\epsilon}_2^{n+1} - \bar{\epsilon}_1^{n+1}) \tag{6.3}$$

This work can be characterized as the internal work W_i. The external work is supplied by external forces (P_e) or pressures (p_e), which are acting over a certain travel distance (l_e); that is, the externally supplied work can be written as

$$W_e = P_e l_e = p_e A_e l_e \tag{6.4}$$

where e refers to "external" and A_e is the cross-sectional area over which the external forces/pressures act.

By equating (6.1) and (6.4), the external forces or pressures can be estimated:

$$P_e l_e = p_e A_e l_e = \int_V \int_{\bar{\epsilon}_1}^{\bar{\epsilon}_2} \bar{\sigma} \, d\bar{\epsilon} \, dV \tag{6.5}$$

If the power necessary is to be determined, the velocity (v_e) with which the force or pressure is supplying the work or the time (t_e) during which it is supplied must be known. Consequently, the power required is given by

$$N = P_e v_e = \frac{W_e}{t_e} \tag{6.6}$$

where N is the power. If an external moment M_e is acting with angular velocity ω_e, the power is correspondingly given by $N_e = M_e \omega_e$. These methods for estimating forces and energies are only approximate, as they are based on the assumption of homogeneous deformation. A homogeneous deformation is the most efficient way to carry out a deformation and requires the smallest possible load or force, since frictional work and redundant work (caused by friction and geometrical constraints) are neglected.

This means that the work [Eqs. (6.1), (6.2), and (6.3)] and the force [of the type $P_{max} = \sigma_0 A_{max}$ and Eq. (6.5)] are the lower limits for any process, producing the same final deformations. The agreements with the actual work and force are, in general, reasonable for very low coefficients of friction and simple geometries, producing minimum internal distortion (note that geometrical constraints provide redundant deformation). This is, however, not the case for many processes, and consequently, to obtain reasonably accurate results, it is often appropriate to introduce empirical correction factors, as discussed later.

6.4.2 Sheet Rolling: Determination of Rolling Force, Moment, and Power

The sheet-rolling process (Fig. 6.4) can be approximately considered as a bar-forging process (Table 6.2), where a bar (the sheet) is produced by forging actions succeeding each other along the bar. The main problem here is that the

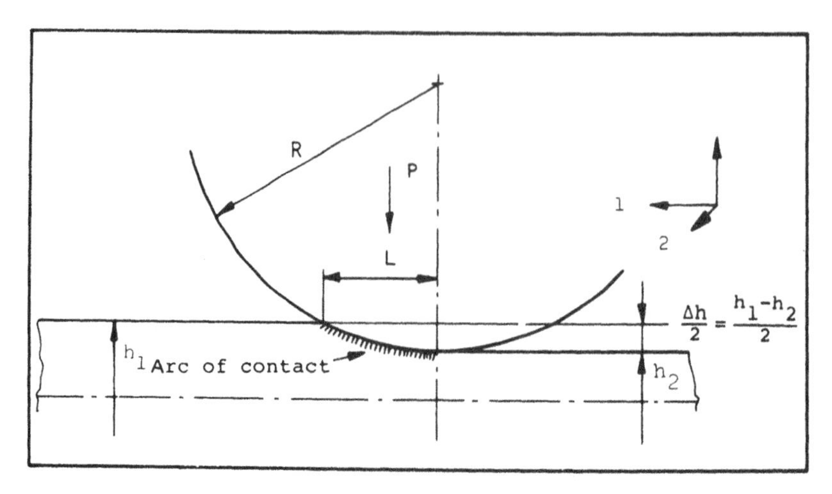

FIGURE 6.4 Sheet rolling.

yield strength of the material increases from the entry (thickness h_1) to the exit (thickness h_2) of the roll gap. It is therefore necessary to use a mean yield stress $\sigma_{3,m}$, where the suffix m refers to a mean value and the suffix 3 refers to the direction of the principal stress (see Fig. 6.4).

Assuming that the curvature of the rolls can be ignored, the rolling load or force can be determined by

$$P = -\sigma_{3,m}A = -\sigma_{3,m}wL \text{ (}P\text{ defined positive)} \tag{6.7}$$

where $\sigma_{3,m}$ is the mean yield strength in direction 3 of the principal stresses, A the area of deformation, w the width of the sheet, and L the longitudinal projection of the arc of contact (i.e., the chord of contact).

Since for a close approximation the width of the sheet can be considered constant during the deformation ($\Delta w = 0$), the state of strain is plane ($\epsilon_1 = -\epsilon_3$, $\epsilon_2 = 0$).

The state of stress is given by (see Chapter 4, Example 5): $\sigma_1 \simeq 0$ (no external longitudinal forces); $\sigma_2 = (\sigma_1 + \sigma_3)/2 = \sigma_3/2$; σ_3. Consequently, von Mises' yield criterion for the plane state of strain gives $\sigma_1 - \sigma_3 = -\sigma_3 = (2/\sqrt{3})\sigma_0$, where σ_0 is the uniaxial yield strength. If mean values are used, this becomes

$$-\sigma_{3,m} = \frac{2}{\sqrt{3}}\sigma_{0,m} = 1.15\sigma_{0,m} \tag{6.8}$$

where the suffix $0,m$ designates the mean uniaxial yield strength. The value of $\sigma_{0,m}$ can be found from the stress–strain curve of the material:

$$\sigma_{0,m}(\bar{\epsilon}_2 - \bar{\epsilon}_1) = \int_{\bar{\epsilon}_1}^{\bar{\epsilon}_2} \bar{\sigma}\, d\bar{\epsilon}$$

or, more simply, from $\sigma_{0,m} = 0.5(\sigma_{0,1} + \sigma_{0,2})$.

From Fig. 6.4 the chord of contact L is found to be given by

$$L^2 = R^2 - \left(R - \frac{\Delta h}{2}\right)^2 \simeq R\Delta h \tag{6.9}$$

where Δh is the reduction in thickness and R is the radius of the rolls. Consequently, from substituting (6.8) and (6.9) in Eq. (6.7), the rolling force becomes

$$P = \frac{2}{\sqrt{3}} \sigma_{0,m} w (R\Delta h)^{1/2}$$

The frictional contribution to P has been empirically determined and is estimated to be an average of 20%; that is, the total rolling force P^* becomes

$$P^* = 1.2P = 1.2\frac{2}{\sqrt{3}} \sigma_{0,m} w (R\Delta h)^{1/2}$$

The rolling force per unit width is, consequently,

$$\frac{P^*}{w} \cong 1.35\sigma_{0,m}(R\Delta h)^{1/2} \tag{6.10}$$

This expression gives reasonably good results.

The moment to drive one of the rolls can be approximated by assuming that P is acting in the middle of the length L. Thus

$$M = P^*\frac{L}{2} \tag{6.11}$$

The necessary power per roll is given by

$$N = M\omega = P^*\frac{L}{2}\,\omega \tag{6.12}$$

where ω is the angular velocity of the rolls.

6.4.3 Extrusion: Determination of the Extrusion Pressure or Force

In an extrusion process, the necessary pressure is very important, as it determines the size of the extrusion press or the maximum cross-sectional area that can be extruded on a given press. Figure 6.5 illustrates the extrusion process. It is not possible here to use the simple method to determine the pressure/force as in rolling, since the external force does not act directly on the deforming cross-sectional area. Consequently, the work method or work formula must be used. The work per unit volume w is given by

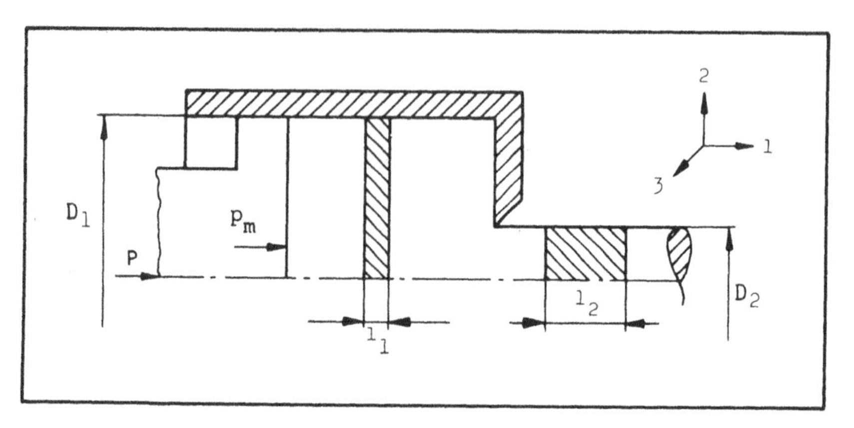

FIGURE 6.5 Extrusion of bars.

$$w = \int_{\bar{\epsilon}_1}^{\bar{\epsilon}_2} \bar{\sigma}\, d\bar{\epsilon}$$

The state of strain in the extrusion is given by

$$(\epsilon_1, \epsilon_2 = \epsilon_3 = -0.5\epsilon_1)$$

The corresponding effective strain, Eq. (4.37), becomes

$$\bar{\epsilon} = \left[\frac{2}{3}(\epsilon_1^2 + \epsilon_2^2 + \epsilon_3^2)\right]^{1/2} = \epsilon_1$$

ϵ_1, and consequently $\bar{\epsilon}$, is determined by (Fig. 6.5):

$$\epsilon_1 = \bar{\epsilon} = \ln\frac{l_2}{l_1} = \ln\frac{A_1}{A_2} = 2\ln\frac{D_1}{D_2}$$

where the suffixes 1 and 2 refer to the states before and after deformation, l is a length, A is a cross-sectional area, and for circular bars D is a diameter.

Assuming a homogeneous deformation (i.e., each volume element in the deforming material is supplied with the same amount of work), $\bar{\epsilon}_1 = 0$, and assuming that the material follows the stress–strain curve $\bar{\sigma} = c\bar{\epsilon}^n$,

$$W = V\frac{c}{n+1}\left(\ln\frac{A_1}{A_2}\right)^{n+1} \left[= V\frac{c}{n+1}\left(2\ln\frac{D_1}{D_2}\right)^{n+1}\right] \tag{6.13}$$

Considering the production of the extruded volume $V = A_2 l_2$, which is equal to the original volume (volume constancy) $V = A_1 l_1$ [$= (\pi/4)D_1^2 l_1$ for circular bars], the work supplied by the external forces can be expressed as

$$W_e = Pl_1 = p_m A_1 l_1$$

where P is the extrusion force and p_m is the mean extrusion pressure. Equating W_e and Eq. (6.13) gives

$$P = p_m A_1 = A_1 \frac{c}{n+1}\left(\ln \frac{A_1}{A_2}\right)^{n+1}$$

or

$$p_m = \frac{c}{n+1}\left(\ln \frac{A_1}{A_2}\right)^{n+1} \tag{6.14}$$

For circular bars this becomes

$$p_m = \frac{c}{n+1}\left(2 \ln \frac{D_1}{D_2}\right)^{n+1}$$

Since the extrusion process involves a high degree of internal distortion for the extrusion ratios (A_1/A_2) normally used, the assumption of a homogeneous deformation is rather poor. Allowing for both internal distortion and friction, p_m (6.14) must be increased about 50% on average, giving

$$p^*_m = 1.5 \frac{c}{n+1}\left(\ln \frac{A_1}{A_2}\right)^{n+1} \tag{6.15}$$

where p^*_m is the corrected mean extrusion pressure.

If an ideal-plastic material is used, $\bar{\sigma} = \sigma_0$ and the corrected extrusion pressure becomes

$$p^*_m = 1.5\sigma_0\,\bar{\epsilon} = 1.5\sigma_0 \ln \frac{A_1}{A_2} \tag{6.16}$$

For circular bars

$$p^*_m = 3\sigma_0 \ln \frac{D_1}{D_2}$$

6.4.4 Wire Drawing: Determination of Drawing Force and Maximum Reduction of Area in One Pass

As in extrusion, the work formula must be used to determine the force in wire drawing, as this does not act directly on the deforming material. The state of strain for wire drawing corresponds to that for extrusion, giving ϵ_1, $\epsilon_2 = \epsilon_3 = -0.5\epsilon_1$, that is,

$$\bar{\epsilon} = \epsilon_1 = \ln \frac{l_2}{l_1} = \ln \frac{A_1}{A_2} \quad \left(= 2 \ln \frac{D_1}{D_2} \text{ for circular wires}\right)$$

If an ideal-plastic material is assumed, $\bar{\sigma} = \sigma_0$ (the mean yield strength) and Eq. (6.2) gives

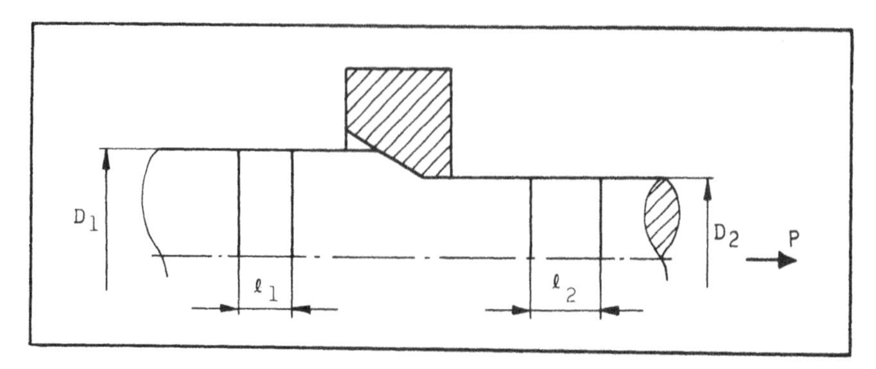

FIGURE 6.6 Wire drawing, circular wires. $V_1 = (\pi/4)D_1^2 l_1 = (\pi/4)D_2^2 l_2$.

$$W = V\sigma_0\bar{\epsilon} = V\sigma_0 \ln \frac{A_1}{A_2} \tag{6.17}$$

The work supplied by the external force in drawing a final wire length l_2 with the cross-sectional area A_2 (see Fig. 6.6) is $W_e = Pl_2$. Equating this and Eq. (6.17) gives

$$P = \frac{V}{l_2}\sigma_0 \ln \frac{A_1}{A_2}$$

Since the volume considered is $V = A_2 l_2 = A_1 l_1$, the drawing force becomes

$$P = A_2\sigma_0 \ln \frac{A_1}{A_2} \tag{6.18}$$

For circular wires

$$P = \frac{\pi}{4}D_2^2\sigma_0 \ln \frac{A_1}{A_2} = \frac{\pi}{2}D_2^2\sigma_0 \ln \frac{D_1}{D_2}$$

If a mean yield stress $\sigma_{0,m}$ is used for σ_0 in Eq. (6.18), reasonable results can be obtained.

When the stress–strain curve can be expressed as $\bar{\sigma} = c\bar{\epsilon}^n$, the drawing force can be found in a similar manner. Thus

$$P = A_2 \frac{c}{n+1}\left(\ln \frac{A_1}{A_2}\right)^{n+1} \tag{6.19}$$

where A_1 is the original and A_2 is the final cross-sectional area of the wire.

The reduction in area r is defined by

$$r = \frac{A_1 - A_2}{A_1} = 1 - \frac{A_2}{A_1}$$

From this the ratio A_1/A_2 can be found:

$$\frac{A_1}{A_2} = \frac{1}{1 - r} \tag{6.20}$$

The maximum possible reduction of area in one pass is limited by the tensile or ultimate strength of the drawn wire, as this must transmit the drawing force. For heavily cold worked materials, the yield stress and the ultimate stress will be nearly equal; that is, the maximum stress in the drawn wire can be approximated by $\bar{\sigma} = \sigma_0$ (ideal-plastic materials, where σ_0 is the mean yield stress). Equation (6.18) gives

$$P = \sigma_0 A_2 = A_2\sigma_0 \ln\frac{A_1}{A_2} = A_2\sigma_0 \ln\frac{1}{1 - r_m}$$

where r_m is the maximum reduction of area; consequently,

$$\ln\frac{1}{1 - r_m} = 1 \Rightarrow r_m = 1 - \frac{1}{e} = 63\% \tag{6.21}$$

In practice, only values around $r_m \simeq 50\%$ are obtainable, owing to internal distortion and friction.

If the drawing force [Eq. (6.18)] is corrected for internal distortion and friction by a 50% increase, Eqs. (6.18) and (6.19) become

$$P = 1.5A_2\sigma_0 \ln\frac{A_1}{A_2} \qquad (\bar{\sigma} = \sigma_0)$$

$$P = 1.5A_2 \frac{c}{n + 1}\left(\ln\frac{A_1}{A_2}\right)^{n+1} \qquad (\bar{\sigma} = c\bar{\epsilon}^n)$$

6.4.5 Explosive Forming: Determination of the Necessary Charge of Explosives

In this section explosive free forming of sheets is considered (see Fig. 6.7). This means that the final shape is created by the induced stress fields determined by the process conditions.

The charge is placed in water at a stand-off distance L from the sheet blank, and the energy is transmitted to the blank as a shock wave created by the detonation of the high explosive. To avoid too heavy a loss in energy, the water head H above the charge must exceed a certain value, which is approximately equal to the blank diameter.

To estimate the charge size necessary (the amount of high explosives), the work formula can be used; this involves a determination of the work of deformation necessary to shape the blank (internal work) and a determination of the external work (i.e., the work/energy delivered from the charge to the blank). Equating the internal and the external work allows a determination of the charge size required.

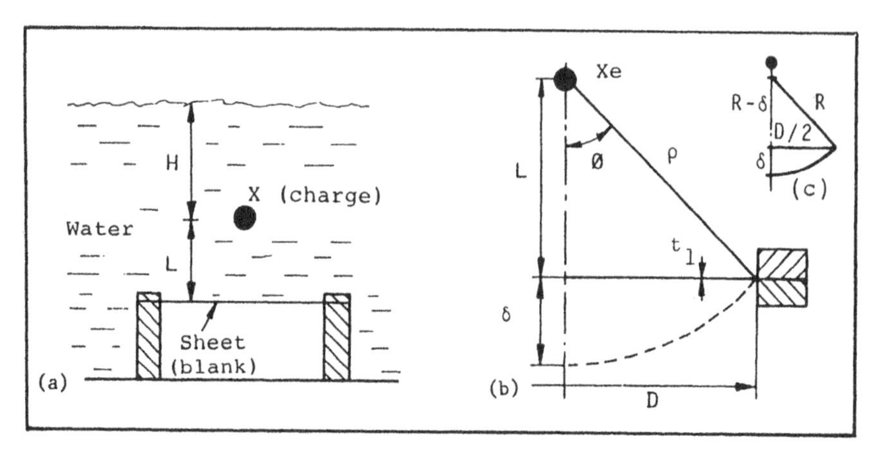

FIGURE 6.7 Explosive free forming of sheets.

Work of Deformation

It is assumed that the blank is firmly clamped in the tooling rings (i.e., the deformation involves pure stretching) and that the desired shape can be approximated by a spherical segment with radius R and a deflection δ (Fig. 6.7c).

The surface area of the spherical segment is

$$A_s = 2\pi R\delta$$

where δ is the maximum deflection. Volume constancy gives

$$\frac{\pi}{4}D^2 t_1 = t_2 2\pi R\delta$$

that is,

$$\frac{t_2}{t_1} = \frac{D^2}{8R\delta} \tag{6.22}$$

From

$$R^2 = (R - \delta)^2 + \frac{D^2}{4}$$

it is found that

$$2R\delta = \delta^2 + \frac{D^2}{4}$$

which, when substituted in Eq. (6.22), gives

$$\frac{t_2}{t_1} = \frac{D^2}{4(\delta^2 + D^2/4)} = \frac{1}{1 + 4(\delta/D)^2} \tag{6.23}$$

The thickness strain is, consequently,

$$\epsilon_t = \ln\frac{t_2}{t_1} = -\ln\left[1 + 4\left(\frac{\delta}{D}\right)^2\right] \tag{6.24}$$

At the pole of the segment, the radial ϵ_r and circumferential ϵ_0 principal strains in the plane of the segment are equal and thus $\epsilon_r = \epsilon_0$.

Volume constancy gives the state of strain:

$$(\epsilon_t,\ \epsilon_r = \epsilon_0 = -\tfrac{1}{2}\epsilon_t)$$

Consequently, the effective strain becomes

$$\bar{\epsilon} = \left[\frac{2}{3}(\epsilon_t^2 + \epsilon_r^2 + \epsilon_0^2)\right]^{1/2} = -\epsilon_t = \ln\left[1 + 4\left(\frac{\delta}{D}\right)^2\right] \tag{6.25}$$

If it is assumed that the material follows the stress-strain curve $\bar{\sigma} = c\bar{\epsilon}^n$, the work of deformation is given by Eq. (6.3), when $\bar{\epsilon}_1 = 0$ and $\bar{\epsilon}_2 = \bar{\epsilon}$:

$$W = V\frac{c}{n+1}\bar{\epsilon}^{n+1}$$

Substituting Eq. (6.25) in this expression gives

$$W = \frac{\pi}{4}D^2 t_1 \frac{c}{n+1}\left\{\ln\left[1 + 4\left(\frac{\delta}{D}\right)^2\right]\right\}^{n+1} \tag{6.26}$$

For other geometries, the same procedure for the determination of the work of deformation can be used.

Charge Energy/Work Supplied to the Blank

The explosive is assumed to contain an amount of energy per unit mass of a (J/kg); that is, the energy contained in a charge with the mass X is equal to Xa.

Only a portion of this energy hits the blank, corresponding to the solid angle ω, which subtends the blank (see Fig. 6.7b):

$$\omega = \frac{2\rho\pi(\rho - L)}{4\pi\rho^2}$$

where ρ is the radius of the sphere determining ω. This expression can be reduced to

$$\omega = \frac{1}{2}\left(1 - \frac{L}{\sqrt{L^2 + D^2/4}}\right) = \frac{1}{2}(1 - \cos\phi) \tag{6.27}$$

The energy E'_l directed toward the blank is consequently

$$E'_L = \frac{Xa}{2}(1 - \cos \phi)$$

Of this energy, only a portion is converted into useful mechanical energy. If the coefficient of efficiency is η, the useful amount of energy can be expressed as

$$E_L = \eta\frac{Xa}{2}(1 - \cos \phi) \tag{6.28}$$

The efficiency of η normally varies between 0.5 and 0.25. If $\eta_{max} = 0.5$ is assumed, the useful energy becomes

$$E_L \leq \tfrac{1}{4}Xa(1 - \cos \phi) \tag{6.29}$$

Equating (6.26) and (6.29) yields

$$W = E_L \leq \tfrac{1}{4}Xa(1 - \cos \phi)$$

or

$$X \geq \frac{4W}{a(1 - \cos \phi)} \tag{6.30}$$

Substituting W in this expression, X is found:

$$X \geq \frac{\pi D^2 t_1 c}{(n + 1)a(1 - \cos \phi)}\left\{\ln\left[1 + 4\left(\frac{\delta}{D}\right)^2\right]\right\}^{n+1} \tag{6.31}$$

where a is the energy per unit mass of the explosive and X is the total mass of the explosive (i.e., the charge size).

In other applications a similar procedure can be followed.

6.5 SUMMARY

In this chapter the mass-conserving processes based on solid materials and plastic deformation as primary basic process have been discussed. The discussion has elucidated the main elements in the three basic flow systems for material, information (shape), and energy. This means that this chapter and the morphological model described in Chapter 1 constitute a fundamental and general background, allowing a primary evaluation of those processes capable of acquiring a desired geometry (i.e., it allows an evaluation of the possibilities and limitations concerning the geometries, materials, surfaces, and tolerances).

It is important that the engineer be able to apply the basic principles described by the flow systems in a new and imaginative manner to obtain technical and economical advances in production situations.

7

Solid Materials: Mass-Reducing Processes

7.1 INTRODUCTION

Mass-reducing processes are extensively used in the manufacturing industry. In these processes the size of the original workpiece is sufficiently large that the final geometry can be circumscribed by it, and the unwanted material is removed as chips, particles, and so on (i.e., as scrap). The chips or scrap are necessary to obtain the desired geometry, tolerances, and surfaces. The amount of scrap may vary from a few percent to 70–80% of the volume of the original work material.

Most metal components have, at one or another stage, been subjected to a material removal process. Many other materials (e.g., plastics and wood) are frequently subjected to material removal processes.

Owing to the rather poor material utilization of the mass-reducing processes, the anticipated scarcity of materials and energy, and increasing costs, development in the last decade has been directed toward an increasing application of mass-conserving processes. However, die costs and the capital cost of machines remain rather high; consequently, mass-reducing processes are, in many cases, the most economical, in spite of the high material waste, which only has value as scrap. Therefore, it must be expected that the material removal processes will maintain their important position in manufacturing for many years to come. Furthermore, the development of automated production systems has progressed more rapidly for mass-reducing processes than for mass-conserving processes.

TABLE 7.1 Classification of Mass-Reducing Processes in Terms of Basic Processes and Fundamental Methods of Material Removal

Category of basic process	Fundamental removal method	Examples of processes
Mechanical	I 	Cutting: Turning Milling Drilling Grinding, etc.
	II 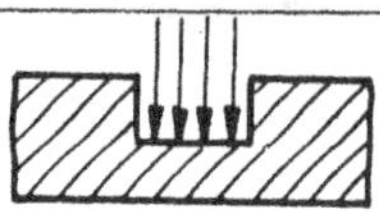	Water jet cutting Abrasive jet machining Sand blasting, etc.
	III 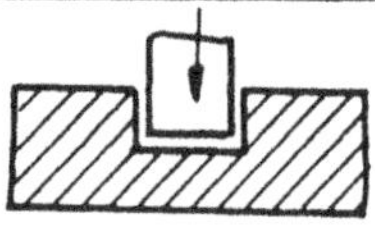	Ultrasonic machining
	IV 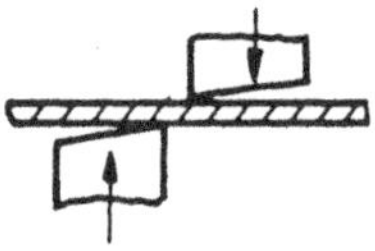	Blanking Punching Shearing
Thermal	II 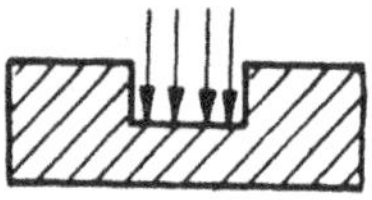	Thermal cutting (melting) Electron beam machining Laser machining
	III 	Electrodischarge machining
Chemical	II 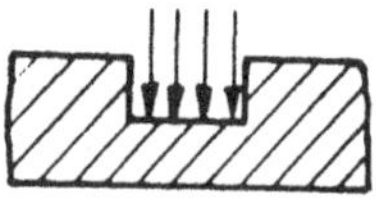	Etching Thermal cutting (combustion)
	III 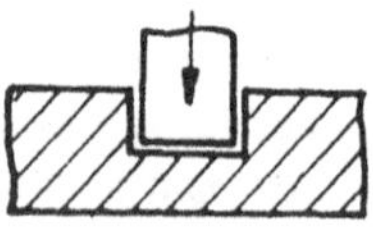	Electrochemical machining

In this chapter the following topics are discussed: the characteristics of mass-reducing processes, the geometrical possibilities, typical examples of mass-reducing processes, and the determination of forces and power.

7.2 CHARACTERISTICS OF MASS-REDUCING PROCESSES

In this section the fundamental conditions concerning mass-reducing processes will be discussed. As an introduction, the basic principles of the processes are described and fundamental definitions given, after which the chip formation and the process conditions are elucidated.

7.2.1 Basic Principles

As discussed in Chapter 1, material removal can be based on four fundamental removal methods, which illustrates the relationship between the imprinting of the information and the energy supply (see Table 7.1).

In removal method I, the imprinting of information is carried out by a rigid medium of transfer (the tool), which is moved relative to the workpiece, and the mechanical energy is supplied through the tool. The final geometry is thus determined from the geometry of the tool and the pattern of motions of the tool and the workpiece. The basic process is mechanical: actually, a shearing action combined with fracture.

In removal method II, the imprinting of information is carried out by an energy source, having certain characteristics, which is moved relative to the workpiece (i.e., the final geometry is a result of the properties of the energy source and the pattern of motions). The medium of transfer can be fluid, gaseous, granular, or combinations thereof. The basic process can be mechanical, thermal, or chemical.

In removal method III, the imprinting of information is established by the motion of a rigid transfer medium containing elements of the desired geometry (the negative picture) relative to the workpiece (i.e., the final geometry is a product of the geometry of the rigid medium and the pattern of motions). The energy supply is provided through a coupling medium, which is necessary to establish and control the basic process. The state of the medium can be fluid, gaseous, granular, or combinations thereof, and the basic process can be mechanical, thermal, or chemical.

In removal method IV, the imprinting of information is established through two rigid media (i.e., the final geometry is a result of the geometry of the two media and the pattern of motions). This method is similar to method I concerning the basic process, which is mechanical and fracture specifically caused by shearing. The energy is supplied primarily through one of the two media.

An evaluation of the geometrical accuracies obtainable for the four fundamental removal methods shows that methods I, III, and IV have the potential of providing fairly accurate components, due to the rigid media of transfer. Method II does not have quite the same potential because the final geometry relies entirely on the interaction between a geometrically unconfined energy supply through a fluid, gaseous, or granular medium and the workpiece.

Removal method I, which represents the traditional cutting processes and thus has the major industrial importance, is discussed in detail in the following sections; the other removal methods are described only as examples in Section 7.4. This is also the case for such important processes (included in method IV) as blanking, punching, and shearing, where two tools work against each other, so that the material can be separated to produce various patterns (curved or straight edges, closed contours, etc.).

7.2.2 Fundamental Definitions

As mentioned previously, the unwanted material in mass-reducing processes—based on removal method I—is removed by a rigid cutting tool, so that the desired geometry, tolerances, and surface finish are obtained. Examples of processes in this group are turning, drilling, reaming, milling, shaping, planing, broaching, grinding, honing, and lapping.

Most of the cutting or machining processes are based on a two-dimensional surface creation, which means that two relative motions are necessary between the cutting tool and the work material. These motions are defined as the *primary motion*, which mainly determines the cutting speed, and the *feed motion*, which provides the cutting zone with new material.

In turning the primary motion is provided by the rotation of the workpiece, and in planing it is provided by the translation of the table; in turning the feed motion is a continuous translation of the tool, and in planing it is an intermittent translation of the tool.

Cutting Speed. The cutting speed v is the instantaneous velocity of the primary motion of the tool relative to the workpiece (at a selected point on the cutting edge).

Figures 7.1 through 7.3 show fundamental definitions, designations, and so on, for turning, drilling, and milling. The cutting speed for these processes can be expressed as

$$v = \pi dn \quad \text{m/min} \tag{7.1}$$

where v is the cutting speed in m/min, d the diameter of the workpiece to be cut in meters, and n the workpiece or spindle rotation in rev/min. Thus v, d, and n may relate to the work material or the tool, depending on the specific kinematic pattern. In grinding the cutting speed is normally measured in m/s.

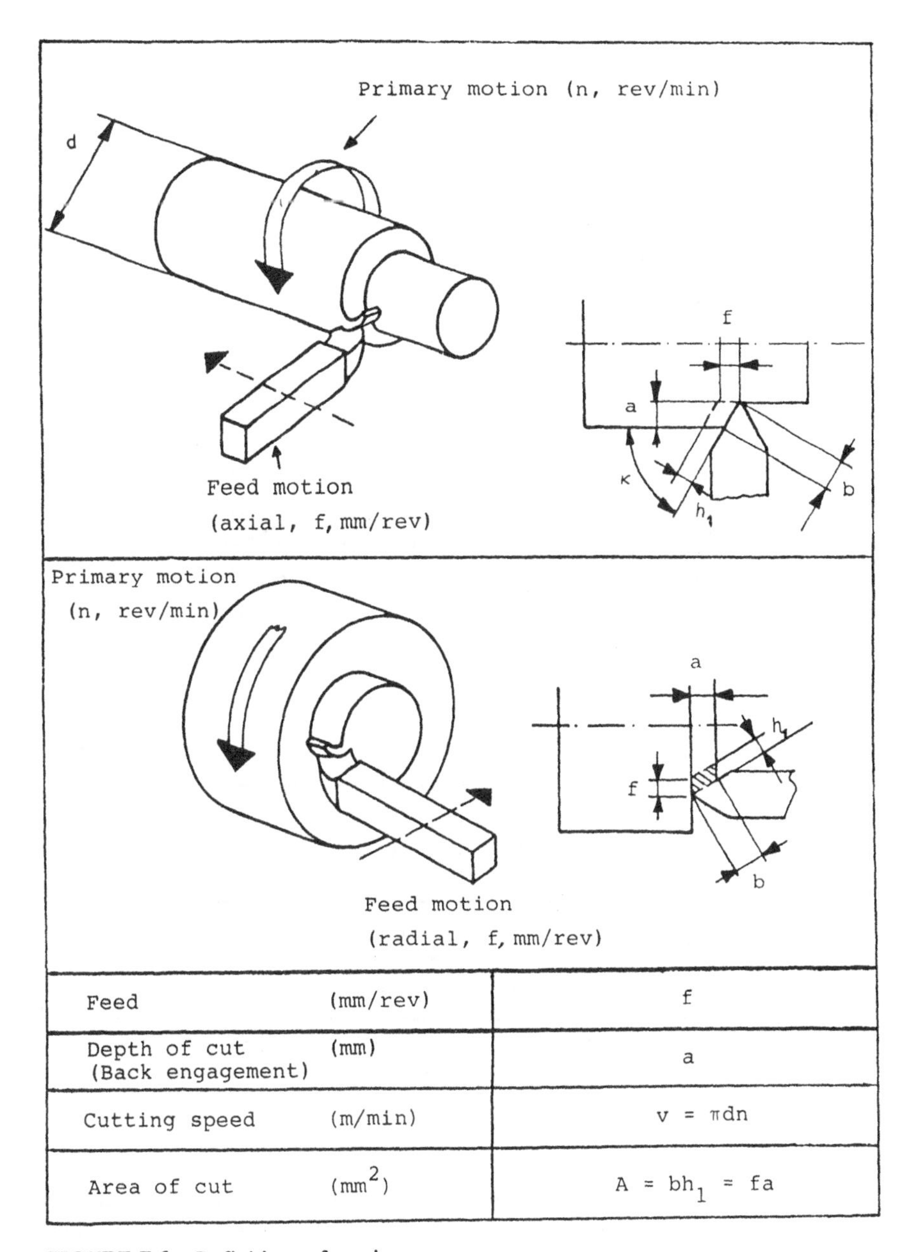

Feed	(mm/rev)	f
Depth of cut (Back engagement)	(mm)	a
Cutting speed	(m/min)	$v = \pi dn$
Area of cut	(mm^2)	$A = bh_1 = fa$

FIGURE 7.1 Definitions of turning.

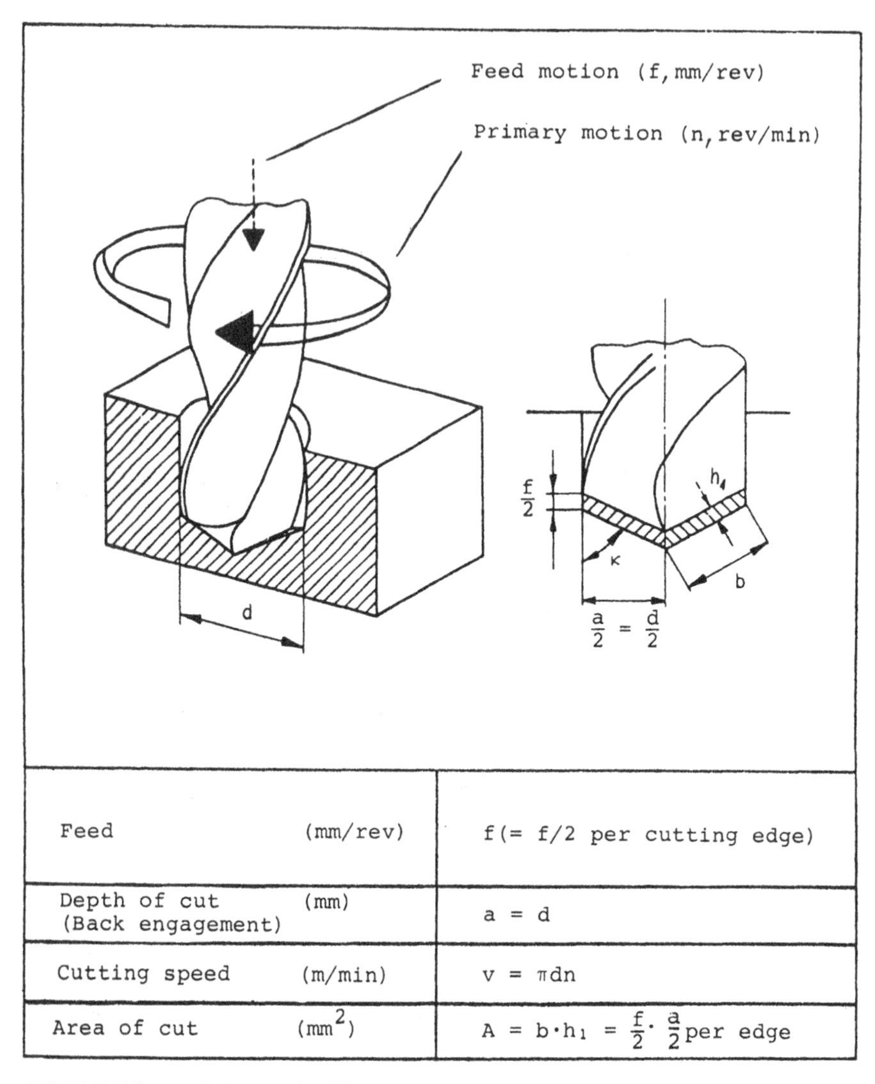

Feed	(mm/rev)	f(= f/2 per cutting edge)
Depth of cut (Back engagement)	(mm)	a = d
Cutting speed	(m/min)	$v = \pi dn$
Area of cut	(mm^2)	$A = b \cdot h_1 = \frac{f}{2} \cdot \frac{a}{2}$ per edge

FIGURE 7.2 Definitions of drilling.

It should be mentioned here that the basic SI units in many cases are not used in the field of machining, but the units used are recognized by ISO (International Organization for Standardization) and applied in the relevant standards for example ISO 229, ISO 3002/1, and ISO 3685. ISO 1000 describes the SI-unit system.

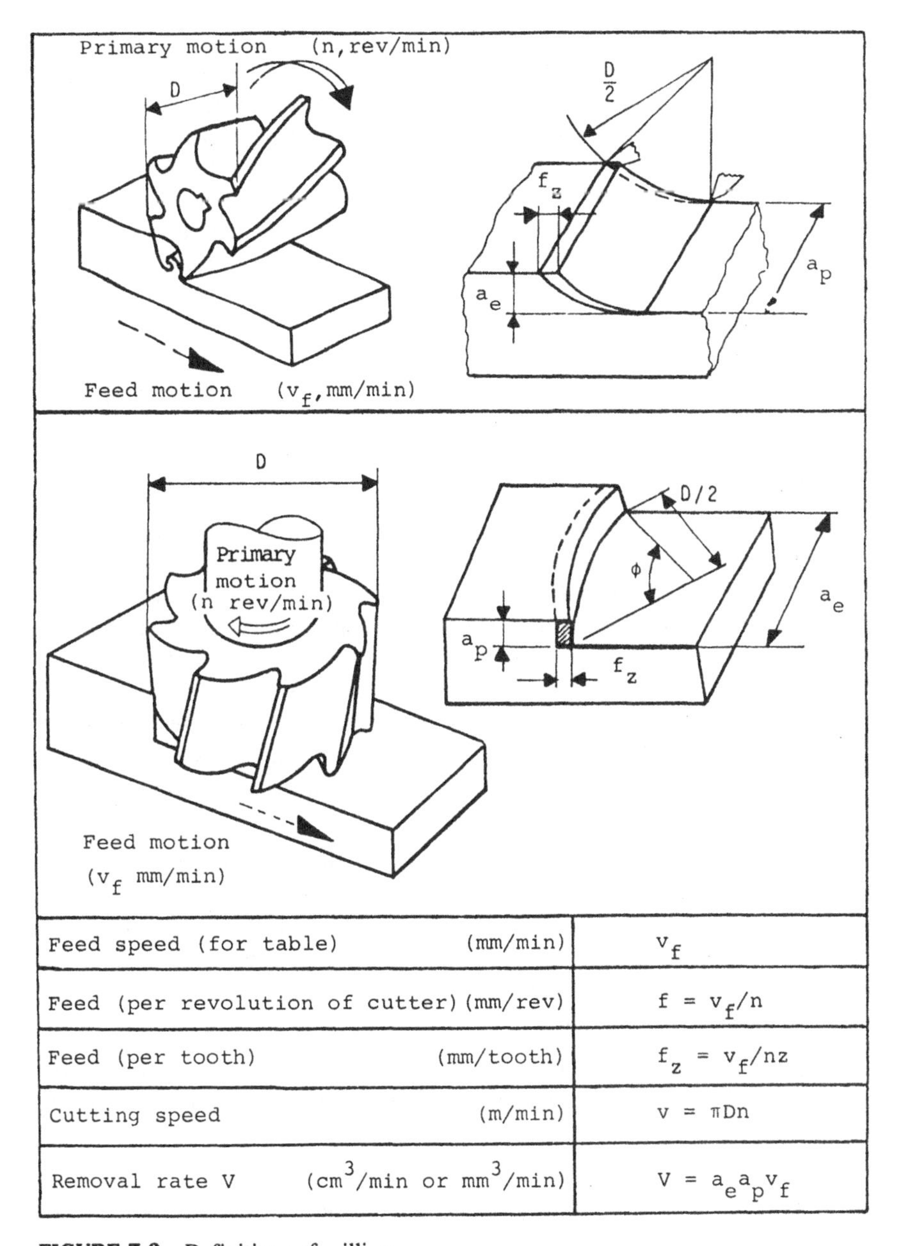

Feed speed (for table) (mm/min)	v_f
Feed (per revolution of cutter) (mm/rev)	$f = v_f/n$
Feed (per tooth) (mm/tooth)	$f_z = v_f/nz$
Cutting speed (m/min)	$v = \pi Dn$
Removal rate V (cm^3/min or mm^3/min)	$V = a_e a_p v_f$

FIGURE 7.3 Definitions of milling.

Feed. The feed motion f is provided to the tool or the workpiece and, when added to the primary motion, leads to a repeated or continuous chip removal and the creation of the desired machined surface. The motion may proceed by steps or continuously. The feed speed v_f is defined as the instantaneous velocity of the feed motion relative to the workpiece (at a selected point on the cutting edge).

For turning and drilling, the feed f is measured per revolution (mm/rev) of the workpiece or the tool; for planing and shaping f is measured per stroke (mm/stroke) of the tool or the workpiece. In milling the feed is measured per tooth of the cutter f_z (mm/tooth); that is, f_z is the displacement of the workpiece between the cutting action of two successive teeth. The feed speed v_f (mm/min) of the table is therefore the product of the number of teeth z of the cutter, the revolutions per minute of the cutter n, and the feed per tooth ($v_f = nzf_z$) (see Fig. 7.3).

A plane containing the directions of the primary motion and the feed motion is defined as the working plane, since it contains the motions responsible for the cutting action.

Depth of Cut (Engagement). In turning (Fig. 7.1) the depth of cut a (sometimes also called back engagement) is the distance that the cutting edge engages or projects below the original surface of the workpiece. The depth of cut determines the final dimensions of the workpiece. In turning, with an axial feed, the depth of cut is a direct measure of the decrease in radius of the workpiece and with radial feed the depth of cut is equal to the decrease in the length of workpiece. In drilling (Fig. 7.2) the depth of cut is equal to the diameter of the drill. For milling, the depth of cut is defined as the working engagement a_e and is the radial engagement of the cutter. The axial engagement (back engagement) of the cutter is called a_p (see Fig. 7.3).

Chip Thickness. The chip thickness h_1 in the undeformed state is the thickness of the chip measured perpendicular to the cutting edge and in a plane perpendicular to the direction of cutting (see Figs. 7.1 and 7.2). The chip thickness after cutting (i.e., the actual chip thickness h_2) is larger than the undeformed chip thickness, which means that the cutting ratio or chip thickness ratio $r = h_1/h_2$ is always less than unity.

Chip Width. The chip width b in the undeformed state is the width of the chip measured along the cutting edge in a plane perpendicular to the direction of cutting (see Figs. 7.1 and 7.2).

Area of Cut. For single-point tool operations, the area of cut A is the product of the undeformed chip thickness h_1 and the chip width b (i.e., $A = h_1 b$). The area of cut can also be expressed by the feed f and the depth of cut a as follows:

$$h_1 = f \sin \kappa \qquad \text{and} \qquad b = \frac{a}{\sin \kappa} \tag{7.2}$$

where κ is the major cutting-edge angle (i.e., the angle that the cutting edge forms with the working plane).

Consequently, the area of cut is given by

$$A = fa \tag{7.3}$$

Removal Rate. For single-point tools the removal rate V (cm^3/min or mm^3/min) is the product of the area of cut and the cutting speed:

$$V = Av = fav \tag{7.4}$$

For milling (Fig. 7.3) the removal rate is given by

$$V = a_e a_p v_f \tag{7.5}$$

After these general definitions, the tool geometry is discussed. Only tools with well-defined cutting edges are described. Cutting tools with geometrically undefined cutting edges are mentioned later.

Major Cutting Edge. The major cutting edge is that cutting edge for which a portion is responsible for the major part of the cutting (see Fig. 7.4a, b, and c).

Minor Cutting Edge. The minor cutting edge is that cutting edge for which a portion is partly responsible for generating the surface on the workpiece (see Section 7.2.4).

Corner Radius. The corner radius r is the transition curve between the major cutting edge and the minor cutting edge (see Fig. 7.4a).

Tool Face. The tool face (or rake face), the slope of which is established by the cutting edge inclination and the normal rake, is the face over which the chip flows (see Fig. 7.4b).

Flank. Two flanks exist. The *major flank* is the flank adjacent to the major cutting edge; that is, the major cutting edge is formed by the intersection of the face and the major flank (see Fig. 7.4b). The *minor flank* is the flank adjacent to the minor cutting edge; that is, the minor cutting edge is formed by the intersection of the face and the minor flank.

Cutting-Edge Inclination. The cutting edge inclination λ is the angle between a plane perpendicular to the working plane and the major cutting edge (Fig. 7.4c).

Clearance. The clearance angles are necessary to prevent the major and minor flanks from rubbing against the workpiece. The most important is the normal clearance α (see Fig. 7.4c and d), which is the angle between the cutting direction and the flank.

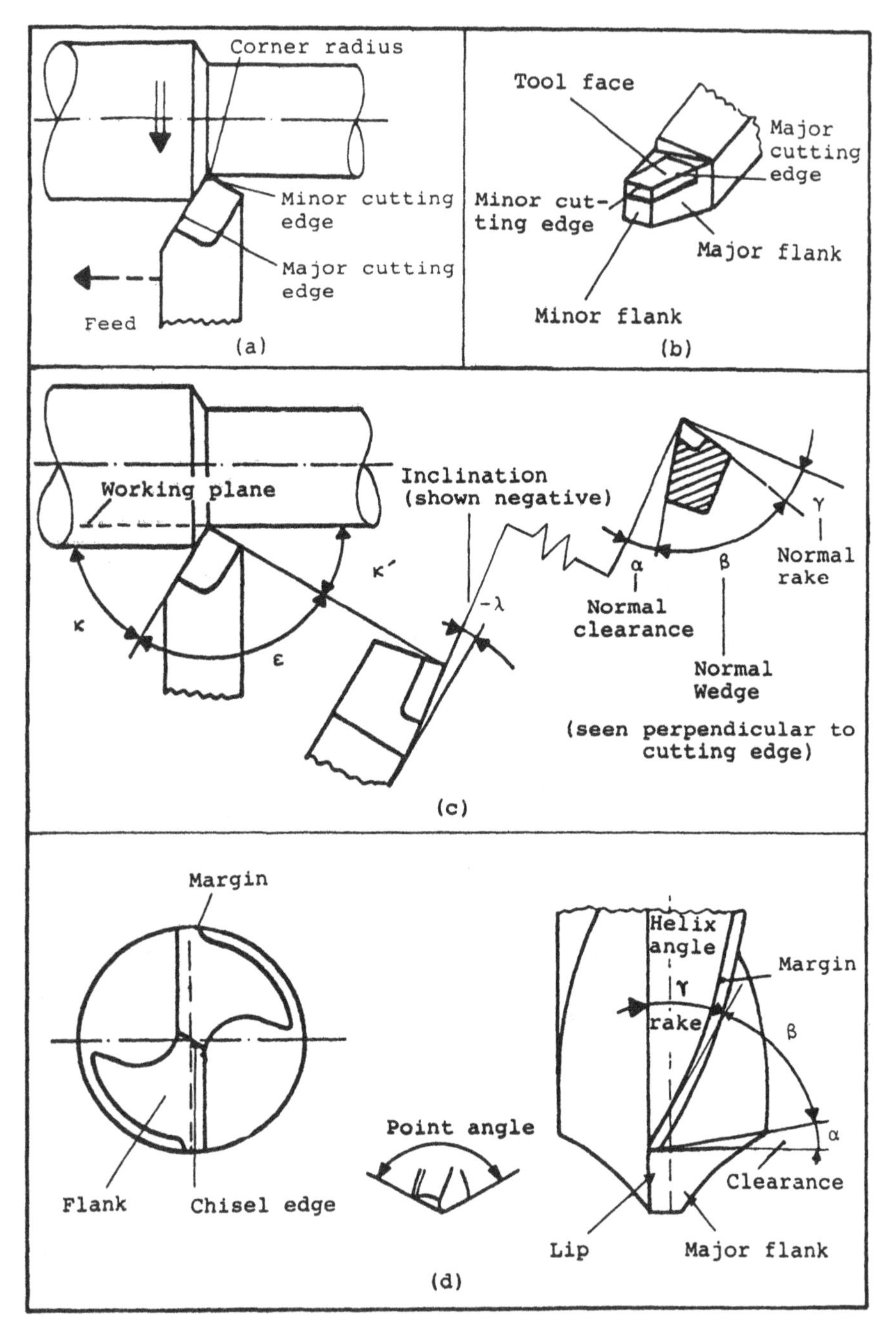

FIGURE 7.4 Definitions of faces and angles: (a–c) for turning; (d) for drilling.

Major Cutting Edge Angle. The major cutting edge angle κ is the angle between the major cutting edge and the working plane, measured in a plane perpendicular to the working plane (see Fig. 7.4c).

Minor Cutting Edge Angle. The minor cutting edge angle κ' is provided to prevent the minor cutting edge from contacting the workpiece over the whole of the minor cutting edge (see Fig. 7.4c).

Normal Rake. The normal rake γ is the angle between the normal to the cutting direction and the tool face, measured in a plane perpendicular to the cutting edge (see Fig. 7.4c and d). The rake is the most important angle of the cutting process when considering the mechanisms of the process.

The Wedge Angle. The normal wedge angle β is the angle between the face and the flank, measured in a plane perpendicular to the cutting edge (see Fig. 7.4c and d).

The sum of the normal clearance α, the wedge angle β, and the normal rake γ is equal to 90°:

$$\alpha + \beta + \gamma = 90° \tag{7.6}$$

γ is shown positive in Fig. 7.4.

In drilling, the tool (a twist drill) has two major cutting edges and the angle between these is called the *point angle* (see Fig. 7.4d).

Included Angle. The included angle ϵ is the angle between the major and minor cutting edges measured in a plane perpendicular to the working plane and parallel to the direction of the feed motion (see Fig. 7.4c).

More detailed descriptions of single-point cutting tool geometry can be found in ISO 3002/1.

7.2.3 Chip Formation

A cutting process is a controlled interaction among the workpiece, the tool, and the machine. This interaction is influenced by the selected cutting conditions (cutting speed, feed, and depth of cut), cutting fluids, the clamping of the tool and workpiece, and the rigidity of the machine. Figure 7.5 illustrates this interaction [14]. The clamping of the tool and the workpiece are not discussed here, and it is assumed that the machine will possess the necessary rigidity and power to carry out the process.

The main factors in chip formation are: the tool (material and geometry), the work material (material, geometry, rigidity), and the cutting conditions. These factors are designated with an asterisk in Fig. 7.5.

To reduce the number of parameters in the study of the chip formation mechanism, orthogonal cutting is considered, which approximates a plane strain

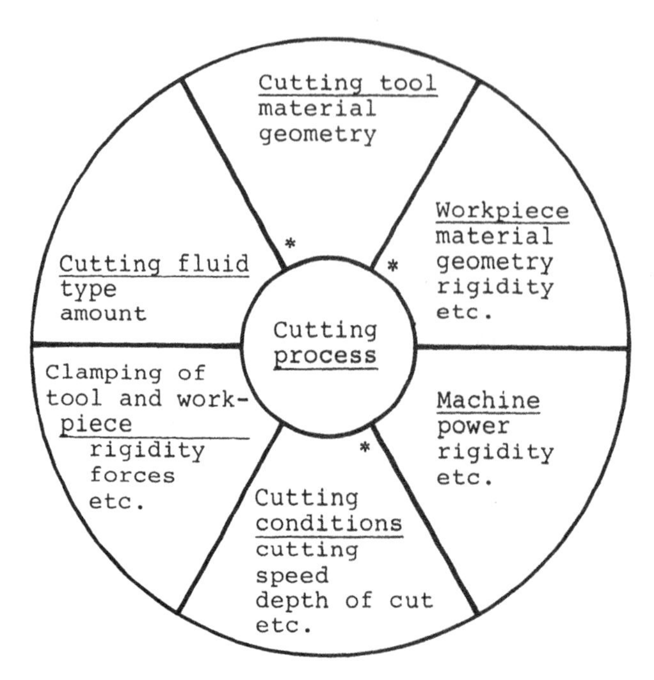

FIGURE 7.5 Main factors affecting the cutting process [14].

problem. In orthogonal cutting, the major cutting edge is perpendicular to the working plane; consequently, the depth of cut a becomes equal to the chip width b and the undeformed chip thickness h_1 becomes equal to the feed f (see Fig. 7.6).

Chip Formation Mechanism

From Fig. 7.6a it can be seen that the shear deformation in the model is confined to the shear plane AB, extending from the tool cutting edge to the intersection of the free surfaces of the workpiece and chip. In practice, shearing is not confined to the plane AB, but in a narrow shear zone. At low cutting speeds the thickness of the zone is large, but at practical speeds its thickness is comparable to that shown in Fig. 7.6b and c and can be approximated to a plane. The angle ϕ that the shear plane forms with the machined surface is called the *shear angle*.

The chip can be considered as built up of thin layers, which slide relative to each other (see Fig. 7.6d). These layers can be compared to a stack of cards pushed toward the tool face. High normal pressures exist between the chip and the tool, causing high frictional forces resulting in a chip with a smooth rear surface (see Fig. 7.6b and c). The influence of friction is not shown in Fig. 7.6d.

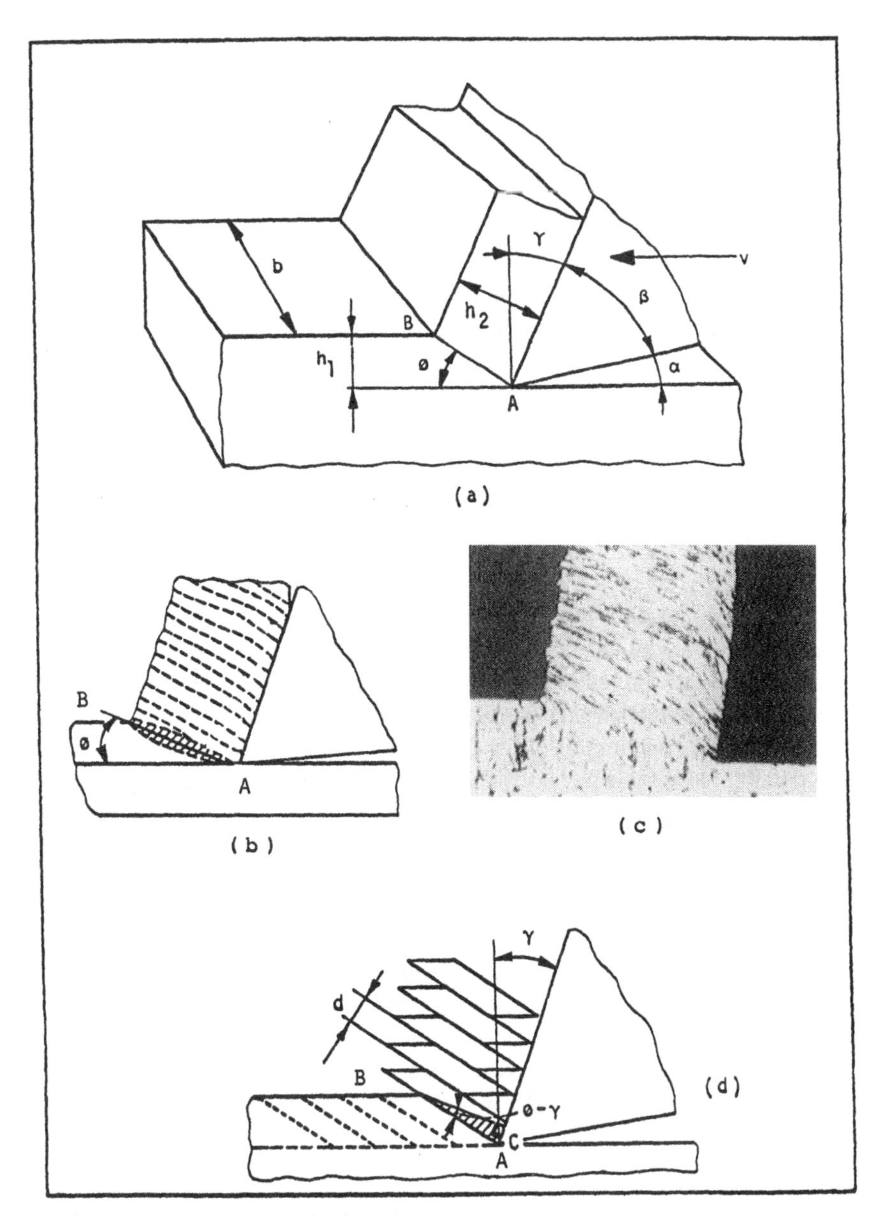

FIGURE 7.6 Orthogonal cutting. (c) from Ref. 15, © 1960, Addison-Wesley Publishing Company, Inc., Chap. 3, p. 32, Fig. 3–2. Reprinted with permission.

In the cutting process, the properties of the tool and the work material and the cutting conditions (h_1, γ, and v) can be controlled, but the chip thickness h_2 ($> h_1$) is not directly controllable. This means that the cutting geometry is not completely described by the chosen parameters.

The cutting ratio or chip thickness ratio, which is defined by

$$r = \frac{h_1}{h_2} \quad (<1) \tag{7.7}$$

can be measured and used as an indicator of the quality of the cutting process.

The shear angle ϕ can be expressed by the rake angle γ and the inverse cutting ratio λ_h (Fig. 7.6d):

$$\lambda_h = \frac{1}{r} = \frac{h_2}{h_1} = \frac{AB \cos(\phi - \gamma)}{AB \sin \phi} \tag{7.8}$$

Solving this equation gives

$$\tan \phi = \frac{\cos \gamma}{\lambda_h - \sin \gamma} \tag{7.9}$$

The inverse cutting ratio (also called the chip compression) and the rake angle determine the shear angle ϕ. The smaller ϕ, the larger h_2, which means that the shear zone increases in length (i.e., the force and power requirements increase). Consequently, a large shear angle will give the best utilization of the supplied power. The chip compression ($\lambda_h = 1/r$) must thus be kept as small as practically possible, since this increases the shear angle and, consequently, decreases the power consumption.

Hard work materials give lower chip compression values than do soft materials but require higher cutting forces. Friction increases the chip compression and can be reduced by introducing suitable cutting fluids.

The chip compression can be reduced further by increasing the cutting speed or the feed. These increases in cutting speed and feed have an upper limit, however, because the tool life decreases, which might have a greater economic effect than the resulting increases in material removal rate. Different theoretical models have been developed (e.g., see Cook [15]) to predict the shear angle, but these are not discussed here. The actual shear angle ϕ can be determined experimentally by measuring h_2 [Eq. (7.9)].

Types of Chip

From the appearance of the chip, much valuable information about the actual cutting process can be gained, as some types of chip indicate more efficient cutting than others. The type of chip is determined mainly by the properties of the work material, the geometry of the cutting tool, and the cutting conditions.

It is, in general, possible to broadly differentiate three types of chip: (1) the discontinuous (segmental) chip, (2) the continuous chip, and (3) the continuous chip with built-up edge.

The Discontinuous Chip. In this case, which represents the cutting of most brittle materials such as cast iron and cast brass, the stresses ahead of the cutting edge cause fracture. This is because the actual shear strain exceeds the shear strain at fracture in the material in the direction of the shear plane, so that the material is removed in fairly small segments (see Fig. 7.7a). Fairly good surface finish is, in general, produced in these brittle materials, as the cutting edge tends to smooth the irregularities.

Discontinuous chips can also be produced under certain conditions with more ductile materials such as steel, causing a rough surface. These conditions may be low cutting speeds or low rake angles in the range 0–10° for feeds greater than 0.2 mm. Increasing the rake angle or the cutting speed normally eliminates the production of a discontinuous chip.

The Continuous Chip. This type of chip, which represents the cutting of most ductile materials that permit the shearing to take place without fracture, is produced by relatively high cutting speeds, large rake angles ($\gamma = 10–30°$), and low friction between the chip and the tool face (see Fig. 7.7b and c).

Continuous and long chips may be difficult to handle and, consequently, the tool must be provided with a chip breaker, which curls and breaks the chip into short lengths. The chip breaker can be formed by grinding a stop or a recess in the tool or brazing/screwing a chip breaker onto the tool face.

The Continuous Chip with Built-Up Edge. This type of chip represents the cutting of ductile materials at low speeds where high friction exists on the tool face. This high friction causes a thin layer of the underside of the chip to shear off and adhere to the tool face. The chip is similar to the continuous chip, but it is produced by a tool having a nose of built-up metal welded to the tool face. Periodically, portions of the built-up edge separate and escape on the chip undersurface and on the material surface, resulting in a rough machined surface (see Fig. 7.8a). The built-up edge effectively increases the rake angle and decreases the clearance angle (see Fig. 7.8b).

The cutting speed influences the size of the built-up edge and, consequently, the final surface roughness (see Fig. 7.8c). Some materials do not exhibit the minimum shown in roughness for low cutting speeds.

At sufficiently high cutting speeds, the built-up edge normally disappears, and this upper limit is called the free machining cutting speed. A hard material will generally have a lower free machining speed than a soft material. At increasing feed, the curve (Fig. 7.8c) will shift to the left. In most processes, cutting speeds above the free machining speed are chosen, but for broaching, for example, it is sometimes necessary to approach the minimum (Fig. 7.8c).

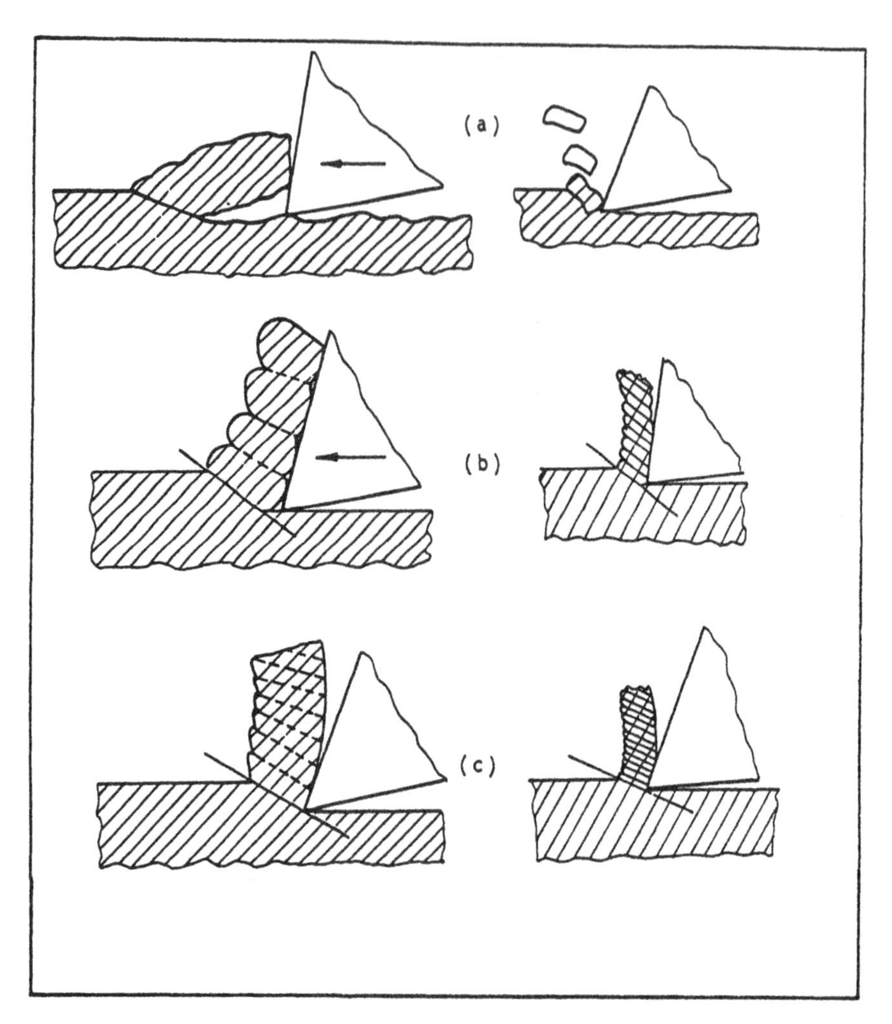

FIGURE 7.7 Basic types of chip: (a) the discontinuous chip; (b), (c) the continuous chip.

The built-up edge decreases the tool life. Summarizing, it can be said that the development of the built-up edge can be reduced or eliminated through increased rake angles, higher cutting speeds, higher feeds, and good lubrication.

7.2.4 Conditions of the Cutting Process

As mentioned previously, the main factors in the cutting process are the tool, the workpiece, and the machine that establishes the interaction between the tool and

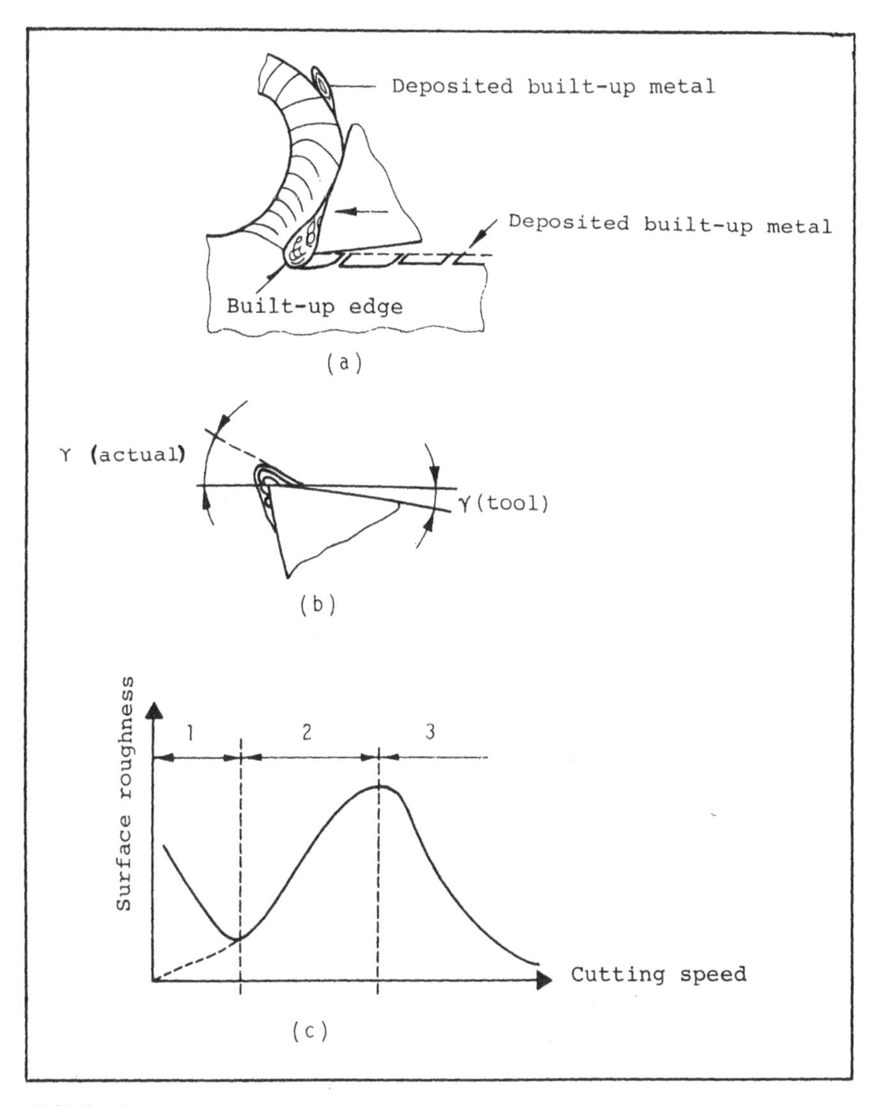

FIGURE 7.8 The effect of a built-up edge on (a) the work material; and (b) the tool geometry. (c) The built-up edge and, consequently, the surface roughness are functions of the cutting speed.

the workpiece. The machine (i.e., type, size, stability, rigidity, and state) and the support (i.e., rigidity, accuracy, etc.) have a major influence on the efficiency of the actual cutting process. It is assumed in the following that the machine is well maintained and that the clamping corresponds to standard (good) workshop practice.

The interaction between the tool and the workpiece is governed by the properties of the work material, the tool material, the tool geometry, and the cutting data (cutting speed, feed, depth of cut). The selection of tools (material and geometry) and cutting data depend on the work material, the requirements of workpiece geometry, tolerances and surface finish, the desired (economical) tool life, chip formation, the cutting forces, and power. The cutting data (based on workpiece requirements, machine, and tool material) must be selected so that an economical cutting process is obtained.

In the following, the tool (material, geometry, wear, and tool life) and the selection of cutting data are discussed. The machines are not described in detail, but the basic principles are illustrated in later sections.

The Tool Material

Chip formation involves high local stresses, friction, wear, and high temperatures; consequently, the tool material must combine the properties of high strength, high ductility, and high hardness or wear resistance at high temperatures. The most important tool materials are carbon tool steels (CTS), high-speed steels (HSS), cemented or sintered carbides (CC), ceramics (C), and diamond (D).

Carbon Tool Steel. Plain carbon steels of about 0.5–2.0%C when hardened and tempered have a high hardness and strength, and can be used as hand tools for cutting softer materials at low speeds. The wear resistance is relatively low, and cutting-edge temperatures must not exceed about 300°C. This material is used now only for special purposes and has generally been replaced by the materials described below.

High-Speed Steel. High-speed steels are alloyed steels that permit cutting-edge temperatures in the range 500–600°C. The typical alloying elements are tungsten, chromium, vanadium, and cobalt: for example, 22%W, 4.7%Cr, 1.4%V, and 0.75%C. The permissible higher cutting-edge temperatures make it possible to increase the cutting speed by about 100% compared to that used with carbon tool steels—hence the name high-speed steels. This steel is used quite extensively in twist drills, milling cutters, and special-purpose tools and is, in fact, the most common tool material.

Sintered Carbide. Sintered (or cemented) carbides are produced by powder metallurgical processes. Sintered carbides of tungsten carbide (WC) with cobalt (Co) as a binder are hard and brittle and are used in cutting cast iron and bronze. If titanium carbide (TiC) is added or used as the main constituent, the strength and toughness can be increased, and these types can be used in cutting hard materials. A large variety of sintered carbides exist, and each is generally developed to fulfill the requirements of effective cutting of different material groups.

Sintered carbides are very hard, and they permit an increase in cutting speeds of about 200–500% compared to high-speed steel tools. But it must be remembered that they have a relatively low ductility and, consequently, care must be taken to avoid high-speed impacts such as those that occur during interrupted cutting operations.

Sintered carbides are, in general, used as "throwaway" inserts supported in special holders or shanks. The inserts may have from three to eight cutting edges, and when one edge becomes dull, the insert is indexed to a new cutting edge. This procedure continues until all edges are used, when a new insert is substituted.

During recent years, coated sintered carbide tools have been developed, allowing both higher cutting speeds and consequently higher temperatures. Production rate increases of about 200% are obtainable compared to conventional sintered carbides. Titanium carbide, titanium nitride, aluminum oxide, and so on, can be used as coating materials to prolong the life of the tool.

Ceramics. Ceramic tool materials have been developed within the last couple of decades. The material most frequently used is aluminum oxide, which is pressed and sintered. For light finishing cuts the cutting speeds obtainable are two to three times larger than the cutting speeds for sintered carbides. They are used mainly where close tolerances and high surface finish are required. The ceramics are produced as throwaway inserts or tips.

Diamond. Diamond is the hardest of all tool materials and is used mainly where a very high surface quality is required as well as close tolerances.

Cutting Fluids

The unwanted effects of high friction and high cutting-edge temperatures can be reduced by introducing suitable cutting fluids. The purposes of introducing a cutting fluid are in general:

Reduction of friction and wear (increased tool life)
Cooling of the cutting edge
Protection of the new surface against corrosion
Flushing away of chips

Some other advantages are decreased tendency to produce a built-up edge on the tool at low cutting speeds, the possibility of increasing the cutting speed, and improved component surface finish, accuracy, and so on. A wide variety of cutting fluids are on the market, and among the factors affecting the selection are the work material, the particular process, the properties of the fluid as coolant and lubricant, its stability in use and storage, and whether it is nontoxic or will in any way affect the health of the operating personnel.

The Tool Geometry

The most suitable tool geometry depends on the properties of the tool material and the work material, the most important parameters being the normal rake angle γ, the wedge angle β, the normal clearance angle α, and the corner radius r.

The *normal rake angle* γ affects the ability of the tool to shear the work material and form the chip. It can be positive or negative. Positive rake angles lead to smaller-cutting forces, and consequently smaller deflections of the workpiece, toolholder, and machine.

The normal rake angle must not be too large, as the strength of the tool is reduced as well as its capacity to conduct heat. In machining hard work materials, γ must be small, even negative for carbide and diamond tools: the higher the hardness, the smaller γ. For high-speed steels, γ is normally chosen in the range 0–30°, depending on the type of tool (turning, planing, end milling, face milling, drilling, etc.) and the work material.

For carbide tools, γ is normally chosen in the range $-8°$–25°, depending on the type of tool and work material. Inserts for different work materials and toolholders can be supplied with several standard values of γ; $\gamma = -6°$ or $\gamma = +6°$ are often used.

In general, the power consumption is reduced by approximately 1% for each 1° increase in γ.

The *wedge angle* β determines the strength of the tool and its capacity to conduct heat, and depends on the values of γ and α.

The *clearance* α mainly affects the tool life and the surface quality of the workpiece. To reduce the deflections of the tool and the workpiece and to provide good surface quality, larger α values are required. For high-speed steel α values in the range 5–10° are normal, the smaller values being for the harder work materials. For carbides, the α values are slightly lower to give added strength to the tool.

Other important angles include the major cutting-edge inclination λ, which affects the direction of chip flow. The major cutting-edge angle κ characterizes the type of tool and directly influences the chip thickness. The higher the rigidity of the support, the smaller the values of κ that are applicable, and vice versa. The angle κ is often chosen in the range 20–60°. For small-diameter workpieces in turning, the value $\kappa = 90°$ is used.

The *corner radius* r has a major influence on the surface finish of the workpiece. Increasing r decreases the wear rate and provides better surfaces; it is often chosen in the range 0.5–3 mm. Inserts of carbides can be supplied with

r = 0.2, 0.4, 0.8, 1.2, 1.6, and 2.4 mm. For further information on tool geometry, see the literature [16–18] and ISO standards.

Tool Wear and Tool Life

During the cutting process, the tool wears and eventually fails to perform satisfactorily, causing loss in dimensional accuracy, increased roughness and power consumption, and even breakage of the cutting edge. The wear rate depends on the tool temperature, which depends mainly on the cutting speed, the hardness and type of tool material, type and condition of the work material, the tool geometry, the dimensions of the cut, and so on.

Tool wear is caused by the workpiece and the chip rubbing against the tool surfaces. The wear from the workpiece (i.e., the abrasive contact between the tool and the machined surface) is called the *flank wear* and is measured by the width of the land *VB* (see Fig. 7.9a). The wear from the chip (i.e., abrasive contact between the tool face and the chip that slides over it) is called *crater wear* or *cratering* and is measured by the crater depth *KT* and the crater width *KB* (see Fig. 7.9b).

The tool may also fail due to overheating, chipping of the cutting edge, plastic deformation, and thermal cracking. In general, most tools fail due to gradual wear of the flank or crater type. When the wear has reached a specified amount, determined from the workpiece requirements and the cutting performance, the tool must be reground or changed.

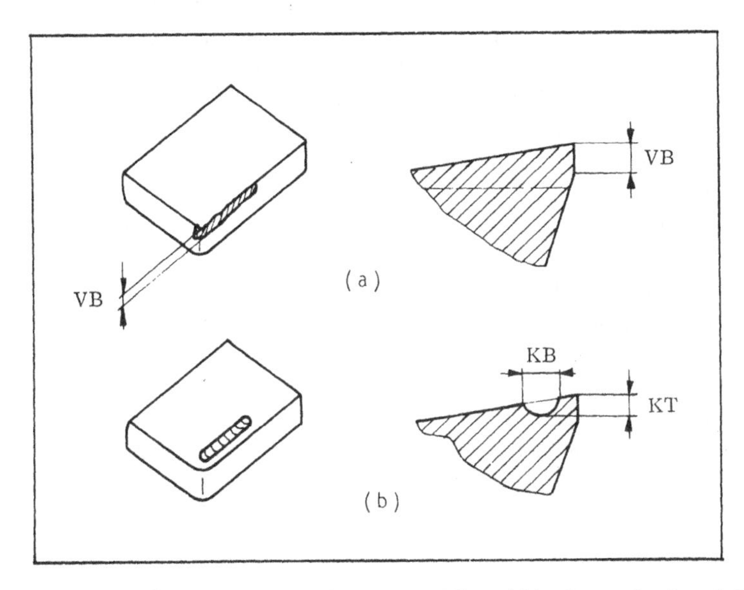

FIGURE 7.9 Tool wear: (a) flank wear (VB, width of wear land) and (b) crater wear (KT, depth; and KB, width of crater).

Wear or failure criteria may be defined as a maximum value of the width of the flank wear land (VB_{max} or a maximum value of the crater wear (KT_{max}, KB_{max}). The value VB_{max} is predominantly used as a wear criterion, but at high cutting speeds in the machining of materials that form continuous chips, it is sometimes necessary to include the crater wear in the criteria.

The way in which the flank wear varies with time is very important since it is the basis for the determination of the optimal (economical) cutting conditions.

If the flank wear VB is plotted against the cutting time T at different cutting speeds (v_1 through v_5) the curves shown in Fig. 7.10a are obtained for fixed values of tool geometry, feed, depth of cut, and corner radius. Using the flank wear criterion $VB = VB_{max}$, it can be seen that increasing cutting speeds yield decreasing tool life until tool failure. The time until $VB = VB_{max}$ is obtained is called the *Tool life T* corresponding to the cutting speed v. A logarithmic plot of tool life against cutting speed usually shows an approximate straight line within the practical range of cutting speeds (see Fig. 7.10b). This line can be expressed by the Taylor equation

$$vT^{-1/k} = C \tag{7.10}$$

where v is the cutting speed in m/min, T the tool life in minutes, k a material constant that defines the slope of the tool-life curve, and C is a constant.

If a certain tool life is desired, the permissible maximum cutting speed is found from the graph. For single-point tools Eq. (7.10) is influenced by the feed f, the corner radius r, the major cutting-edge angle κ, and the depth of cut a.

Figure 7.10c shows that to obtain a certain tool life, increasing the feed will necessitate a reduction of cutting speed. The speed, and to a lesser extent the feed, have a major influence on the tool life, and the effect of the corner radius, the major cutting-edge angle, and the depth of cut can be introduced as corrections to the cutting speed determined from Fig. 7.10d, *e*, and *f*. The corrections are generally relatively small and, consequently, as a first approximation, it is acceptable to neglect them.

Among other factors affecting the tool life are surface effects (strain hardening, etc.) of the work material and the type and condition of the tool material.

For different steels, the steel manufacturers normally provide data for Eq. (7.10) for various feeds together with correction tables for other influencing factors [18]. Today, several research institutions are continuously carrying out experimental work to supply the industry with accurate information, which is very important in the effective utilization of modern, high-capital-cost production equipment [17].

The Choice of Cutting Data

We shall now discuss how the choice of cutting data (cutting speed, feed, and depth of cut) can be made. The criterion normally used is cost (economical cut-

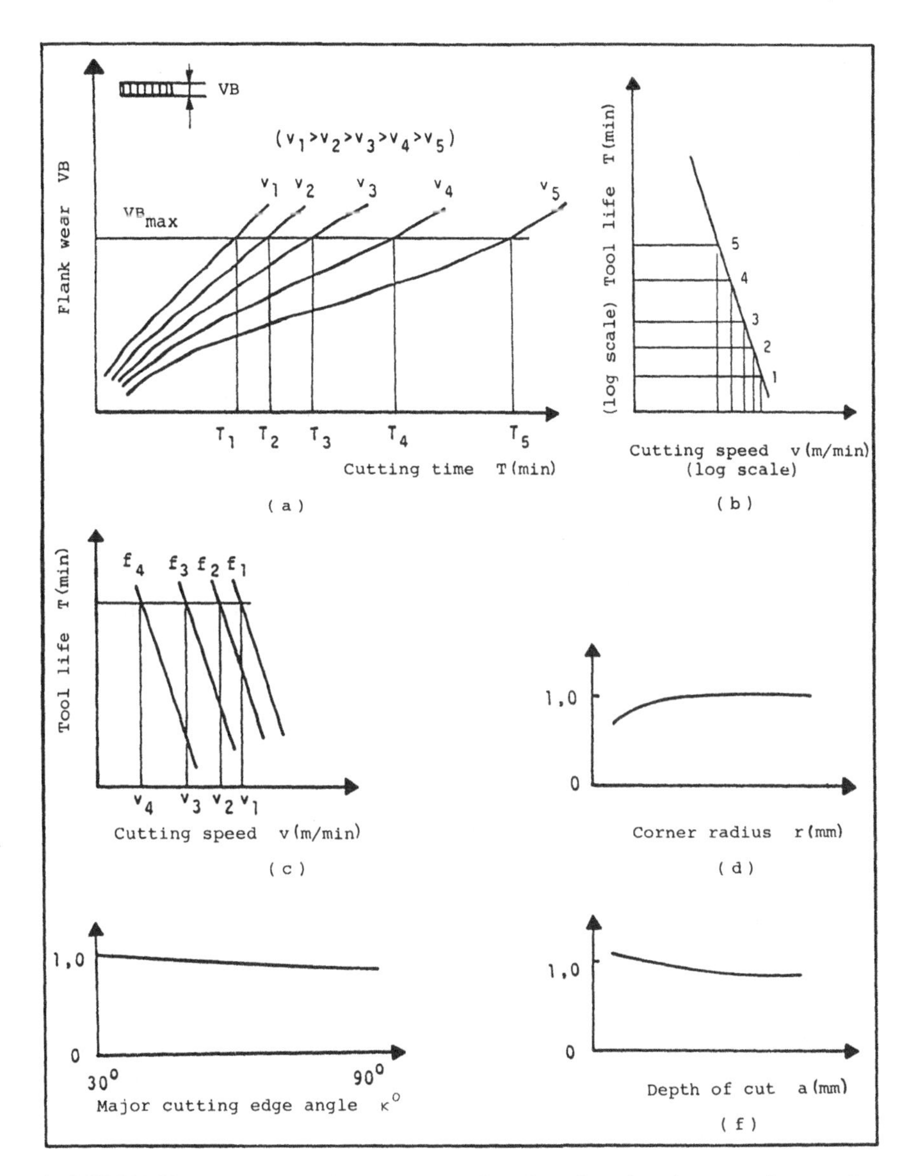

FIGURE 7.10 (a) Width of flank wear as a function of cutting time at various cutting speeds; (b) the tool life as a function of cutting speed; (c) the tool life as function of cutting speed at different feeds; (d), (e), and (f) corrections to the cutting speed, part (c), showing the influence of corner radius, major cutting-edge angle, and depth of cut.

ting data), but maximum production rate determined by the power available can also be used and does not give the same cutting data as the criterion of cost.

The economical cutting data are based on an economical tool life T_e determined by minimum-cost considerations. This can be expressed as the maximum of metal removed per unit cost. The metal removed can be expressed as [see Eq. (7.3)]

$$Q = AvT \tag{7.11}$$

where A is the area of cut, v the cutting speed, and T the tool life.

The cost can be written as

$$E = MT + Mt_{ct} + c_t \tag{7.12}$$

where M is the machine and operation rate, T the tool life, t_{ct} the tool changing time, and c_t the tool cost (including regrinding).

The metal removed per unit cost is

$$U = \frac{Q}{E} = \frac{AvT}{MT + Mt_{ct} + c_t} \tag{7.13}$$

By substituting $CT^{1/k}$ for v [the Taylor equation, (7.10)] and differentiating with respect to T, the economical tool life T_e is found:

$$T_e = -(1 + k)\left(\frac{c_t}{M} + t_{ct}\right) \tag{7.14}$$

k is often in the range -2 to -7.

The determination of T_e does not automatically lead to the desired cutting data. From Fig. 7.10 it can be seen that different combinations of cutting speed and feed can give any desired tool life T_e. The combination yielding the highest removal rate ($V = fav$ cm^3/min) is then chosen. If the removal rate V is plotted against the feed, the feed and cutting speed corresponding to the maximum removal rate for the desired T_e can be determined (see Fig. 7.11a). The subscript e for f and v refers to the economical tool life. Another approach is shown in Fig. 7.11b, where the cutting speed is plotted on a logarithmic scale against the feed for a desired tool life T_e. In the diagram straight lines are drawn, each representing a constant removal rate (V_1, V_2, . . .). The point where a removal rate line is tangential to the v_e, f_e curve determines the economical data.

This method cannot always be applied, because of low power available, insufficient rigidity of the tool and its support, surface requirements, and so on. If the power available is the limiting factor, it is generally preferable to reduce the cutting speed.

The machining of a component is often carried out in one or more roughing cuts and a finishing cut. The cutting data for the roughing cuts are usually determined from economical considerations, and the finishing cut is determined by the required surface quality (see pages 202–203).

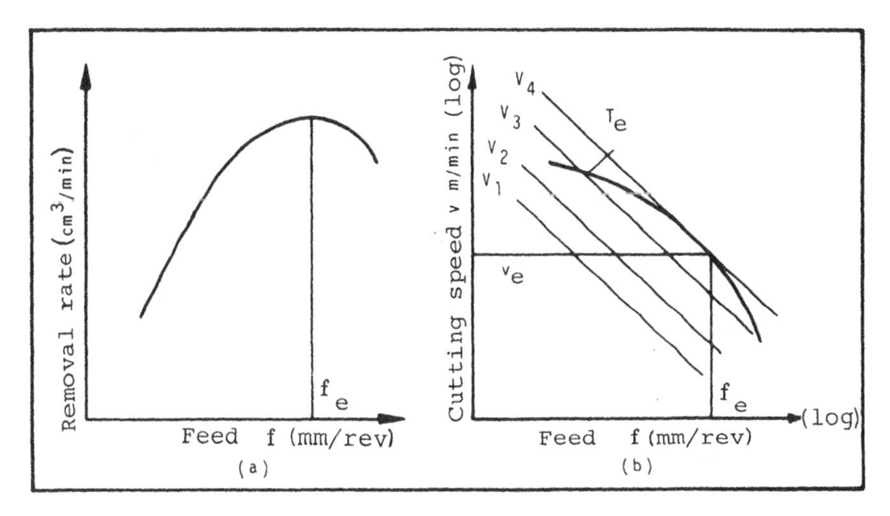

FIGURE 7.11 (a) The removal rate as a function of feed and (b) the cutting speed as a function of feed corresponding to the economical tool life T_e.

To show the general trends, typical values of cutting speed and feed are suggested next.

Cutting Speed (v m/min). For high-speed-steel tools, the cutting speeds for machining steel are in the range 20–50 m/min, depending on the feed and the desired tool life. For carbides, the cutting speeds for machining steel are in the range 80–350 m/min, depending on the quality of the carbides, the feed, and the desired tool life.

Feed (f mm/rev, mm/tooth). For most processes the feed is below 1.0 mm. The lower values are used in finish machining and the higher in rough machining. Large variations exist from process to process; for example, in single-point operations the feed might be 0.1–1 mm, in milling 0.01–0.1 mm, and in grinding less than 0.01 mm.

The Work Material

When an economical machining operation is to be established, the interaction among the geometry, the material, and the process must be appreciated. As mentioned previously, it is not sufficient to choose a material that fulfills the required functional properties; its technological properties describing the suitability of the material for a particular process must also be considered. In mass-conserving processes, the material must possess a certain ductility (formability), and in mass-reducing processes, it must have properties permitting machining to take place in a reasonable way. The technological properties describing the suitability of a material for machining processes are collectively called its *machinability*.

Machinability cannot be completely described by a single number, as it depends on a complex combination of properties which can be found only by studying the machining process in detail. The term *machinability* describes, in general, how the material performs when cutting is taking place. This performance can be measured by the wear on the tool, the surface quality of the product, the cutting forces, and the types of chip produced. In most cases, tool wear is considered the most important factor, which means that a machinability index can be defined as the cutting speed giving a specified tool life. Experiments show that machinability defined this way is different in turning, drilling, milling, and so on, and means that a machinability index must be ascribed for a particular process. When a component is to be machined by several processes, the machinability index corresponding to the process most used is chosen.

Machinability tests are carried out under standardized conditions (i.e., specified quality of tool material, tool geometry, feed, and depth of cut). The tool life for which the machinability index is quoted is generally 30 min (i.e., T_{30}) and the wear criterion is VB_{max} = 0.30 mm (see ISO Standard 3685).

The machinability of a material greatly influences the production costs for a given component. Poor machinability results in high costs, and vice versa. In Table 7.2 the machinability for the different material groups is expressed as the removal rate per millimeter depth of cut when turning with carbides [19]. The table can be used only as a general comparative guideline; in actual situations accurate values must be obtained for the particular material.

The machinability of a particular material is affected primarily by its hardness, composition, and heat treatment. For most steel materials, the hardness

TABLE 7.2 Removal Rate per Millimeter Depth of Cut for Different Groups of Materials When Turning with Carbides

Material	Removal rate per mm depth of cut (mm^2/min)
Constructional steel	47,000–63,000
Tool steel (annealed)	15,000–37,000
Stainless steel	17,000–43,000
Cast steel	20,000–27,000
Cast iron	13,000–23,000
Copper alloys	50,000–63,000
Brasses	60,000–70,000

Source: From Ref. 19.

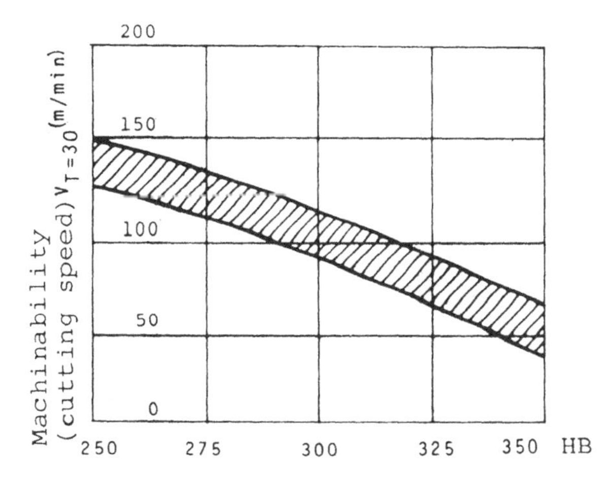

FIGURE 7.12 The influence of hardness on the machinability of hardened and tempered alloyed steel (0.35%C) in turning with carbides [18].

has a major influence on the machinability. A hardness range of HB from 170 to 200 is generally optimal. Low hardnesses tend to lead to built-up edge formation at low speeds. High hardnesses above HB = 200 lead to increased tool wear, as seen in Fig. 7.12, which gives the machinability as the cutting speed for a tool life of 30 min (T_{30}) for hardened and tempered alloyed steel.

Sometimes, it is preferable to accept a lower tool life when machining hard materials (HB from 250 to 330) instead of annealing and rehardening the material.

The composition of the work material has a direct influence on the machinability since the strength properties are affected. It has been found that some alloying elements that do not significantly affect the mechanical properties have a favorable effect on the machinability. The addition of, for example, 0.2% sulfur will increase the machinability of steel significantly. In some cases, however, because of functional requirements of high-temperature strength, for example, a sulfur content of only 0.01–0.05% is acceptable. Other alloying elements, such as lead, tellurium, selenium, and bismuth, have a similar influence on the machinability, but the disadvantage is that these materials are more difficult to produce.

The heat treatment of the work material can have a significant influence on its machinability. A coarse-grained structure generally has a better machinability than does a fine-grained structure. The distribution of pearlite and cementite has an influence, too, but this will not be discussed here. It should be mentioned, however, that hardened, plain carbon steels (>0.35%C) with a martensitic structure are very difficult to machine. Inclusions, hard constituents, scale, oxides, and so on, have a deteriorating effect on the machinability, as the abrasive wear on the cutting tool is increased.

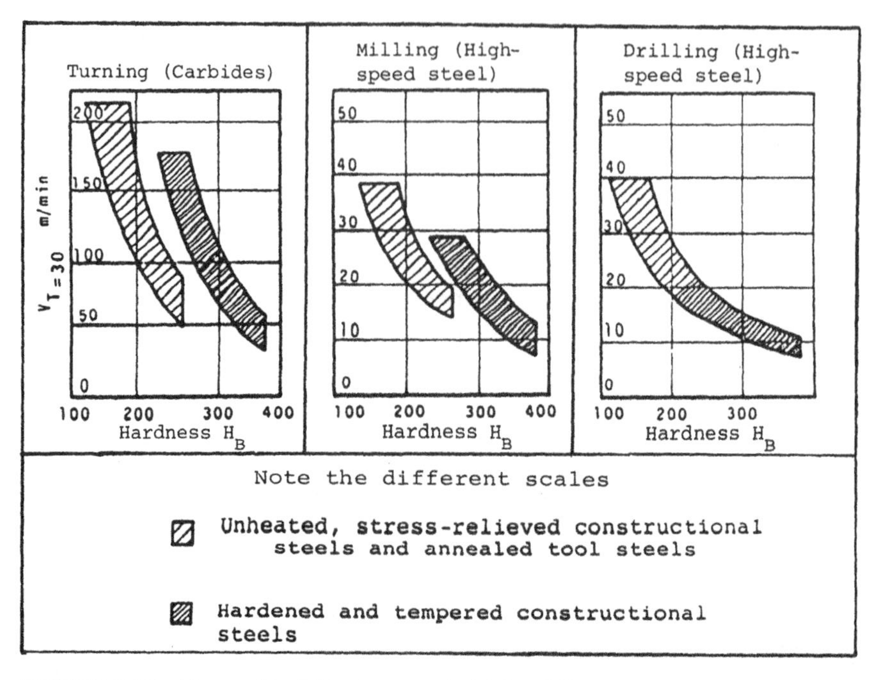

FIGURE 7.13 The machinability ($v_{T=30}$ m/min) for different material groups and processes [18].

Figure 7.13 shows machinability as a function of hardness for different material groups. The machinability is again defined as the cutting speed giving a tool life of 30 min. From the figure it can be seen that hardened and tempered materials—in spite of their higher hardness—have machinabilities approximately as high as the softer materials in turning and milling. In drilling, an increased hardness results in a poorer machinability.

The Surface Quality

In a machining process, a specific geometry is produced, which also implies that a surface of satisfactory quality must be produced. In this section a few definitions concerning the description of a surface are given, enabling an evaluation of the surface possibilities of the different processes.

A machined surface always deviates from the theoretical surface. The real surface looks like a mountain landscape (see Fig. 7.14).

The most important terms used in specifying a surface are waviness, lay, and roughness. *Waviness* is the recurrent deviations from an ideal surface and of relatively large wavelength (greater than 0.1 mm, for example). Such deviations generally result from deflections of the tool, workpiece, or machine, vibration,

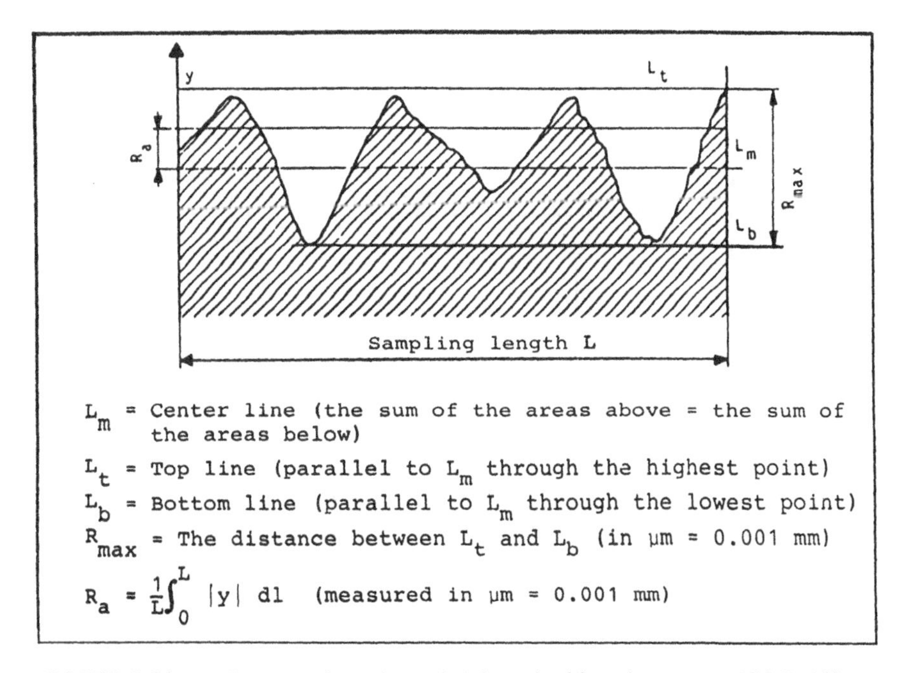

FIGURE 7.14 Definitions of roughness height and arithmetic average, ISO/R 468.

or warping, and means that the tool and the workpiece should be held rigidly with as little overhang as possible in order to minimize waviness. The *lay* is the direction of the predominant surface pattern produced by feed marks.

Roughness refers to the finely spaced irregularities or irregular deviations, and is shown in Fig. 7.14. The roughness is affected by the tool shape and the feed as well as the machining conditions.

Figure 7.14 gives a few definitions, but more detailed descriptions can be found, for example, in ISO/R 468.

Here the roughness will be described by the maximum height of the irregularities R_{max} and the arithmetical mean value R_a. The maximum height R_{max} is the maximum peak-to-valley height within the sampling length. The arithmetical mean value R_a is the average of the numerical deviations from the mean line of the surface within the sample length (see Fig. 7.14).

For surfaces with triangular irregularities, a relationship exists between R_{max} and R_a as follows:

$$R_a \simeq \frac{R_{max}}{4} \tag{7.15}$$

The relation can be used for approximate calculations.

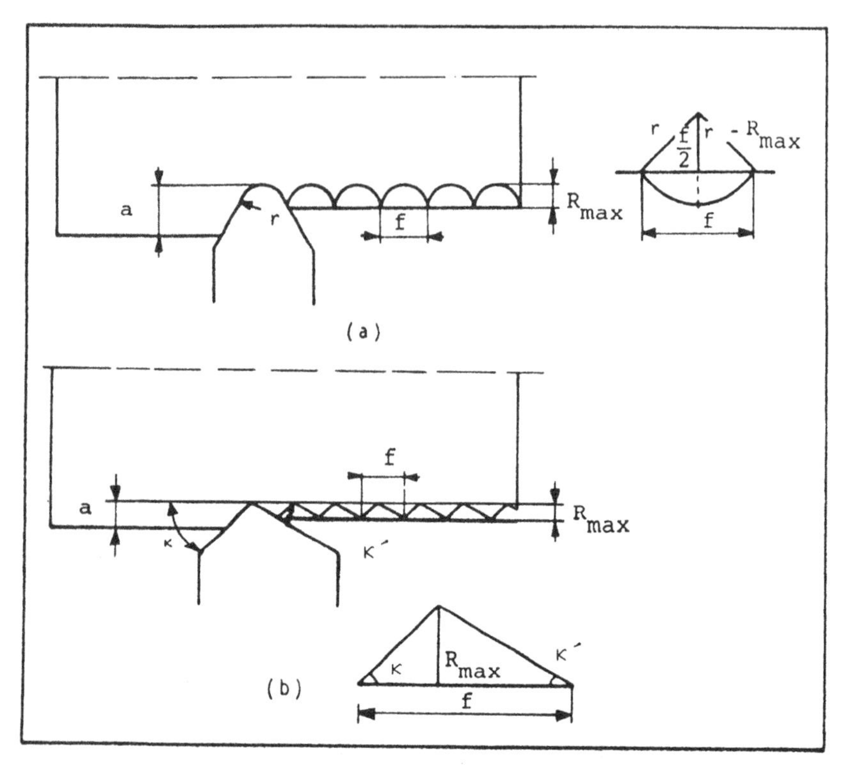

FIGURE 7.15 Turning: (a) the corner radius r of the tool and the feed determine the surface roughness; (b) the tool nose radius is so small that the major and minor cutting-edge angles determine the surface roughness.

The roughness of a machined surface is dependent primarily on the tool geometry and the feed. Figure 7.15 shows a turning process. In Fig. 7.15a the corner radius r of the tool and the feed f determine the surface roughness, and in Fig. 7.15b the roughness is determined by the major and the minor cutting-edge angles. From Fig. 7.15a, we have

$$R_{max} \simeq \frac{f^2}{8r} \tag{7.16}$$

By using the approximate relationship given by Eq. (7.15), R_a is found to be

$$R_a \simeq \frac{f^2}{32r} \tag{7.17}$$

Considering the situation shown in Fig. 7.15b, the calculation for R_{max} leads to

$$R_{max} = \frac{f}{\cot \kappa + \cot \kappa'} \tag{7.18}$$

And from Eqs. (7.15) and (7.18), R_a can be found:

$$R_a = \frac{f}{4(\cot \kappa + \cot \kappa')} \tag{7.19}$$

From these equations it can be seen that decreasing the feed gives a lower roughness value. A large corner radius r gives a low roughness.

Superimposed on these purely geometrical considerations are the effects of the cutting process, including the possible existence of a built-up edge on the tool. A built-up edge results in a rough surface. Since the tendency to produce a built-up edge is decreased for increasing cutting speeds, it might be expected that the surface roughness decreases by increasing cutting speed, and this is indeed the case. Furthermore, an effective cutting lubricant can reduce the surface roughness because it reduces the built-up edge.

The tool material has some influence, and ceramics and diamonds give the best surfaces.

Summarizing, it can be concluded that the roughness decreases (i.e., the surface quality improves) when the feed is decreased, the nose radius is increased, and the major cutting-edge angle κ and the minor cutting-edge angle κ' are reduced. Furthermore, increasing cutting speeds and effective cutting lubricants can improve the surface quality.

The preceding description related only to turning. Figure 7.16 shows a face milling operation where the diameter of the cutter is D and the feed is f per revolution. A calculation gives

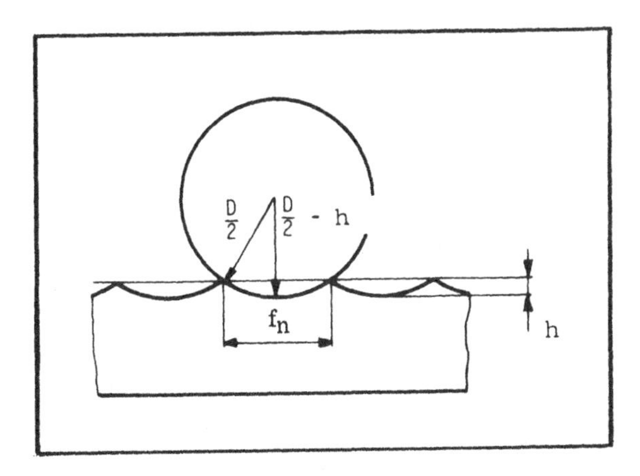

FIGURE 7.16 Roughness height in face milling.

TABLE 7.3 Typical Roughness Values (Arithmetical Mean Value R_a) for Different Processes

Process	Roughness (R_a) (μm)
Turning	3–12
Planing	3–12
Drilling	3–25
Milling	1–10
Grinding	0.25–3

$$R_{max} = \frac{f^2}{4D} \tag{7.20}$$

The calculation is based on the fact that only one tooth is responsible for the workpiece. As illustrative information, Table 7.3 gives typical roughness values (R_a) for different processes corresponding to normal workshop practice.

7.3 GEOMETRICAL POSSIBILITIES

An evaluation of the geometrical possibilities of a process must, as discussed in Chapter 1, be based on the surface creation principle (TF, ODF, TDF, FF), the pattern of motions for the work material and the tool, and the fundamental characteristics of the particular removal method (energy, medium of transfer, basic process). In this section only mass-reducing processes based on removal method I are discussed (i.e., the traditional machining processes). Examples of processes based on removal methods II, III, and IV are given in Section 7.4. An evaluation of the geometrical possibilities can be carried out considering the above-mentioned factors.

For the machining processes (removal method I), the tools (media of transfer) may be classified into two major groups:

1. Tools with well-defined edge geometry
   a. Single-point tools (one cutting edge)
   b. Multipoint tools (more than one cutting edge)
2. Tools with undefined edge geometry (multipoint random-edge-geometry tools)
   a. Grinding tools

A description of cutting edges was given in Section 7.2.2. Tools with well-defined cutting edges according to Section 7.2.2 are, in general, classified as tools with one major cutting edge (turning, planing, shaping tools, etc.), that is, single-point tools, and tools with more than one major cutting edge (drills, milling cutters, etc.), that is, multipoint tools.

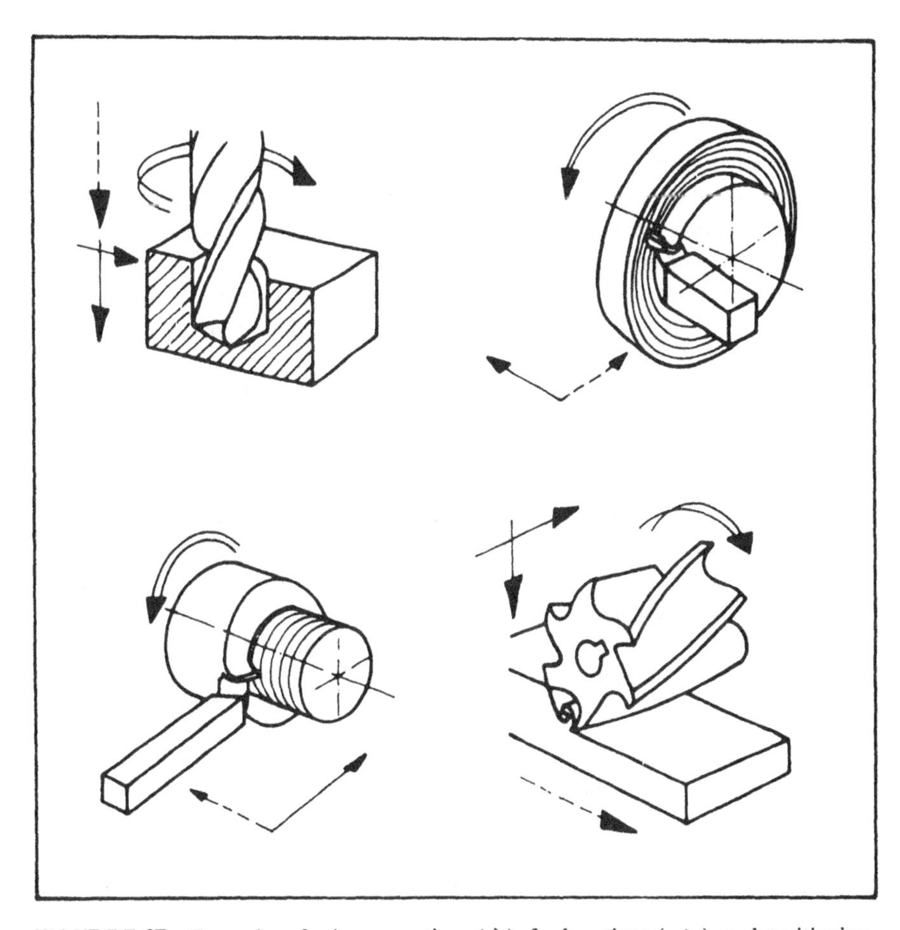

FIGURE 7.17 Examples of primary motions (⇒), feed motions (--->), and positioning or adjustment motions (→).

In tools with multipoint random-edge geometry, the description in Section 7.2.2 is not valid. Considering a grinding tool (or grinding wheel) which consists of many abrasive particles bonded together by a suitable bonding material (clays/ceramics, silicate of soda, shellac, rubber, resins, etc.) in the desired tool shape, the dull particles will fracture or be broken out, exposing new and sharp edges with random geometries. This occurs either during the machining process or during a special dressing process.

The different types of tools are described later in this section.

The surface having the desired component geometry may be produced by

Total forming (TF)
One-dimensional forming (ODF)

Two-dimensional forming (TDF)
Free forming (FF)

The applicable surface creation principle depends on the tool geometry and the characteristics of the removal method. Most machining processes are based on two-dimensional forming (TDF), but some are based on one-dimensional forming (ODF). When the geometry or structure of the tool has been chosen, the pattern of motions for the tool and the work material must be established. It must be emphasized here that the design of a process has an iterative character; that is, the tool geometry, the pattern of motions, and so on, to produce the desired geometry must be selected iteratively.

In the machining processes, three types or categories of motions exist: (1) the primary motion, (2) the feed motions, (3) the positioning or adjustment motions. If the primary motion is a rotation, the feed motion is continuous, and if the primary motion is linear, the feed motion is discontinuous. The positioning or adjustment motions enable the workpiece and the tool to be brought into the correct positions for the cutting process; this means setting the depth of cut. Depending on the type of machine, one or more of the positioning movements are integrated into the primary and feed motions.

Figure 7.17 shows examples of the three types of motions without relating them to machinery, which is done later. The types of motions available are:

Translations
Rotations
Combinations of translations and rotations
Stationary

Considering the primary and feed motions, the following general combinations are possible:

Pattern of motions			
Workpiece		Tool	
Primary	Feed	Primary	Feed
T	T	T	T
R	R	R	R
T/R	T/R	T/R	T/R
O	O	O	O

Every one of the four columns in this table has four different types of motion, yielding a total number of possible patterns of $4^4 = 256$. Quite a few of these are impractical, reducing the number somewhat, but generally the table allows for an imaginative utilization of the motion principles, enhancing the possibilities of new and favorable applications.

The geometrical possibilities in cutting with single-point tools, multipoint tools with well-defined edge geometry, and multipoint random-edge geometry tools are discussed separately in the following sections.

7.3.1 Single-Point Cutting Tools

Single-point tools are characterized by one major cutting edge with a well-defined geometry. The major cutting edge ends in a corner which constitutes the transition curve to the minor cutting edge. The major portion of the cutting process is carried out by a portion of the major cutting edge, while the corner and a portion of the minor cutting edge have a major influence on the final surface roughness.

The edge of a single-point tool normally contacts only a short line on the desired surface and, consequently, surface creation must be based on two-dimensional forming.

For single-point tools, the diagram in Table 7.4 shows examples of the patterns of motions of the workpiece and tool, the geometrical possibilities, and the processing machines. The pattern of motions is selected from the practical applications of combined motions in the table discussed above. These patterns of motions are provided by the particular processing machines or machine tools. It should be remembered that the machines possess positioning or adjustment motions in addition to the functional motions and defining the limits of geometry and size of the machined components.

Table 7.4 must not be considered a complete description of all the practical possibilities of machining with single-point tools, but only as an illustration of the principles based on important industrial examples.

Figures 7.18 and 7.19 show schematically the main elements and the basic pattern of motions for planers and shapers. These motions and the many possible shapes of the cutting tools provide or describe the geometrical possibilities of the processes. The tools can be of a wide variety of shapes, restricted only by the selection of the correct angles and faces for the particular material and sufficient strength of the tool and rigidity and power of the machine to produce a satisfactory component.

The different appearance of shaping and planing machines is due to the fact that shaping is intended for the machining of relatively small parts, whereas planing is intended for the machining of large parts. It is impractical to build shapers with ram strokes and multiple table motions long enough to machine large parts.

TABLE 7.4 Pattern of Motions, Geometrical Possibilities, and Machine Tools for Single-Point Tools

Pattern of motion[a]				Cutting with single-point tools[b]	Processing machine (machine tool)
Workpiece		Tool			
Primary	Feed	Primary	Feed		
—	—	T	(T)		Chisel (hand)
T	—	—	T	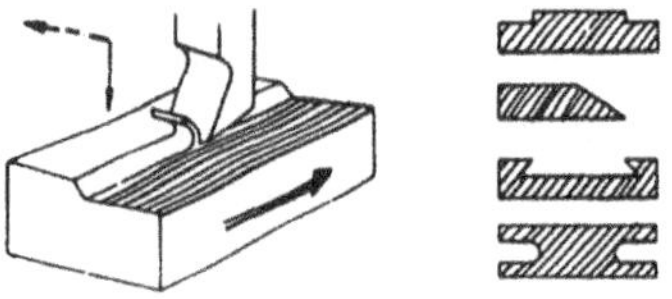	Planer
—	T	T	—	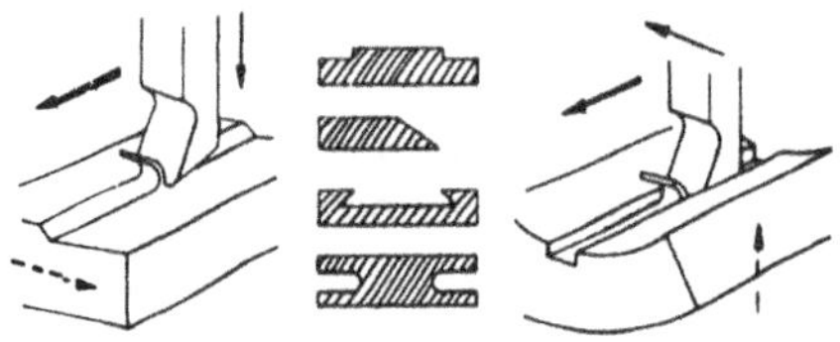	Shaper Slotter
—	—	R	T	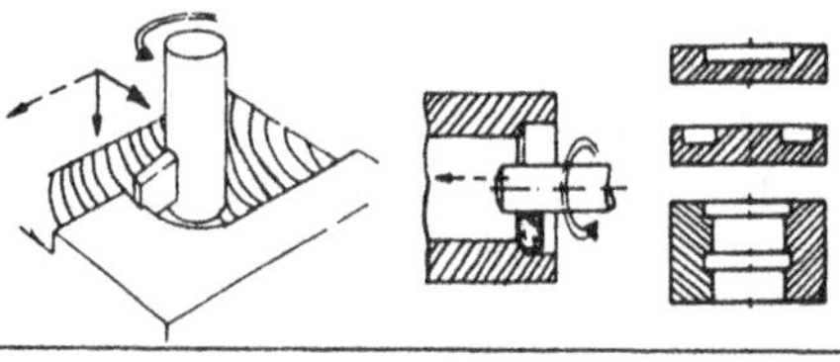	Horizontal boring machine
—	T	R	—	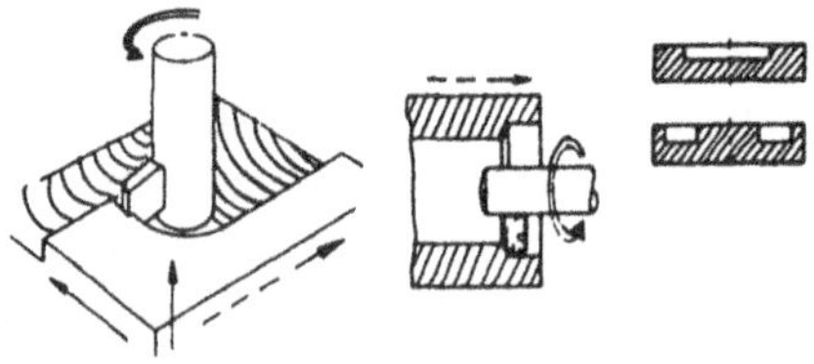	Horizontal boring machine
R	—	—	T	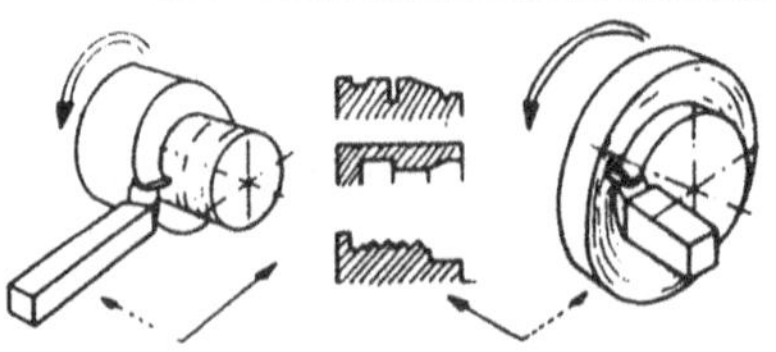	Lathe Boring mill

[a]R, rotation; T, translation.

[b]⇒, primary motion; -->, feed motion; →, adjustment motion.

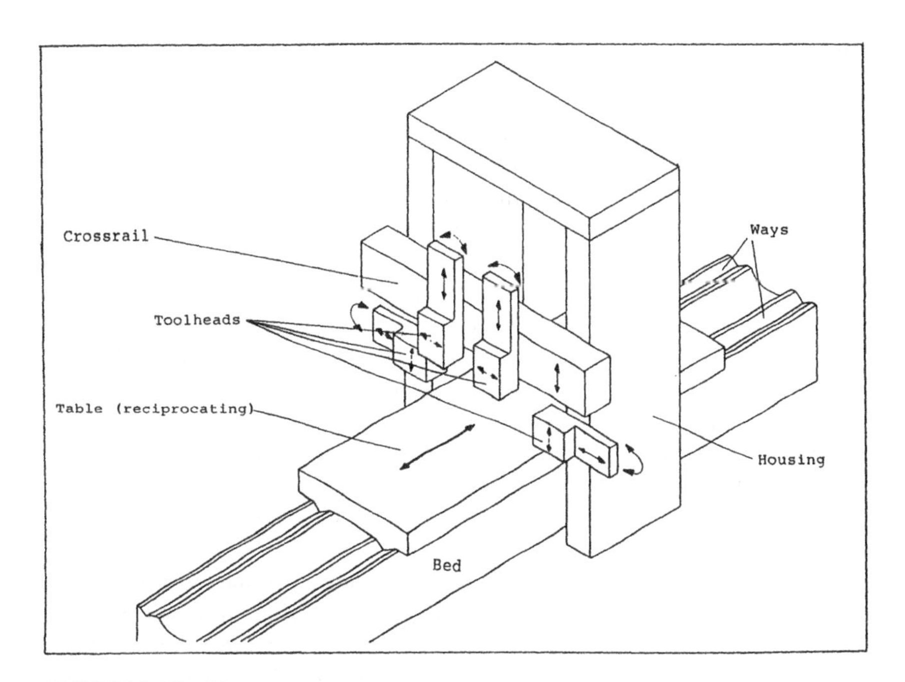

FIGURE 7.18 Planer.

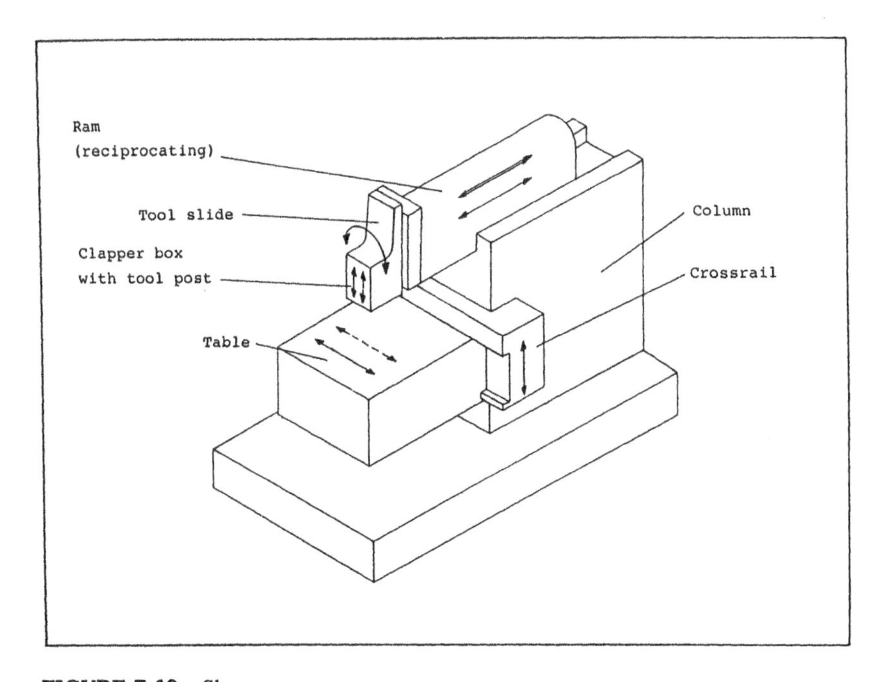

FIGURE 7.19 Shaper.

The examples in Table 7.4, where a horizontal boring machine is the appropriate machine tool, are discussed later in this section as well as the pattern of motions of the horizontal boring machine (Fig. 7.27).

The last example in Table 7.4, where the workpiece has a rotary primary motion and the tool a translatory feed motion, represents turning. Turning is carried out on a lathe, and the pattern of motions is shown in Fig. 7.20.

A special variant of the lathe is the vertical boring mill, the pattern of motion for which is shown in Fig. 7.21. Lathes (Fig. 7.20) are the most frequently used machines in industry, and they are available in a wide range of sizes. A selection of typical turning tools is shown in Fig. 7.22. Cutting takes place on the side of right-hand, left-hand, and facing tools. For round-nose, cutoff, finishing, and threading tools, cutting takes place near the corner of the tool. The tool materials are most often carbides (including the coated ones) and ceramics.

The pattern of motions (Fig. 7.20) and the geometry of the tools (Fig. 7.22) yield a tremendous range of geometrical capabilities for the turning process. A transverse adjustment motion of the tailstock permits tapers to be turned. Short tapers can be produced by swiveling the compound rest the desired angle, locking it, and feeding the rest manually.

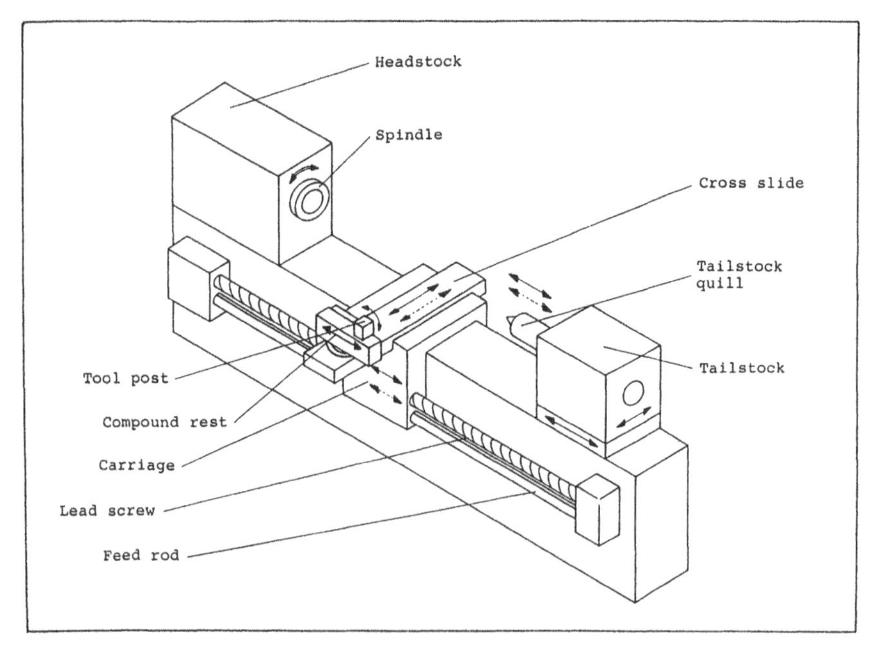

FIGURE 7.20 Lathe.

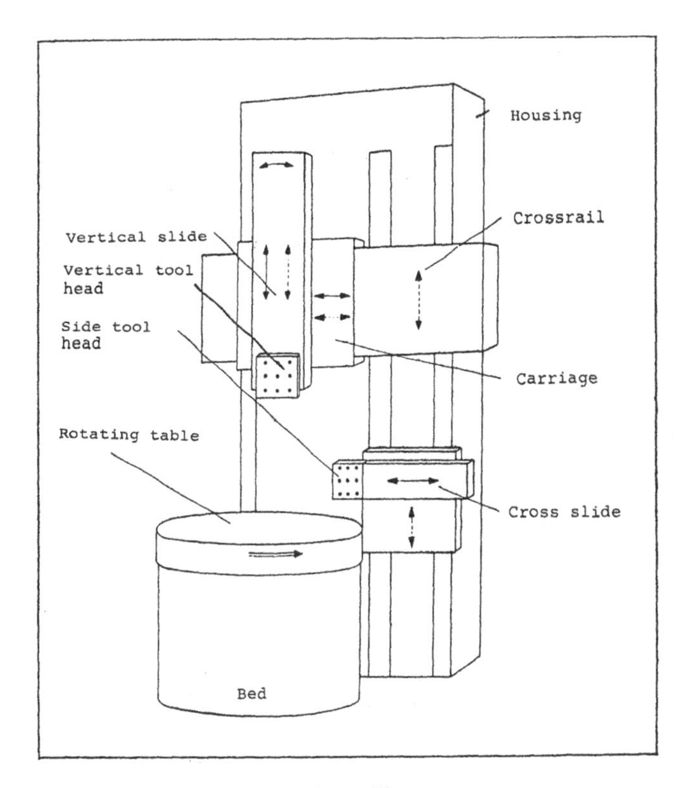

FIGURE 7.21 Vertical boring mill.

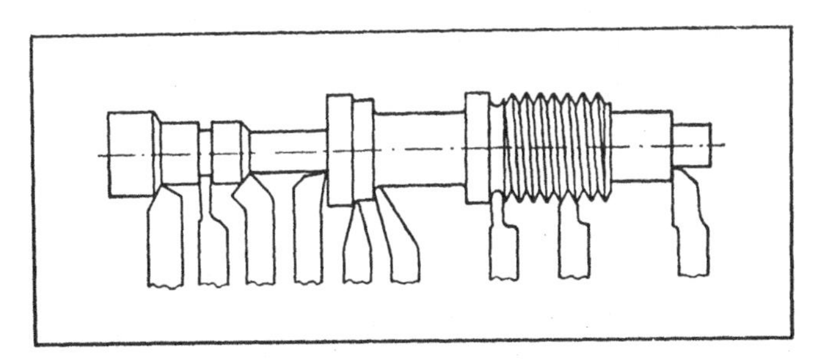

FIGURE 7.22 Examples of typical turning tools.

If heavy and large workpieces are to be machined, the horizontal lathe is impractical. Therefore, the vertical boring mill, which can be considered as a vertical lathe, has been developed (see Fig. 7.21).

A wide range of special machines utilizing single-point cutting tools are available, some permitting simultaneous cutting with several tools. A description of all the different types of machinery is beyond the scope of this book, but it is strongly recommended that journals and magazines be studied and tours around workshops taken, to acquire broad knowledge of the various types of machine tools.

7.3.2 Multipoint Cutting Tools

Multipoint cutting tools have more than one major cutting edge of well-defined geometry. These edges can be arranged in different geometrical patterns giving a wide spectrum of tool or cutter shapes (see Fig. 7.23).

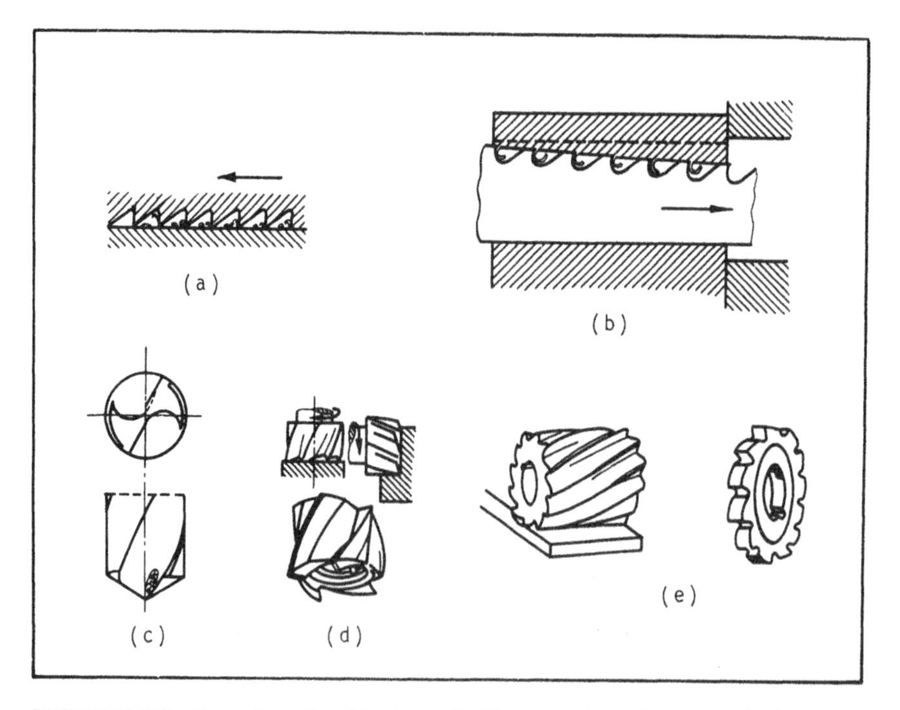

FIGURE 7.23 Examples of multipoint tools illustrating how the geometrical arrangement of the cutting edges allows the design of a wide spectrum of cutter shapes: (a) files and saw blades; (b) broaching tools; (c) twist drills; (d) end mills; (e) plain milling cutters, metal-slitting cutters, and so on.

If the cutting edges are arranged after each other in a straight line, files, saw blades, and so on, are obtained (see Fig. 7.23a). The tool angles must be appropriate to the particular work material or group of materials. The feed is provided to the work material or to the tool.

When cutting edges are arranged in a straight line, inclined at a certain angle to the primary motion (i.e., successive edges protruding farther from the line than the previous ones), which means that the feed is built into the tool, the tool is a broaching tool (see Fig. 7.23b). In practice, the feed (per tooth) varies from about 0.15 mm for roughing teeth to 0.02 for finishing teeth. The feed depends on several factors: for example, the strength of the teeth, rubbing tendencies, and the work material.

The major cutting edges can be arranged at the end of a circular bar, inclined at a certain angle to the axis of the bar, yielding twist drills (see Fig. 7.23c).

If the cutting edges are arranged at the end and around the circumference of a cylindrical bar, plain end mills, shell end mills, side milling cutters, and so on, are obtained (see Fig. 7.23d). If the edges are arranged only on the circumference of the cylindrical body, plain milling cutters, metal slitting saws, T-slot cutters, and so on, are obtained (see Fig. 7.23e).

The examples in Fig. 7.23 illustrate principles only, and, as one can imagine, a huge variety of tool shapes are available, many of them as standard tools. It should be mentioned that the use of inserted-tooth tools for throwaway tungsten or titanium carbide (eventually coated) inserts is widespread and increasing. In the determination of the practical pattern of motions for the production of specific components, the orientation of the cutting edges must be appropriate for the intended direction of cutting. But, as the examples show, the geometrical capabilities determined by the shape of the tools and the pattern of motions are tremendous.

Table 7.5 shows, for multipoint tools, examples of the patterns of motion for the workpiece and tool, the geometrical possibilities, and the processing machines or machine tools. In the first three examples the cutting edges are arranged in a straight line either parallel to or inclined to the cutting direction, and in the remaining examples the tool body is cylindrical with the cutting edges arranged in a variety of ways.

The machine tools are basically grouped into drilling machines (Fig. 7.24), plain column-and-knee type milling machines (Fig. 7.25), and horizontal boring machines (Fig. 7.27). Vertical drilling machines (Fig. 7.24) are typical for heavy-duty drilling, but many variants exist, including bench type, radial, multiple-spindle, deep-hole, and transfer drilling machines. In all these machines, different tools—requiring the same pattern of motions and range of power as twist drills—can be used.

Figure 7.25 shows the pattern of motions for plain column-and-knee-type milling machines, which are general-purpose machines. They are primarily

TABLE 7.5 Pattern of Motions, Examples of Geometrical Possibilities, and Machine Tools for Multipoint Tools

Pattern of motion[a]					
Workpiece		Tool			
Primary	Feed	Primary	Feed	Cutting with multipoint tools[b]	Processing machine (machine tool)
—	—	T	—	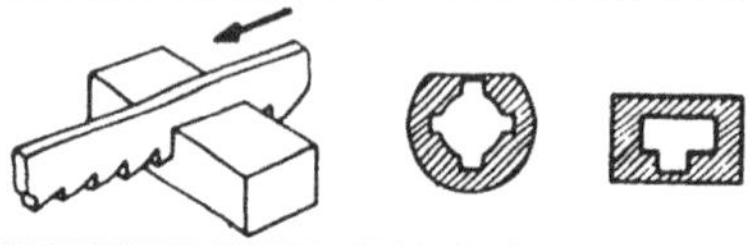	Broaching machine
—	—	T	T	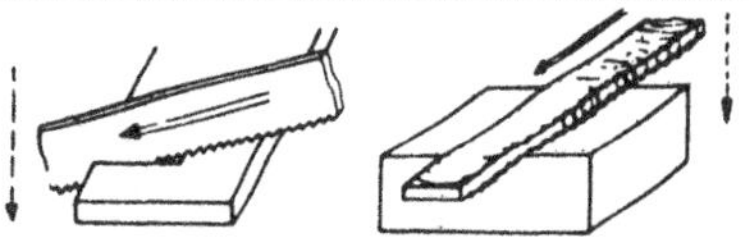	(Hand) Saw File
—	T	T	—	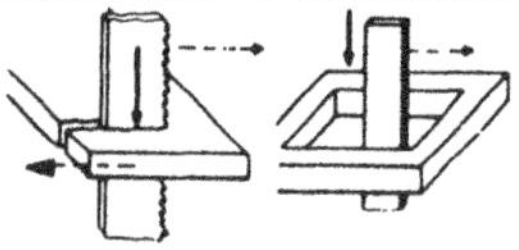	Band-sawing machine Filing machine
R	—	—	T	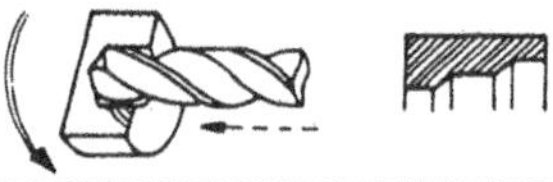	Lathe Boring mill
—	—	R	T	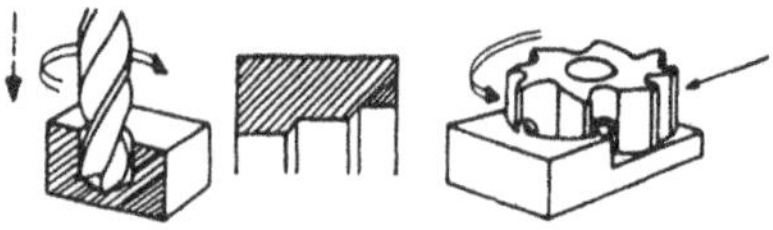	Horizontal boring machine Drill press
—	T	R	—	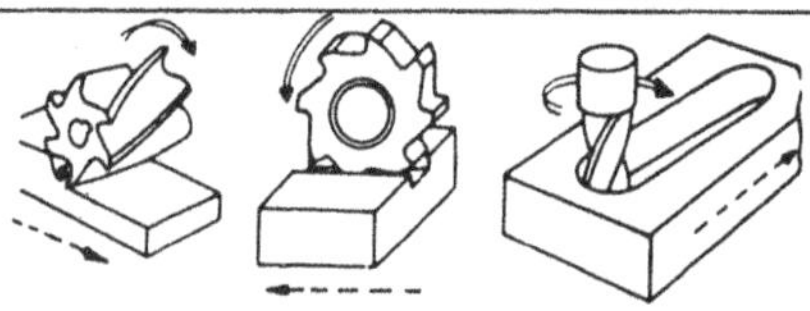	Milling machine Rotary cutoff saw
—	R	T	—	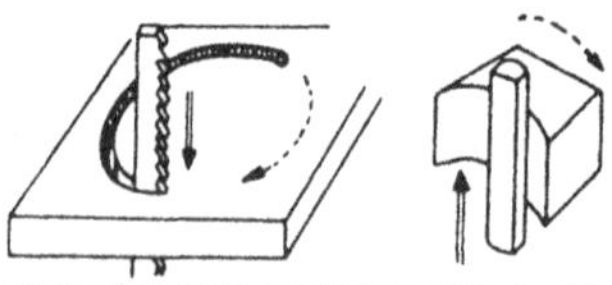	Band saw Filing machine
—	R	R	—	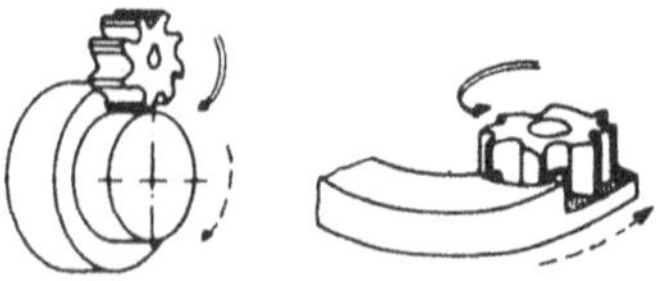	Milling machines

[a]R, rotation; T, translation.

[b]⇒, cutting motion; --->, feeding motion; →, adjustment motion.

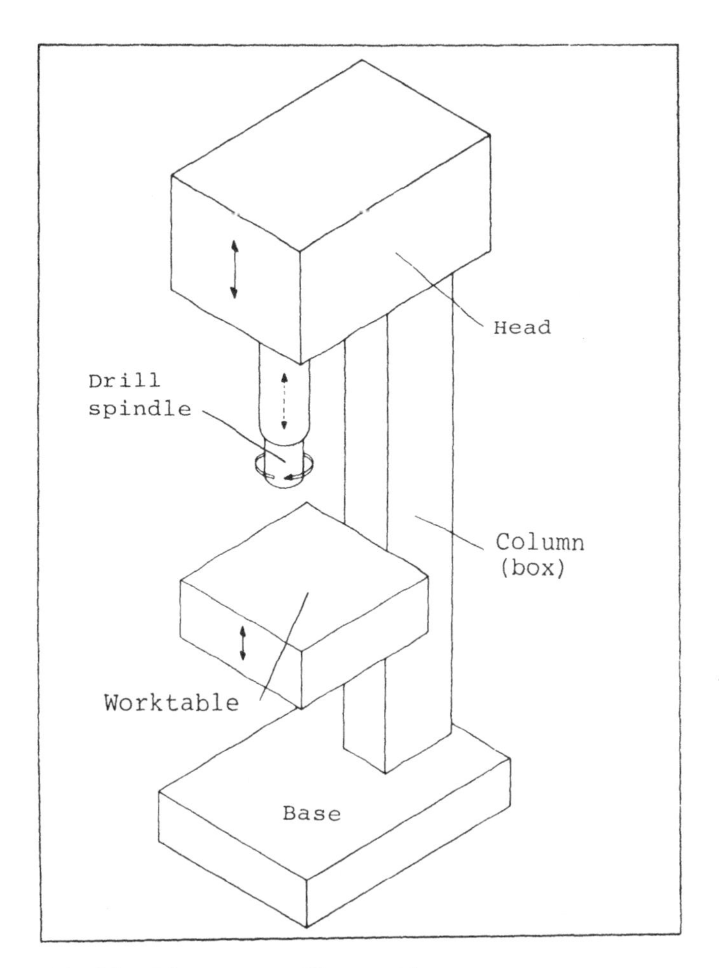

FIGURE 7.24 Vertical drilling machine.

designed for arbor-mounted cutters (see Fig. 7.26a) and have a high degree of versatility and control as well as good productivity. Shank-mounted cutters can also be used to a limited extent in these machines (see Fig. 7.26b). Figure 7.26 shows typical milling cutters, but they represent only a small portion of the many shapes available. Milling machines can also be used for drilling and boring. Many different milling machines are on the market; examples are universal column-and-knee-type milling machines (plain column-and-knee-type milling machines supplied with a swivel on the saddle, enabling helices to be cut when swivelling the work table), ram-type milling machines, bed-type milling machines, and planer-type milling machines. Milling machines are among the most important machine tools, as they can produce a wide variety of machined surfaces.

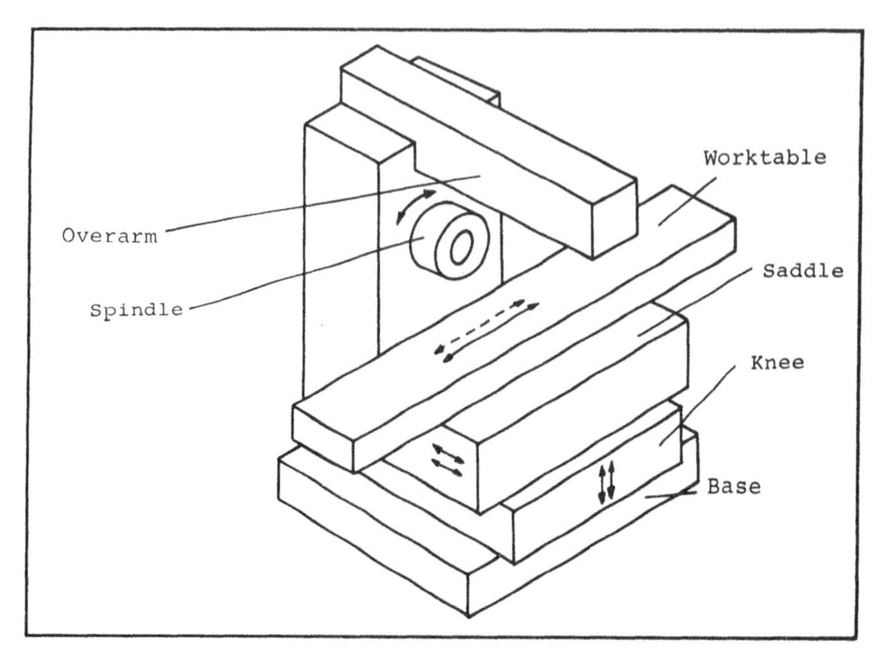

FIGURE 7.25 Plain column-and-knee-type milling machine.

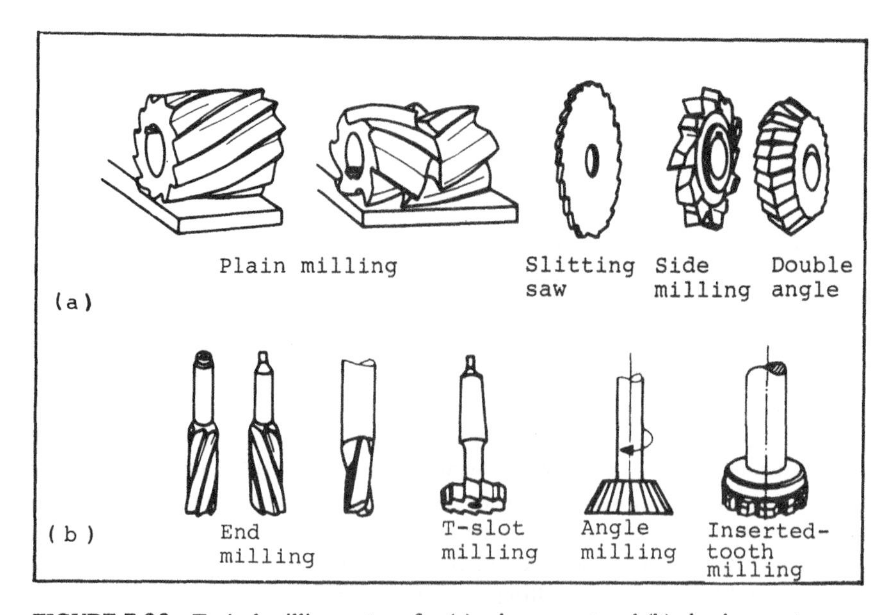

FIGURE 7.26 Typical milling cutters for (a) arbor-mount and (b) shank-mount.

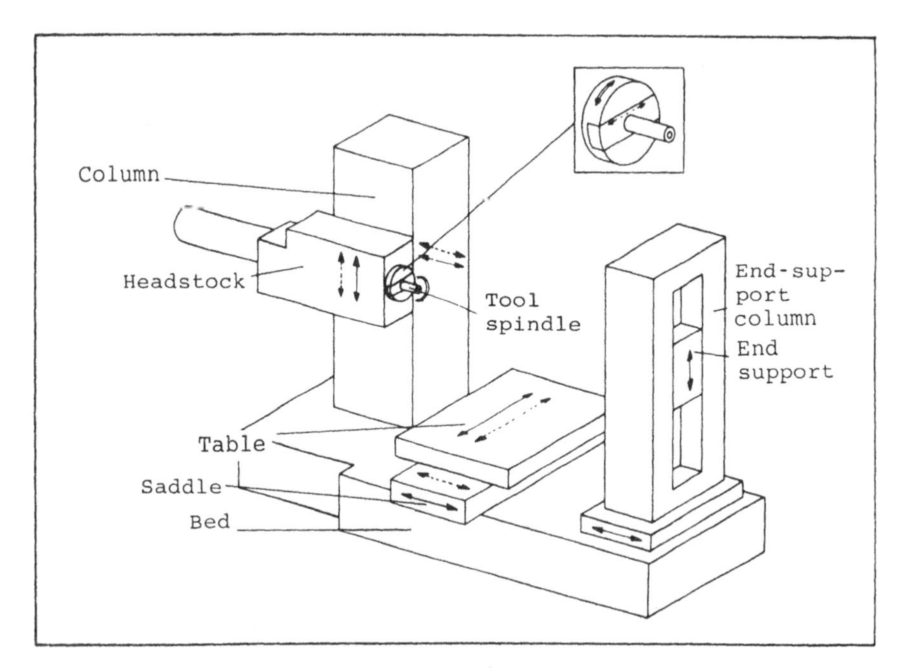

FIGURE 7.27 Horizontal boring (drilling and milling) machine.

Figure 7.27 shows the pattern of motions for horizontal boring, drilling, and milling machines. Boring is carried out by a single-point tool, mounted either in a stub-type bar or a long boring bar supported in a bearing in the end support. Most shank-mounted milling cutters can be used in this machine, but because of the spindle construction, drills and milling arbors can be mounted, permitting arbor-mounted cutters to be used.

It is possible to mount a turret head on the spindle, permitting a quick tool change, since the turret head can contain many tools.

Several types of boring machines are available, but a description of these will not be given here. However, in recent years, many types and sizes of numerically controlled machining centers have become available. These centers are variants of or substitutes for the versatile boring, drilling, and milling machines, and they are equipped with a storage magazine that may hold up to 50 or more different tools, enabling quick tool changes (5–10 s). These machines are generally used for small and medium lot sizes.

7.3.3 Multipoint Random-Edge-Geometry Tools

As described previously, the random-edge-geometry tools consist of abrasive particles embedded in a suitable bonding material in the desired tool shape.

TABLE 7.6 Pattern of Motions, Examples of Geometrical Possibilities, and Machine Tools for Grinding with Multipoint Random-Edge Geometry Tools

Pattern of motion[a]					
Workpiece		Tool			
Primary	Feed	Primary	Feed	Cutting (grinding) with multipoint random-edge geometry tools	Processing machine (machine tool)
	T	T		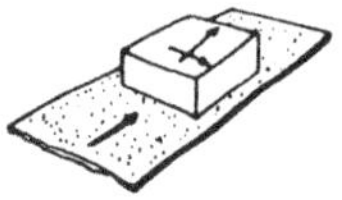	Belt grinder
	T	R		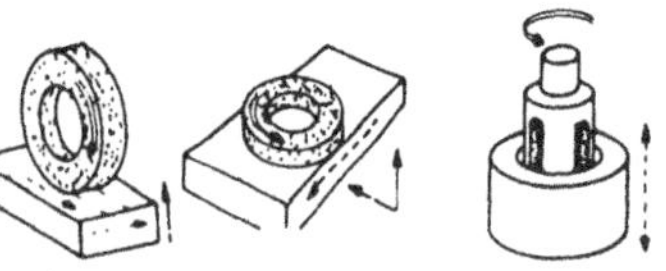	Surface grinder (horizontal or vertical) Honing machine
	R	R			Surface grinder (vertical) Lapping machine (rotary)
	R/T	R		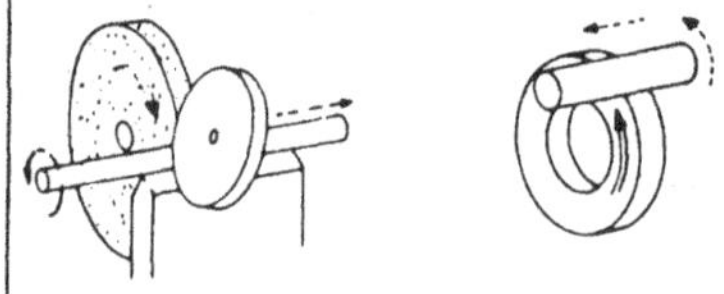	Centerless grinding machine Cylindrical grinding machine (plain center type)
	R	R	T		Surface grinder Internal grinder
	R	T		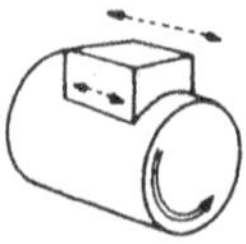	Superfinishing machine

[a]R, rotation; T, translation.

One of the most characteristic processes based on random-edge-geometry tools is grinding. Many edges are cutting simultaneously and have very short cutting times.

The grinding tools are available in many geometrical shapes, such as wheels, segments, bands, and so on. The most frequently used grinding tool is the grinding wheel used for cylindrical or plain grinding. Grinding offers close dimensional control and fine surface finishes and has become extremely important in recent years, because of the increasing demands of high accuracy and surface quality. Formerly, grinding was used only for finishing operations, but a rapid development is taking place with regard to roughing (high-speed) grinding, which may substitute for turning, for example.

Table 7.6 shows examples of the pattern of motions (of workpiece and tool), the geometrical possibilities, and the machine tools. Figure 7.28 shows the pattern of motions of a plain center-type cylindrical grinder. The many other types available will not be described here. It is recommended that relevant literature sources be studied and visits made to machine shops.

The geometry of grinding tools and the pattern of motions of the machines provide many geometrical possibilities in obtaining high dimensional accuracy and fine surface qualities.

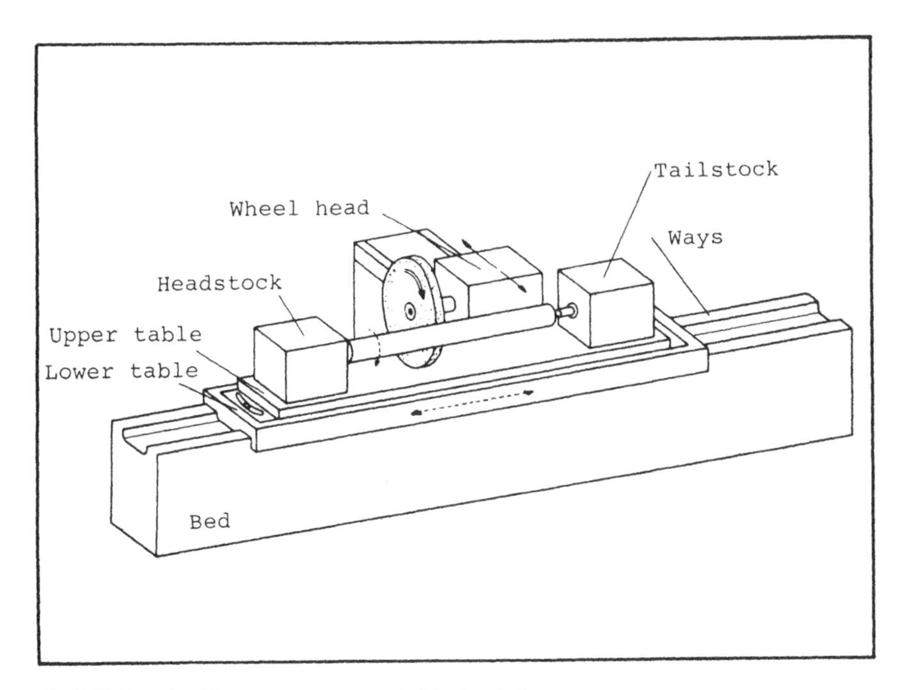

FIGURE 7.28 Plain center-type cylindrical grinder.

7.4 EXAMPLES OF TYPICAL MASS-REDUCING PROCESSES

In this section short descriptions of some of the most frequently encountered mass-reducing processes used in industry are given. The processes will be classified in a similar way to that used in Section 6.3 according to the category of basic process, type of energy, transfer medium, surface creation principle, and state of stress. The abbreviations used are the same as in Section 6.3.

The field of mass-reducing processes is huge, containing the conventional machining processes, blanking and punching, shearing, electrodischarge machining, electrochemical machining, and so on. It is beyond the scope of this book to describe them all and, consequently, only typical processes will be described. Further information must be sought in the literature. It should, however, be emphasized that examples illustrate only the basic principles discussed in Chapter 1, providing a background for imaginative and practical applications.

PROCESS 1: Turning (M, Me, Ri, TDF, Sh)

Description. The turning process is characterized by solid work material, two-dimensional forming, and a shear state of stress. The workpiece (*W*) is supported [e.g., clamped in a chuck (*C*) and supported by a center] and rotated (the primary motion, *R*). Through the primary motion (*R*) and the translatory feed (T_a = axial feed for turning and T_r = radial feed for facing) of the tool (*V*) the workpiece is shaped.

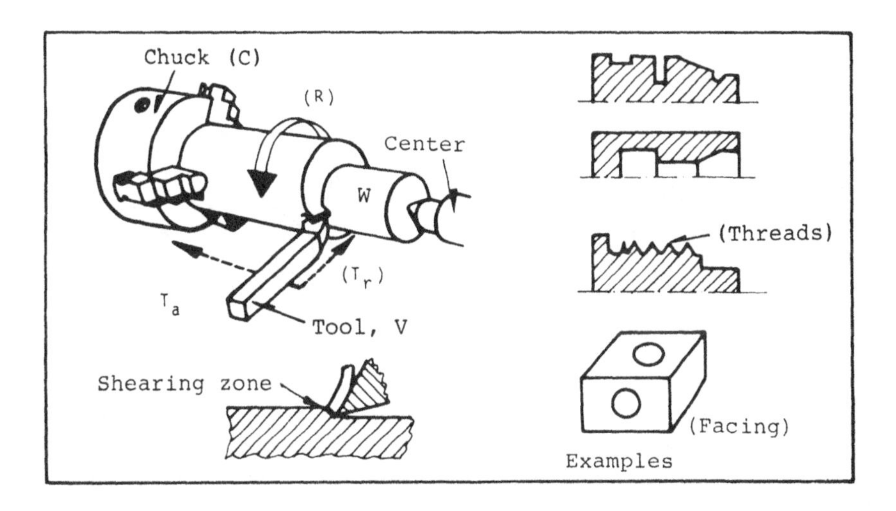

Applications. Turning is used primarily in the production of various cylindrical components with a nearly unlimited number of external and internal axial cross-sectional shapes (including tapers, threads, etc.). Facing is used for both regular and irregular shapes. Turning is the most extensively used industrial process.

Material Requirements. The material should not be too hard (HB $<$ 300) and should possess a minimum of ductility to confine deformation mainly to the shear zone.

Tolerances/Surfaces. Turning provides close tolerances, often less than ± 0.01 mm. Tighter tolerances may be obtained. The surface quality is good, normally in the range $3 \leq R_a \leq 12$ μm.

Machinery. A wide variety of lathes are on the market: for example, the engine lathe, the turret lathe, single- and multispindle screw machines, automatic lathes, and NC lathes.

PROCESS 2: Milling (M, Me, Ri, TDF, Sh)

Description. The milling process is characterized by solid work material, two-dimensional forming (one-dimensional forming may be used in a few cases), and a shear state of stress. The workpiece (*W*) is clamped on the table (*B*), which is given a translatory feed (*T*), that together with the primary motion (*R*) of the cutter (*V*) provides the many geometrical possibilities. Many shapes of cutters are available.

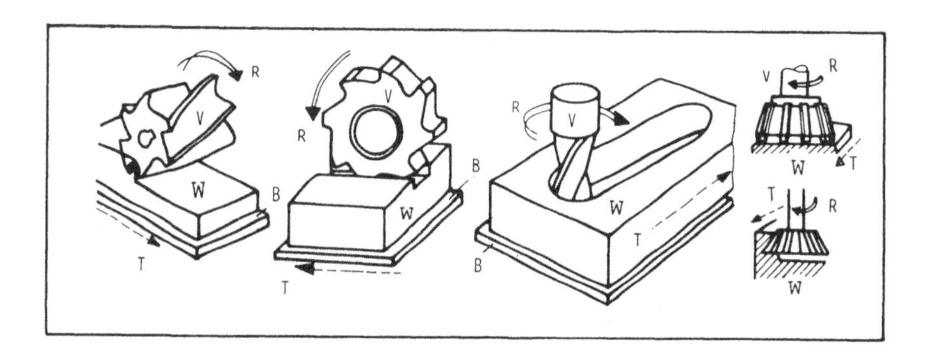

Applications. The milling process—through the various types of cutters and the wide variety of machines—is a versatile high-production process. Through various accessories (dividing head, attachments, etc.) many different special shapes can be produced. The milling process comes close to turning in extensive industrial use, since the geometrical possibilities are enormous and the removal rate high.

Material Requirements. The hardness of the material should not be too high (HB < 250–300) and a minimum of ductility is advisable.

Tolerances/Surfaces. The obtained tolerances are normally good ($\simeq \pm 0.05$ mm) and the surface quality high, $1 \leq R_a \leq 10$ μm.

Machinery. A wide variety of milling machines are available: for example, the plain column-and-knee type (general purpose), universal column-and-knee type, bed-type, and planer type.

PROCESS 3: Drilling (M, Me, Ri, TDF, Sh)

Description. The drilling process is characterized by solid work material, two-dimensional forming, and a shear state of stress. The workpiece (*W*) is clamped on a table (*B*) and the tool (*V*) is given a rotation (the primary motion, *R*) and a translatory feed (*T*). In drilling on lathes, the workpiece is rotated and the feed is applied to the tool.

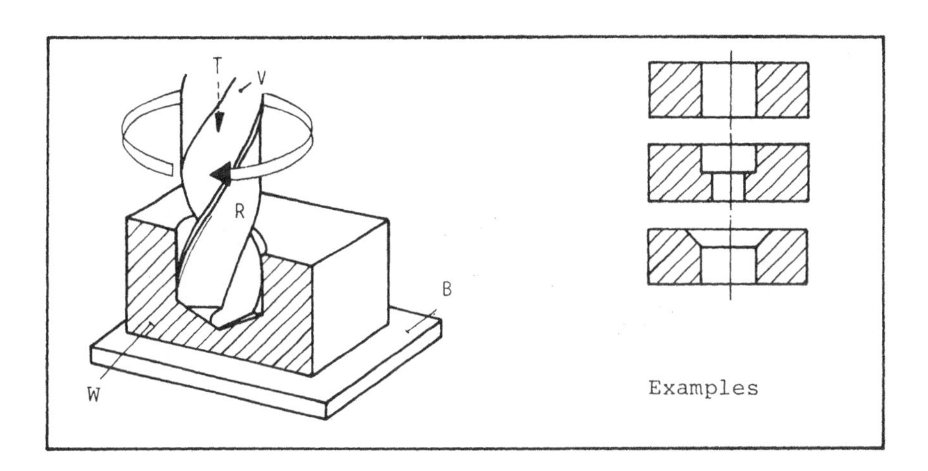

Applications. The drilling process is primarily used to produce interior circular, cylindrical holes. Through various tools (twist drills, combination drills, spade drills, etc.) different hole shapes can be produced (cylindrical holes, drilled and counterbored, drilled and countersunk, multiple diameter holes, etc.). Drilling is an important industrial process.

Material Requirements. The hardness of the material should normally not exceed HB = 250.

Tolerances/Surfaces. For diameters less than 15 mm, the normal tolerance is around ±0.1 mm. Larger holes often have tolerances around ±0.3 mm. Finer tolerances may be obtained, but finishing is often carried out by a special reaming process. The surface roughness is typically $3 \leq R_a \leq 25$ μm.

Machinery. Many types of drilling machines are available: for example, bench, upright, radial, deep-hole, and multispindle drilling machines.

PROCESS 4: Planing (M, Me, Ri, TDF, Sh)

Description. The planing process is characterized by solid work material, two-dimensional forming (one-dimensional forming may be used), and a shear state of stress. The workpiece (W) is clamped on the table (B), which is given a translatory primary motion (T_B), and the tool (V) is given a translatory feed (T_V), providing the geometrical possibilities.

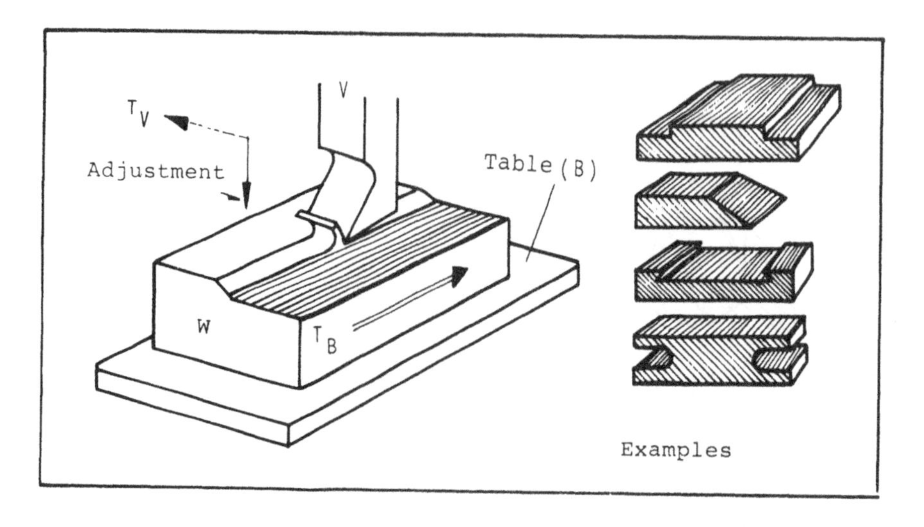

Applications. The planing process is, in general, used to produce large horizontal, vertical, or inclined flat surfaces (e.g., ways, beds, etc.).

Material Requirements. The hardness should generally not exceed HB = 300, and a minimum of ductility is advisable.

Tolerances/Surfaces. Normally, quite good tolerances can be obtained, ±0.05 to ±0.10 mm. The surface roughness is in the range $3 \leq R_a \leq 12$ μm.

Machinery. Different types of planers are available: for example, double housing planers, open-side planers, edge or plate planers, and pit-type planers.

PROCESS 5: Shaping (M, Me, Ri, TDF, Sh)

Description. The shaping process is characterized by solid work material, two-dimensional forming (one-dimensional forming may occur), and a shear state of stress. The workpiece (*W*) is clamped on the table (*B*), which is provided with the feed (T_B), and the tool (*V*) is provided with the primary motion (T_V), giving the geometrical possibilities.

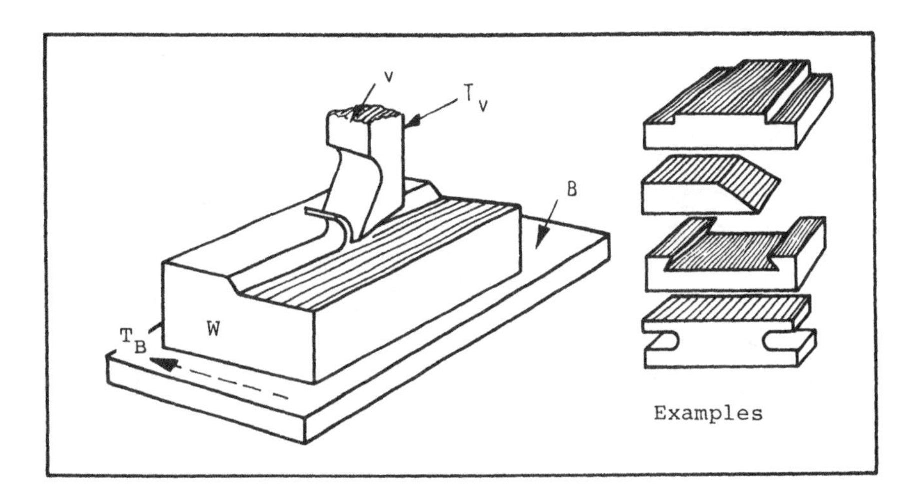

Applications. The shaping process is primarily used to produce smaller horizontal, vertical, or inclined flat surfaces (curved surfaces may be produced) mostly in toolrooms. Planing and shaping are, in many cases, substituted for by the more productive milling process.

Material Requirements. The hardness should not exceed HB = 300, and a minimum of ductility is advisable.

Tolerances/Surfaces. Good tolerances can be obtained (±0.1 mm to ±0.05 mm). The surface roughness will normally be in the range $3 \leq R_a \leq 12$ μm.

Machinery. Different types of shapes are available, mechanical or hydraulically powered. Shaping is gradually being replaced by milling.

PROCESS 6: Grinding (cylindrical and surface) (M, Me, Ri, TDF, Sh)

Description. The grinding process is characterized by solid work material, two-dimensional forming (one-dimensional forming may occur), and a shear state of stress. The workpiece (*W*) is supported between centers (*P*) or clamped on a table (*B*) and given a rotary (*R*) and translatory (T) feed. The tool *V* (the grinding wheel) is given a rotary primary motion (R_V) and, depending on the particular process, sometimes a feeding motion also.

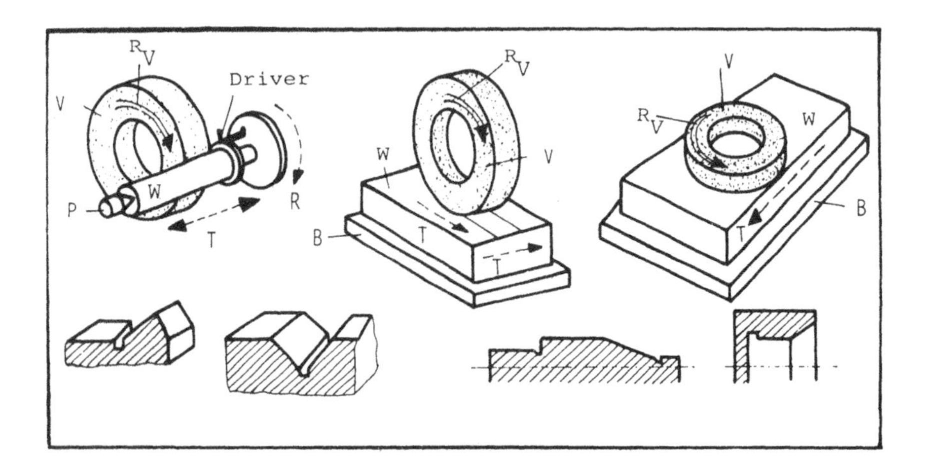

Applications. The grinding processes are used primarily in finishing cylindrical or flat surfaces which have been produced by various other processes. Today, roughing grinding, including profile grinding at high cutting speeds, can sometimes substitute for turning, milling, or planing.

Material Requirements. The material has no limit in hardness provided that it is less than the hardness of the grains. Also, high ductility gives difficulties.

Tolerances/Surfaces. Grinding is normally a finishing process, and tolerances around ± 0.001 mm are obtainable and very fine surfaces can be produced, $0.25 < R_a < 3$ μm. The grinding processes have a low material removal rate.

Machinery. Various types of grinding machines are available; for example, cylindrical grinders, centerless grinders, and surface grinders.

PROCESS 7: Electrical discharge machining, EDM (T, El, Fl)

Description. The electrical discharge machining process is characterized by solid work material, total, one-dimensional, or two-dimensional forming, and an unspecified state of stress, as the material is removed by melting and evaporation. The surface creation principle depends on the shape of the tool and the pattern of motions. The workpiece (*W*) is placed in a dielectric fluid (*D*) and the tool (electrode, *V*) is fed toward the material. The electrical discharges, taking place successively at the changing positions of minimum distance, melt and evaporate the material in small craters, so that the surface is composed entirely of extremely small craters. The discharges occur when the potential difference between the tool and the workpiece is large enough to cause an electrical breakdown in the dielectric medium.

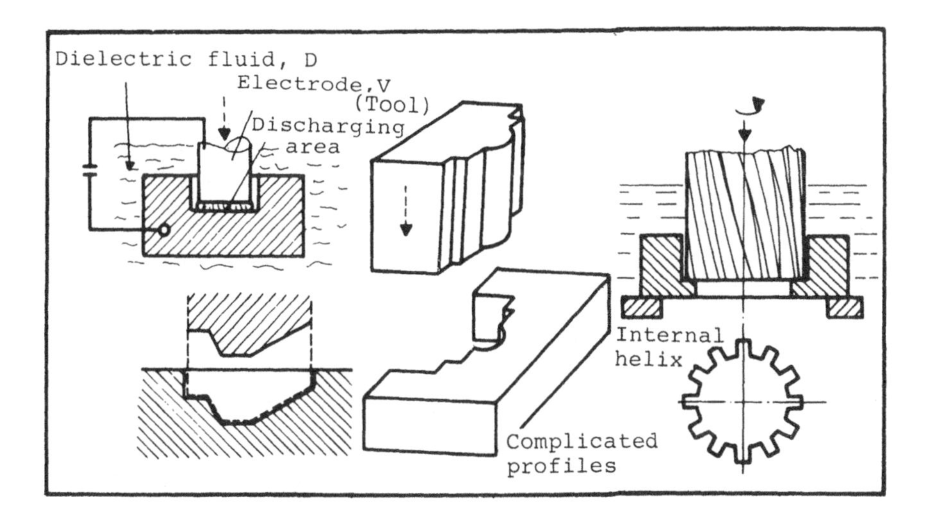

Applications. The EDM process is used primarily to produce cavities and dies, blanking and punching tools, cutting small holes, and so on. The geometrical possibilities are unlimited within the constraints of electrode geometry and pattern of motions.

Material Requirements. Electrically conductive materials can be shaped independent of hardness and strength (i.e., also in the heat-treated state).

Tolerances/Surfaces. The removal rate is rather low, but the tolerances fine, $\pm$ 0.02 to $\pm$ 0.005 mm, and the surface quality high, $0.1 \leq R_a \leq 10$ μm.

Machinery. Various types of electrical discharge machines are available, of different sizes and capabilities.

PROCESS 8: Shearing (M, Me, Ri, TDF, Sh)

Description. The shearing process is characterized by solid work material, two-dimensional (sometimes one-dimensional) forming, and a shear state of stress. The workpiece (*W*) can either be clamped so that the tool (*V*) is carrying out both the shearing and the feed motions, or moved (fed) so that the tool is carrying out only the shearing motion. The metal is cut between two shearing blades or rolls with a clearance of 5–10% of the thickness of the work material.

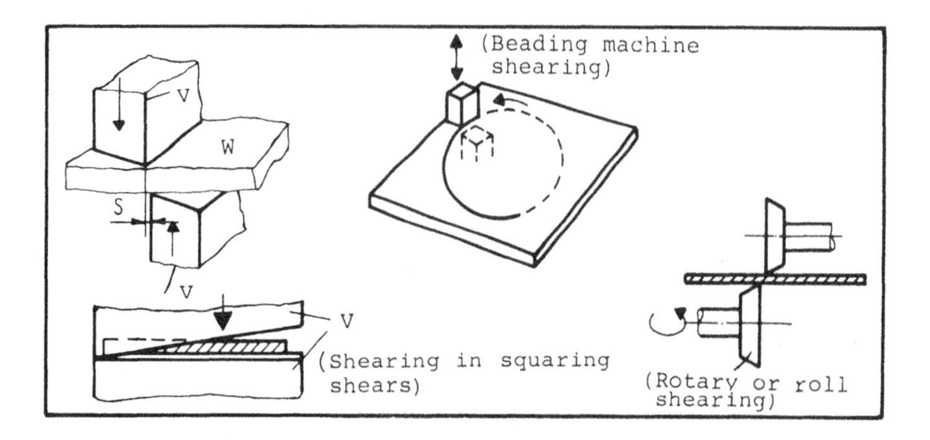

Applications. The shearing process is used extensively in industry to cut sheets and plates. When the blades are straight, the process is called shearing. With curved blades, the process often is given special names.

Material Requirements. The hardness should not be too high (HB < 200) and considerably lower than the hardness of the shears to avoid heavy wear.

Tolerances/Surfaces. The tolerances depend on the actual pattern of motion but are normally in the range 0.1–1.0 mm. The roughness of the cut surface is in the range $10 \leq R_a \leq 100$ µm.

Machinery. A wide variety of machines are available: for example, squaring shears, roll shears, slitting machines, and blading machines.

PROCESS 9: Blanking and Piercing (M, Me, Ri, ODF, Sh)

Description. The blanking and piercing processes are characterized by solid work material, one-dimensional forming, and a shear state of stress. The material (*M*), in the form of sheets or strips, is placed on the die (*D*) and the punch (*P*) is driven toward and through the material. If the piece punched out is the workpiece, the process is called blanking, and if the piece punched out is scrap, the process is called piercing.

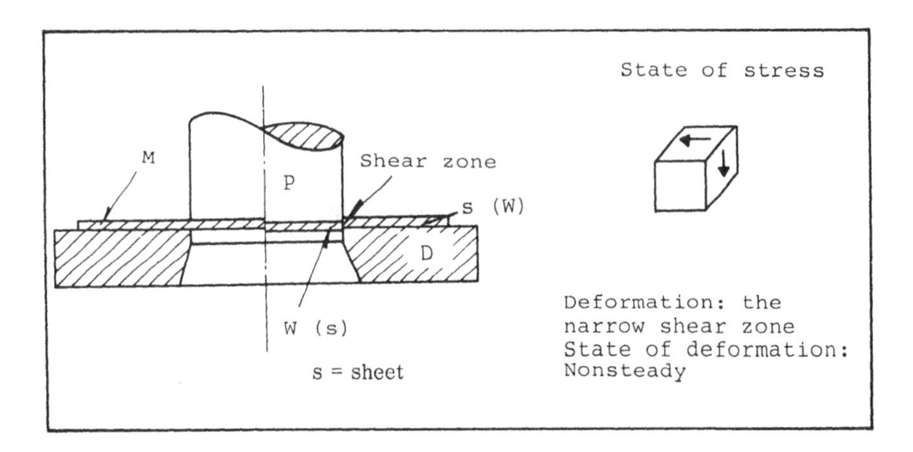

Applications. The blanking and piercing processes are used extensively in the electrical and mechanical industries to produce parts for electric motors, electrical equipment, household appliances, typewriters, and so on. It is typically a mass production process.

Material Requirements. Limited hardness, HB $\leq$ 200, and sufficient ductility to permit controlled shearing.

Tolerances/Surfaces. Depending on the clearance and the conditions, the tolerances vary considerably, ±0.05–0.5 mm. The fine tolerances are obtained in fine blanking, where the work material is held by a pressure plate. The surface roughness is normally in the range $5 \leq R_a \leq 50$ μm.

Machinery. The machinery is usually a mechanical or hydraulic press, but the dies are the important part of the process. They may be single-operation dies or progressive dies.

7.5 DETERMINATION OF FORCES AND POWER

All manufacturing processes involve, as discussed in Chapter 1, material flow, information flow, and energy flow. In the previous sections of this chapter, the material and information flow in mass-reducing processes have been elucidated. Based on this, and on the elementary theory of plasticity, the main elements in the energy flow system can be determined. The energy system supplies the energy necessary—through a suitable medium of transfer—to carry out the processes according to the planned material and information flow.

Forces and power are major elements in the detailed specification of the energy system, since they set the requirements of the media of transfer and the machinery.

In the following sections, approximate methods are described, but more refined theories can be found in the literature.

7.5.1 General Background (the Work Method)

The energy per unit volume w necessary to carry out a cutting process is composed of

1. The shear work w_s consumed in the shearing zone (see Fig. 7.6)
2. The frictional work w_f consumed when the chip slides over the tool face

This means that

$$w = w_s + w_f \tag{7.21}$$

For most cutting processes, the frictional work is approximately equal to 20–30% of the total work and, consequently, Eq. (7.21) can be modified to

$$w \simeq (1.25 \text{ to } 1.50)w_s \tag{7.22}$$

The shearing work per unit volume can be found from the shear stress–shear strain curve (τ–γ_s) of the material,

$$w_s = \int_0^{\gamma_s} \tau \, d\gamma_s \quad \left(= \int_0^{\bar{\epsilon}} \sigma \, d\bar{\epsilon}\right) \tag{7.23}$$

which can be shown to be identical to Eq. (4.42).

If a mean value of the shear stress τ_m is used, Eq. (7.23) becomes

$$w_s \simeq \tau_m \gamma_s \tag{7.24}$$

For most cutting processes γ_s is in the range 2 to 4, which means that Eq. (7.24) can be written

$$w_s \simeq (2 \text{ to } 4)\tau_m \tag{7.25}$$

Tresca's yield criterion for pure shear (Section 4.4.1) gives $\tau_m = \sigma_0/2$ or $\sigma_0 = 2\tau_M$, where σ_0 is the uniaxial yield stress. The yield stress σ_0 can be related to the hardness of the material through

$$HB \simeq c\sigma_0 \tag{7.26}$$

[Section 2.3.2, Eq. (2.14), relates σ_{uts} and HB but the same type of relation is also valid for the yield stress] where c is a constant for a given material. This means that HB $\simeq 2c\tau_m$. Equation (7.25) consequently becomes

$$w_s \simeq \frac{(2 \text{ to } 4)\text{HB}}{2c} = (1 \text{ to } 2)\,\frac{\text{HB}}{c} \tag{7.27}$$

Equation (7.27) substituted in Eq. (7.22) leads to

$$w \simeq (1.25 \text{ to } 3)\,\frac{\text{HB}}{c} \tag{7.28}$$

For steel (cold worked) c is approximately $c = 0.3$ when σ_0 is measured in N/mm^2, which substituted in Eq. (7.28) gives

$$w \simeq (4 \text{ to } 10)\text{HB} \quad \text{N/mm}^2 \tag{7.29}$$

Equation (7.29) shows that, as a first approximation, the work per unit volume necessary to carry out a cutting process can be determined from the hardness of the material. The lower value 4 of the factor in Eq. (7.29) corresponds to favorable cutting conditions, which means that it is reasonable to estimate the necessary work in cutting as

$$w \simeq 10\text{HB} \quad \text{N/mm}^2 \tag{7.30}$$

Experiments show that this equation gives acceptable results to within 20% [15].

It must be remembered that most of the equations traditionally used in machining are not homogeneous in units and require special attention.

The work w necessary is dependent on the cutting conditions, especially the normal rake angle γ and the feed f. The influence of the rake angle can be formulated as follows: The necessary work w decreases about 1% for every degree the rake angle is increased. The influence of the feed f on w can be expressed by the empirical relation [15]

$$\frac{w_1}{w_2} = \left(\frac{f_2}{f_1}\right)^{0.2} \tag{7.31}$$

where w_1 and w_2 are the amounts of the work necessary to carry out cutting at feeds f_1 and f_2 with other parameters kept constant.

To be able to use these two relationships for the influence of γ and f, a reference value of w must be defined corresponding to known values of γ and f. If such a reference value cannot be found, the approximate value for w given by Eq. (7.30) must be used independent of γ and f. Usually, w is measured under controlled conditions (given values of v, f, γ, etc.) when accurate calculations are necessary. If w is known, the power consumption N can be determined as w times the removal rate [Eq. (7.4)]:

$$N = wAv = wfav \tag{7.32}$$

where A is the area of cut (equal to the feed times the depth of cut) and v is the cutting speed.

7.5.2 Cutting Forces and Power in Turning

In a turning process, the total force P acting on the tool can be resolved into three components (see Fig. 7.29)

P_t = tangential force (cutting force)
P_a = axial force
P_r = radial force

Experimentally, it has been shown that the magnitudes of these forces are very approximately given by

$$P_t : P_r : P_a = 4:2:1$$

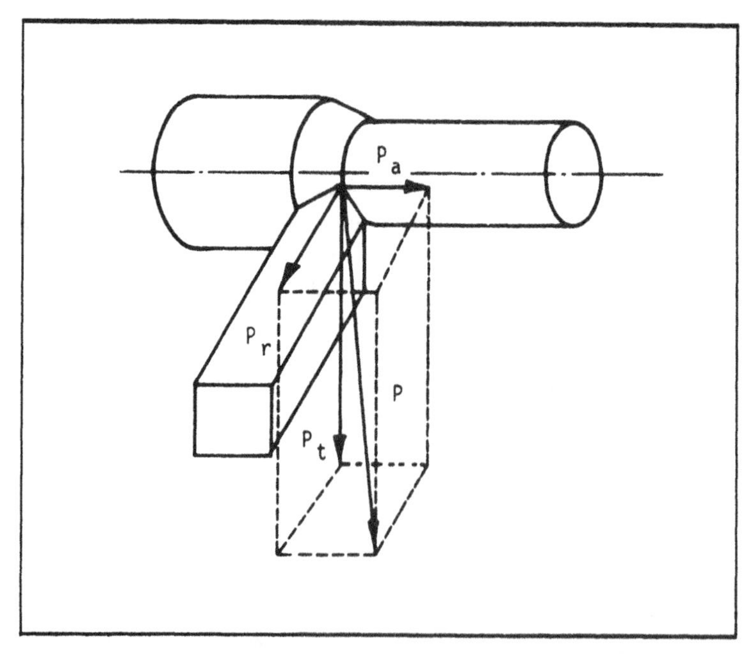

FIGURE 7.29 Forces in turning: P_t = the tangential force (cutting force), P_r = the radial force, and P_a = the axial force.

The tangential force P_t is the external force component which must supply the necessary (internal) work:

$$P_t = wA \tag{7.33}$$

Equation (7.33) can be modified to

$$P_t = wfa \tag{7.34}$$

where the area of cut A is substituted by $A = fa$ [Eq. (7.3)]. Often, P_t is measured experimentally and, consequently, the work necessary w can be calculated [Eq. (7.34)]:

$$w = \left(\frac{P_t}{fa}\right)_{f,\gamma} \tag{7.35}$$

where the suffixes f, γ indicate the actual values of f and γ used in the test. In many cases only f is known, which means that the influence of γ on w must be neglected.

The work w per unit volume may also be interpreted as the specific cutting pressure (i.e., the cutting force per unit area of the area of cut) and in the literature this is often designated by k_s. Table 7.7 shows, for different materials, typical values of k_s (or w), and the corresponding values of feed f and rake angle γ, at which k_s is determined. Furthermore, mean hardness values are listed for the different materials, enabling a comparison of w, determined from Eq. (7.30), with the measured values of k_s. It can be seen that Eq. (7.30) gives reasonable results as a rough approximation.

If the values of k_s (or w) in Table 7.7 are used at values of f and γ other than the reference values, corrections must be carried out according to Eq. (7.31), and the rule used that k_s decreases 1% for every degree γ is increased. The tangential force P_t determines the power consumption,

$$N = P_t v \tag{7.36}$$

If Eq. (7.34) is substituted into Eq. (7.36), the power is given by

$$N = wfav = k_s fav \tag{7.37}$$

where $w = k_s$ is the specific cutting pressure, f the feed, a the depth of cut, and v the cutting speed. From Eq. (7.37) it can be seen that w or k_s can also be interpreted as the specific power (i.e., the power required to remove unit volume of work material in unit time).

If the power N is measured in KW, f in mm/rev, a in mm, v in m/min, and k_s in N/mm^2, Eq. (7.37) becomes

$$N = \frac{k_s fav}{60{,}000} \text{ kW} \tag{7.38}$$

TABLE 7.7 Typical Values of $w = k_s$ for Various Materials[a]

Material	HB	$w = k_s$ (N/min^2)	f (mm/rev)	γ_0
Plain carbon steel				
Mild steel (0.15%C)	120–150	1950	0.4	−6
Steel (0.35%C)	140–160	2100	0.4	−6
Steel (0.50%C, stress relieved)	170–210	2250	0.4	−6
Alloyed steel				
Steel (0.4%C, 3%Cr, 0.5%Mo, hardened and tempered)	375	2350	0.4	−6
Steel (0.35%C, 1.4%Cr, 1.4%Mo, hardened and tempered)	230–330	2250	0.4	−6
Steel (0.5%C, 1.2%Cr, 2.4%W, annealed)	230	2300	0.4	−6
Stainless steel				
0.2%C, 17%CR, 2%Ni, hardened and tempered	250–310	2200	0.4	−6
0.08%C, 17.5Cr, 8.5% Ni, annealed	160	2050	0.4	−6
Cast steel	180–200	1900	–	–
Cast iron (gray)	200	1300	–	–
Copper	–	1100	–	–
Brass	80–120	850	–	–
Aluminum	–	550	–	–
Aluminum alloys	–	700	–	–

[a]Evaluated from various sources.

Equation (7.38) determines the power consumption at the cutting edge. If the efficiency of the lathe is η, the required motor power is given by

$$N = \frac{k_s fav}{60{,}000\eta} \text{ kW} \tag{7.39}$$

Often, η has a value from 0.7 to 0.8.

It should be remembered that equations such as (7.38) and (7.39) are a result of an incorrect use of the SI system. But here it has been decided to use the units traditionally used in the machining field.

The radial force P_r has no influence on the power consumption, but has a major influence on the deflection of the workpiece and consequently on the accuracy obtained.

The axial force p_a is relatively small, and its contribution to the power consumption ($<5\%$) is negligible, since the feed speed normally is very low.

7.5.3 Forces and Power in Shaping and Planing

As in turning, the total force P acting on the tool can be resolved into three components. Only the tangential force P_t parallel to the direction of cutting is considered. In general, P_t is given by $P_t = wfa$. Practical experience shows that the specific cutting pressure in shaping and planing normally lies 15–20% above the value for turning, so that P_t can be approximated to

$$P_t \simeq 1.18 k_s fa \tag{7.40}$$

where k_s can be found from Table 7.7.

The power necessary (when f is measured in mm/stroke, a in mm, k_s in N/mm^2, and v in m/min) is given by [modification of Eq. (7.39)]

$$N \simeq 1.18 \frac{k_s fav}{60{,}000\eta} \tag{7.41}$$

The efficiency is normally 0.7–0.8

7.5.4 Moment and Power in Drilling

In drilling with twist drill, two cutting edges are working simultaneously. The total force can be divided into the separate forces P, equal in size, acting on each major cutting edge. For each edge P can be resolved into three components, P_t, P_a, and P_r (see Fig. 7.30).

The tangential force P_t can be found from Eq. (7.34):

$$P_t = wfa = k_s \frac{f}{2}\frac{D}{2} \tag{7.42}$$

where k_s has the reference feed $f_z = f/2$ (the feed per cutting edge), f the total axial feed of the drill (see Fig. 7.2 and 7.30a), and D the diameter of the drill.

The torque acting on the drill is given by (see Fig. 7.30a)

$$M = P_t \frac{D}{2} = k_s \frac{f}{2}\frac{D}{2}\frac{D}{2}$$

that is,

$$M = k_s \frac{fD^2}{8} \tag{7.43}$$

This means that the required moment increases with the feed and the diameter D of the drill. The change in k_s with f is given by Eq. (7.31).

The total axial force P_A can be expressed as (see Fig. 7.30b)

$$P_A = 2P_a = 2P'_a \sin\left(\frac{120}{2}\right)$$

where P'_a is the feeding force acting perpendicular to the edge. P'_a can be calculated as

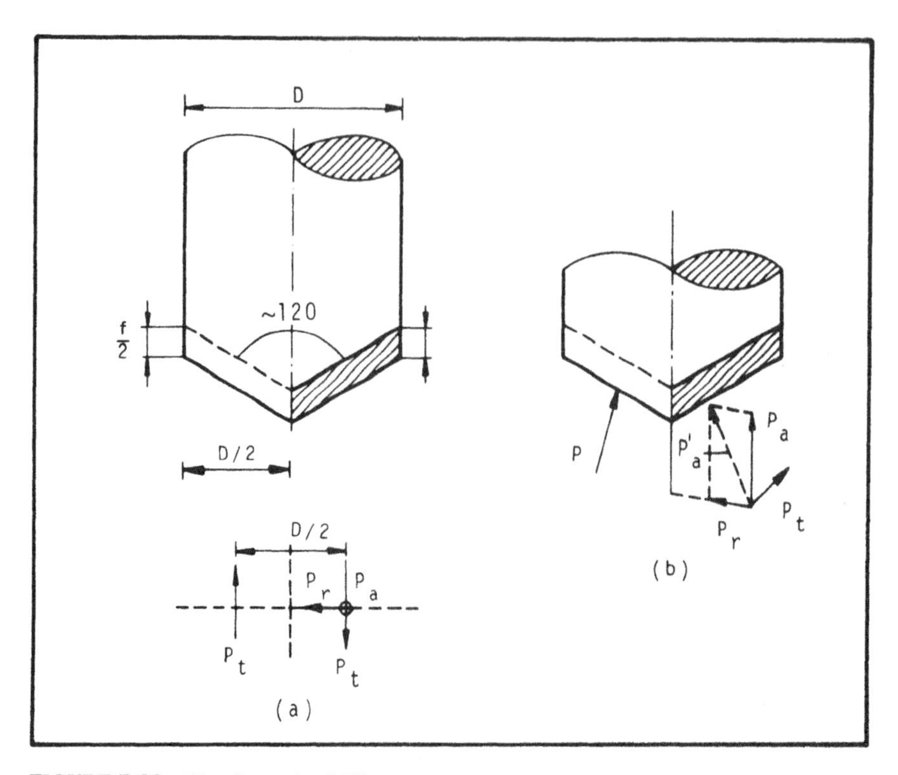

FIGURE 7.30 The forces in drilling.

$$P'_a = k\frac{fD}{22} = \frac{kfD}{4}$$

where k is the specific cutting pressure to press the edge into the material (area of cut $f/2$ times $D/2$). If k is approximated by k_s, P'_a becomes

$$P'_a = k_s\frac{fD}{4} = P_t$$

Substituting in the equation for P_A,

$$P_A = 2P_a = 2P_t \sin\left(\frac{120}{2}\right)$$

that is,

$$P_A = \sqrt{3}P_t = \frac{\sqrt{3}}{4} k_s fD \tag{7.44}$$

If P_A is related to the moment (7.43), it is found that

$$P_A = 2P_a = 2\sqrt{3}\,\frac{M}{D} \tag{7.45}$$

The power required at the cutting tool is given by

$$N = M\omega = \frac{k_s f D v}{4(60{,}000)} \tag{7.46}$$

where k_s is measured in N/mm^2, f and D in mm, and $v = \pi D n$ in m/min (n = rev/min).

The power required to provide the feed is normally only 1–5% of the total power and, consequently, it is neglected.

If the efficiency of the drilling machine η is included, the motor size required is

$$N = \frac{k_s f D v}{\eta 4(60{,}000)} \tag{7.47}$$

Values of k_s can be found from Table 7.7.

7.5.5 Power Consumption in Milling

In milling, the cutting process is more complicated than in turning and drilling, but calculations similar to the preceding ones based on the specific cutting pressure may be carried out. Since the chips are comma-shaped, it is necessary to define a mean thickness.

In most cases it is acceptable to consider only the specific power consumption V_s. This can be based on the specific removal rate and is defined as the volume of material removed per unit power per unit time. Thus V_s is often measured in cm^3/kW·min; that is, V_S indicates how many cubic centimeters of the work material can be removed per kilowatt per minute.

If V_s, as well as the cutting conditions a_e = working engagement, mm, a_p = back engagement (axial depth of cut), mm, and v_f = table feed speed, mm/min, are known (see Fig. 7.3), the required power can be calculated. The removal rate V is

$$V = a_e a_p v_f \quad \text{mm}^3/\text{min} \tag{7.48}$$

The required power is

$$N = \frac{V}{V_s} = \frac{a_e a_p v_f}{V_s} \tag{7.49}$$

Table 7.8 shows typical values for V_s *where the efficiency of the machine is included.*

If it is necessary to know the tangential force acting on the cutter or workpiece to estimate the required strength of the support, this can be found from Eq. (7.49), since the power, in general, is given by (V_s measured in cm^3/kW·min)

TABLE 7.8 Typical Values of the Specific Removal Rate V_s

Material	V_s (cm^3/kW·min)	
	End milling	Face milling
Carbon steel, $\sigma_{uts} < 600$ N/mm^2	14–17	12–15
Carbon steel, $\sigma_{uts} > 600$ N/mm^2	12–15	11–14
Alloyed steel, $\sigma_{uts} < 900$ N/mm^2	11–14	10–12
Alloyed steel, $\sigma_{uts} > 900$ N/mm^2	9–11	8–10
Cast Iron, HB < 250	24–30	22–28
Cast iron, HB > 250	18–22	16–20
Bronze	20–28	18–24
Brass	40–50	35–45
Light metal	50–80	45–70

Source: From Ref. 20.

$$N = \frac{1}{\eta}\frac{P_t v}{60{,}000} = \frac{a_e a_p v_f}{V_s} \times 10^{-3}\ \text{kW} \tag{7.50}$$

where v is the cutting speed and η is the efficiency of the machine (0.7–0.8).

If V_s is unknown, the work defined by Eq. (7.32) (where $w = 10\text{HB N/mm}^2$) can be used in rough calculations. An empirical relationship exists relating V_s and HB:

$$V_s\text{HB} \cong 1800 \tag{7.51}$$

where V_s is measured in cm^3/kW·min.

Considering a steel with 0.35%C, Table 7.7 gives HB = 150, which substituted in Eq. (7.51) leads to

$$V_s \cong \frac{1800}{150} = 12\ \text{cm}^3/\text{kW·min}$$

From Table 7.8 it can be seen ($\sigma_{uts} < 600$ N/mm^2) that V_s for end and face milling together are in the range 12–17 cm^3/kW·min, which means that Eq. (7.51) gives reasonable results. In these approximate calculations, end and face milling cannot be distinguished.

8

Solid Materials: Joining Processes

8.1 INTRODUCTION

In previous sections the production of components in one solid piece has been discussed. Often, it is more economical, or in some cases the only practical possibility because of size or geometry, to build up the desired component by joining two or more elements produced by the methods formerly discussed. In this section the fundamental processes that can be used to join elements or components are described.

If the elements (components) *A* and *B* in Fig. 8.1 are to be joined, this can be done in three principal ways:

1. The elements can be joined permanently through localized coalescence based on cohesion and/or adhesion between the elements.
2. The elements can be joined through a geometrical locking of the elements based on elastic or plastic deformation.
3. The elements can be joined through the use of special joining elements or fasteners (rivets, bolts, etc.).

In the first joining method, coalescence is brought about through cohesion or adhesion:

Cohesion, which means that the elements form common metallic crystals, is established through a suitable combination of temperature and pressure to

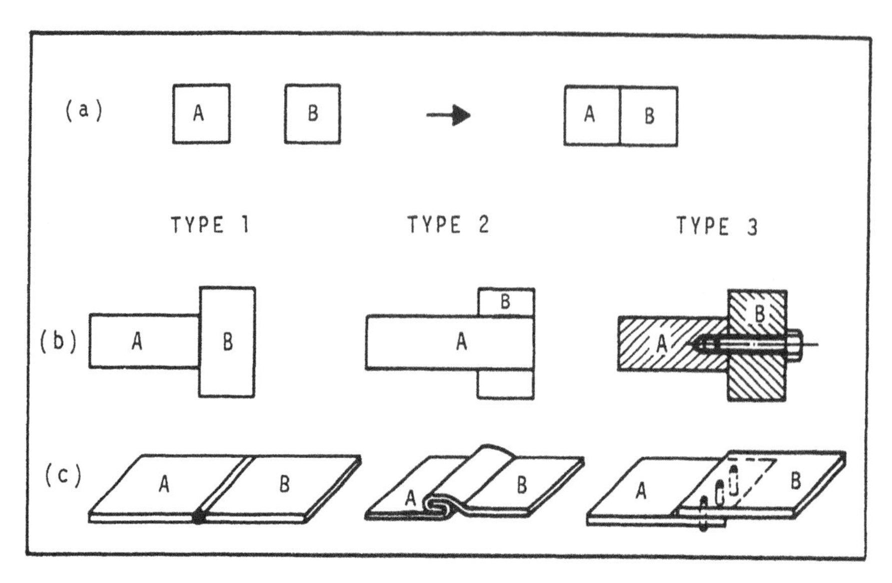

FIGURE 8.1 Joining of two elements *A* and *B*: (a) illustration of the problem; (b), (c) examples of joints based on the three fundamental principles (cohesion/adhesion, geometrical locking, joining elements).

create sufficient proximity and activity (i.e., the coalescence is based on atomic bonding forces)

Adhesion means that the elements are bonded together through surface forces of physical, electrical, or chemical nature (i.e., the elements do not form common structures)

Coalescence based on cohesion requires that the two elements have the same basic structure. This is not a requirement for adhesion; that is, different types of materials can be joined.

In many joining processes it is necessary to apply filler materials, which means that the cohesion or adhesion is established between the filler material and the element materials (see Fig. 8.1b and c, assembly type 1).

Assemblies based on geometrical locking are obtained through stresses generated by elastic deformation (assembly type 2b) or by plastic deformation (assembly type 2c). The elements must be shaped according to the chosen assembly method. Assembly type 2b is called shrinkage and may, for example, be established by producing the shaft (*A*) oversize for the hole (*B*) (interference fit). By cooling the shaft and/or heating the housing, the two members can be assembled. After cooling to room temperature, high stresses arise between the elements locking them together. If the interference is not too large, the elements may be pressed together cold.

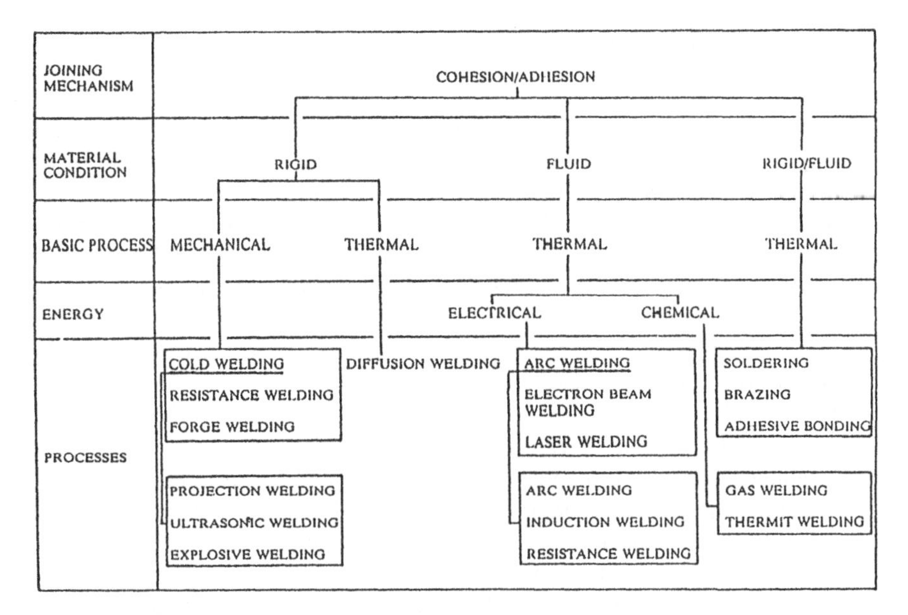

FIGURE 8.2 Structuralization of welding methods.

Assembly methods based on the use of joining elements (Fig. 8.1b and c, type 3) could mean riveting or bolting, for example. In many cases this type of assembly is designed to produce stresses in localized areas of the joins and, consequently, the result may be compared to shrinkage. The action of a screw in the thread, spring elements, paper clips, wedges, and so on, are examples.

Before the joining principle can be selected, the functional requirements of the joint must be defined. The joint can be permanent or nonpermanent. Permanent joints are not intended to be taken apart again, whereas nonpermanent joints are intended to be separated at some future date.

In this chapter only those joining methods based on a specific process are discussed. This means, that only type 1 assemblies involving cohesion and/or adhesion are considered. Figure 8.2 shows a detailed classification of the different methods, naming the processes most commonly used in industry.

8.2 CHARACTERISTICS OF THE JOINING PROCESSES

To achieve a satisfactory joint or bond based on coalescence—cohesion and/or adhesion—two basic requirements must be fulfilled:

1. The surfaces involved must be free of oxide layers, adsorbed gas, and other contaminants.

2. The surfaces involved must be brought into intimate contact so that the bonding forces (atomic forces or surface forces) can be activated.

These basic conditions can be fulfilled in different ways, resulting in the various joining methods. Referring to Fig. 8.2 it should be pointed out that in fusion weldings (state of material: fluid in the vicinity of the joint) the conditions necessary to obtain a satisfactory bonding are fulfilled by a localized melting along the edges of the two parts to be joined, often using a filler material because of the need to fill gaps in the joints. These gaps are made by chamfering the edges before welding in order to facilitate the conveyance of heat, particularly when welding heavier gages. The filler material is the main ingredient in some processes: brazing, soldering, and gluing, for example.

In the family of pressure weldings (material condition: solid) the two conditions are fulfilled through pressure applied to the joint. The surface oxides and other impurities are removed by plastic deformation of the surfaces, so that virgin metallic surfaces are created. Depending on the amount of deformation and the geometry of the specimens, some of the oxide layer becomes distributed in the joint causing weaker bonding.

Brazing, soldering, and gluing are processes based on the introduction of a filler material having a lower melting point than those of the materials to be joined. The bonding conditions are fulfilled through cleaning (mechanical such as wirebrushing and/or chemical) and proper design of the joint, so that capillary action can take place. The filler material is metallic in brazing and soldering, non-metallic in gluing, and the bonding mechanism is mainly adhesion.

Fusion weldings are characterized by a high temperature, and pressure weldings by a high pressure. Between these limits, many suitable and useful combinations of temperature and pressure exist, which is the background for the development of a large number of industrially important processes.

8.3 FUSION WELDING

In fusion welding, localized melting of the joining surfaces is established to create cohesion-based bonding, Fig. 8.3. Fusion welding thus requires a suitable energy source to produce the melting. In addition to that, a suitable method to protect or shield the molten metal from contamination is needed because the melt has a tendency to react with the oxygen and nitrogen in the surrounding atmosphere resulting in a vesicular, porous weld with inferior mechanical properties.

The most frequently used types of energy are electrical (the major type) and chemical, although other types may be used. Because of the highly differing characteristics of electrical and chemical energy, a short account of important conditions is given in the following section.

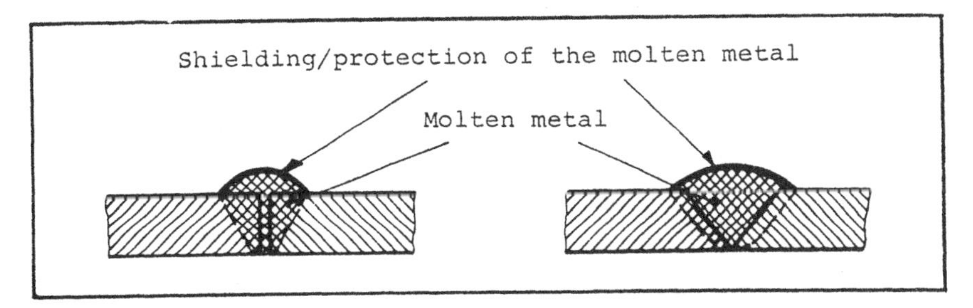

FIGURE 8.3 The molten pool of metal in fusion welding.

8.3.1 Electrical Energy Released Through Arcing (Arc Welding)

Under the joint designation arc welding, a number of processes are found where the necessary energy is induced by means of an electric arc. With few exceptions, the arc is established between a metallic electrode and the specimen. The arc is thus a primary condition for any arc welding process.

Theory

An electric current is a stream of electrons. If an air gap is created in the circuit, the movement of the electrons is broken and the current ceases to flow. The flow can be reestablished by ionizing the air gap, making electron movement possible again. During the ionization, other electrons are moved from their stationary positions; that is, electrons not emitted from the atom change orbits instead. Some of them reach orbits of lower energy levels, releasing energy in the form of electromagnetic waves with a wavelength in the visible area. Consequently, a light is created in the gap between the electrodes—an arc—whose characteristics in several important respects are decisive for all arc welding processes.

For precautionary reasons, the maximum allowable voltage in arc welding equipment is of an order of magnitude well below 100 V, which is not sufficient to create an arc by direct flashover, which demands about 5000 V/mm air gap. Under normal conditions, therefore, only an extremely small number of electrons are emitted. To increase the emission to usable proportions, so-called thermionic emission is utilized. This covers the fact that electrons are liberated easily from a metal surface heated to a very high temperature because of the additional energy given to them. The electrons gather in a cloud around the cathode (see Fig. 8.4). If a suitable voltage is applied, such as the no-load voltage of the welding machine, the electrons will move toward the anode (see Fig. 8.4).

When passing through the applied electric field across the air gap, the electrons accelerate and, after a certain distance a_1, they possess sufficient kinetic

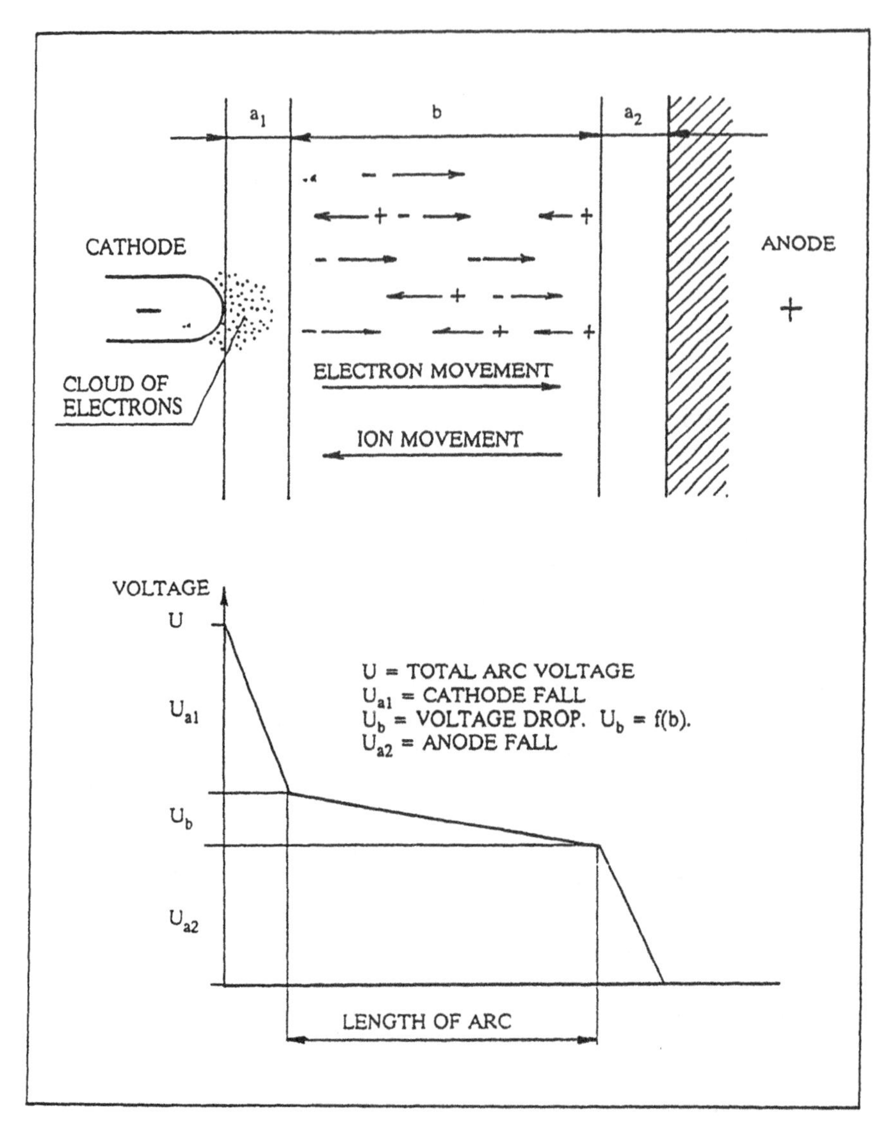

FIGURE 8.4 Generation of an electric arc. (From *Sandvikens Handbook*, 0.34 Sv. Sandviken, Sweden.)

energy to destroy the structure of the atoms in the gap, causing the atoms either to ionize (i.e., lose at least one electron) or to have their electrons change orbits, resulting in a change in the energy content of the atoms. This is called ionization by collision and takes place over the distance b. The electrons speed on toward

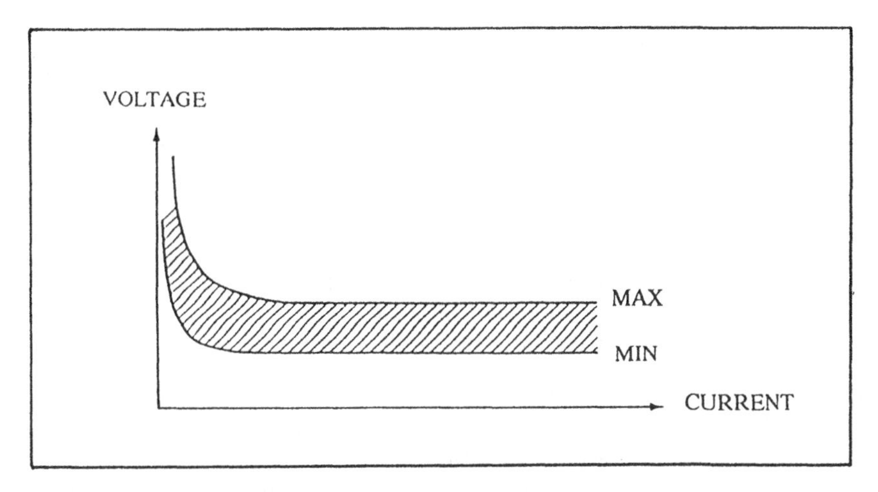

FIGURE 8.5 Arc characteristics at maximum and minimum. (From *Sandvikens Handbook*, 0.34 Sv. Sandviken, Sweden.)

the anode where they collide, converting their energy to heat. Correspondingly, the ions that also are created over the distance b accelerate toward the cathode, whose temperature—and thus its ability to emit new electrons—is maintained. The electron bombardment is fiercer than that of the ions, causing the anode to be heated more than the cathode when using direct current. When alternating current is used, the heat distribution is, of course, equal.

In the cathode and anode areas a_1 and a_2 respectively, voltage drops are observed. Their sum is the absolute minimum, below which an arc cannot exist. The total arc voltage is also dependent on the length of the arc (voltage drop across b). Increasing arc length is therefore, up to a point, synonymous with increasing arc voltage, physical conditions in practice limiting the arc length to 10–12 mm. Above this maximum the arc breaks and extinguishes itself. Figure 8.5 shows a general outline of the curves for maximum and minimum arc length. The area between the curves is the field of activity for this particular arc.

It has been mentioned that a hot cathode is necessary to start the flow of electrons. In practice, this is achieved by briefly short-circuiting the anode and the cathode. Because of the roughness of the parts, short-circuiting does not come into being all over the cross section but is limited to discrete points, resulting in very high local temperatures and melting or vaporization here. The heating causes the cloud of electrons, whereupon the short circuit is broken, by lifting the welding electrode a short distance away from the workpiece, causing a rise in the voltage, which in turn accelerates the emitted electrons, and so on, as described earlier.

The temperature in the arc is of an order of magnitude of 5000–7000°C, and the working voltage is between 18 and 35 V, with 50–80 V needed to initiate arcing. The current intensity depends on—among other things—the gage of the workpiece, typical values being 100–600 A. Alternating as well as direct current may be used, each possessing advantages and disadvantages such as:

All types of electrodes may be used with dc (but not with ac).
Welding of light gages is more difficult using ac
Ac welders are cheaper to purchase than dc equipment. Ac welders have a better efficiency and lower no-load losses.
Magnetic distortion of the arc is practically nonexistent on ac welders.

Dc welding machines have, until recently, dominated the welding field, but because of the cheap and effective ac machines developed in the last few years, they are now used extensively in industry.

In general, the arc can be established in either of two ways. First, it can be established between a consumable metal electrode, which is gradually melted during the process, supplying the necessary filler metal to the joint; that is, the electrode corresponds to the work material and has a lower melting point than the arc temperature. Second, the arc can be established between the work material and a nonconsumable tungsten electrode which has a melting point above the arc temperature, and the necessary filler metal, corresponding to the work material, must be supplied separately.

As mentioned earlier, it is necessary to shield the molten pool against contamination to obtain a weld with good mechanical properties. This is often done by coating the electrodes with a suitable fluxing material that melts and/or vaporizes during the process, forming a protective slag as well as stabilizing the arc by aiding the ionization of the air gap. Table 8.1 gives compositions for the three main groups of electrodes, of which the low-hydrogen or basic type is the one most commonly used.

Instead of coating the electrodes, the arc can be covered with granulated fluxing powder of a composition corresponding to Table 8.1, or it can be protected by inert gases forming a shielding atmosphere around the welding zone.

Although the physical phenomena behind fusion welding are fairly simple, their practical utilization gives rise to a number of industrially important processes to be described briefly next.

Processes

Metal–Electrode Arc Welding. Here the arc is maintained between a bare consumable metal electrode and the work material (see Fig. 8.6a and b). To compensate for the continuous melting of the electrode, it must be moved toward the work material to keep a constant arc length along the weld. In welding with bare electrodes, the arc tends to become unstable and, consequently, the

TABLE 8.1 Types of Electrode Coatings

	Low-Hydrogen	Rutile (%)	Neutral (%)
CaO_2	27	5	4
CaF_2	34	—	—
SiO_2	15	20	36
TiO_2	9	45	—
MnO	5	8	26
FeO	4	5	21
Al_2O_3	—	5	4
MgO	—	5	2
BALANCE	6	7	7

process is used only in special applications. The best known application is stud welding, where a metal stud is joined to a workpiece. The arc is established between the workpiece and the stud and maintained until a sufficiently high temperature is produced, and the stud is pressed against the workpiece to provide coalescence. Special equipment in the form of stud-welding guns has been developed for this process.

Shielded Metal-Arc Welding. This process (see Fig. 8.6c) is used extensively in industry. Here the shielded electrodes consist of metal wires or rods (2–10 mm in diameter) upon which is extruded a special coating. During welding the coating provides a gas shield around the arc and forms a protective slag coating that prevents oxidation and other contamination. At the same time it prevents too rapid cooling of the molten metal. The coating, in general, fulfills several purposes, some of which are: It provides a protective atmosphere; it stabilizes the arc; it performs metallurgical refinement; it provides slag to accumulate impurities; it prevents oxidation and slows down the cooling rate; it adds alloying elements; it increases deposition rate by powdered metal in the coating; and so on. Many types of electrodes are available, specially developed for the different applications, and the manufacturer's recommendation should always be followed. About 80–90% of all manual welding is done with coated electrodes.

Submerged-Arc Welding. In this process the metal arc is shielded by a granular fusible flux (see Fig. 8.6d). Otherwise, the process is very similar to shielded metal-arc welding. The electrode, in the form of coiled wire, is copper coated to provide good electrical contact and is fed into the granular material, supplied ahead of or around it. The arc is thus completely submerged by the flux, and high-quality welds are produced. The process is generally automatic, and high welding speeds can be obtained when welding thick plates. It is widely used in large-volume welding such as in ships, large tanks, tubes, and so on.

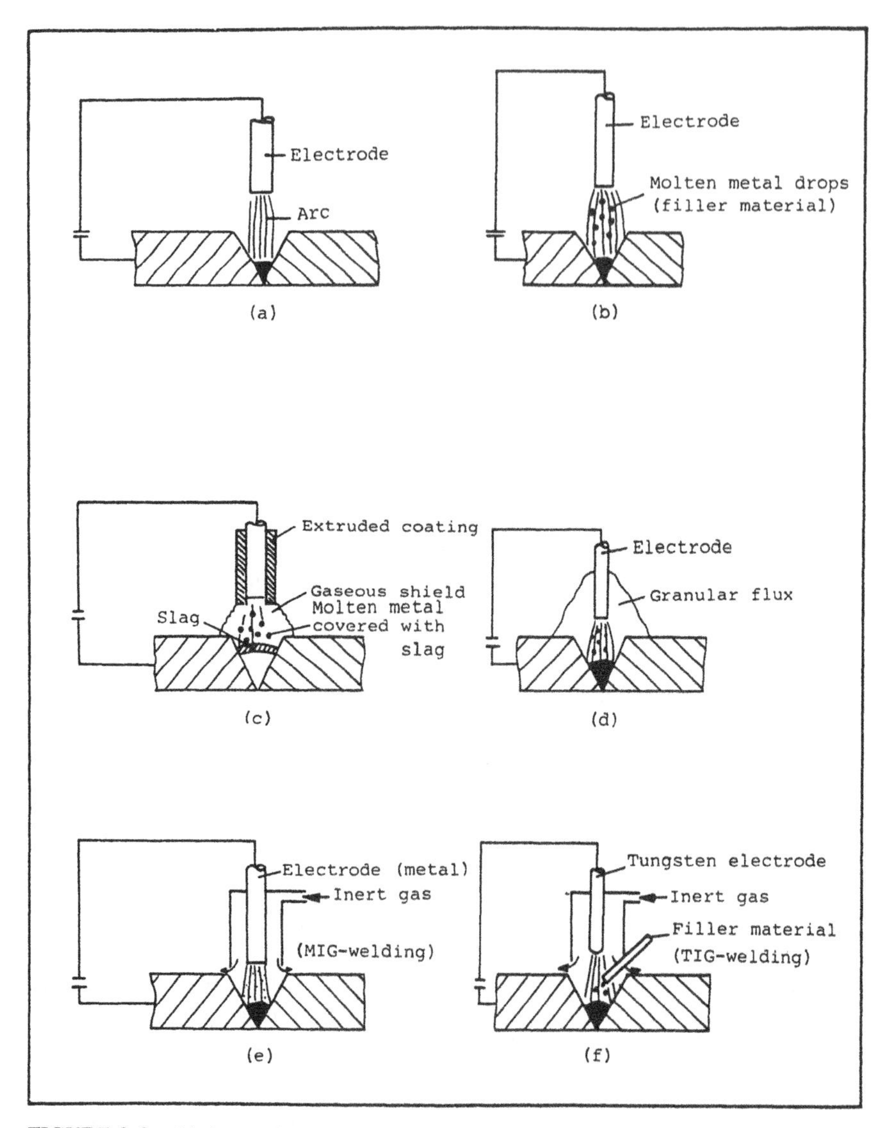

FIGURE 8.6 (a) Arc welding; (b) metal–electrode arc welding; (c) shielded metal-arc welding; (d) submerged-arc welding; (e) gas metal-arc welding; (f) gas tungsten-arc welding.

Gas Metal-Arc Welding. Here a bare metal electrode is used and both the arc and the molten metal are shielded by an atmosphere of inert gas. This process is called MIG welding (metal-inert gas) (see Fig. 8.6e). The electrode is fed through a special gun, which also supplies the protecting gas. The following gases can be used: argon, helium, or carbon dioxide. Carbon dioxide is used extensively to weld steel as it is the cheapest gas. For heavier weldings, small amounts of flux may be supplied through a hollow electrode, a process that also has many applications.

The more expensive gases, such as argon and helium, are mainly used in the welding of aluminum, magnesium, and stainless steel. Gas metal-arc welding is being used at a growing rate, giving high-quality welds at high speeds even with thick plates. The process can be carried out manually or automatically.

Gas Tungsten-Arc Welding. In this process, a nonconsumable electrode of tungsten is used, the filler material is supplied separately and the shielding established by an inert gas, for example, argon, helium, or carbon dioxide for steel welding (see Fig. 8.6f). The process, often called TIG welding (tungsten-inert gas), is used mainly in light-gage work and is not competitive with other welding methods for heavier gages of metal.

More detailed descriptions of processes and equipment can be found in the literature [2,21–24].

8.3.2 Fusion Welding Based on Electrical Energy Through Electron and Laser Beams

Electrical energy can be utilized in fusion welding in other ways than through arcing. The latest developments are electron-beam welding and laser-beam welding. These two processes are described next.

Electron-Beam Welding

In electron beam welding (see Fig. 8.7), the heat necessary for coalescence is obtained from the bombardment of the workpieces with a high-intensity electron beam consisting of high-velocity electrons. The process is carried out in a vacuum chamber to avoid heavy loss of energy and to shield the molten metal against oxidation. The electron beam is produced by a special and rather expensive electron gun. The need to use a vacuum chamber to perform welding imposes serious limitations on the size of the workpieces that can be welded. In the past few years, much effort has been put into the development of a type of electron-beam welding machine wherein the workpiece remains outside the vacuum chamber. Here, special precaution must be taken, for example, by shielding by inert gases, to avoid oxidation.

Electron-beam welding is usually carried out in a vacuum chamber. No filler material is applied, the penetration power is high (>100 mm) and the heat-affected zone is narrow. The process can be used to join not only common materials, but also materials difficult to weld by other processes: refractory metals, oxidizable metals, and super alloys, for example. Because of the high cost of this welding process, it is used primarily when other welding processes cannot be applied.

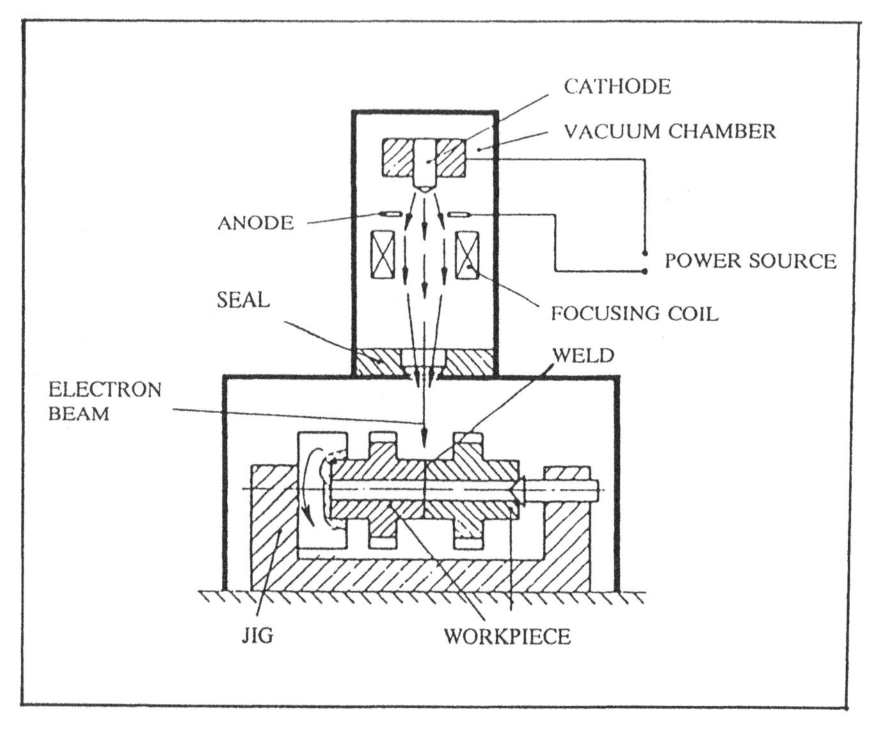

FIGURE 8.7 Electron beam welder. (From J. Flimm, Hanser Verlag, Munich.)

Laser-Beam Welding

In laser-beam welding (see Fig. 8.8), the heat is provided by a high-intensity light beam (10^9–10^{12} W/m^2). This beam is created in special laser (*L*ight *A*mplification by *S*timulated *E*mission of *R*adiation) media through direct supply of electrical energy to the media or indirectly through, for example, flash lamps. Depending on the laser media, different wavelengths are produced. Today, the CO_2 laser especially has found many industrial applications in welding. Commercial lasers with power in the range up to 20 kW are available. In welding, the power is mostly utilized in pulses rather than as a continuous beam. The penetration is shallow, 10–12 mm in steel with a 2-kW laser, but the heat-affected zone is very small. Laser beam welding is used in the electronics industry and at a rapidly increasing rate in the mechanical industry for light-gage welding, welding of small components, and other uses. The laser equipment is relatively expensive, but because of the production rates obtainable and the versatility (no vacuum), it is becoming economical in an increasing number of cases.

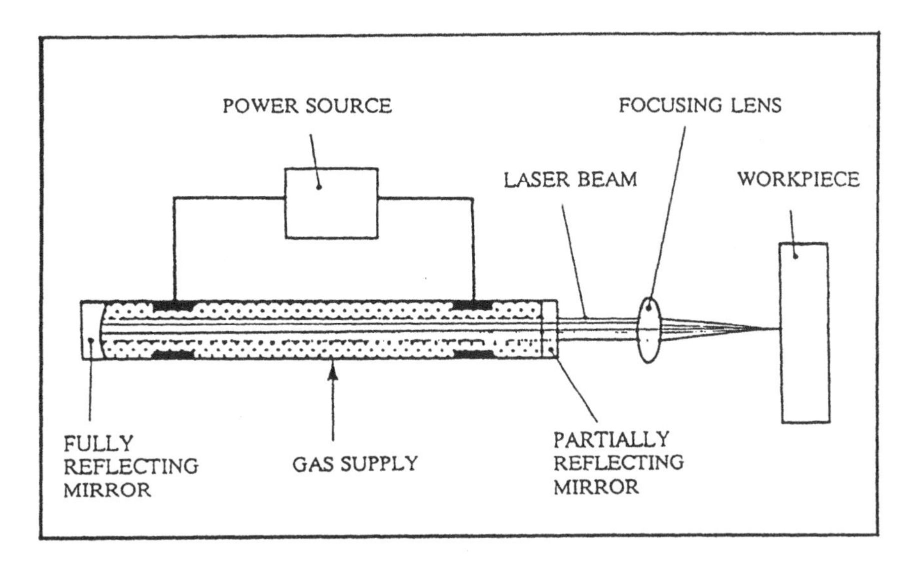

FIGURE 8.8 Gas laser.

8.3.3 Thermo-Chemical Welding Processes

Welding processes in which the parts to be joined are fused by the heat of chemical reactions are of two types—one employing flames, or gas welding as it is frequently called, and one in which exothermic processes such as the aluminothermic reaction are used. This method of joining is known as Thermit welding.

Gas Welding

In the gas welding process heat is transferred from a flame to the work by forced convection and by radiation. The flame is produced by supplying nearly equal volumes of oxygen (O_2) and acetylene (C_2H_2) to a torch, whose function is to bring together correct volumes of the fuel gas and oxygen, mix them efficiently, and pass them through a nozzle to form a flame with characteristics suitable for welding (see Fig. 8.9). Primary combustion takes place at the base of the flame in a thin shell-like region that surrounds the cone (Fig. 8.10). The reaction is as follows:

$$C_2H_2 + O_2 \rightleftharpoons 2CO + H_2 + 106{,}500 \text{ cal}$$

Maximum temperature—about 3500°C—is reached just beyond the apex of this combustion zone. The products of this first reaction form the bluish region of the flame called the reducing zone. Because this region is most closely in contact with the work in welding, it largely determines the characteristics of the flame from the welding point of view. The outer zone, or envelope, of the flame preheats the material and provides some shielding against oxidation of the mol-

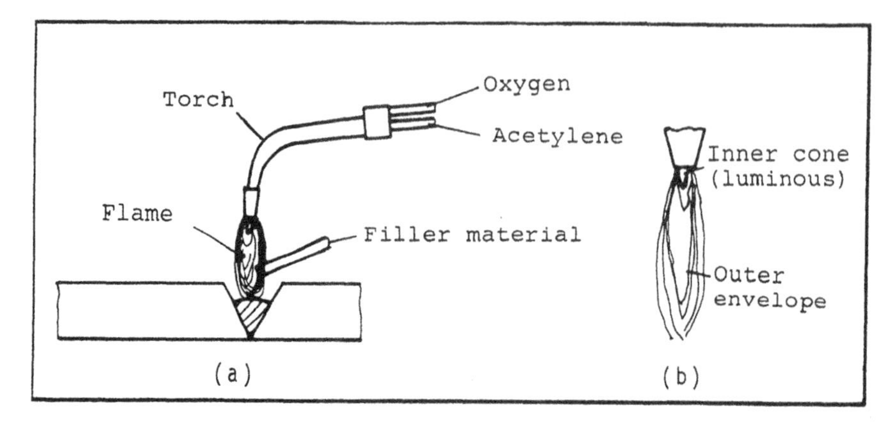

FIGURE 8.9 Gas welding: (a) principle; (b) neutral torch flame.

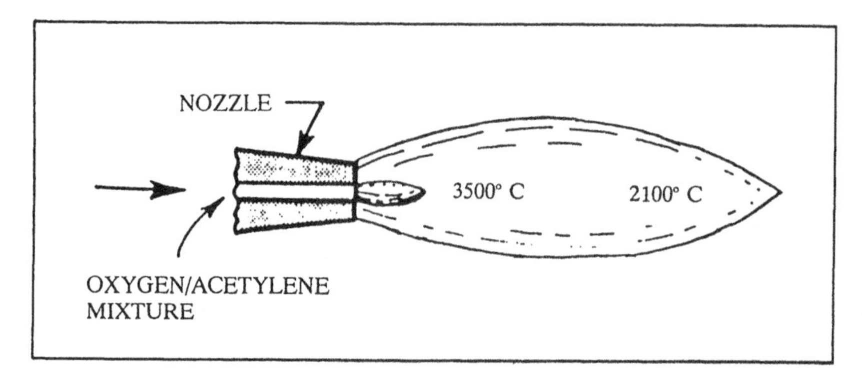

FIGURE 8.10 Temperature distribution in a neutral oxy-acetylene welding flame.

ten metal, owing to the fact that some of the oxygen in the air is used in secondary combustion, and that motions of the flame create a favorable airflow.

The chemical characteristics of the flame can be altered to suit the requirements of the welding process by changing the ratio of acetylene and oxygen. For most applications the so-called neutral flame is used. This is essentially the flame as discussed above, but because it is less desirable to have a slightly oxidizing flame than one which is reducing, it is usual to arrange for a slight excess of acetylene. Correct adjustment is indicated by a slight white flicker at the end of the cone. Any further increase in the volume of acetylene results in the flame becoming carburizing.

Carburizing flames are used when carbon must be added to the material being welded. When such a flame is used on mild steel the surface layers pick up carbon, resulting in a reduced melting point so that only the surface fuses. This technique is useful for hard surfacing, where deep fusion is to be avoided. Oxidizing flames are used with alloys containing zinc (e.g., brass). The zinc is oxidized on the surface of the pool where the oxide layer inhibits further reaction. With a normal flame, zinc is volatilized from the pool continuously and oxidizes in the atmosphere.

The essential process requirements of surface protection and deoxidation are performed mainly by the CO and H_2 in the reducing zone of the flame. With ferrous materials the reactions

$$FeO + CO \rightarrow Fe + CO_2$$
$$FeO + 2H \rightarrow Fe + H_2O$$

proceed at welding temperature, so that clean welds are produced. The alumina film on aluminum alloys, however, is not reduced by either CO or H_2 so a flux must be used with these alloys. Such fluxes are mainly halides of the alkali metals, and because they are corrosive they must be removed from the joint afterward. Designs of joints that might trap flux must be avoided. Fluxes are also required with copper and some copper-base alloys and these may be based on boric acid.

Acetylene is produced by the reaction of water on calcium carbide. The carbide is formed by fusing anthracite or coke with limestone at high temperature in an electric furnace:

$$3C + CaO \rightleftharpoons CaC_2 + CO$$

After crushing, the calcium carbide is reacted with water to produce acetylene, which is then purified to free it from traces of sulfur and phosphorus:

$$CaC_2 + 2H_2O \rightleftharpoons Ca(OH)_2 + C_2H_2$$

Acetylene is unstable at pressures above 30 psi and cannot be compressed directly into cylinders. Cylinders for acetylene are therefore packed with a porous filler saturated with acetone. The porous mass divides the space in the cylinder into many small cells, making the propagation of an explosion impossible. Acetone absorbs 25 times its own volume of acetylene for each atmosphere of applied pressure and this permits acetylene to be compressed safely up to 250 psi. Between the supply of gas and the torch it is necessary to have a pressure regulator and a gauge.

Metallurgical control of the weld deposit is aided by the optimum choice of filler rod, which often contains extra deoxidants to control the oxygen content of the molten pool. With steel this is performed largely by silicon but also by manganese. The correct ratio of these elements is necessary not only to control weld chemistry but also to impart to the pool the most suitable fluidity. During welding the products of deoxidation form a thin film on the metal surface which has a dominant effect on the fluidity and stability of the molten bead.

Gas welding once ranked equal in importance with metal-arc welding. Since the introduction of the inert-gas methods (particularly tungsten-arc welding, with which it has most in common from the view of welding technique), its use has declined, and it is now primarily used for repair work and in the field where other equipment cannot be transported or where electric energy is not available.

Thermit Welding

A number of metal oxides can be reduced by reaction with finely divided aluminum with the liberation of considerable heat, so that the products of reaction are molten. Molten, superheated iron produced this way can be poured between two parts of a joint to produce a weld. The reaction is obtained with any of the iron oxides but ferric oxide produces the highest temperature, up to 2450°C being reported:

$$Fe_2O_3 + 2Al \rightleftharpoons Al_2O_3 + 2Fe$$

A charge of 1000 g of Thermit produces 476 g of slag, 524 g of iron, and 181,500 cal. The ferric oxide is prepared from mill scale to which is added other materials to control the reaction of the Thermit and the ultimate analysis of the metal produced. Thermit powder will not ignite below 1300°C, so the reaction is started with a small quantity of special mixtures and an ignitor. The reaction is completed in about 30 s to 1 or 2 min regardless of the size of the charge. The quality and soundness of the metal is improved by adding small pieces of scrap steel or alloys to the powder. In this way excessive heating is also avoided, and the molten steel produced has a temperature of approximately 2100°C—that is, a superheat of 600°C.

The main application of the process is for joining steel, primarily rails, and for welding joints with large, compact cross sections, such as rectangles or rounds. It has also been used in heavy construction, shipbuilding, joining reinforcing bars, and repair welding. The use of Thermit welding in heavy construction is now receiving competition from the more recently developed electroslag process. Nonferrous aluminothermic mixtures have been used for joining copper conductors.

8.3.4 Joints in Fusion Welding

Figure 8.11a shows typical joints for arc welding and gas welding, and Fig. 8.11b shows a few applications. The fusion welding processes can be used for the welding of all gages of materials, the heavy gages necessitating welding from both sides. Many national standards for welding exist as well as quality control procedures. Important welds must be carefully inspected, for example, by X-rays.

The quality of a weld depends on many factors, such as materials, electrodes, the production of the weld, and the geometry of the design. The welding procedure chosen has a major influence on the distortion of the product.

TABLE 8.2 Principle of Pressure Welding and Some Pressure Welding Process Groups

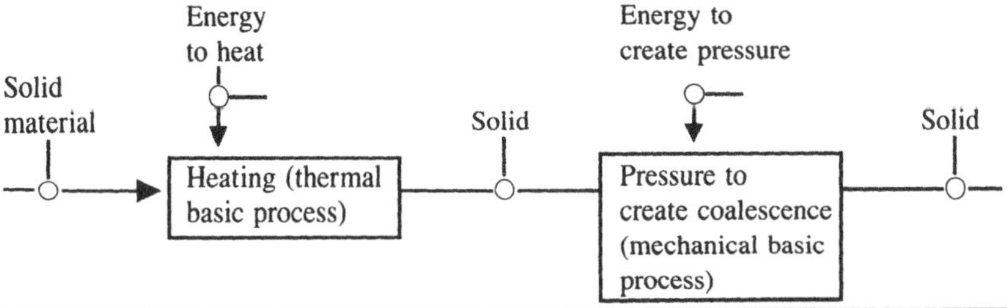

Solid material → [Heating (thermal basic process); Energy to heat] → Solid → [Pressure to create coalescence (mechanical basic process); Energy to create pressure] → Solid

Type of energy for heating	Type of energy to create pressure	Process groups (examples)
—	Mechanical	Cold welding
Electrical	Mechanical	Resistance welding
Mechanical	Mechanical	Friction welding
Chemical	Mechanical	Pressure gas welding
—	Chemical[a]	Explosive welding
—	Mechanical[a]	Ultrasonic welding

[a]Also creates heat in the welding zone.

8.4 PRESSURE WELDING

The term *pressure welding* covers all the processes where coalescence is obtained by a combination of pressure and temperature, thus fulfilling the two basic requirements (see Table 8.2). Pressure welding ranges from high pressure at low temperature, known as cold welding, to low pressure combined with a high temperature slightly below the melting point of the workpieces. The processes are thus a combination of two basic processes: a mechanical basic process involving plastic flow to remove oxides and other contaminants and to create sufficient proximity and activity, and a thermal basic process to facilitate the welding process by lowering the yield stress of the workpieces. The removal of oxides and other contaminants may also be done by mechanical or chemical cleaning prior to welding.

8.4.1 Cold Welding

In cold welding or cold pressure welding, coalescence is created by pressure alone. The pressure causes the workpiece to deform plastically, providing the necessary intimacy between the virgin metal structures. To obtain reasonably high strength welds, the surfaces must be cleaned, usually mechanically by wire brushing. The remaining oxide layers will be dispersed in the welding zone as islands and will decrease the strength of the weld. To obtain good welds, the

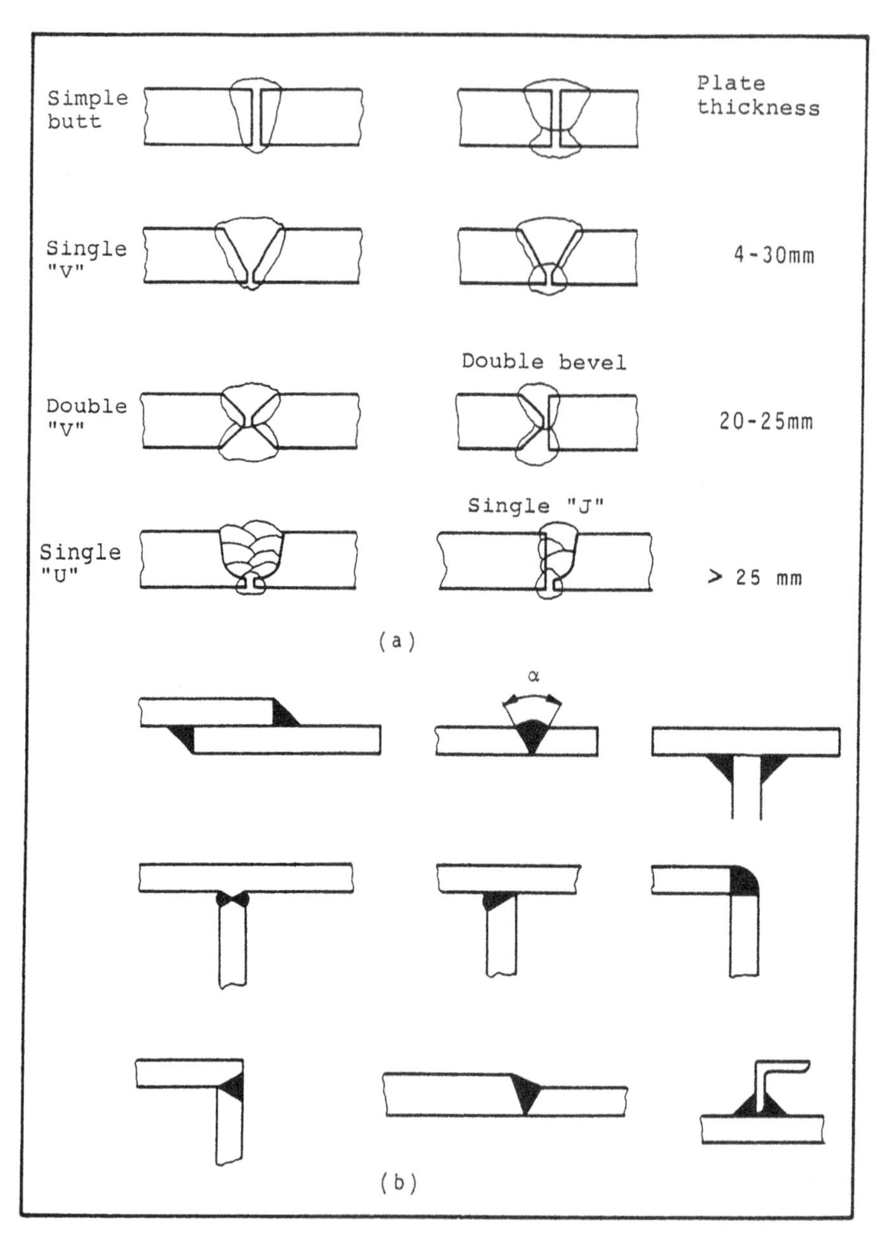

FIGURE 8.11 Examples of (a) welded joints and (b) applications.

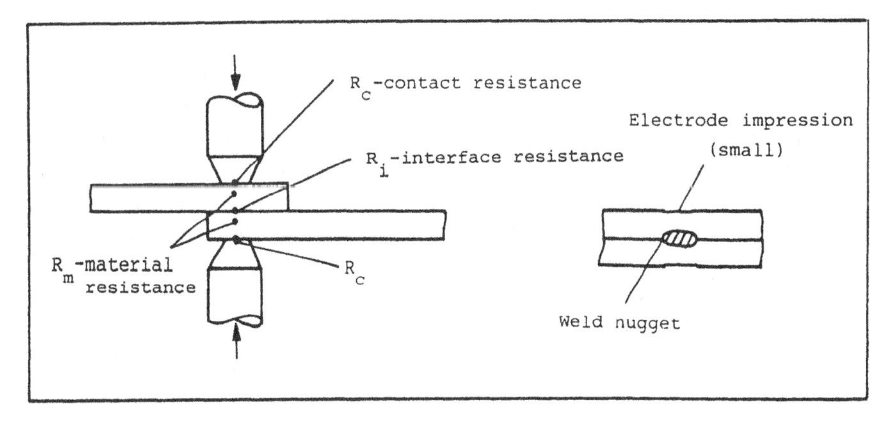

FIGURE 8.12 Resistance welding.

surface expansion caused by plastic deformation must be in the range of 50–90%. The cold-welding process is generally used to join relatively small parts (wires, rods, sheets, rings, etc.) as butt or lap (inclusive seam) welds. Materials most frequently cold-welded are aluminum, copper, lead, nickel, and zinc. A common example is the cold-welded electrical connection.

8.4.2 Resistance Welding

To reduce the pressure necessary to obtain sufficient coalescence of the work materials, they are often heated to temperatures corresponding to the forging range. Since no melting occurs, resistance welding is solid-state welding. Heating is usually accomplished by passing an electrical current through the area to be welded. Most heat develops where the resistance is greatest—in the interface of the two members (see Fig. 8.12). The total resistance is $R = 2R_c + 2R_m + R_i$, where R_c is the contact resistance between the electrodes and the work materials, R_m the resistance of the material, and R_i the resistance between the surfaces. Heat liberation is controlled by the relation $W = kRI^2t$, where k is a constant less than 1 to compensate for heat loss, R the total resistance, I the current, and t the time. The process involves two water-cooled electrodes that are pressed against the work materials, and a current is passed through the materials. The interface resistance is, as mentioned, the greatest and depends on pressure, surface quality, cleanliness, and material. The contact resistance R_c is minimized by appropriately shaping the electrodes of high-electrical-conductivity material with copper tips and applying a suitable pressure. The resistance of the electrode material is, in general, small compared to that of the work materials.

Resistance welding includes processes such as spot welding, seam welding, projection welding, and butt and flash welding.

Spot Welding

Spot welding is the most extensively used resistance welding process and is suitable for joining two or more metal sheets (see Fig. 8.12). The electrodes are normally conical with end diameters of about $5\sqrt{h}$ (where h is the thickness of the sheet) to give a reasonable size of the weld nugget, and they are water cooled to keep the temperature low. If different materials having different electrical conductivity are to be welded, the contact areas of the electrodes must be inversely proportional to the conductivity.

The main parameters of the process are current, time, and pressure, which must be mutually adjusted and coordinated depending on material and geometry. For mild steel, typical values are 200–400 A, 0.15–1.0 s, and 70–100 N/mm^2, respectively.

Spot welding is used extensively, in both small and large companies, to join many different materials and combinations of materials. The upper thickness limit is about 3–4 mm. The welding quality is dependent on the cleanliness of the materials, so they must be free of dirt, scale, and other contaminants, and have reasonably smooth surfaces.

Seam Welding

In many applications it is essential to obtain continuous welding of the sheets. This can be accomplished in seam welding, where a continuous series of spot welds are made. The electrodes can be rotating disks (see Fig. 8.13b). When the material is moved by and between the disk electrodes, the current is turned on and off to form the overlapping spot welds. Cooling of both the electrodes and the work area is provided directly by water streams. Seam welding is used primarily in the manufacture of liquid or pressure-tight vessels, tanks, radiators, and so on. By notching the disk electrodes, irregularities in the workpiece can be accommodated. In the manufacture of pipes and other structural shapes from flat sheets, resistance seam welding is applied as a butt welding process. After heating, which is provided by a high-frequency current, the butting surfaces are successively pressed together. In seam welding, the current must be higher than in spot welding to compensate for the flow of current through the previously made weld.

Projection Welding

In projection welding, the current is concentrated in suitable projections made on one of the sheets, allowing flat electrodes to be used (see Fig. 8.13c). The projection welding process is thus similar to spot welding, but the location of the spot is determined by the projections made on the sheet.

Many projections can be welded simultaneously, resulting in cost reductions compared to spot welding. The projections can be natural, that is, they are directly associated with the geometry of the components, or artificial, that is, they

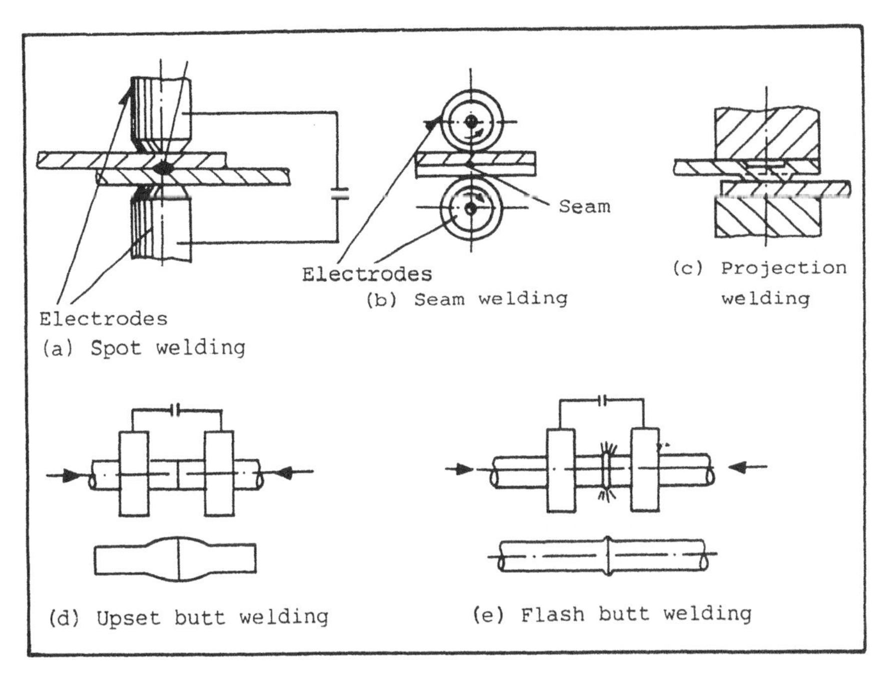

FIGURE 8.13 Resistance welding: (a) spot welding; (b) seam welding; (c) projection welding; (d) upset butt welding; (e) flash butt welding.

are made especially for the welding process by embossing, bending, and so on. The projections have a diameter corresponding to the thickness of the sheet and a height of about 60% of the thickness of the sheet.

Projection welding has many industrial applications, including welding together flat or curved sheets, tubes to flat sheets or cylindrical components, and nuts and bolts for sheet components. Many components are available ready for projection welding; that is, they are supplied with suitable projections.

Upset and Flash Butt Welding

The butt welding of bars can be carried out as upset or flash welding (see Fig. 8.13d and e). In upset butt welding, the mating surfaces are brought into light contact, an appropriate current flowing across the interface to heat the surfaces, which are kept under a slight pressure, and after heating the pressure is increased to form the upset. Some oxides from the surfaces are normally distributed in the welded zone, resulting in a slight reduction of the weld strength.

In flash butt welding, which requires more complicated equipment, the mating surfaces are brought together at a slow, controlled speed. When the highest points or asperities approach each other, the large current causes these to melt and boil away; when the next highest points approach each other, the process is

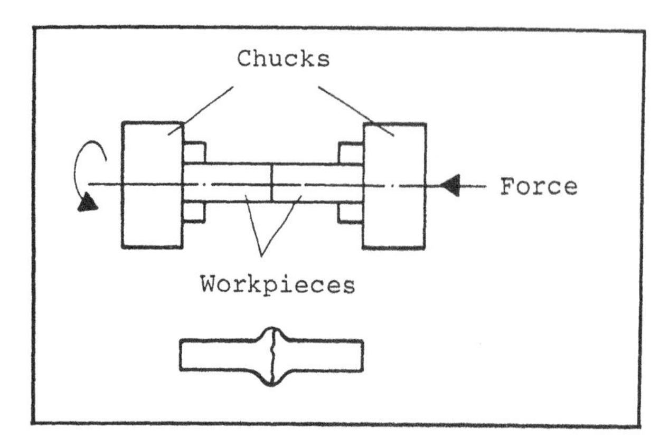

FIGURE 8.14 Principle of friction welding.

repeated; and so on. The melted surfaces are expelled from the interface. After a short time, the flashing has completely removed a thin layer of the materials at the interface, and they are now pressed together, causing a small upset. The quality of the weld is better than that for upset butt welding. Rods, bars, tubes, and structural shapes of uniform section can be welded by these methods.

8.4.3 Other Pressure Welding Processes

The processes described below are listed in Table 8.2.

Friction Welding

In friction welding, the heating is provided by mechanical friction established by relative motion, under pressure, between the surfaces to be welded (see Fig. 8.14). One of the parts to be welded is stationary and the other part rotates. In the first phase, the relative motions under an applied axial pressure create sufficient heat to reduce the yield stress considerably. A temperature up to the fusion temperature is reached. In the second phase, the rotation is stopped and the pressure is maintained or even increased until the weld is completed. The flash formed due to the deformation is removed after welding, if necessary. The process is used to weld circular bars, tubes, and so on. In the past few years, the number of applications for friction welding has increased considerably.

Pressure Gas Welding

Pressure gas welding is used to butt join bars, tubes, rails, and so on. The heat is supplied through suitable water-cooled oxyacetylene torches, and when the temperature reaches the fusion temperature the workpieces are pressed against each other at pressures, depending on the material, up to 50 N/mm^2. The equipment for this process is relatively cheap.

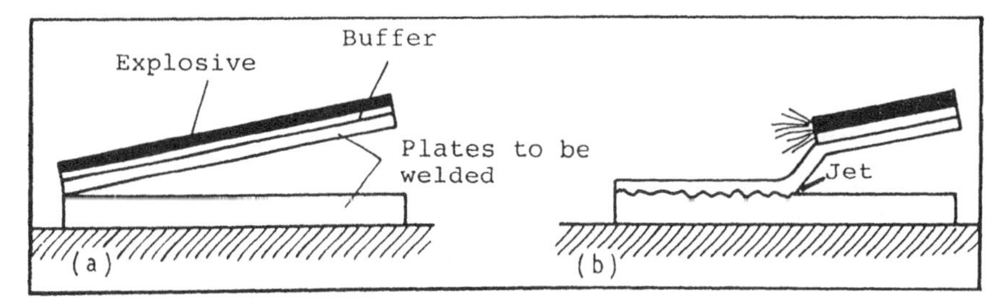

FIGURE 8.15 Explosive welding.

Explosive Welding

In explosive welding (see Fig. 8.15), high explosives are used to accelerate the top plate to a high velocity, providing both a cleaning action and high pressures when it collides with or impacts on the bottom plate. The cleaning action, the removal of oxides, and so on, is provided by a high-speed jet emitted from the collision point.

Explosive welding is used in the manufacture of metallic compound materials, cladding, welding of tubes to tube plates, and so on. It is a specialized process performed by only a few companies [25].

Ultrasonic Welding

Ultrasonic welding is used to weld similar or dissimilar materials, including plastics, in overlap joints. Vibrations parallel or perpendicular to the surface of the weldment are generated by an ultrasonic transducer attached to one of the clamping tools, which supplies the necessary clamping or welding force perpendicular to the surface of the weldment. The oscillatory shear or normal stresses break up and remove the oxide layers or contaminants, so that perfect coalescence can be established through the clamping pressure. Ultrasonic welding is mainly used to join sheets, foils and wires, and so on, and can be of the spot or seam type.

More detailed descriptions of the various welding processes can be found in the literature.

8.5 JOINING PROCESSES BASED ON FILLER MATERIALS WITH $T_f < T_w$—BRAZING, SOLDERING, AND ADHESIVE BONDING

This group of assembly processes (group 3 in Fig. 8.2) are based mainly on adhesion forces between the filler material and the work materials. The melting temperature of the filler material T_f is lower (often considerably lower) than the melting points of the work materials T_w. The processes can be divided into two groups, based on metallic filler materials (brazing and soldering) and non-metallic filler materials (adhesive bonding).

8.5.1 Soldering and Brazing

In soldering and brazing, a permanent joining of similar or dissimilar metallic materials is obtained by the application of a nonferrous filler metal with a melting point below the melting points of the work materials (i.e., no melting of these occurs). The liquid filler material is distributed in the joints by capillary action.

The joining process is based primarily on adhesion between the filler metal and the work materials, but in brazing, cohesion also occurs. This permits virtually all metallic materials to be joined by these processes. Because of the lower joining temperature compared to welding, distortion of the assembly causes little difficulty.

To obtain a high-quality joint, the following requirements must be fulfilled:

Clean surfaces
Correctly shaped joints (i.e., the gaps or clearances must be small enough to permit capillary action)
Correct joining temperature

Depending on the temperature (i.e., the melting point of the filler material) the following processes can be identified:

Soldering (<450°C)
Brazing (>450°C)
Braze welding (>450°C)

Before a short description of these processes is given, the heating sources usually applied will be mentioned. The primary basic process is mechanical (flow of the filler material into the joint), but the secondary process of heating is thermal. Table 8.3 shows the types of heat source and their practical applications. The words in parentheses indicate whether the heat source is used for soldering or brazing or both.

Soldering

In soldering, a solder is used with a melting point below 450°C. The most extensively used solders are alloys of tin and lead with small amounts (less than 0.5%) of antimony. A typical example would be solder with 40% tin and 60% lead. Higher contents of tin increase the fluidity, cost, and strength of the solder. The working temperature is about 250°C. For the soldering of light metals, a solder consisting of tin, zinc, and cadmium is often used, with a working temperature of about 300°C. A wide variety of solders are available with working temperatures between 100 and 400°C. As mentioned previously, the surfaces to

TABLE 8.3 Heat Sources Used for Soldering and Brazing

Type of energy	Practical heating principle
Chemical	Torch (flame) (soldering/brazing)
Electrical	Resistance (soldering/brazing)
	Induction (soldering/brazing)
Thermal	Soldering iron
	Furnace (brazing)
	Salt bath (brazing)
	Metal bath (dip soldering and brazing)

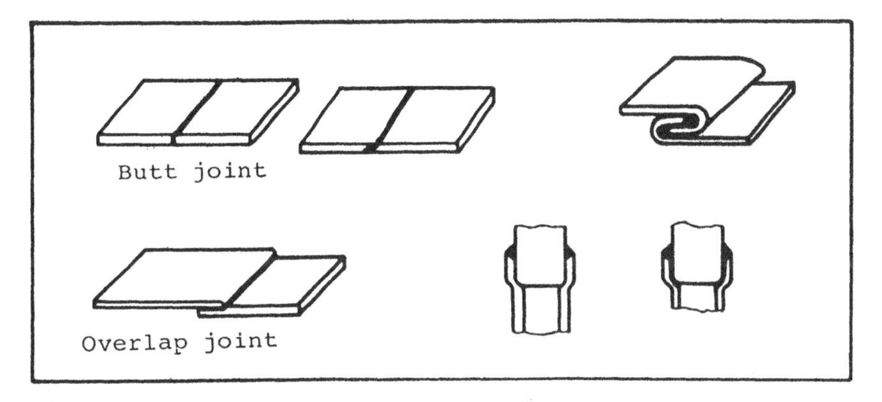

FIGURE 8.16 Common soldered joints.

be soldered must be clean. This is accomplished by the use of fluxes, which can be corrosive or noncorrosive. After soldering, the fluxes must be removed. Some fluxes can also act as temperature indicators, as their color changes with temperature.

Soldering is used extensively in industry for many different applications. The shear strength is low (25–50 N/mm^2). Figure 8.16 shows typical joints. The clearances must be in the range of 0.25–0.025 mm, depending on solder, flux, and material. The desired strength and the geometry of the joint also influence the choice of clearance. Most of the heating methods are the same as those used in brazing (see Table 8.3) and are mentioned briefly in the next section.

Brazing

In brazing, brazing metals with melting points above 450°C are used. Table 8.4 shows typical groups of brazing metals and examples of their application. Many brazing metals are available, and the right one for the specific application must be carefully selected.

TABLE 8.4 Examples of Brazing Metals and Their Applications

Brazing metal	Brazing temperature (°C)	Application
Copper/copper alloys (Cu, Ni, Co, Cr)	850–1100	Steels, carbides, high-speed steels, etc.
Copper with phosphorus	750–850	Copper, copper alloys, etc.
Brass (Cu, Zn, Mn, Ni)	850–1000	Steel, cast iron, copper, nickel, etc.
Silver alloys (Ag, Cu, Zn, Cd)	600–850	Copper, copper alloys, steel, etc.
Aluminum alloys (Al, Si)	500–600	Aluminum, aluminum alloys

To obtain a high-quality braze, the surface oxides must be removed, no oxidation should take place during heating, and the surface tension of the brazing metal reduced. This is accomplished by a suitable fluxing agent. Before applying the flux, the work materials must be clean (i.e., free of dirt, oil, etc.).

The flux must be selected on the basis of the base metals used. For furnace, induction and dip brazing, paste fluxes are used, and for torch brazing, paste or powder fluxes can be used. The paste can be brushed onto the surfaces.

Since the fluxes are corrosive, the brazed assemblies must be cleaned carefully. As the liquid brazing metal is distributed in the joint by capillary action, the clearances or gaps must be of a suitable size and have parallel walls. For copper brazing, the clearances are generally in the range 0.02–0.05 mm; for silver brazing, 0.05–0.20 mm; for brass brazing, 0.2–0.5 mm; and for aluminum brazing, 0.1–0.3 mm. The brazing metals can be supplied in the form of wires, rods, thin shims, bands, foils, and so on. In the brazing of complex components, jigs or fixtures may be used to assure proper alignment. It is often possible to provide fits, staking, and so on, in the design of the individual components, facilitating alignment. Figure 8.17 shows some typical brazed joints.

The heating source for brazing were listed in Table 8.3. Torch brazing is used for repair work and small batch production. The equipment necessary is simple and cheap. Furnace brazing, where preloaded assemblies, possibly held in jigs or fixtures, are heated in a controlled atmosphere, is used mainly for mass production.

Salt bath brazing, where the preloaded and fastened assemblies are dipped into molten salt kept at a temperature slightly above the melting point of the brazing metal, is used primarily for larger components and where very thin or very thick sections are brazed. Induction brazing, where high-frequency induction currents heat the preloaded assemblies, is a fast production method that lends itself to some automation. It is widely used in industry, especially for assemblies with a good surface finish. Resistance brazing is mainly used in the electrical industry to braze conductors, connections, and so on, and the heat is supplied through electrodes of graphite. Metal bath or dip brazing, where the

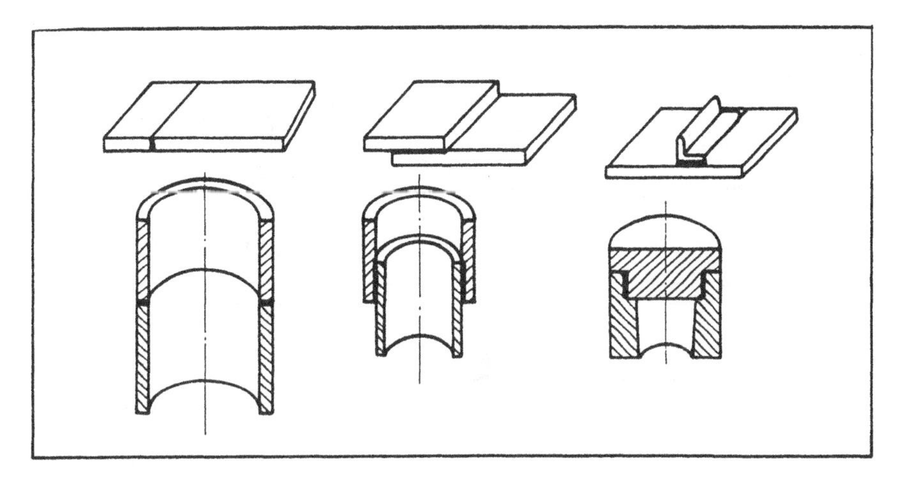

FIGURE 8.17 Common brazed joints.

assemblies are dipped in a molten bath of the brazing metal, is used principally for small components, typically for the fastening of wires in the electrical industry. In soldering, the metal bath or dip heating is used extensively, and the application of induction heating is increasing. Much soldering is done with electrical soldering irons, which is a versatile and cheap method.

The shear strength of brazed assemblies is rather high, normally 100–150 N/mm^2, but values as high as 300 N/mm^2 may be obtained.

Braze Welding

Braze welding is used primarily for the repair of gray and malleable cast-iron components, and sometimes for steel parts. It is generally preferred to welding because of the longer heating time. Braze welding differs from ordinary brazing only in the way that the filler material is supplied. In brazing, capillary forces distribute the liquid filler metal, whereas in braze welding using vee or double-vee joints, as in welding, the material is distributed by gravity. An oxyacetylene torch is predominantly used as a heating source.

8.5.2 Adhesive Bonding

In adhesive bonding, a nonmetallic adhesive material is used to join similar or dissimilar materials. Bonding, which takes place between the adhesive material and the workpieces, not between the workpieces themselves, is based on adhesive forces of a physical, electrical, and chemical nature. The primary basic process is mechanical, involving the flow or the placing of the adhesive material in the joint. The secondary basic process is chemical, resulting in a hardening of the adhesive.

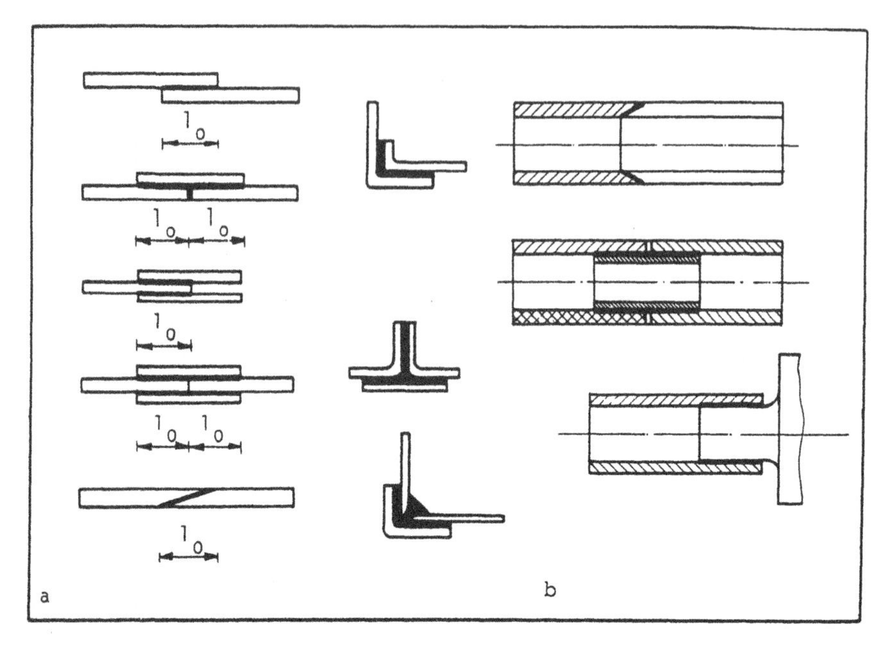

FIGURE 8.18 Examples of (a) joints and (b) components in adhesive bonding.

Nearly all types of materials can be joined by adhesive bonding. Joining of metals by adhesives has increased very rapidly in the past 10 years, and many suitable adhesive materials have been developed.

To obtain a high-quality adhesive bond, the recommendations or prescriptions given by the adhesive manufacturer must be followed carefully. Selection of the right adhesive material is dependent on the actual work materials, the functional requirements (loading: mechanical, thermal, chemical, surrounding media, etc.), and the joining procedure practical (or desired). A short description will be given of adhesive materials, joint geometries, and joining procedures [26,27].

A wide variety of adhesive materials are available offering a large spectrum of properties. An adhesive material consists in general of the following ingredients: the base material, the solvent, the filler material, and the hardener. Depending on the base material and the desired properties of the adhesive, one or more of the last three ingredients may be missing.

The base material gives the adhesive the desired adhesion properties. The solvent gives a suitable viscosity and releases the cure of the adhesive. A filler material may be added to increase strength, to reduce shrinkage or thermal expansion, and so on. The hardener activates the adhesive.

The base material can be thermoplastic and thermosetting resins and artificial elastomers. Other types of materials, such as ceramics, can also be used as adhesives. Thermoplastic base materials can be polyamides, vinyls, and nonvulcanized rubbers. Thermosetting base materials can be epoxies, phenolic rubber, and vinyl. Depending on the setting temperature, the thermosetting adhesives can be classified into cold-setting adhesives (normally low-molecular-weight thermosetting resins) and hot-setting adhesives requiring heat to produce cross-linking. Since the thermoplastic adhesives soften and lose strength when the temperature increases, they cannot be used at elevated temperatures. Thermosetting and thermoplastic resins may be combined.

As mentioned, many types of adhesives are available. Selection of the proper one for a specific application must be done carefully and the recommendations of the manufacturer followed. Common joints used in adhesive bonding are shown in Fig. 8.18. An important characteristic of a joint is the thickness of the adhesive and the overlap length l_0. The thickness is normally in the range 0.05–0.3 mm and the overlap length is generally 5–10 times the thickness of the workpiece.

The shear strengths obtainable in adhesive bonding are in the range 10–50 N/mm^2, but special procedures and adhesives may result in higher strengths. As mentioned, rapid development of excellent adhesives in the past few years has increased considerably the use of adhesive bonding of metals.

8.6 SURVEY OF THE JOINING METHODS

When selecting a joining method for a specific application, many factors must be considered, including the following:

Functional requirements
Materials
Design (geometry)
Dimensions
Production conditions and rates
Available methods
Economy

It should be remembered that soldering, brazing, adhesive bonding, and resistance welding are normally used to join thin sheets (<3 mm thick).

Table 8.5 shows the most common joining methods and the suitable materials.

TABLE 8.5 Schematic Survey of Joining Methods and Suitable Materials

Process	Steel	Stainless steel	Cast iron	Aluminum + aluminum alloys	Copper + copper alloys
Gas welding	X		X	X	X
Shielded metal-arc welding	X	X	X	X	X
Submerged-arc welding	X	X			
MIG welding	X	X	X	X	
TIG welding	X	X		X	X
CO_2 welding	X				
Spot welding	X	X		X	
Projection welding	X	X			
Friction welding	X	X		X	X
Brazing	X	X	X	X	X
Soldering				X	X
Braze welding	X	(X)	X	(X)	(X)
Adhesive bonding	X	X		X	X

8.7 EXAMPLES OF TYPICAL JOINING PROCESSES

In this section short descriptions of some of the most frequently used joining processes are given. The processes are classified in a way similar to that used in Sections 6.3 and 7.4: according to the category of basic process, type of energy, transfer medium, material condition, and surface creation principle. The abbreviations used are the same as in Section 6.3. Because the field of welding covers a very wide area, only those processes most frequently used in industry are described. Further information may be sought in the literature.

Figures are provided by courtesy of The Manufacturing Consortium, Brigham Young University, Provo, Utah.

PROCESS 1: Shielded Arc Welding (Manual) (T, El, Ga, ODF)

Description: Localized melting is obtained using the heat from an electric arc established between a consumable metal electrode and the work material. The electrode, which is coated, acts as filler material. The coating decomposes in the arc, providing a gas shield around it, as well as forming a protective slag that prevents oxidation and other contamination of the weld. Further, the coating helps stabilize the arc and performs metallurgical refinements. The length of the electrode is 350–450 mm, and the melting rate is 200–250 mm/min. Since the process is intermittent, a welder will usually have to stop and fit a new electrode several times in the course of making a weld.

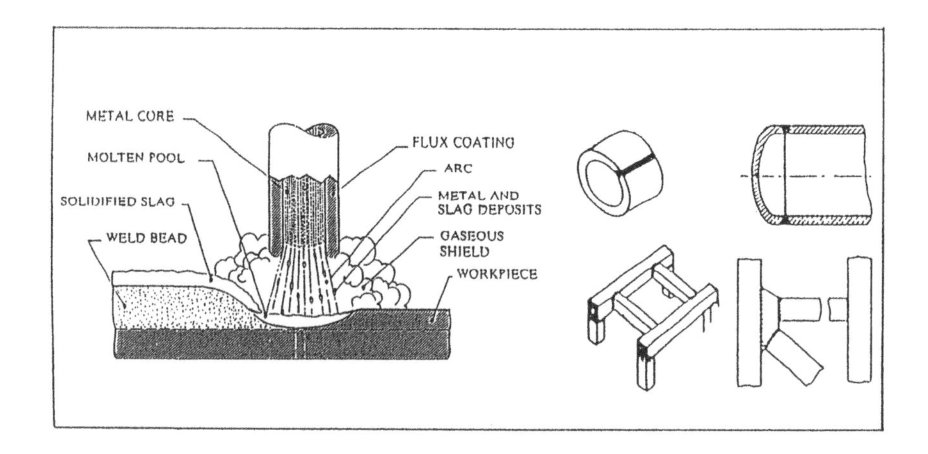

Applications: Manual metal-arc welding is the most used welding process. Because it can be used in all welding positions, the process is widely used in shipbuilding and structural and general engineering. Almost ten times as many mild steel electrodes are used as all other types put together. It is a low capital cost process which uses portable equipment and can be applied to a wide range of joint types. Thickness of material typically 3–30 mm, with thicknesses above 4 mm requiring chamfering the edges of the two parts to be joined and a number of weld runs to fill the joint.

Tolerances/Surfaces: Light gages with few welds, ~1 mm. Heavy gages with few welds, ~5 mm, with many welds, ~50 mm.

Machinery: Transformer (ac) or generator (dc), electrode holder, cables.

PROCESS 2: Submerged-Arc Welding (T, El, Ga, ODF)

Description: The heat for coalescence is provided by an electric arc struck between the workpiece and the bare-metal, consumable electrode, which also delivers the filler material. Shielding is ensured by a blanket of granular flux deposited over the area to be welded. The flux is of the same composition and fulfills the same conditions as in shielded arc welding. Since the arc is completely covered by the flux, extremely high welding currents can be used without spatter or entrainment of air. Currents from 200 to 2000A are commonly used, resulting in a very high productivity.

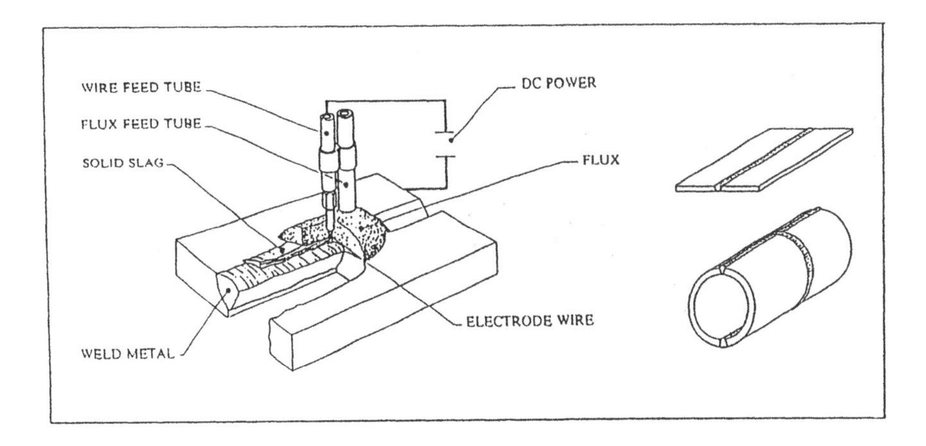

Applications: Submerged-arc welding has found its main applications with mild and low alloys steels, although it has been used for other materials as well. It is a method used mainly for downhand automatic/semiautomatic welding of thicknesses above 5 mm where the welds are long and straight. Flat butt welds or fillet welds are the most common joints produced. Ships, pressure vessels, large steel constructions, and storage tanks are typical fields of applications.

Tolerances/Surfaces: Highly dependent on the fashioning of the parts to be welded, as well as their setup. Typical values are a few millimeters per meter length of workpiece. The weld metal has a smooth almost ripple-free surface.

Machinery: Semi- or fully automatic welding machines, welding speeds from 10 to 50 m/hour. For heavy gages tandem electrodes are often used, offering higher production rates.

Equipment using continuous *covered* electrodes is also available.

PROCESS 3: Gas Metal-Arc Welding (MIG) (T, El, Ga, ODF)

Description: This process is an electric arc welding process in which an arc is struck between a consumable wire electrode and the workpiece. Shielding is provided by an inert gas. MIG is an acronym for *M*etal *I*nert *G*as. The electrode is fed through a special gun, which also supplies the protecting gas. The following gases are used: argon, helium, and carbon dioxide. CO_2 is used extensively to weld steel as it is the cheapest gas. Ar and He are mainly used in the welding of aluminum, magnesium, and stainless steel. MIG welding is always used with direct current and usually with the electrode positive.

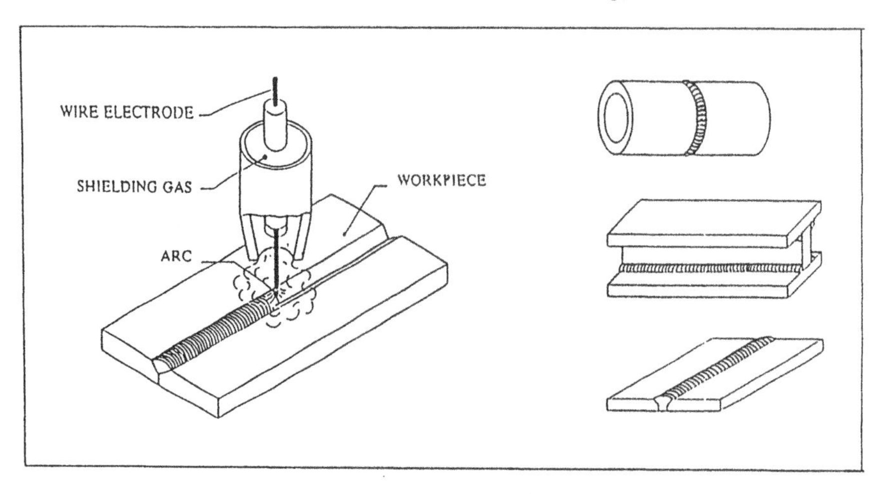

Applications: Gas metal arc welding is a very versatile process. With currents of less than 250 A, it is used for all-position welding with semiautomatic equipment. Using special techniques, nonferrous metal can be welded from sheet gauges up to massive thicknesses with suitable choice of edge preparation. For steels the competitive process at currents below 250 A is manual metal-arc. Higher welding currents than 250 A are used only with downhand welding, and when the joints are straight or simple circumferential seams, automatic equipment may be used. Maximum current is about 700 A.

The industries served include shipbuilding, general engineering (pressure vessel tanks, pipes), and automotive industries.

Tolerances/Surfaces: Dependent on size and geometry of the workpiece and on the number of welds. Typical values are 1–5 mm.

Machinery: Welding machines that provide a regulated dc source of power for melting of the electrode, a wire feeder which advances the electrode as it melts, shielding gas, a control system to regulate gas flow and electrode feed, and a welding gun, which carries the electrode and gas to the workpiece.

PROCESS 4: Tungsten Inert Gas Welding (TIG) (T, El, Ga, ODF)

Description: A metal joining process in which an arc is struck between a tungsten electrode and the workpiece. The tungsten electrode is not consumed because of its extremely high melting point. Shielding of the molten metal and the electrode is provided by an inert gas, typically Ar or He. Filler material, matched to the base material, is supplied separately as a wire or rod. It is not usually required for thin materials. With mechanized welding, filler wire from a coil is fed automatically into the leading edge of the weld pool.

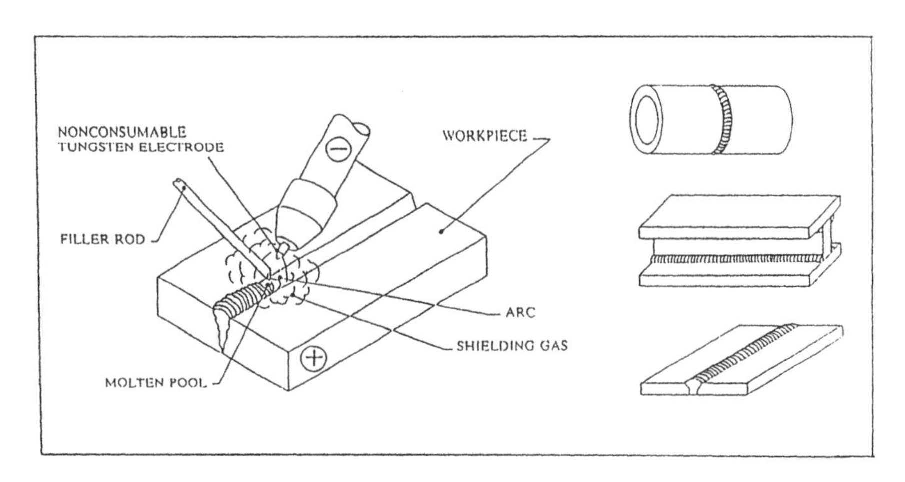

Applications: Although high welding currents (up to 700 or 800 A) are possible, permitting the welding of thick metal, TIG welding is primarily a process for welding sheet metal or small parts. Before a surface can be welded, it must be clean, because the inert gas does not provide any cleaning or fluxing action. The process is at its best when welding single-pass or double-sided close butt joints, edge joints, or outside corner joints. Because it is so easily mechanized and gives high-quality welds, the process is greatly favored for precision welding in the aircraft, atomic, and instrument industries. Various automatic devices are available.

Tolerances/Surfaces: Similar to MIG welding. The finished weld is usually very clean, smooth, and uniform.

Machinery: The equipment for TIG welding consists of a power supply, a water supply to cool the torch, shielding gas, a welding gun, and a filler rod. Power supplies can be ac/dc straight polarity or dc reverse polarity. Dc reverse polarity is seldom used except occasionally for welding aluminum and magnesium.

PROCESS 5: Electron beam welding (T, El, Va, ODF)

Description: Electron beam welding is defined as a metal joining process where melting is produced by the heat of a concentrated stream of high-velocity electrons. The kinetic energy of the electrons are changed into heat upon impact with the workpiece. Energy density at the work can amount to 5×10^{12} W/m^2, a much higher figure than for any other arc welding process. In most applications no filler material is used.

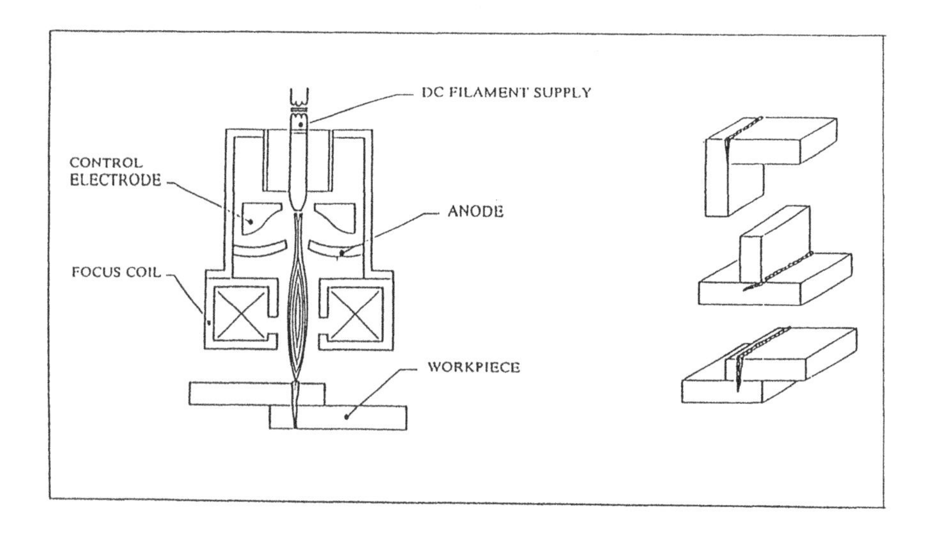

Applications: Because of the requirement of a vacuum chamber, the possible workpiece geometries may be somewhat limited. The chamber also makes it difficult to position the work and align the joint precisely under the beam. The high density of the electron beam as a heat source results in a deep, narrow penetration with a very narrow heat-affected zone. Typical welding thickness is between 5 and 100 mm, and gap widths are typically in the range of 0.02 to 0.3 mm. Joints must be machined and fitted accurately.

Tolerances/Surfaces: The characteristic narrow beam and narrow heat-affected zone permit welding on finished parts to exacting tolerances, typical values being 0.1–0.5 mm.

Machinery: Important features of the equipment are the electron gun, the focusing and beam-control system, and the working chamber. Because of the high cost of the equipment, the process is used primarily for high-quality components and when other welding processes cannot be applied.

PROCESS 6: Laser Beam Welding (T, El, Ga, ODF)

Description: Laser beam welding is a metal joining process that produces melting of materials with the heat obtained from a narrow beam of coherent, monochromatic light. This beam can travel long distances without attenuation and may be focused through lenses to produce spots in which the energy density amounts to over 10^{12} W/m^2 and is equaled only by the electron beam. No vacuum chamber is required for the generation and delivery of the beam, which is generated in a laser medium (gas: CO_2; solids: Nd-YAG), each type having specific characteristics. The laser output can be pulsed or continuous. Shielding gas blown through a nozzle most often coaxially with the beam protects the weld. Typically, no filler material is used.

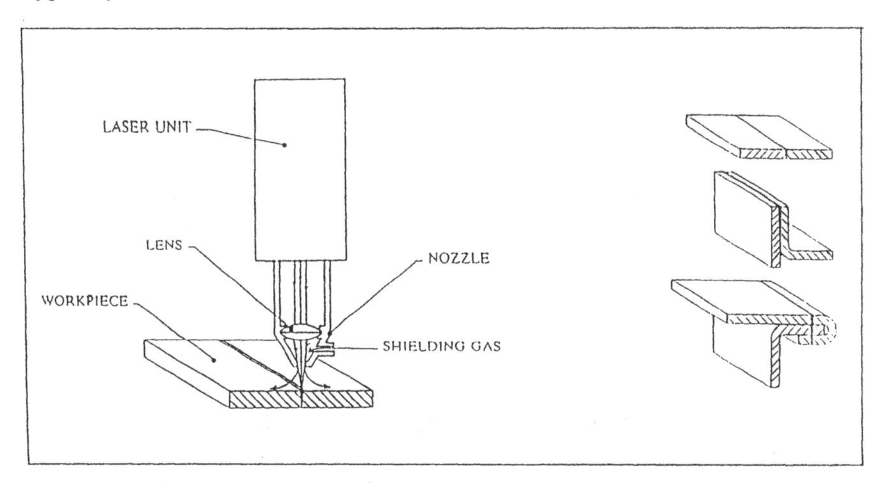

Applications: Laser welding is a relatively new technology in the manufacturing industry but is seeing wider use because of its pinpoint heat and penetration capabilities. Typical welding thicknesses are between 0.3 and 10 mm. Butt joints, edge flange joints, and various lap joints are typical workpiece geometries. Because of the limited dimensions of the beam, high precision is required of the components to be welded. Micro-spot welding of electronic components, particularly where minimum spread of heat is required, is a growing field of application. Very complicated welds may be produced by guiding the laser beam via lenses and mirrors to any desired position in space without significant loss of energy.

Tolerances/Surfaces: As for electron beam welding (see Process 5).

Machinery: The essential equipment for laser beam welding includes the laser gun (often CO_2 or Nd:YAG, the workholding device, and the power supply. The equipment is expensive.

PROCESS 7: Spot Welding (T, El,Ri, TF)

Description: Spot welding is a process in which contacting metal surfaces are joined by the heat obtained from resistance to electric current flow. The electric current generating the heat is introduced to the work through electrodes in contact with the work, resulting in a weld nugget. Pressure is required to place the parts in contact and is an important process parameter. The diameter of the electrodes directly affects the size and shear strength of the weld and must be adjusted according to the thickness of the work.

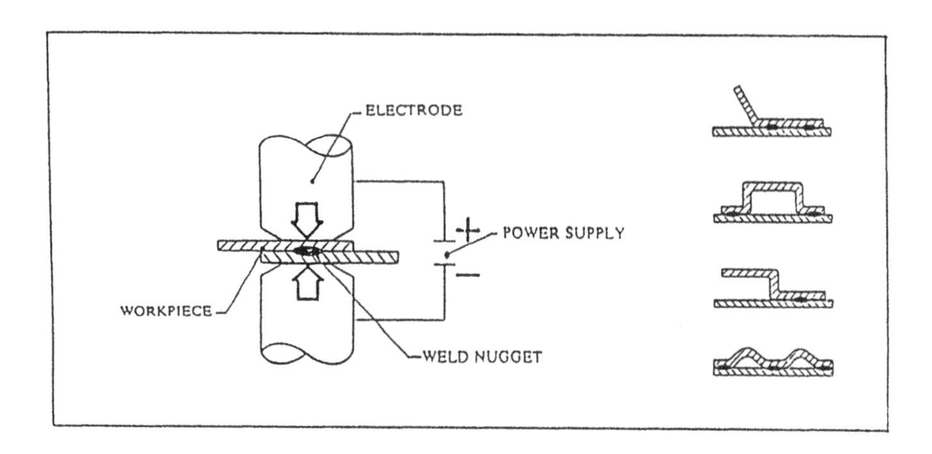

Applications: Spot welding is extensively applied in the joining of sheet metal (thickness ≤3 mm) in mild and stainless steels, heat-resisting alloys, aluminum and copper alloys, etc. The high speed of operation, ease of mechanization, the self-jigging nature of the lap joint, and the absence of edge preparation or filler metal are attractive features of the resistance spot welding process. Typical welding parameters for mild steel are: current 200–400A, welding time per nugget 0.15–1 s, welding pressure 70–100 N/mm^2.

Tolerances/Surfaces: The process as such introduces no serious inaccuracies. Size, geometry, shaping, and holding of the workpieces determine the resultant tolerances, typical values being about 2 mm.

Machinery: Because of the wide variety of applications, there are many different special types of spot-welding machines, all containing a power supply (transformer); arrangements for the setting of current, time, and pressure; electrode holders; and water supply for cooling of the electrodes.

PROCESS 8: Gas Welding (T, Ch, Ga, ODF)

Description: In this process heat is transferred from a flame to the work by convection and radiation. The welding flame is produced by combustion of oxygen (O_2) and acetylene (C_2H_2) in a torch, the function of which is to bring together correct volumes of gas and oxygen, mix them, and form a flame with suitable characteristics. Acetylene and oxygen are supplied in separate steel cylinders in compressed form, necessitating pressure regulators and gauges between the supply cylinders and the torch. The outer zone of the flame provides some shielding against oxidation of the weld. Filler material is supplied from a separate rod.

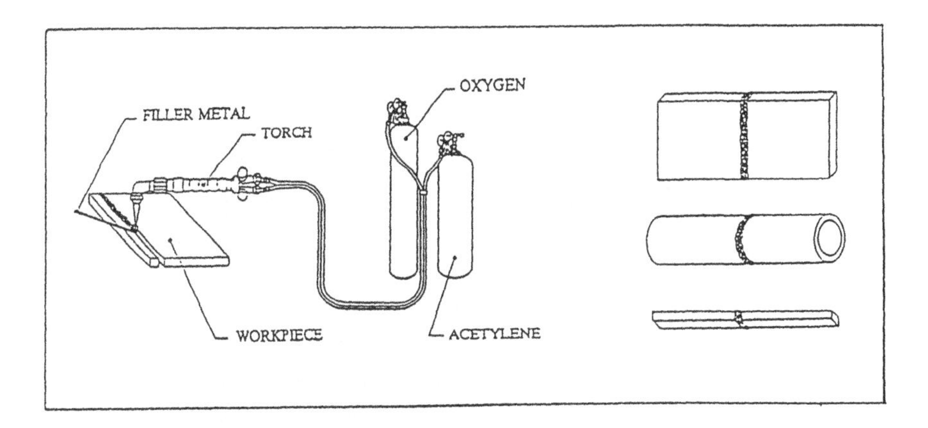

Applications: The gas welding process has about the same fields of applications as the more efficient Shielded Arc and MIG processes; and since their introduction, its use has declined. Today, it is used primarily for repair work and in places where electric power is not available.

Tolerances/Surfaces: Depending primarily upon size, geometry, and holding of the workpieces as well as on the number of welds, typical values being 2–5 mm.

Machinery: The necessary equipment is fairly cheap and consists of a welding torch, pressure bottles for oxygen and acetylene, regulators, gauges, and connecting hoses.

PROCESS 9: Soldering/Brazing (T, El/Ch, Ri/Ga, ODF/TF)

Description: In soldering and brazing the gap between the parts to be joined is bridged by adding a liquid metal having a lower melting point than that of the work, which is not melted. Coalescence is thus brought about primarily through adhesion, permitting joining of similar or dissimilar metallic materials. Heat may be supplied through electric, chemical, or thermal energy. Surfaces to be joined must be clean and shaped correctly so that capillary action can distribute the filler material in the joints. Filler metals are available in a wide variety of shapes, sizes, and consistencies.

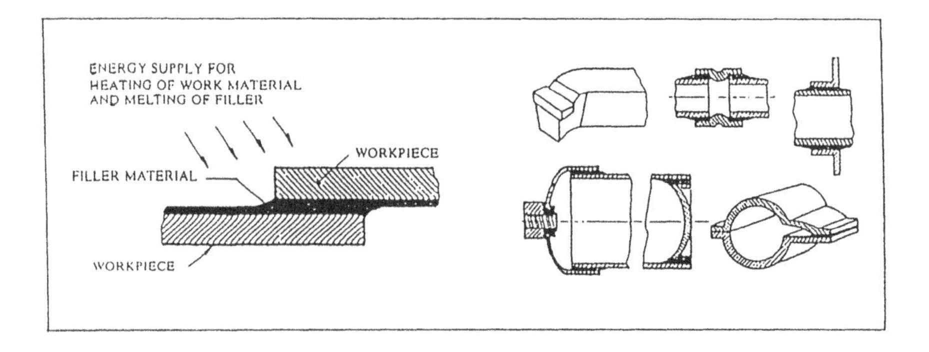

Applications: Soldering and brazing are industrially important joining methods, primarily in the mass producing industries. Because of the lower joining temperature compared to welding, distortion of the assembly causes little difficulty.

When filler materials with melting temperatures below 450°C are used, the joining process is called soldering. Such fillers may be Pb-Sn alloys, which are much used in the electronics industries.

The name brazing is used for the process when employing filler materials with melting temperatures above 450°C. Ag-Cu-Zn and Cu-Zn alloys are examples often used for sheet or tube joining.

Tolerances/Surfaces: As for gas welding, tolerances depend mainly on size, geometry, and holding of the work. Typical values are 0.5–1 mm.

Machinery: A heat source such as a gas welding torch or an electric soldering iron for single piece production or repair work. Salt baths, electrical resistance furnaces, and so on, are used in mass producing industries.

PROCESS 10: Adhesive Bonding (M, T/Ch, Ri/Fl, . . .)

Description: Adhesive bonding is a process in which a substance in a liquid or semiliquid state is applied to adjoining workpieces to provide permanent bonding due to curing through the action of a catalyst, pressure, and/or heat of a nonmetallic bonding material. Similar or dissimilar materials may be joined since the bonding takes place between the adhesive material and the workpieces, not between the workpieces themselves. Adhesives come in a wide variety of types and forms, and joining of metals by adhesives has increased rapidly in the past few years. The parts to be joined must be clean, precisely aligned, and kept together by a source of force to aid the adhesion of the workpieces. Curing at controlled time and temperature is usually necessary.

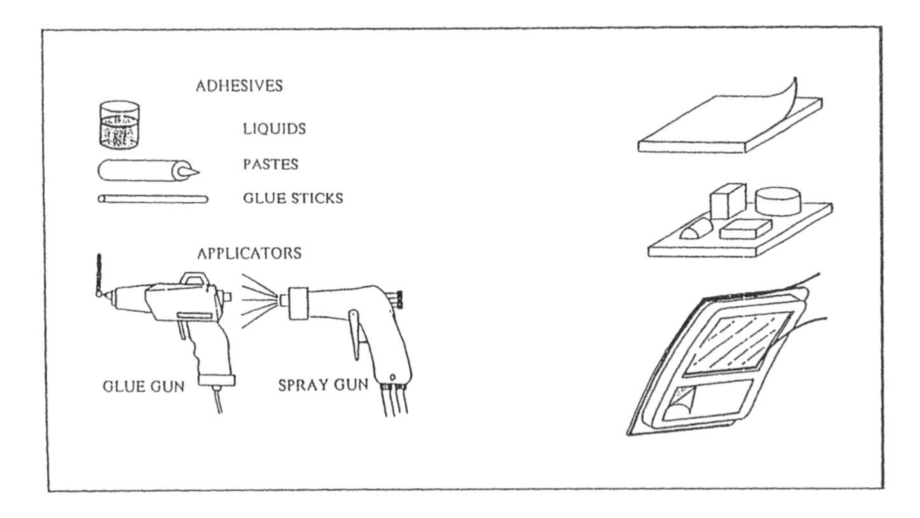

Applications: The geometrical possibilities for adhesive bonding are virtually unlimited, with applications in such areas as the automotive, construction, aerospace, and medical industries as well as in making joints for wood and other relatively lightweight fabrication materials. Bonding pressure may range from 0.1 to 10 N/mm^2 while resultant shear strength, depending on the adhesive used may range from 5 to 100 N/mm^2. Typical geometries include thin sheets or foils joined to heavy gauge workpieces, mounted component parts, and fabrics to solids. Several automated systems for applying the adhesive are available.

Tolerances/Surfaces: Depends primarily on the size, geometry, and holding of the workpieces, 0.2–1 mm being a typical range.

Machinery: The essential equipment is varied, including hot glue guns and caulking guns, both of which are used to apply adhesives essentially in a paste or semiliquid form. Spray applicators are used to apply liquid adhesives and can be automated, whereas brushes are manipulated manually.

9

Granular Materials: Powder Metallurgy

9.1 INTRODUCTION

As described in Chapter 1, the forming or shaping of materials can be done from the solid, granular, or liquid state—referring to the state of the work material in the shaping phase prior to stabilization of the material. Shaping and stabilization can sometimes be integrated.

The manufacture of a product from the granular or particle state covers, in general, a broad spectrum of materials and components or products, such as carbide tools (sintered or cemented tool inserts), metal powder components, sand molds, ceramics, concrete, tablets, and bread.

A granular material is a mixture of solid grains or particles possibly of varying sizes. Each grain or particle may be a combination of smaller units, for example, the crystals in metal grains. Granular materials are generally used for one or more of the following reasons:

The particular material is only available or can only be produced in the granular state.

The desired properties (porosity, combination of materials, etc.) can only be obtained from granular materials.

Manufacture of the product is cheaper than by other methods.

Small components are difficult to produce by other methods.

Production of components from granular materials will generally follow the same pattern:

Production of the granular material
Conditioning or preparation for shaping and stabilization
Shaping
Stabilization of the shape
Finishing operations

Depending on the material and the requirements of the component, these phases involve different basic processes. In this context, only the production of metal powder components will be discussed; this area is, in general, called powder metallurgy.

Within the last decade, the production of components from metal powders has increased rapidly, with a probable yearly expansion of 10–20% in the years to come. This is due to one or more of the following reasons:

The production of simple or complicated geometries can be performed in one operation and with high dimensional accuracy.
A high (nearly 100%) material utilization.
The final properties, even if they are not on the same level as corresponding solid materials, are satisfactory for most applications.
The production of components that can be produced by other methods only with difficulty.
Powder metallurgy is the most economic process in many situations.

Powder metallurgy competes primarily with casting, hot and cold forging, and cutting. These methods normally involve several operations. In the following sections the characteristics of powder metallurgical processes (powders, preparation, compacting, sintering, post-sintering treatments, etc.), properties and applications are discussed [28–31].

9.2 CHARACTERISTICS OF THE POWDER METALLURGICAL PROCESSES

Production of a component by powder metallurgy techniques normally involves the following stages or phases:

Production or selection of powder
Preparation, including the mixing and blending
Pressing or compacting
Sintering or heat treatment
Post-sintering treatment, if necessary

9.2.1 Metal Powders

Different methods for the production of metal powders have been developed. The most important are the reduction of ores, atomizing, and electrolytic deposition. In conventional powder metallurgy, powders produced by the reduction of ores are used extensively, but in recent years the application of powders produced by atomization has increased rapidly. Powders produced by electrolytical deposition are used only for special purposes, and the market for these powders is decreasing. The types and properties of the powders have a major influence on the final properties of the component, so that a fundamental knowledge of powders is important.

Powders Produced by the Reduction of Ores

The reduction process is used primarily to produce iron powders. The purity of the powder is directly related to the purity of the ore. In the production of iron powders, pure and dry iron ore is heat-treated in sealed drums together with coal dust, coke, gravel, and chalk at about 1200°C for about 90 h (the Hoeganaes method). After this reduction, the resulting iron sponge cake is crushed, ground, and heat-treated in a hydrogen atmosphere to provide a reduction of the oxides and to anneal the powder particles. The powders contain the impurities from the ores, and a single grain has many internal pores, making the powder unsuitable for pressing to very high densities, as it will require enormously high pressures to close these internal pores. Depending on the production conditions, the number and size of the internal pores vary, and the general shape of the grains is irregular. In addition to iron powder, nonferrous metal, cobalt, molybdenum, and tungsten powders can be produced by the reduction of ores.

Powders Produced by Atomization

In atomization, the powders are produced from the liquid state, which gives great freedom in the choice of materials and in the alloying process. The purity of the powders is directly related to the raw materials and the melting and refining processes. The shapes and sizes of the particles can be varied within wide limits, depending on the process parameters.

A flow of liquid metal through an orifice is broken up by a jet stream of gas (air or inert gases), water steam, or water. Gas atomization gives spherical and large particles; water atomization gives smaller and irregular grains without internal pores.

Atomization can be used to produce powders of iron, steels (including stainless), lead, zinc, aluminum, bronzes, brasses, and so on.

The use of powders produced by atomization has increased rapidly due to the purity obtainable, the alloying possibilities, and the powder properties. It should be mentioned that in the past, the price for atomization powders has been higher than for reduction powders, but now these prices are comparable.

Powders Produced by Electrolytic Deposition

After electrolytic deposition, the metal is crushed and ground by mill grinding to the desired grain sizes. Iron powders produced by electrolytic deposition are more expensive than those produced by reduction or atomization. The electrolytic powders are used only where their special properties (including high purity, density, and compressibility) can be utilized.

As mentioned previously, several other powder manufacturing processes exist. They will not be described here, as the powders produced by these processes are used only for special applications.

A metal powder can be characterized by:

Chemical composition
Particle-size distribution
Particle shape (spherical, irregular)
Surface characteristics
Internal structure (pores, etc.)
Flow ability (ASTM 213.48/212.48)
Compressibility
Green strength (strength after compaction)
Sintering properties or sintering abilities (change of dimensions, strengths, etc.)

To describe these characteristics, several testing methods have been developed (SAE/ASTM/MPI standards or recommendations).

Powder manufacturers supply all the necessary information about their powders, and these should be studied carefully before selecting a powder for a specific application. Most of the listed characteristics or properties of the powder influence the pressing and sintering processes as well as the green and final strengths. In the present context, powders will not be described further, but additional information can be found in the literature [30].

9.2.2 Preparation of the Powder

An important stage in the production of metal powder components is the preparation of the powder for compaction and sintering. The preparation of a powder consists mainly of mixing or blending to obtain a uniform distribution of the base powder and alloying elements. The mixing process must be carried out carefully. Too heavy mixing may cause strain hardening, wear of the particles against each other, layering, and so on. Recommendations from the manufacturer must be followed.

Lubrication can be provided as internal or external lubrication. In internal lubrication, a lubricant (zinc stearate or stearic acid, 0.25–1% by weight) is mixed with the powder, increasing its compressibility and decreasing its green

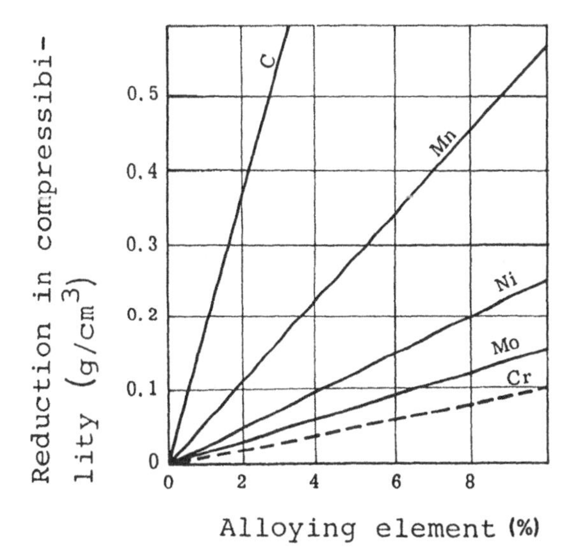

FIGURE 9.1 The influence of alloying elements on the compressibility of steel powders [28].

strength. After pressing, the lubricant is driven out by heat treatment (in air at 375–425°C), before sintering in a controlled atmosphere. In external lubrication, only the die walls are lubricated, avoiding the heat treatment necessary to drive out the lubricant, but this method does not provide the improved flow and compaction properties.

Considering the alloying elements, a distinction between metallic and nonmetallic alloying elements must be made, since they have quite different diffusion rates and thus require varying sintering times to obtain a homogeneous structure.

In general, iron powder should contain very small amounts of carbon and other nonmetallic alloying elements, as these increase the hardness and decrease the compressibility (see Fig. 9.1). The compressibility is measured as the density obtained for a compacting pressure of 400 N/mm^2. The preferred method is to mix the powder with graphite (1% graphite results in a steel with 0.8%C after sintering) so that good compressibility is retained.

Alloying by mixing the powder with the right amounts of metallic elements will require a very long sintering time to achieve a homogeneous structure. To reduce the sintering time required, prealloyed powders are preferred, as the reduction in compressibility is not very severe (see Fig. 9.1). If reduction powders are used, alloying with metallic elements is normally based on partial alloying, which requires a special alloying stage. The pure powder is mixed with the alloying elements and heat-treated so that an incomplete diffusion into the base powder has taken place. During the final sintering, after compaction, the diffusion is completed. If atomization powders are used, a regular or true alloying is obtained, as the alloyed powder is produced directly from the liquid state. Such powders are called prealloyed powders.

In the past decade, a rapid increase in the consumption of partial alloyed (reduction) powders and prealloyed (atomization) powders has taken place. By using these powders, excellent mechanical properties of the components can be obtained with tensile strengths in the range 400–1000 N/mm^2, and even values of 1500 N/mm^2 can be reached with special and more expensive powders. The alloying elements are primarily Cu, Ni, Mo, and Mn. Stainless steel powders are being employed at an increasing rate. The development of powders leading to high-strength components considerably increases the potential market for powder metallurgy.

The discussion above focused on iron and steel powders, but it must be emphasized that a wide spectrum of nonferrous metal powders are available, including brass, bronze, aluminum, nickel and zinc powders.

9.2.3 Pressing or Compacting of Powders

The technology of pressing or compacting is a broad and complicated subject, requiring a high degree of engineering ingenuity. Therefore, the description given here must be considered as elementary.

Background

In this section the background for the effect of the different powders on die design is discussed briefly before the various die design principles (i.e., compacting methods) are discussed.

The component is specified by the desired density, strength, tolerances, and so on, and the powder is specified by its compressibility curve, that is, its density as function of compacting pressure (double-action pressing) (see Fig. 9.2a). The apparent or filling density is 2.4 g/cm^3 and for a compaction ratio of 2 (resulting in half the original height), the density will be 4.8 g/cm^3 For practical purposes, the compaction ratio must be in the range 2.5–2.8, corresponding to densities in the rage 6–7 g/cm^3. Figure 9.2b shows the compression ratio versus compacting pressure. From Fig. 9.2 a new curve showing the punch motion within the die as a function of the compacting pressure can be constructed based on the desired density and for 100% punch motion. Such a curve shows that about 85% of the motion is already carried out at compacting pressures of 200 N/mm^2. For the remaining 15% of the motion, the compaction pressure must increase from 200 to about 800–1000 N/mm^2, which means that the compaction press is required only to provide high compacting pressures over a very short travel length.

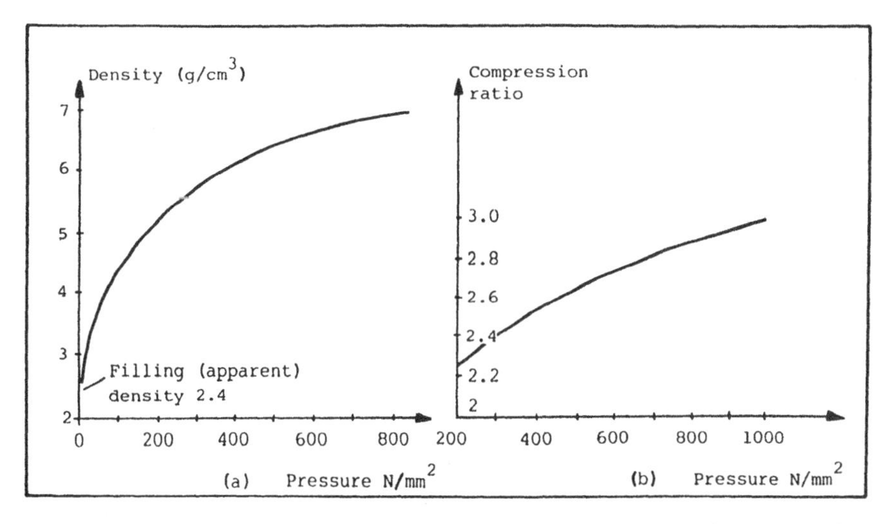

FIGURE 9.2 Density and compression ratio as a function of compacting pressure for iron powder [30].

Approximately 90% of the industrially applied powder components have densities in the range 5.7–6.8 g/cm^3, but in the past few years, the application of components having densities in the range 7.0–7.2 g/cm^3 has increased; these components have excellent mechanical properties. This has become economical because of the development of better die materials with higher wear resistance and powders with high compressibility.

The filling of the die cavity is generally achieved through volume dosing where the powder flows into the cavity and excess powder is scraped off, giving tolerances of $\pm 1\%$. If higher accuracy is required, dosing by weight must be used, but this is more tedious.

When the powder has been compacted, it must be ejected from the die. The ejection phase must be considered carefully, since fracture may arise at weak points or weak sections, when the elastic energy is released, or when forces act over a small fraction of the powder surface. To obtain optimal production, the designer of the component must analyze both the compaction and ejection phases carefully before deciding on the final geometry.

Compacting or Pressing Principles

The desired component must, in general, have a uniform density distribution throughout. When compacting powder in a cavity with one movable punch (single-action pressing) (Fig. 9.3a), the properties of the powder cause a nonuniform density distribution due to friction between the individual grains and between the particles and the die walls. This means that density decreases with

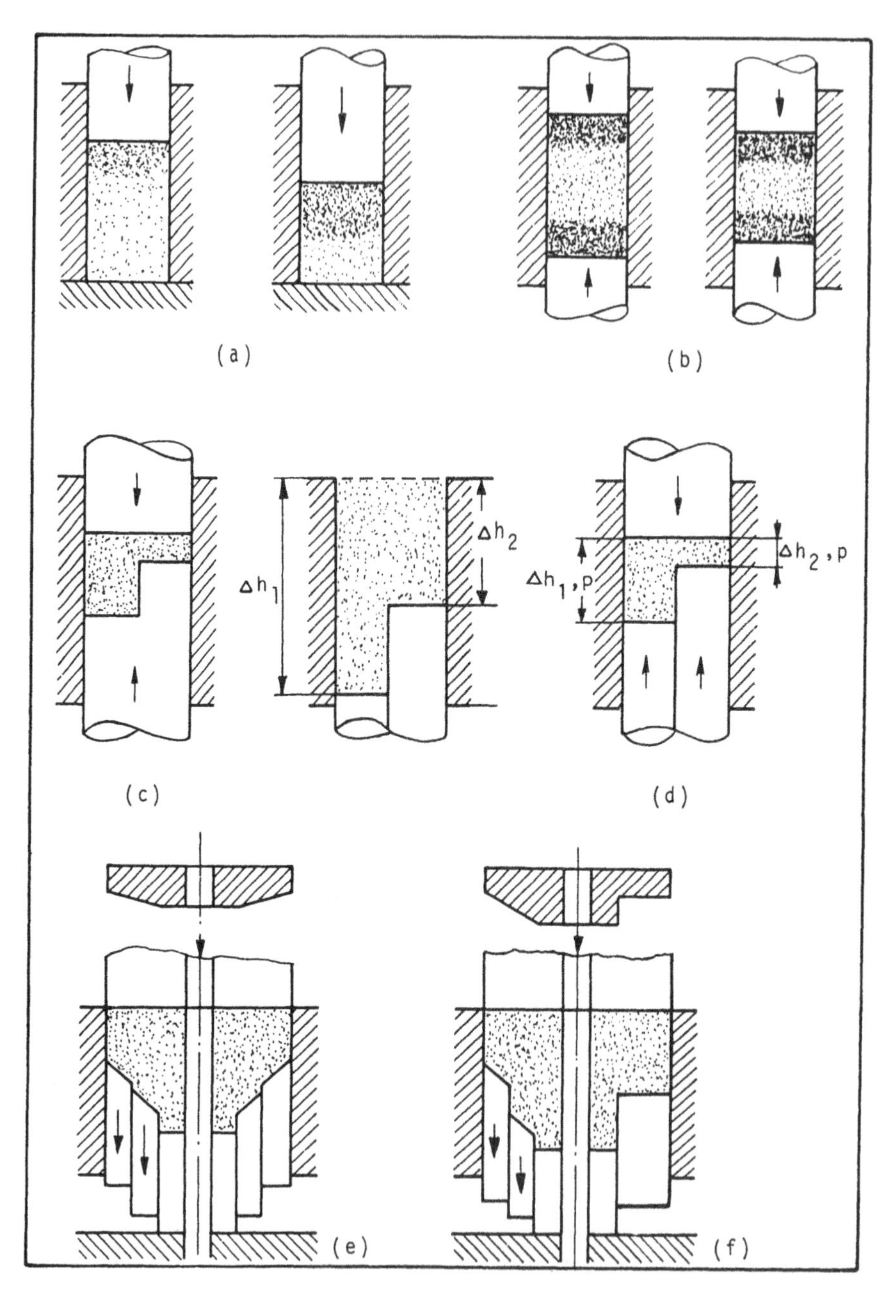

FIGURE 9.3 Powder compaction principles: (a) single-action pressing; (b) double-action pressing; (c), (d) compaction of components with more than one level in height; (e), (f) examples of double-action compaction with sectional punches.

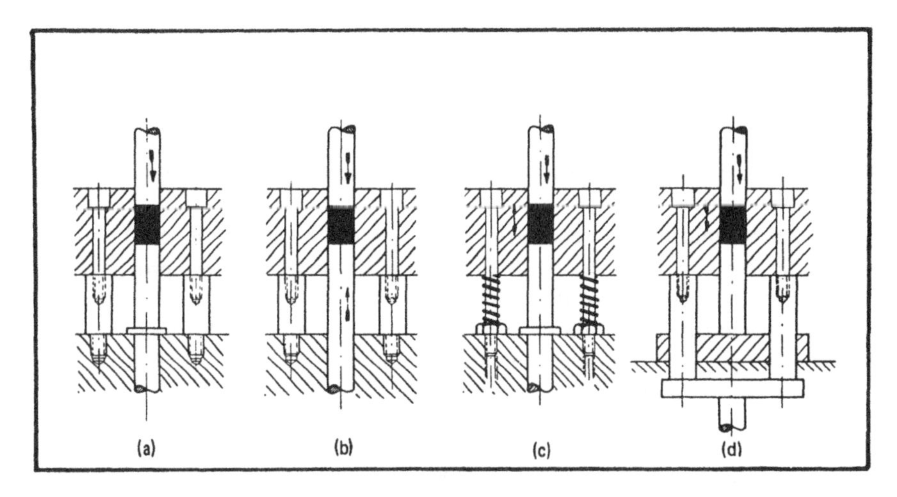

FIGURE 9.4 The four main principles of die design in powder compacting [30].

increasing distance from the punch. Because of this, only nearly plane and thin components can be produced satisfactorily by single-action compaction. By pressing from both sides (double-action pressing) a more uniform density distribution can be obtained (see Fig. 9.3b). To obtain a reasonably good density distribution, the height/width ratio should be kept below 2–2½ whenever possible. For the compaction of components with different levels in height, it is necessary to section the lower punch to obtain the same compression ratio (i.e., density) throughout the component (see Fig. 9.3c and d) ($\Delta h_1/\Delta h_1, p = \Delta h_2/\Delta h_2, p$). Figures 9.3e and f show examples of die design when sectioned punches are necessary.

Figure 9.4 shows the four main die design principles, illustrating how the die design affects the pressing equipment. The single-action compaction, Fig. 9.4a, requires two motions: an active pressing motion, provided by the upper punch, and an ejection motion, provided by the lower punch. As mentioned, only thin and flat components can be produced by this method.

Double-action compaction (Fig 9.4b) requires the same number of motions as single-action compaction, the only difference being that the lower punch is active during compaction, not simply used for ejection. During compaction, the two punches move against each other with the same velocity. About 80% of all powder components in the United States and about 40% in Europe are produced in this way.

Figure 9.4c shows compaction with a floating container or die, where the lower punch is stationary during the compaction phase. An effect similar to that of double-action compaction is obtained by the floating container, which moves down a distance equal to half of the punch travel, because of friction between the powder and the die walls. Ejection may be carried out either by moving the container farther down, with the lower punch stationary until the component is free, or by moving the lower punch upward.

In compacting with a controlled withdrawal motion of the container (Fig. 9.4d), the ejection is carried out by moving the container down until the component is free.

Any mechanical press (eccentric, crank, cam, knuckle joint, etc.) and any hydraulic press can be used with floating and withdrawn dies, provided that there is sufficient room for the tooling.

The geometrical possibilities in powder compaction depend primarily on the principle of powder compaction and the ingenuity of the engineer. More detailed descriptions of die design can be found in the literature [30].

Sizing and Coining

If the dimensional tolerances and/or the mechanical properties obtained after the sintering (or heat treatment) of the compacts produced by one of the methods described above are unsatisfactory, a sizing or coining operation can be carried out.

The sizing operation, which is carried out at moderate compacting pressures, serves to improve the dimensional accuracy of the product. For a small batch of components, the primary pressing or compacting die can be used to carry out the sizing of the compact. Large batches of compacts are normally sized in a special die using an inexpensive sizing press.

The coining operation serves two purposes: improving the mechanical properties of the product and improving the dimensional tolerances. The mechanical properties can be improved only by increasing the density of the compact, which means high compacting pressures (higher than or equal to the primary compacting pressures). Thus, in general, coining requires a special die for the purpose, often of a higher quality than the primary die, because of the higher pressures and the adverse wear conditions.

When coining is involved, the sintering process carried out between the primary compacting and the coining operation is often incomplete and takes the form of pre-sintering for a short time and at a temperature considerably below the normal sintering temperature but sufficient to anneal the compact. After coining, the compact is fully sintered, producing a component with excellent mechanical properties and dimensional tolerances. If the requirements of the product are exceptionally high, a sizing operation may follow the coining operation.

Details of sizing and coining operations are described in the literature [30].

Various Compacting or Shaping Methods

The *axial pressing* methods described above are by far the most important methods for the production of constructional metal powder components. To produce specialized components, often in materials difficult to shape, various processes have been developed, two of which are described here.

Isostatic Compaction. Here the powder is placed in a deformable container made of plastic, rubber, mild steel, and so on, which is subjected to a high fluid pressure in a heavy pressure chamber. The container is removed after the compaction. High compact densities can be obtained with this method, which is used to produce high-speed steel products and many other specialized components. It is possible to carry out the compaction at high temperatures, and in the last few years the process has become relatively widely used in industry.

Extrusion and Rolling. It is possible to manufacture extruded or rolled products directly from the powder state. Depending on the material and the desired properties, different sequences of heating and processing may be chosen. This field of production is expanding rapidly.

9.2.4 Sintering

The term *sintering* covers both the process, which causes the strength of the green compact to increase and its porosity to decrease, and the practical operations necessary to acquire these changes. The most important factors in sintering are the temperature, the time, and the atmosphere. The properties obtained after sintering are influenced by the powder material, the size and shape of the particles, their surface characteristics, and the compacting pressure applied.

If sintering takes place at temperatures below the melting points of the constituents, solid-state sintering results. If sintering is carried out at temperatures between the melting points of two constituents, sintering with a liquid phase results. Solid-state sintering is used for all structural components; sintering with a liquid phase is used for special products such as carbides and ceramics. Solid-state sintering will be given the most attention here, although liquid-phase sintering will be described briefly.

In solid-state sintering, the particles in the compact form a coherent whole. Two major phases in the sintering process can be identified.

In the first phase, the contact areas (or contact points) between the particles grow and form rounded necks, but porosity is still present in the form of interconnecting channels.

In the second phase, the necks grow and the network of channels gradually reduces to isolated pores. In time, the pores tend to become spherical, the smaller ones vanishing and the larger increasing in size; thus the average pore size increases, but the total porosity decreases only slowly.

In practice, only the first phase is fully completed, and depending on conditions, the second phase is partially completed, resulting in a product with a combination of interconnecting channels and isolated pores.

The sintering process generally takes the form of mass transportation with a minimization of the free surface energy as the driving force. Theoretically, the driving force will not disappear before the surface area reaches a minimum (i.e., equal to the external surface area of the compact). But the time to proceed from the end of the first phase to this ideal situation is so great that it never occurs in practice.

The mass transportation mentioned here is carried out by plastic deformation, evaporation/condensation, surface diffusion, and volume diffusion. Of these, the surface and volume diffusion are the most important. The sintering mechanism is not discussed further here, but detailed descriptions can be found in the literature [30].

The sintering temperature and time have a major influence on the final properties of the product. The density of the component is approximately independent of temperature and time. The shrinkage of the component (0.1–0.3% by volume) is balanced by a loss in weight from the reduction of surface oxides, the evaporation of lubricants, and so on. For iron powders the tensile strength generally increases considerably when the temperature has passed 650°C. This increase in strength slows down when the temperature has exceeded about 900°C. Iron powders are normally sintered in the range 900–1150°C. The sintering time has considerable influence on the final tensile strength. When sintering iron powder at 1150°C, about 85% of the maximum obtainable strength (corresponding to a sintering time of 2 h) is obtained after 15 min. Often, sintering times in the range 0.5–1.5 h are used in practice. The ductility of the final component varies with temperature and time in a way similar to the variation in tensile strength.

During coining, the smallest forces are required when the sintering has not gone too far, in other words when the mechanical strength has not increased too much. For iron powders, pre-sintering is carried out in the range 650–850°C, when the particles have become fully annealed without any significant increase in strength, necessitating higher coining pressures.

Sintering is carried out in a protective atmosphere to prevent oxidation during heating and to reduce existing oxides, to remove gases resulting from the heating of lubricants and other material, and to control the carburization and decarburization of iron and iron-rich compacts. The most commonly used atmospheres (the actual type depends on the material and the purpose) are hydrogen, dissociated ammonia, burned ammonia, exothermic and endothermic gases, and vacuum.

Sintering is carried out in various types of furnaces, which can be of the continuous or batch type.

In recent years, a new process combining compacting and sintering has been developed, the spark sintering process. Here, a high-energy electrical spark discharged from a capacitor bank removes, within a second or two, the oxides and other contaminants from the particles, which are then subjected to a compaction pressure. The current is maintained for about 10 s, and the powder is pressed further between the electrodes to obtain the desired density. This process has been applied to many different materials, but it is not yet widely used in industry.

The foregoing discussion of sintering covers only solid state sintering. When one of the major constituents is liquid during sintering, the process is quite different. Sintering with a liquid phase involves very high volume shrinkages, of 40–60%, causing lower dimensional accuracy than that of solid-state sintering. The production of sintered carbides (WC, TiC, etc., with cobalt as a binder) is an important example.

9.2.5 Post-Sintering Treatments (Finishing Operations)

Depending on the desired properties of the component, it may be necessary to treat a product after sintering to obtain these properties. These finishing operations may be specially developed for particular compacts or may take the form of conventional processes. A few of these processes are described briefly next.

Impregnation. A wide variety of self-lubricating bearings are made from porous compacts impregnated with a type of lubricant, mostly oil. The compact, which is produced with a porosity of 25–35%, is immersed in heated oil under pressure or is vacuum-treated, so that the network of pores is filled with oil. The lubricant is released during service at a rate depending on the load and the temperature. The materials are generally bronze or iron-based.

The impregnation of compacts with plastics, for example, can be carried out to achieve pressure- or airtight components, by closing the pores prior to plating or to produce other desired properties.

Infiltration. In this process, the pores in the compact are closed by filling them with a molten metal of lower melting point than the major constituent of the compact. The infiltration can be carried out either in a special pre-sintering process or during normal sintering. The infiltration material can be applied upon or below the compact as a solid which melts at the actual sintering temperature, or it can be applied as a liquid. The molten metal is drawn into the compact by capillary action.

The strength of a component can be increased by infiltration from 70 to 100%. A typical example is an iron compact infiltrated with copper.

Heat Treatment. Powder compacts can be heat-treated by conventional methods. The higher the density of the compact, the better the results. Hardening processes are applicable for iron and steel compacts; phase transformations, gas carburizing, nitriding, and carbonitriding.

Surface Coatings and Treatments. Compacts with high densities can be plated by standard procedures, but for those having lower densities, impregnation with wax or plastic, for example, may be necessary. Barreling, peening, and so on, may also be used. Phosphate coatings, steam oxidation, and chromizing, as well as the conventional plating methods, can be used to improve the corrosion properties of the compacts.

Machining. Some compacts can be machined by the standard processes, but special attention must be given to the tool materials and geometry. To avoid corrosion, coolants containing water must not be used during machining.

Powder compacts can be welded together to provide complex shapes by the usual welding methods.

9.3 PROPERTIES AND APPLICATIONS

In this section a short description of the properties and the applications of powder metal components is given; more detailed information can be found in the literature.

9.3.1 Properties

The mechanical properties of powder metal components are mainly dependent on the porosity, size, and distribution of pores and the properties of the base metal. Figure 9.5 illustrates the change in the mechanical properties with density. Table 9.1 shows typical values for different material groups. To improve the mechanical properties, the density must be increased (i.e., the porosity decreased), and this can be accomplished by:

Increasing the compaction pressures
Multiple compactions or pressings
Application of powders with high compressibility
Infiltration
Hot pressing/forging

An increase in the compaction pressure to above 600 N/mm^2 will result in heavy die wear. Normally, pressures within the range 300–500 N/mm^2 are used in industry for single pressings.

As mentioned previously, re-pressing or coining can be used to increase the density and thus the mechanical properties of the component. The dimensional accuracy is increased at the same time, but a particular sizing operation can be applied to obtain high accuracy without improving the mechanical properties.

The following procedures for the production of powder metal components can be chosen, depending on the requirements:

1. Pressing + sintering

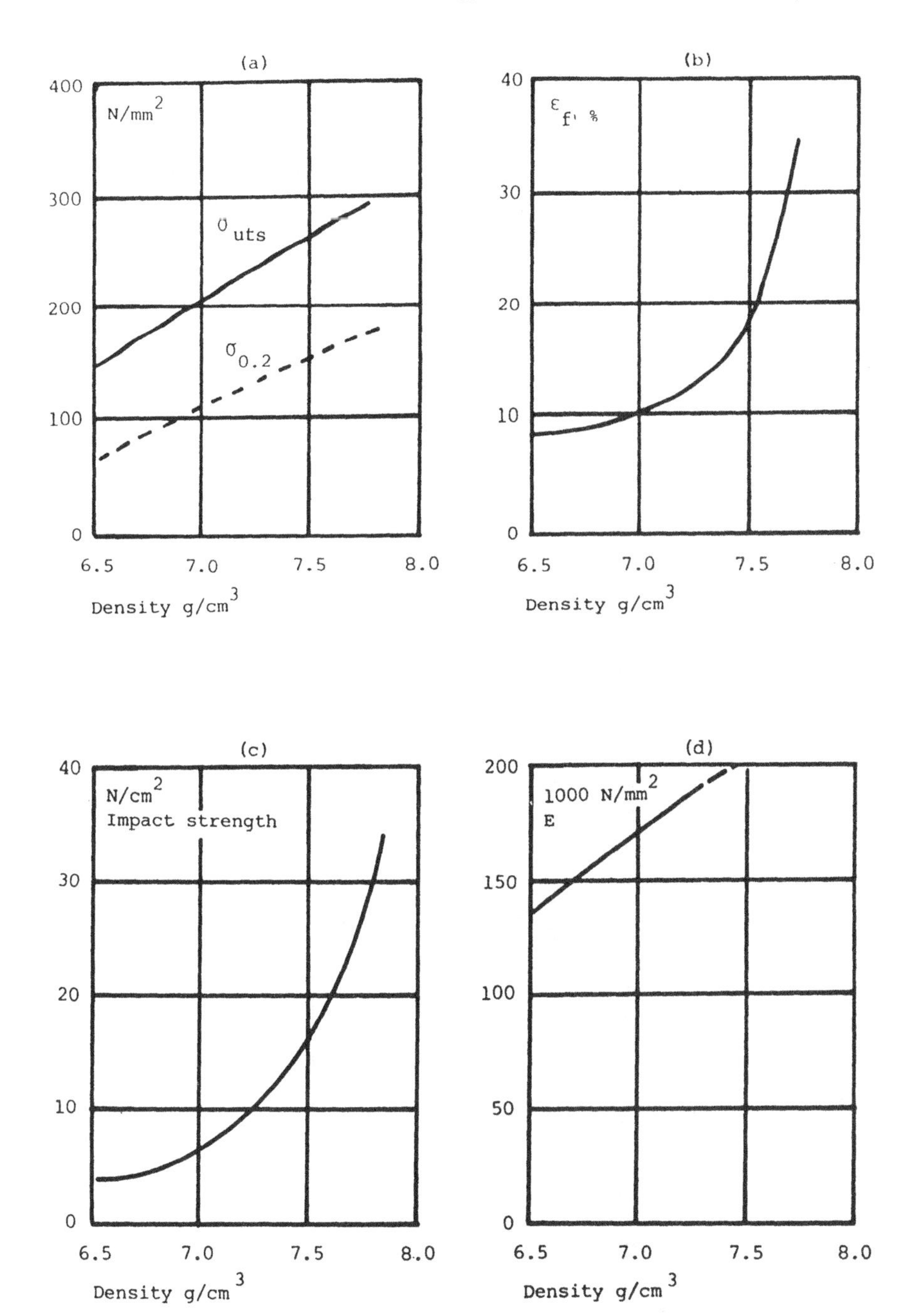

FIGURE 9.5 (a) Tensile strength and yield stress; (b) elongation at fracture; (c) impact strength; (d) modulus of elasticity as function of density for components manufactured from iron powder.

TABLE 9.1 Examples of Properties of Powder Metal Components

Material group	Density (g/cm^3)	Tensile strength (N/mm^2)	Elongation (%)	Examples of applications
Iron and low-alloy compacts	5.2–6.8	5–20	2–8	Bearings and light-duty structural components
	6.1–7.4	14–50	8–30	Medium-duty structural parts, magnetic components
Alloyed steel compacts	6.8–7.4	20–80	2–15	Heavy-duty structural parts, components
Stainless steel compacts	6.3–7.6	30–75	5–30	Components with good corrosion resistance
Bronzes	5.5–7.5	10–30	2–11	Filters, bearings, and machine components
Brass	7.0–7.9	11–24	5–35	Machine components

Source: From Ref. 31.

2. Pressing + sintering + sizing
3. Pressing + pre-sintering + coining + sintering
4. Pressing + pre-sintering + coining + sintering + sizing

In general, the simple procedure 1 is used for about 80% of all applications, but procedure 2 has quite a few applications. The more complex procedures 3 and 4 are used only for special components. This is also due to the development of powders with a high compressibility, giving high-strength-components in one pressing, possibly combined with heat treatment.

As described earlier, infiltration will increase the strength of the compacts by 70–100%, but this process is rather expensive and is used only for special products.

In the past few years, application of hot forging or pressing of powder compacts has increased rapidly. This process, called P/M forging or sinter forging, gives a final relative density of about 98–100% and mechanical properties equal to the corresponding solid material, with even better fatigue properties.

Concerning dimensional accuracy, it can generally be stated that:

Pressing + sintering gives tolerances comparable to common workshop practice in turning, milling drilling, and so on (axial ~ 0.005 mm/mm, radial ~0.002 mm/mm).

Pressing + sintering + sizing (pressing + pre-sintering + coining + sintering) gives tolerances corresponding to grinding (axial ~0.003 mm/mm, radial ~0.001 mm/mm).

Pressing + sintering + heat treatment gives tolerances corresponding to investment casting, die casting, and mean-coarse workshop practice (axial ~0.010 mm/mm, radial ~0.005 mm/mm).

The surface roughness R_a is generally in the range 10–15 μm, but sizing can reduce these figures to 1–4 μm.

The foregoing discussion serves only to illustrate the general capabilities of the powder metallurgical processes, and in specific applications detailed information must be obtained from metal powder manufacturers, powder component producers, and the literature.

9.3.2 Applications

The geometrical possibilities of the powder metallurgical processes can be evaluated from a fundamental knowledge of how metal powders react to the various process parameters and a realization that the surface is created by the total forming (TF). Based on this, some fundamental rules concerning the design of components suitable for production by metal powder compaction can be established. A few of these are:

The compact must be ejectable from the die (no undercuts, grooves, etc.)

The height (length)/diameter (width) ratio should be kept below 2–2.5 (3–4 in thick-walled components)

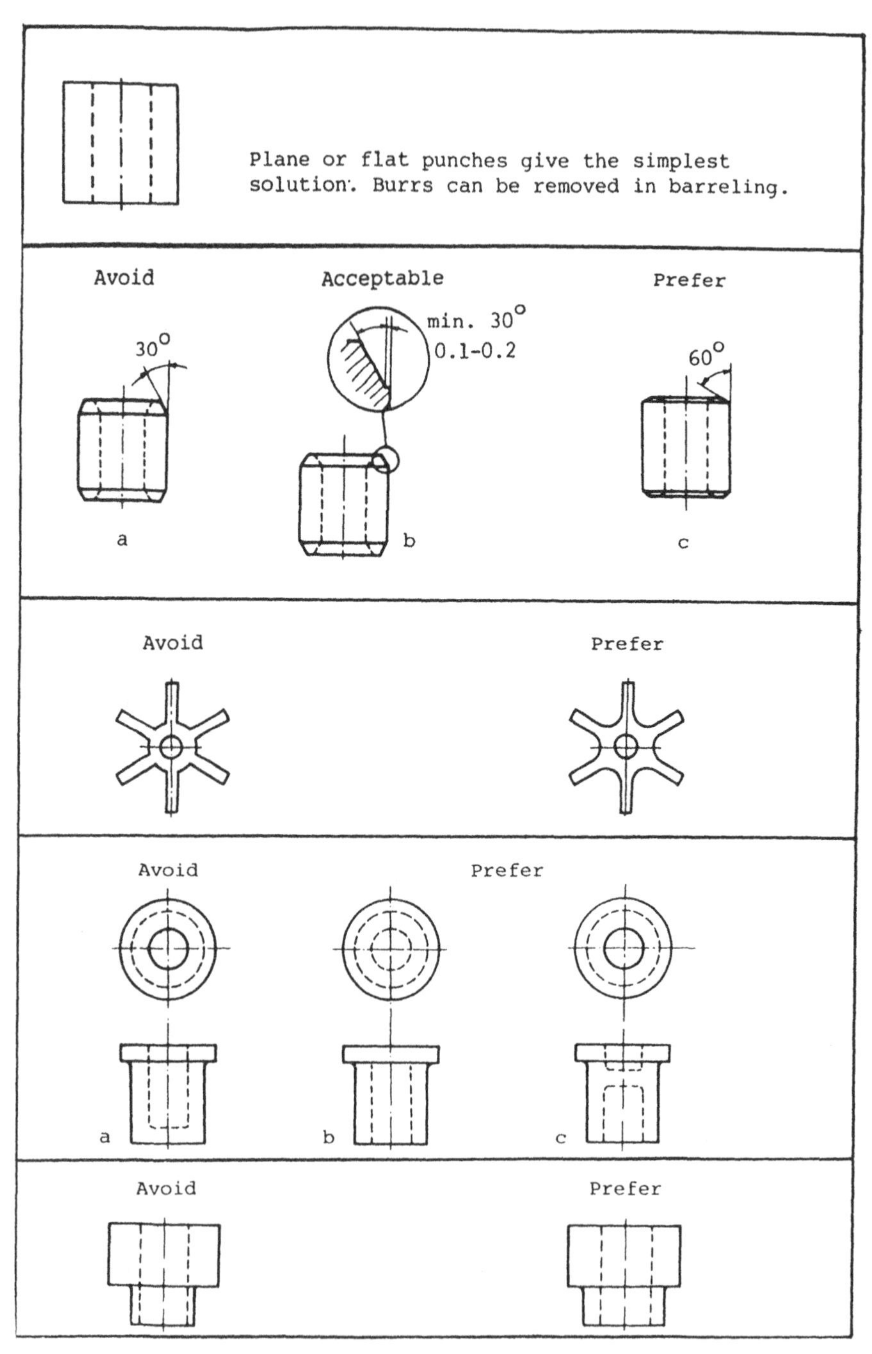

FIGURE 9.6 Examples of the design of acceptable powder metal components [30].

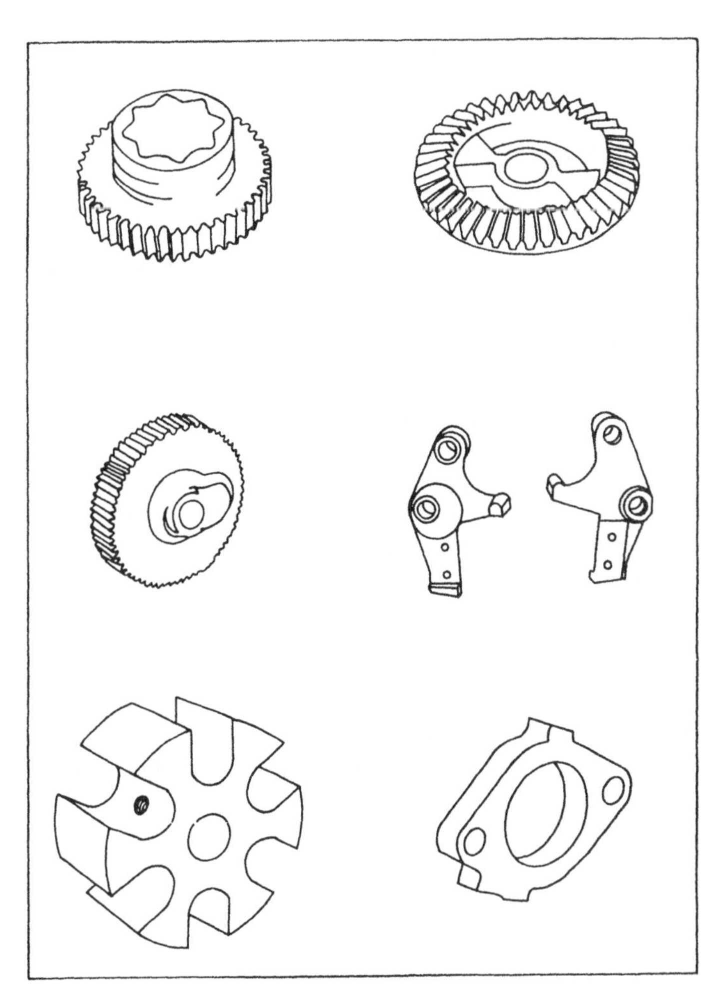

FIGURE 9.7 Typical components produced by powder metallurgy.

Wall thicknesses below 2 mm should be avoided
Sharp edges, corners, and so on, should be avoided
As few variations as possible in wall thickness
In general, the shape must allow the design of strong dies

Some of these rules are illustrated in Fig. 9.6.

The designer must be aware of the fact that powder compaction allows for the production of components having geometries, properties (including porosity), materials, and combinations of materials that cannot be produced by other methods. Before a decision concerning the application of powder metallurgy is made, the designer or production engineer must check that the process is both technically and economically acceptable.

Summarizing, the principle advantages of powder metallurgical processes are: no waste of material, high production rate, production of complicated shapes, elimination of machining, and production of components in various materials or a combination of materials with special properties. The principal disadvantages are: relatively high die costs, lower strengths than the corresponding solid materials, relatively high material costs, and geometrical limitations.

Powder metallurgical parts are used in many fields in industry. Examples include gears, rotors for pumps, bearings, cams, levers, pawls, contact parts, magnets, metallic filters, and sintered carbides (tool inserts). A few examples are shown in Fig. 9.7.

As described previously, the application of extrusion and rolling to produce rods, bars, structural shapes, sheets, and so on, is increasing rapidly, and the same is true for the isostatic compaction of special components.

10

Liquid Materials: Casting Processes

10.1 INTRODUCTION

In previous chapters the shaping or forming of materials in the solid or granular state was discussed. Shaping can also take place in the liquid material state; this is known as casting and is described in this chapter.

In casting, the liquid material is poured into a cavity (die or mold) corresponding to the desired geometry. The shape obtained in the liquid material is now stabilized, usually by solidification, and can be removed from the cavity as a solid component.

Casting is the oldest known process to produce metallic components. The main stages, which are not confined to metallic materials alone but are also applicable to some plastics, porcelain, and so on, are: production of a suitable mold cavity; the melting of the material; pouring the liquid material into the cavity; stabilization of the shape by solidification, chemical hardening, evaporation, and so on; removal or extraction of the solid component from the mold; and cleaning the component.

In principle, no limits exist regarding the size or geometry of the parts that can be produced by casting. The limitations are set primarily by the material properties, the melting temperatures, the properties of the mold material (mechanical, chemical, thermal), and the material's production characteristics (i.e., whether it is used only once or many times).

Normally, the term *casting* is applied to metals, but in general, the principal stages and many of the characteristic problems are the same for most materials which can be shaped from the liquid state. The term *casting* should be treated more broadly, allowing the carryover of new ideas from one field to another (foundry industry, glass industry, plastic industry, etc.). Having a good knowledge of one field allows an easier understanding of another field.

The differences between the many casting processes are due mainly to the mechanical and thermal properties of the work and mold material, the acceptable working temperature of the mold, the cooling method and cooling rate of the workpiece, the radiation of heat from the work and the mold material, the chemical reactions between the molten metal and the mold, the solubility of gas in the work material, and the functional requirements of the component.

In this chapter the discussion will be confined to the casting of metals, but it should be remembered that the principles are generally applicable to most materials that can be melted.

Casting processes are important and extensively used manufacturing methods, enabling the production of very complex or intricate parts in nearly all types of metals with high production rates, average to good tolerances and surface roughnesses, and good material properties. The competitiveness of the casting processes is based primarily on the fact that casting allows the elimination of substantial amounts of expensive machining often required in alternative production methods.

As mentioned, many different casting processes have been developed. The names associated with the processes may be related to the type of mold (nonpermanent, permanent) or to the mold material or the pouring method (gravity, high pressure, low pressure). Furthermore, the application of the names is not always consistent, which sometimes causes confusion. Table 10.1 shows the major casting processes classified according to the different characteristics. The most commonly used names are given, but if doubt about them arises, they can be identified by their characteristics. The individual processes are described later.

Only those casting processes appropriate for specific components are discussed in this chapter. The casting processes used to produce ingots and other materials for use in primary metalworking processes are not described.

10.2 CHARACTERISTICS OF CASTING PROCESSES

Casting processes can be broken down into a few fundamental operations or stages which are common for all casting processes and independent of the work material and of how the mold is produced (see Chapter 1).

Figure 10.1 shows schematically the principal operations or stages in the production of components from the liquid state. From the specifications of the

TABLE 10.1 Some Characteristics of the Major Casting Processes

Type of mold	Mold material	Pouring principle	Pattern material	Process name	Grouping	
Nonpermanent (single-purpose)	Sand (green)	Gravity	Wood, metal, plastics	Green sand, dry sand, core sand casting	Sand casting	
Permanent	Alloy steels	High pressure	—	Die casting	Permanent (metallic) mold casting	
	Graphite, steel, cast iron	Low pressure	—	Low-pressure (permanent mold) casting		
	Cast iron, steel	Gravity		Non-pressure-gravity permanent mold casting		
Nonpermanent (single-purpose)	Nonmetallic (sand, plaster, ceramics, etc.)	Gravity (Low pressure)	Metal	Shell mold casting		
			Wax, plastic, (rubber, metal)	Plaster mold casting		Precision casting
			Wax, plastic, (rubber, metal)	Ceramic shell mold casting	Investment casting	
			Wax, plastic, (rubber, metal)	"Lost wax" casting (investment casting)		
Nonpermanent/ permanent	Nonmetallic/metallic	Centrifugal forces	—	Centrifugal casting	Centrifugal casting	

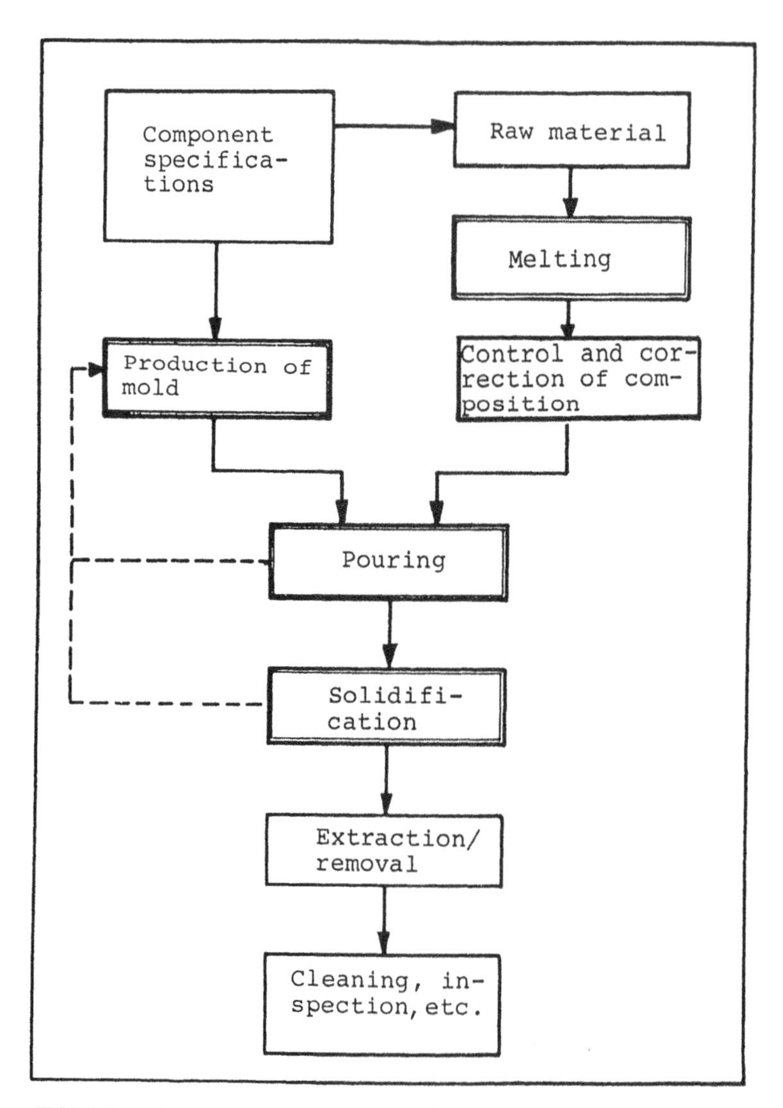

FIGURE 10.1 The main stages or operations in the production of components from the liquid material state (the casting of metals).

desired component (geometry, tolerances, type of material, final material properties, number of components, etc.), the raw material can be chosen. The raw material is melted and the composition controlled and eventually corrected. Depending on the specifications, the casting process can be selected and, consequently, the mold production (mold material and shaping method) can be identified. The molten and refined work material is now inserted or poured into the mold. The shape obtained is stabilized by solidification; a process that is dependent on the work material, the mold, and the external conditions. After solidification, the component is extracted or removed from the mold. The component is finally cleaned and inspected. If approved, machining, heat treatment, and other processes can then be carried out.

Each stage or operation listed in Fig. 10.1 can now be analyzed according to the principles described in Chapter 1 (i.e., material flow, energy flow, and information flow). In this context, only the stages double-lined in Fig. 10.1 will be discussed: melting, mold production, pouring, and solidification.

As mentioned previously, the casting processes can be characterized either by the type of mold, the mold material, or the pouring principle, which makes it difficult to describe mold production, pouring, and solidification without relating these to specific processes. Melting can be described in a general section covering all casting processes. Therefore, in this presentation the melting methods are described first and then a short introduction to the primary characteristics of mold production, pouring, and solidification is given. After this, the casting processes are discussed, including descriptions of molding, pouring, and solidification. In a final section, general design rules and comparisons of the casting processes are presented.

10.3 MELTING (AND CONTROL OF COMPOSITION)

When selecting the melting process (i.e., the furnace equipment) to fulfill the casting specifications and the production requirements, several factors or characteristic features must be considered. Among these are the metal chemistry or the metallurgy, the temperature of the melt, and the melting capacity, including the delivery rate. Before the melting processes are described these factors will be discussed briefly.

10.3.1 Metal Chemistry (Metallurgy)

To be able to fulfill the requirements concerning material properties and to produce a sound casting, the molten metal must have the right composition, with a limited content of metallic and nonmetallic contaminations, including gases. Depending on the raw material (i.e., the metal fed into the furnace), a refining process or correction of the composition of the melt might be necessary. This can involve the removal or addition of elements, the removal of dissolved gases, and so on. When a considerable proportion of uncontrolled scrap is used in the raw material, impurities and contaminants are introduced and the need for a refining process increases.

Concerning gaseous contaminations, it should be mentioned that the molten metal can dissolve these in greater amounts than the solid material; consequently, during freezing, gas is precipitated, causing porosity in the casting. Another effect of the particular gases nitrogen and hydrogen is that they can reduce the ductility of the final casting (i.e., they promote brittleness).

Thus, the final composition of the melt depends on the raw material and all the changes that occur during and after the melting process. The melting procedures can be divided into two categories: (1) melting without refining and (2) melting with refining. In the first category, the raw material and the minor changes caused by the specific melting process determine the final composition of the material. However, small corrections to the compositions can be carried out just before pouring. Examples of the application of this procedure include the melting of alloys with low melting temperatures, the melting of light metals, and the vacuum melting of alloys with high melting points. In the latter case protective atmospheres other than vacuum can be used.

In the second category, melting is followed by a refining phase to obtain the desired composition; in other words, significant corrections are made to the composition of the material. For example, in steel the carbon content as well as the content of other elements are changed during melting either by deoxidation from the atmosphere or by reactions caused by the slag protecting the melt. The oxygen content in the melt must be adjusted frequently by the addition of deoxidation materials.

The various melting processes provide materials with different "hereditary" properties (machinability, cooling properties, etc.). This consideration influences the selection of the melting process, as does the size, shape, and composition of the raw materials. The various melting processes are discussed later.

10.3.2 Temperature of the Melt

To be able to cast complicated shapes, the molten metal must be maintained at high temperatures. If pouring is carried out at too low a temperature, the metal may start to freeze before the mold is filled, causing misruns and other defects. If pouring is carried out at too high a temperature, the metal may react with the mold material, causing gaseous inclusions in the casting. The high thermal loading may also cause detrimental deterioration of the mold in the latter case. The pouring or tapping temperature must be chosen so that these problems are avoided. To assist in specifying the correct tapping temperature, various fluidity tests have been developed, for example, the sand-cast fluidity spiral and the suction-tube method. In the spiral test, the length of the spiral obtained at a given tapping temperature provides an index of relative fluidity. The maximum tapping temperature obtainable from the furnace can be decisive for the selection of the melting process.

10.3.3 Melting Capacity and Delivery Rate

Two factors are of importance here: the rate of the delivery and the mode of delivery. A rate of 5 Mg/h, for example, can be obtained from a batch or a continuous melt. In casting large components, the batch melting processes are normally preferred, as it may take too long a time to accumulate the necessary amount of metal from a continuous-type furnace. In addition, a large furnace is required to maintain the molten metal at the required high temperature. In the production of small castings, the continuous type of furnace is normally preferred.

When selecting melting equipment, both the capital costs and the operating costs must be considered. The calculations that allow the choice to be made will not be discussed here.

10.3.4 Melting Processes/Furnaces

To be able to obtain a sound casting, molten metal of the right composition and temperature must be produced. The furnaces must prevent contamination of the metal and allow corrections of the composition to be made.

Since melting takes place at high temperatures, it is necessary to line the furnace with a refractory material. To protect the molten metal against oxidation, dissolution of gases, to limit or reduce the content of other undesired elements, and so on, it is necessary to cover the metal with a layer of slag.

Depending on the type of furnace lining, the melting process can be classified as acid or basic. In the acid process, the lining consists of fire clay and quartz sand. In the basic process, a lining of magnesite is used. The acid process is often preferred to the basic, as the lining is strong and cheap, giving a lower energy consumption and a slightly higher production. In general, acid linings are used when no refining is necessary and a small quantity of sulfur can be tolerated. The basic process provides low-sulfur iron and allows for carburizing to give higher carbon contents.

The lining and the slag must be of the same type, both acid or both basic. In the acid slag, quartz sand (FeO, MnO, etc.) is used, and in the basic slag, limestone is used. After establishing the melting process, the sources of contaminations must be minimized. Typical sources are:

The atmosphere (O_2, N_2, H_2O, CO_2), for example in unprotected electrical furnaces.

Combustion products (CO_2, CO, H_2O, SO_2) in furnaces fired by oil, gas, coal, coke, and so on. Where solid firing materials are used, the sulfur and phosphor content often contaminates the melt.

The lining may contribute to the contamination if it is not completely inactive. Both metallic and nonmetallic impurities (e.g., Al, Si, O_2, H_2) may be introduced.

TABLE 10.2 Classification of Some of the Most Frequently Used Industrial Melting Furnaces

Type of energy	Furnace	Application
Chemical		
Solid	Cupola	Cast iron, sometimes copper alloys
Granular (coal dust)	Open-hearth furnace (Siemens-Martin)	Steel
Liquid (oil)	Rotary furnace	Cast iron, steel, copper alloys, light metals
Gaseous (gas)	Crucible furnace (gas or oil)	Nonferrous metals and alloys
Electrical		
Arc	Arc furnace Direct Indirect	Steel, cast iron
Induction	Induction High frequency Low frequency	 Steel, cast iron Copper, aluminum alloys
Resistance	Resistance furnace	Alloyed steel, sometimes cast iron, aluminum and copper alloys

Based on the requirements of the melting process, furnace building principles can be developed using the morphological approach described in Chapter 1. In this context, only a classification of the most frequently used furnaces will be given. If the type of energy used to create the thermal basic process is used as a characteristic classification, the summary shown in Table 10.2 results. Within the chemical group, a subdivision according to the state of the firing material is shown. In the last column typical applications of the furnaces are listed.

Cupola. The cupola (Fig. 10.2a) is used primarily to produce cast iron. It is charged continuously with alternating layers of coke and iron; eventually, flux materials and alloying elements are also added. An air blast (cold or hot) is supplied through openings (tuyeres) around the periphery of the lower part of the cupola. The melting rate is directly related to the combustion of the cokes (i.e., to the amount of the air blast). The lining can be either acid or basic. The production rate is typically 8–10 Mg per hour per square meter of the furnace cross-sectional area.

Open-Hearth Furnace. The open-hearth furnace (see Fig. 10.2b) is an open gas- or oil-fired furnace used in steel foundries to produce steel casting. The capacity is often in the range 25–350 Mg per charge, and it takes about 8 h to produce a charge.

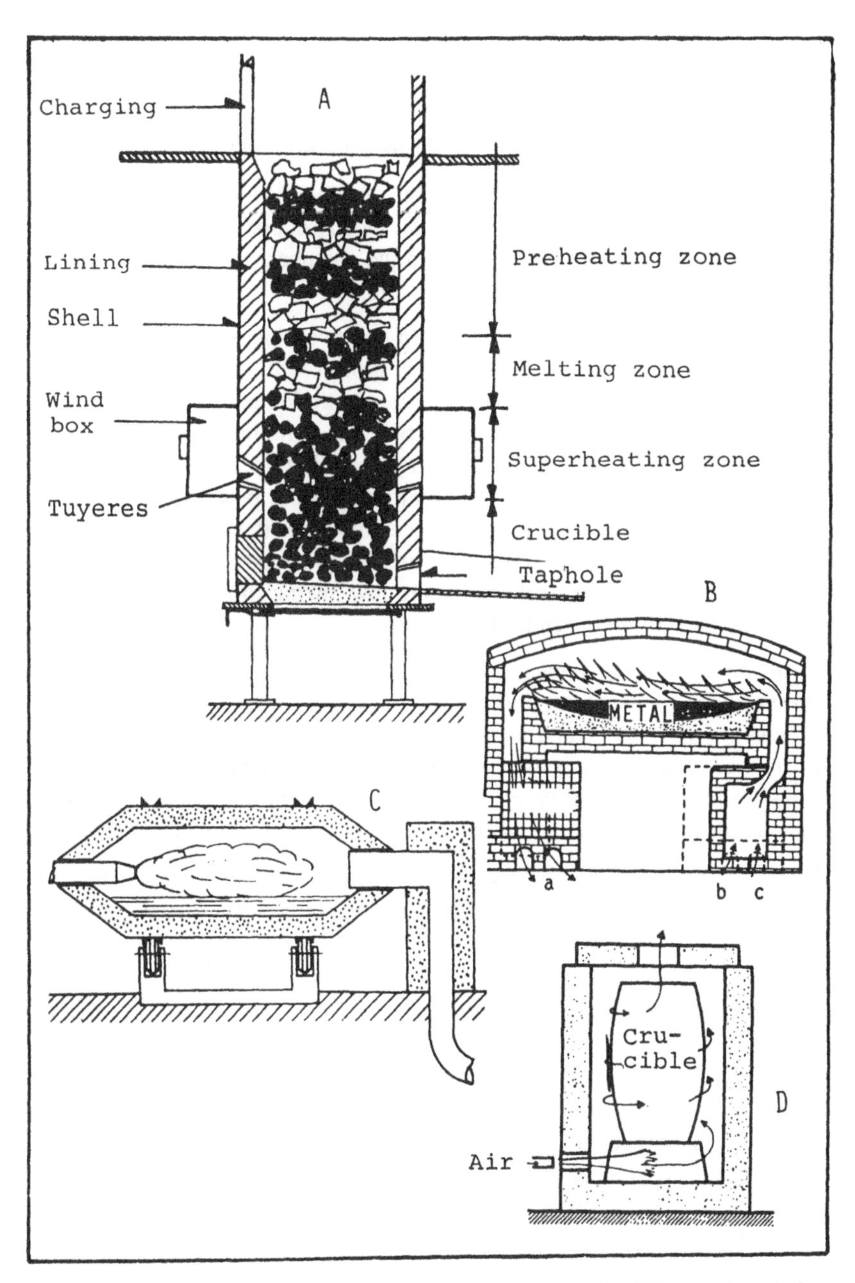

FIGURE 10.2 Furnaces using chemical energy: (A) cupola; (B) open-hearth furnace (a, hot combustion gases; b, hot air; c, preheated gas); (C) rotary furnace; (D) crucible (gas fired).

Rotary Furnace. The rotary furnace (see Fig. 10.2c) is usually gas- or oil-fired. The gaseous combustion products are used to preheat the air. Rotary furnaces have capacities in a wide range, 0.1–1 Mg, for melting of copper alloys and light metals, and in the range 10 Mg for cast iron and steel production.

Crucible Furnace. The gas- or oil-fired crucible furnace (see Fig. 10.2d) is used extensively to melt nonferrous metals, mostly copper alloys and light metals. The burner is arranged tangentially in the chamber, so that the combustion products move in a spiral around the crucible.

Arc Furnace. Arc furnaces are divided into two groups: the direct-arc types and the indirect-arc types. In the direct-arc type (see Fig. 10.3a) the arc is established between the consumable graphite electrodes and the charge. In the indirect-arc type, the arc is established between the graphite electrodes, and the heat from the arc is transmitted to the charge by radiation, conduction, and convection. Arc furnaces are used most commonly in the production of cast iron and steel. The three-phase direct-arc furnace dominates the applications, and the capacity is in the range 2–50 Mg. The indirect-arc type is generally used for smaller capacities, but the applications are few, owing to a low efficiency, excessive wear of the lining, and other factors.

Induction Furnace. Induction furnaces are divided into two groups: the high-frequency (1000–30,000 Hz) crucible-type furnace (see Fig. 10.3c) and the low-frequency (60–180 Hz) channel-type furnace (see Fig. 10.3b). The high-frequency crucible furnace is used mainly in the production of cast iron and steel. The low-frequency channel-type furnace is used for the melting of copper and aluminum alloys. The applications of induction furnaces are increasing rapidly.

Resistance Furnace. Resistance furnaces use the heat produced by the electrical resistance in the graphite heating elements. The heat is radiated to the melt and the furnace. This type of furnace is used mainly to melt aluminum and copper alloys. The resistance furnace is frequently of the crucible type, where the heating elements surround the crucible.

Further information about melting and furnaces can be found in the literature [33,42].

10.4 MOLD PRODUCTION, POURING, AND SOLIDIFICATION

10.4.1 Mold Production

The next step (Table 10.1) is the manufacture of a suitable mold. The mold cavity and mold material must be such that the final component has the desired properties. Figure 10.4 illustrates how the requirements of the mold can be

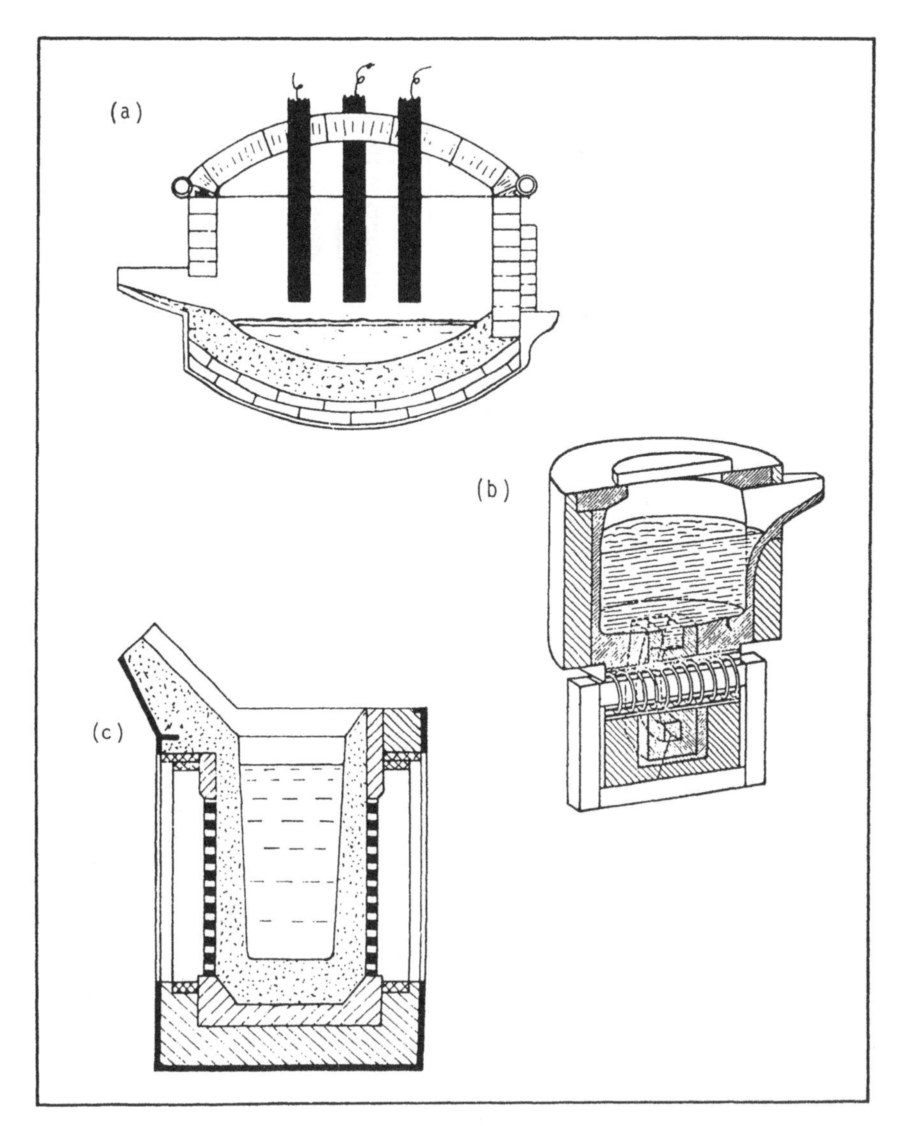

FIGURE 10.3 Furnaces using electrical energy: (a) direct-arc furnace; (b) induction furnace (low-frequency); (c) induction furnace (high-frequency) [32].

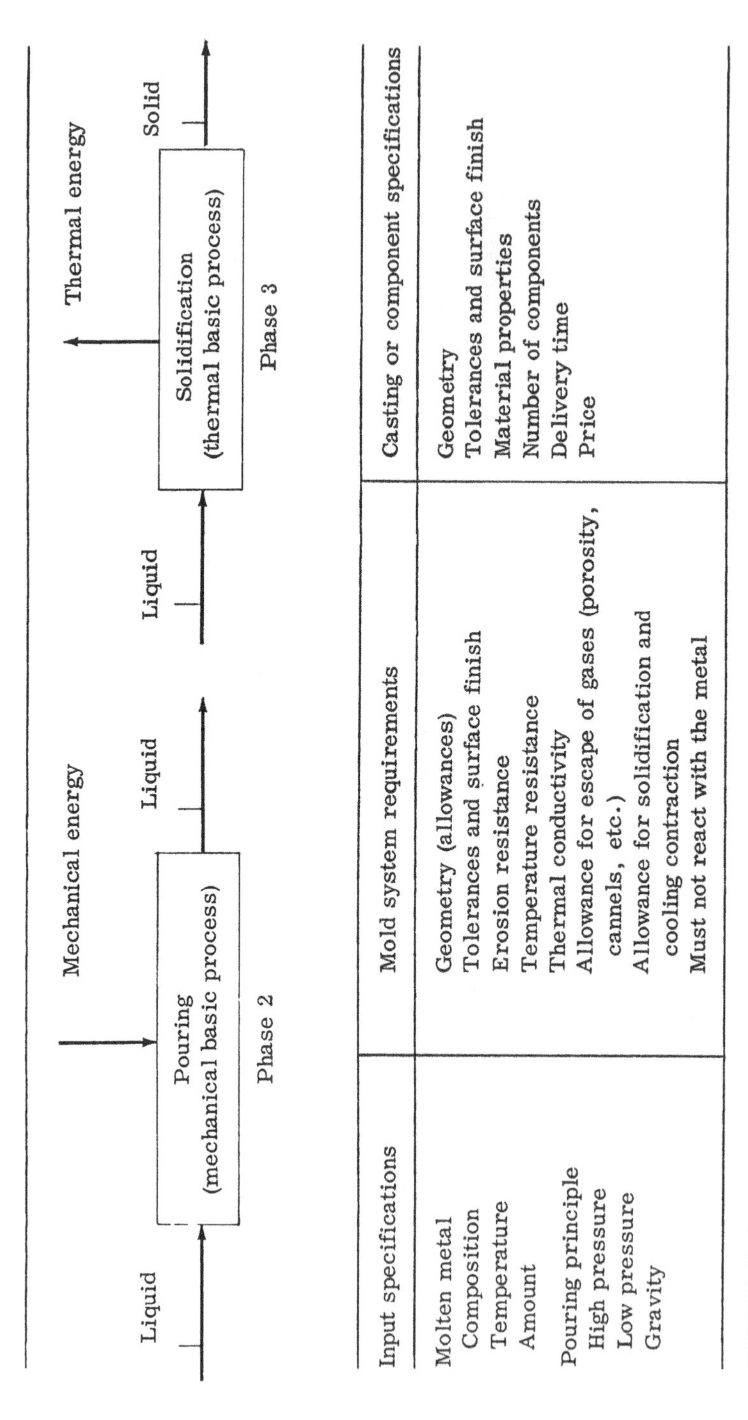

Input specifications	Mold system requirements	Casting or component specifications
Molten metal Composition Temperature Amount Pouring principle High pressure Low pressure Gravity	Geometry (allowances) Tolerances and surface finish Erosion resistance Temperature resistance Thermal conductivity Allowance for escape of gases (porosity, cannels, etc.) Allowance for solidification and cooling contraction Must not react with the metal	Geometry Tolerances and surface finish Material properties Number of components Delivery time Price

FIGURE 10.4 Determination of the requirements of the mold system.

determined. The component specifications lead to specifications of the input material (i.e., the molten metal). Based on this and a fundamental knowledge of metallurgy, melting and solidification, heat transmission, fluid mechanics, materials, and so on, the general requirements of the mold system can be determined. This is, of course, an interactive process.

When the requirements have been defined, the principles of mold design, including selection of mold material and method of mold manufacture, can be analyzed. Here the morphological approach described in Chapter 1 can be utilized. This approach will not be considered in detail, but a brief discussion will reveal the basic characteristics and provide a fundamental understanding of the differences in the casting processes.

Three factors have a major influence on mold design and manufacture:

- Type of mold
  - Permanent (nonexpendable)
  - Nonpermanent (expendable—used once)
- Category of pattern
  - Permanent
  - Nonpermanent
- Pouring principle
  - High pressure
  - Low pressure
  - Gravity

This is illustrated in Table 10.3, where three basic production methods can be identified: I, II, III in the left-hand column.

Method I. The mold is permanent and the mold material solid (i.e., metal, graphite, etc.). The cavity, which is a negative of the desired part, is produced by machining, pressing, and so on. To be able to use the mold, a suitable parting line must be available. This line marks where the two halves of the mold meet.

Method II. The mold is nonpermanent, the material granular (sand, plaster, ceramics, etc.), and a pattern that can be used several times is employed. The two halves of the mold are formed around the pattern. The molds are separated at the parting line and the pattern removed. After reassembly the mold is ready for use. Thus, the basic requirement is that the pattern can be removed.

Method III. The mold is nonpermanent, the material granular (sand, plaster, ceramics, etc.), and the pattern nonpermanent. The pattern is melted, dissolved, or otherwise removed if sufficiently elastic (e.g., rubber patterns), leaving a cavity which can have a complicated shape. The mold is not split into two halves.

TABLE 10.3 Characteristic Features Influencing Mold Design

Type of mold	Mold material	Category of pattern, pattern material	Pouring principles
Permanent I 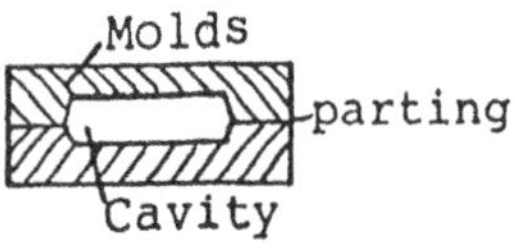 Molds parting Cavity	Metal Graphite etc.	—	High pressure Low pressure Gravity
Nonpermanent II 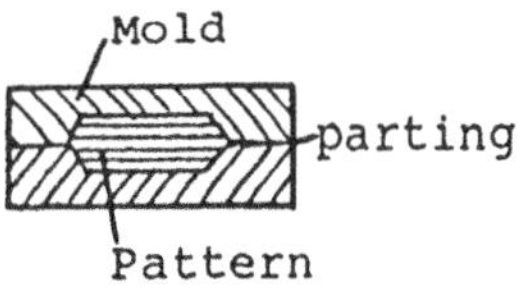 Mold parting Pattern	Sand Plaster Ceramics etc.	Permanent Wood Metal Plastic	Gravity
Nonpermanent III 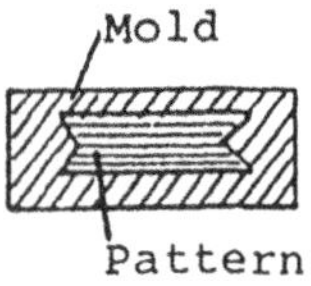 Mold Pattern No parting	Sand Plaster Ceramics etc.	Nonpermanent Wax Plastic Permanent Rubber	Gravity

These three fundamental methods cover the major methods of mold design used today, but the systematic approach, if analyzed carefully, may lead to new concepts. Mold manufacturing methods are discussed further in Section 10.5, where the most commonly used casting processes are described.

10.4.2 Patterns

The mold cavity is not a true negative of the desired component, since it must compensate for some or all of the following:

Shrinkage
- Contraction from casting temperature to solidification temperature
- Contraction during solidification

TABLE 10.4 Patterns Classified by Category, Type, and Material

Category of pattern	Type of pattern	Pattern material
Permanent pattern (removable)	Loose pattern	Wood
	Single piece	Metal
	Split	Plastic
	Gated pattern	Rubber, etc.
	Match-plate pattern	
	Cope and drag pattern	
	Sweep and skeleton pattern	
		Wax
		Plastic (polystyrene, styrofoam)

Contraction in the solid state
Machining to final dimensions, if necessary
The draft necessary to remove the pattern from the mold or the component from a permanent mold

Considering Table 10.3, it can be seen that these allowances must be considered either in the mold (for permanent molds) or in the pattern (for nonpermanent or expendable molds). Patterns, in general, play an important role in casting processes and the principles are described briefly below.

The requirements or specifications for the desired component lead to the selection of a casting process. If a casting process employing nonpermanent or expendable molds is selected, the next question concerns the manufacture of the pattern necessary to produce the mold. Table 10.4 shows a classification of patterns according to category (permanent, nonpermanent), type, and material. The quality of the pattern depends on the number of castings and the casting process. If only a few castings are to be made, soft wood may be employed; for a larger number, hard wood; and for very large numbers, metal will be used. Patterns can thus be divided into different types, according to how they are employed in mold production.

Loose patterns, which are normally used in the casting of up to 100 components, are made as one solid piece (like the desired component with allowances) or split along a plane or irregular surface to facilitate extraction from the mold halves (see Fig. 10.10).

Gated patterns are, in principle, loose patterns with attached gates and runners (see Section 10.4.3). Because of their higher cost, they are used only for small castings and when the number of castings is high (~1000).

Match-plate patterns are split patterns mounted on opposite sides of a wooden or metal plate (known as a match plate) giving the parting surface (see

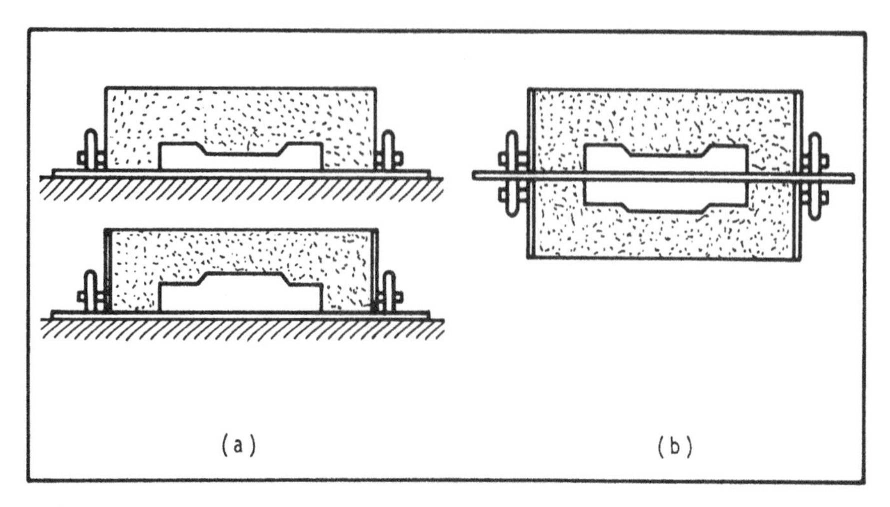

FIGURE 10.5 (a) Cope and drag pattern and (b) match-plate pattern.

Fig. 10.5b). Gates and runners are also mounted on the plate. Locating pins provide an accurate location of the plate between the upper (cope) and lower (drag) flask. A flask is a four-sided metal frame containing the mold halves. The match plate is used in machine molding, allowing the cope and drag mold halves to be made around the plate. After removal of the plate, the two halves are brought together, giving the desired cavity. Because of the high cost of match plates, they are used only when a large number of castings is required.

Cope and drag patterns are split patterns mounted on separate plates (the cope plate and the drag plate), allowing the production of the mold halves on different machines (see Fig. 10.5a).

Sweep and skeleton patterns are wooden cross sections of very large components used in hand forming of a mold. A sweep can be used to form cylindrical shapes by rotation of the section. Skeletons are wooden frames outlining the shape of the casting which serve as guidance for the molds.

As discussed previously, the pattern must include allowances for draft, shrinkage, and machining.

Draft. Draft is provided to enable the pattern to be removed from the mold (see Fig. 10.6a). Normally, split patterns are used, so that half of the pattern is in the cope flask and half in the drag flask. The draft is often in the range 1–2°.

In many cases, the castings are hollow, which necessitates the use of cores (see Fig. 10.6b). The pattern gives the external shape of the casting, and the core the internal shape. Cores are produced by a number of special methods and placed in the cavity when the pattern has been removed. In sand casting, the cores are made of baked sand and can be handled (see Section 10.5.1).

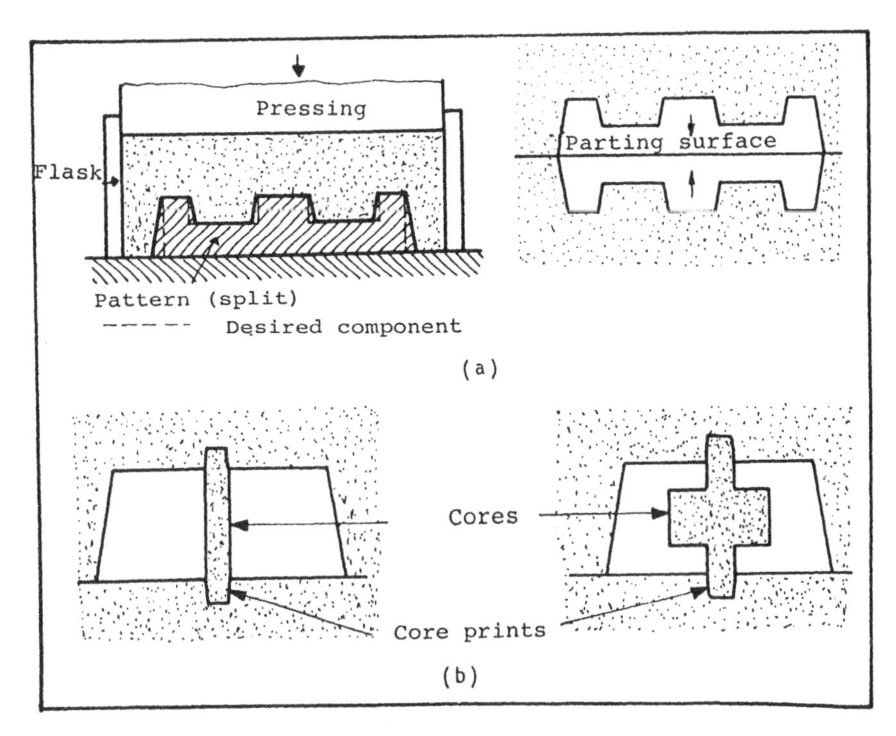

FIGURE 10.6 (a) Pattern draft and (b) cores.

Shrinkage Allowance. To compensate for the shrinkage of the casting during cooling from the casting temperature to room temperature, a shrink rule is used to transfer measurements to the pattern. The linear contraction or shrinkage here must not be confused with the volume shrinkage, which—to some extent—is compensated for by the application of risers.

Machining Allowance. On those surfaces where the casting is to be machined, the pattern must be provided with a reasonable machining allowance, varying between 2 and 10 mm for cast iron, depending on the wall thickness and the tolerances.

Selection of pattern material depends, as mentioned, on the casting process, the number of castings, the tolerances, and other factors. If disposable patterns made of polystyrene or Styrofoam are used, they are not removed from the mold. When pouring the liquid metal into the mold, they evaporate or burn. If wax is used as the pattern material, heating of the mold results in the removal of the wax before the molten metal is poured into the cavity.

Figure 10.7 shows examples of models for sand casting.

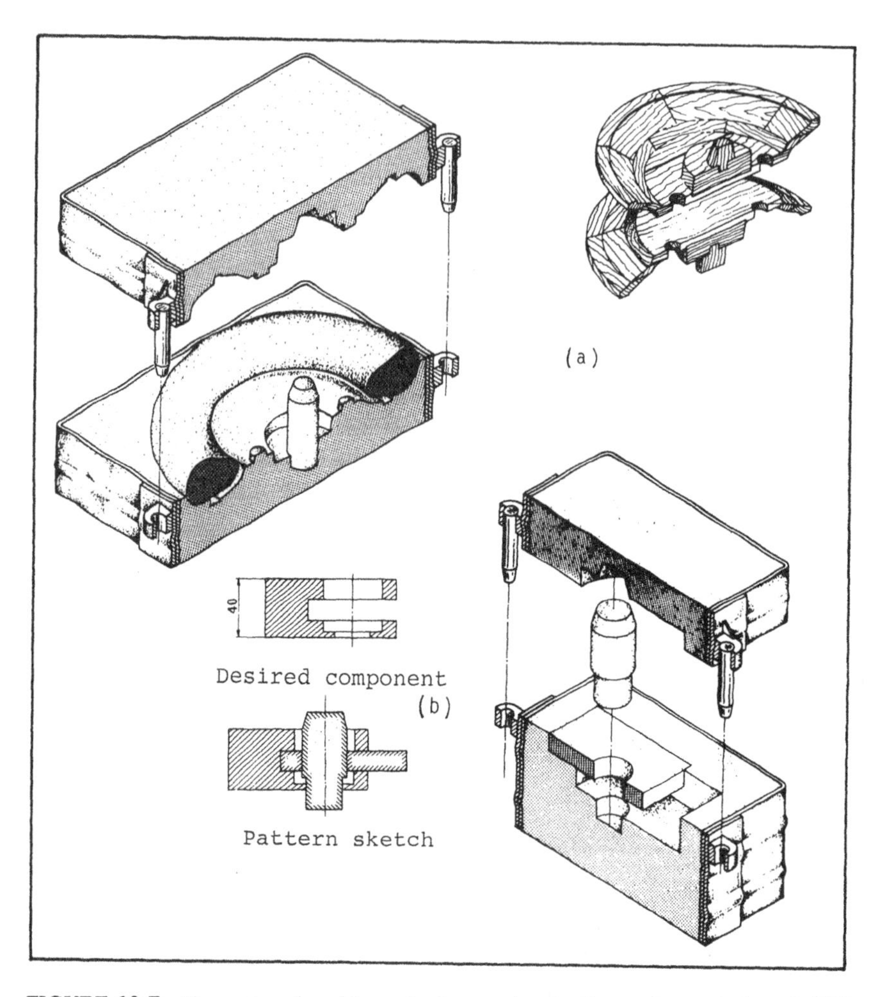

FIGURE 10.7 Examples of molds ready for pouring (gating system not shown) [34].

10.4.3 Pouring

The production of a casting without defects depends on many factors. One of the important factors is the way in which the metal is poured into the mold cavity. This pouring process can be characterized by the filling or pouring pressure and the gating system (channels for delivering the metal). The pouring pressure can be classified as high (2–15 MPa), low (0.12–0.30 MPa), or gravity. The higher pressures permit the casting of thinner sections, higher quality, and so on, but at the same time, there is a requirement for higher-strength mold material. High or

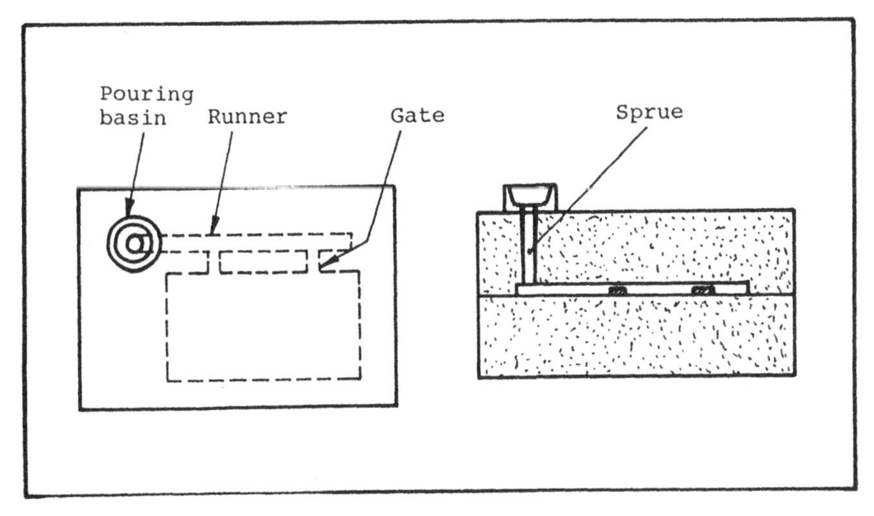

FIGURE 10.8 Schematic illustration of the main elements of the gating system for sand casting.

low pouring pressures normally require perma-nent metal molds, but other materials, such as graphite, can be used for the lower pressures.

The general objective of the gating system is to allow liquid metal to be supplied to the mold cavity at the proper rate and temperature. A poorly designed gating system may cause excessive heat loss (i.e., require too high a pouring temperature influencing the grain structure, porosity, etc.), turbulence in the fluid stream, entrapping of gases, slag, dross, heavy erosion, and so on. The actual gating system depends primarily on the molding method, the mold material, the metal, the geometry, and the pouring or injection pressure. The main elements in the gating system are the same for most casting processes. Thus, the gating systems for sand casting will be described to illustrate these main elements, since this process involves most of the problems met in other casting processes.

A gating system for sand casting (Fig. 10.8) generally includes a pouring basin, a sprue, a runner, and a gate. The liquid metal is poured first into a pouring basin to avoid too heavy erosion at the bottom of the sprue and to prevent slag from entering the cavity. The pouring basin is especially important with large molds. In pressure casting, the basin is replaced by an injection system.

From the pouring basin, the metal flows through the sprue, which should be filled constantly during pouring. The sprue is tapered about 2°, so a higher velocity occurs at the bottom and does not cause aspiration of air through the mold. It is sometimes necessary to place a special well or splash core at the sprue

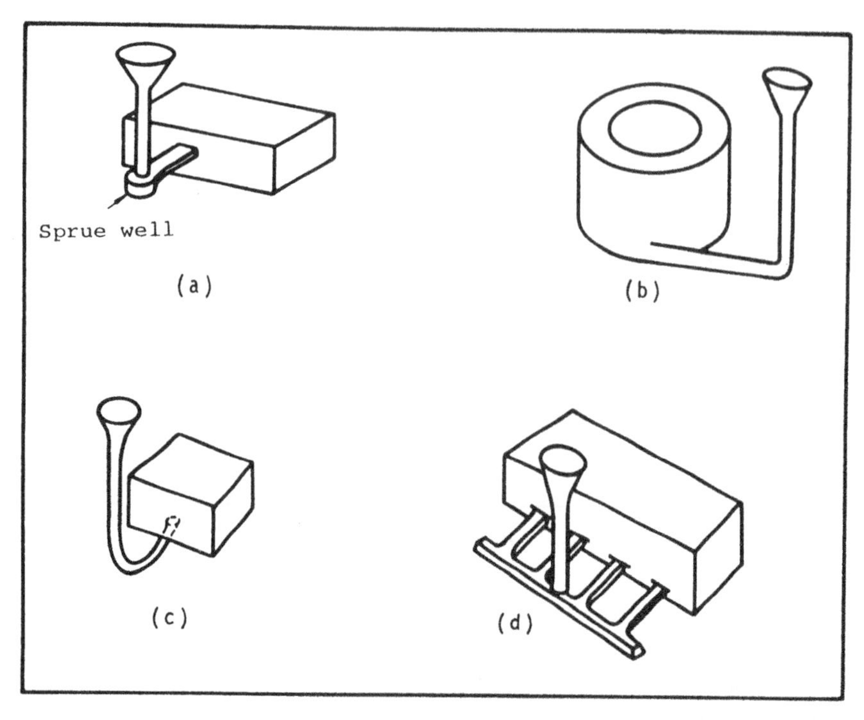

FIGURE 10.9 Typical gating systems: (a) gating system with sprue well; (b) centrifugal gating; (c) horn gating; (d) comb or finger gating.

bottom. If more than one gate is necessary, the metal is distributed to these through a runner. The final flow into the mold cavity takes place in the gate(s).

Figure 10.9 shows a few typical gating systems. Several more exist, but there is no general agreement about the relative advantages and disadvantages of the different systems.

In the designs of gating systems, the cross-sectional area of the sprue is made about 20% larger than the total area of the gates, to obtain filling of the runner from the beginning. The cross-sectional area of the runner is decreased after each branching gate. The importance of proper gating systems has been recognized during the last few years and the design is no longer carried out by the craftsman but by the engineer.

Summarizing, the most important requirements for a well-engineered gating system are:

Prevention of slag and oxides from entering the mold cavity
Prevention of the inclusion of air or gases
Prevention of mold and core erosion
Decreasing the requirements for a high pouring temperature

Leading the liquid metal into the mold at the right place and at the correct rate, resulting in castings with minimized shrinkage voids and distortion
Minimizing the amount of metal used in the gating system

Details can be found in the literature [33,42].

When the gating system has been designed, attention must be given to the avoidance of shrinkage voids. As mentioned, the total contraction or shrinkage during cooling from the pouring temperature to room temperature is made up of the following three contributions:

1. Contraction during cooling from the pouring temperature to the solidification temperature
2. Contraction during solidification
3. Contraction during cooling from the solidification temperature to room temperature

For carbon steel, the solidification shrinkage is in the range 2.5–3%, for aluminum 6–7%, and for cast iron from 1.9% to an expansion of 2.5%, depending on the carbon content.

To compensate for the shrinkage taking place during cooling from the pouring temperature to solidification, a reservoir of molten metal should be attached or connected with the component. To supply liquid metal to compensate for the shrinkage, the reservoir or riser must solidify after the component has solidified. The size and placement of the risers are thus of utmost importance. The riser design can be quite complex and will not be discussed here. Risers should be large in section and located near heavy sections of the casting. If they are placed at the top of the section, gravity assists in feeding the molten metal. Risers may also serve as vents for steam and gases. An example is shown in Fig. 10.10 and further information can be found in the literature [33,42].

10.4.4 Solidification

The mold design and mold material have a significant influence on the solidification pattern. Solidification in sand and shell molds is relatively slow, and a directional solidification is desired, starting at the lowest parts of the casting and continuing up through the mold. The possibilities of obtaining a directional solidification depend on the design of the component.

With permanent molds, solidification starts even before the mold is completely filled and terminates shortly after. This means that thin sections may close to prevent further feeding to other sections. Consequently, the requirement of uniform sections is very important here. In die casting (permanent metal molds), feeding is supplied by the high pressure acting through the gating system. Modifications to the component geometry may be necessary to obtain asound casting.

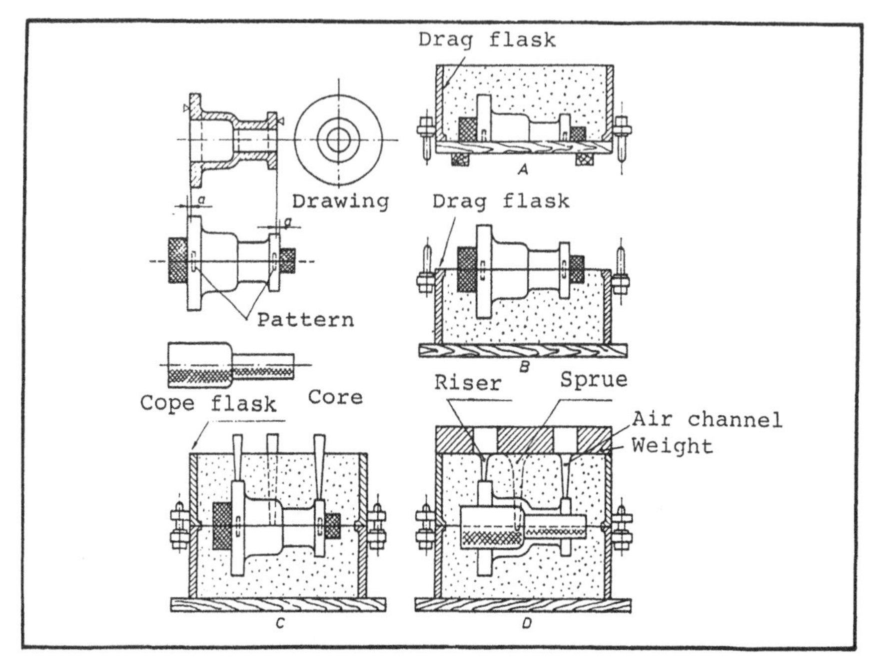

FIGURE 10.10 Typical stages in the production of sand molds using split patterns [35].

10.5 CASTING PROCESSES

In this section the most important casting processes are described briefly.

10.5.1 Sand Casting

Sand casting of cast iron, steel, and metals is extensively used in industry. In this description, molding materials, mold production, and applications are discussed.

Molding Sand

Sand is the most extensively used mold material. To produce a casting of 1000 kg, it is necessary to employ 4000–5000 kg of molding sand; in other words, large amounts of molding sand must be handled and maintained. The fundamental requirements of the molding sand are: temperature resistance, strength to retain a given shape and to withstand the mechanical loading from the liquid metal, permeability (to permit the escape of gases), and collapsibility (to permit shrinkage). The fulfillment of these requirements is obtained in molding sand from the following ingredients:

The sand (giving temperature resistance and permeability)

The binder (giving strength)
Additives (giving collapsibility)
Water (to activate the binder)

The most widespread type of sand used in foundries is silica (SiO_2) or quartz sand. It is readily available and cheap. However, because of its harmful effects on the lungs, it is being replaced to some extent by olivine sand (especially for steel casting). Zircon sand can also be used, but it is more expensive. In most countries, natural sand is being replaced by synthetic sand, which has a specified type of grain and grain-size distribution. The compounding elements are added in the foundries. The binder, which together with water gives strength and formability, can be:

Clay
Cement
Sodium silicate (CO_2 process)
Oils
Resins

In molding sand, clay is generally used as the primary binder. The types of clay are bentonite, kaolinite, or illite. The amounts vary between 5 and 20% of the sand. Cement (portland) can be used as a binder when strong molds are desired. Sodium silicate and oils are used primarily in core production (see later). These, together with the additives, give high strength. The resins are used as primary binders, in both molds (shell molding) and cores. Cereal (flour, starch, dextrin, etc.), wood flour, and sea coal are used as additives to increase strength, permeability, surface quality, and other characteristics, but they will not be discussed in the present context.

The water content in molding sand is usually 4–8% when clay is used as the primary binder. A molding sand based on sand, clay, and water, perhaps with some additives, is called green sand.

If a sand mold is baked in an oven (at 100–300°C) for several hours, a dry sand mold is obtained. When using dry sand molds, usually 1–2% of cereal flour and 1–2% of pitch are added. Dry sand molds reduce gas holes, blows, or porosity in the casting.

The molding sand must be maintained very carefully and tested frequently to control the properties.

The production of sand molds can be analyzed using the morphological approach (Chapter 1), so the molding processes must be considered in the same category as other shaping processes. The shape of the mold is determined by the tool and the pattern of motions. The surface creation principle is generally total forming. The tool, determining the shape of the cavity, is called a pattern and was described in Section 10.4.2.

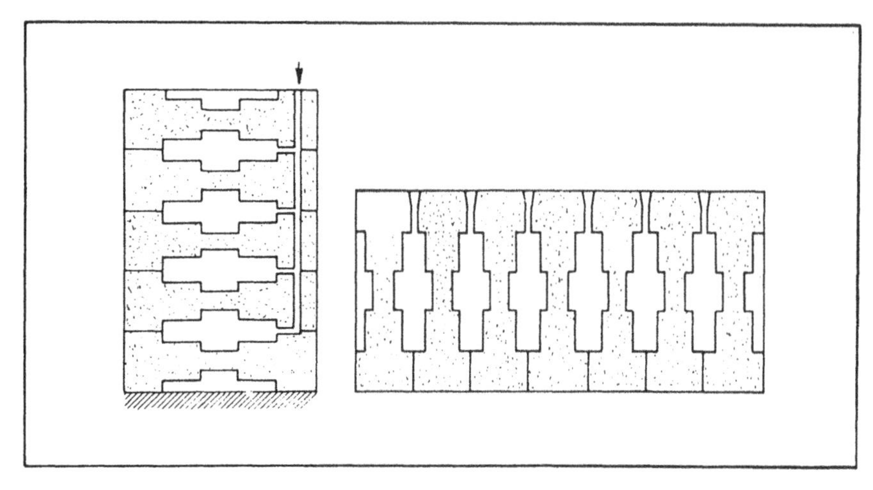

FIGURE 10.11 Arrangement of flaskless molds.

Mold Production

When the pattern has been made (see Section 10.4.2), the mold can be produced, and Fig. 10.10 shows some typical stages in this process. The molds are produced in flasks (i.e., four-sided metal frames), usually consisting of a top flask (cope) and a bottom flask (drag). The flasks are supplied with handles and guide pins for accurate location of the mold halves. To facilitate mold production, many different molding machines have been developed, to facilitate compressing the molding sand to a suitable strength at a high rate. These machines can, for example, be activated by a compression punch combined with a shaking or vibration motion to obtain uniform strength. The sand can also be slung into the flask by sandslings, giving a high production rate.

Much effort has been put into the development of flaskless molding methods. Figure 10.11 shows two different principles in the arrangement of flaskless molds. Figure 10.12 shows an automatic molding machine producing a horizontal string of molds at enormously high speeds. This process, developed by the late V. Aa. Jeppesen at the Technical University of Denmark, has revolutionized the mass production foundry [36]. Further information on molding machinery can be found in the literature.

Applications

Sand casting is used extensively in industry for various cast iron, steel, and metal castings. Casting is, in general, a relatively cheap method of obtaining very complicated components in one stage. Many different types of sand casting processes have been developed to fulfill special purposes. Examples are green sand casting (by far the most widely used process), dry sand casting (similar to

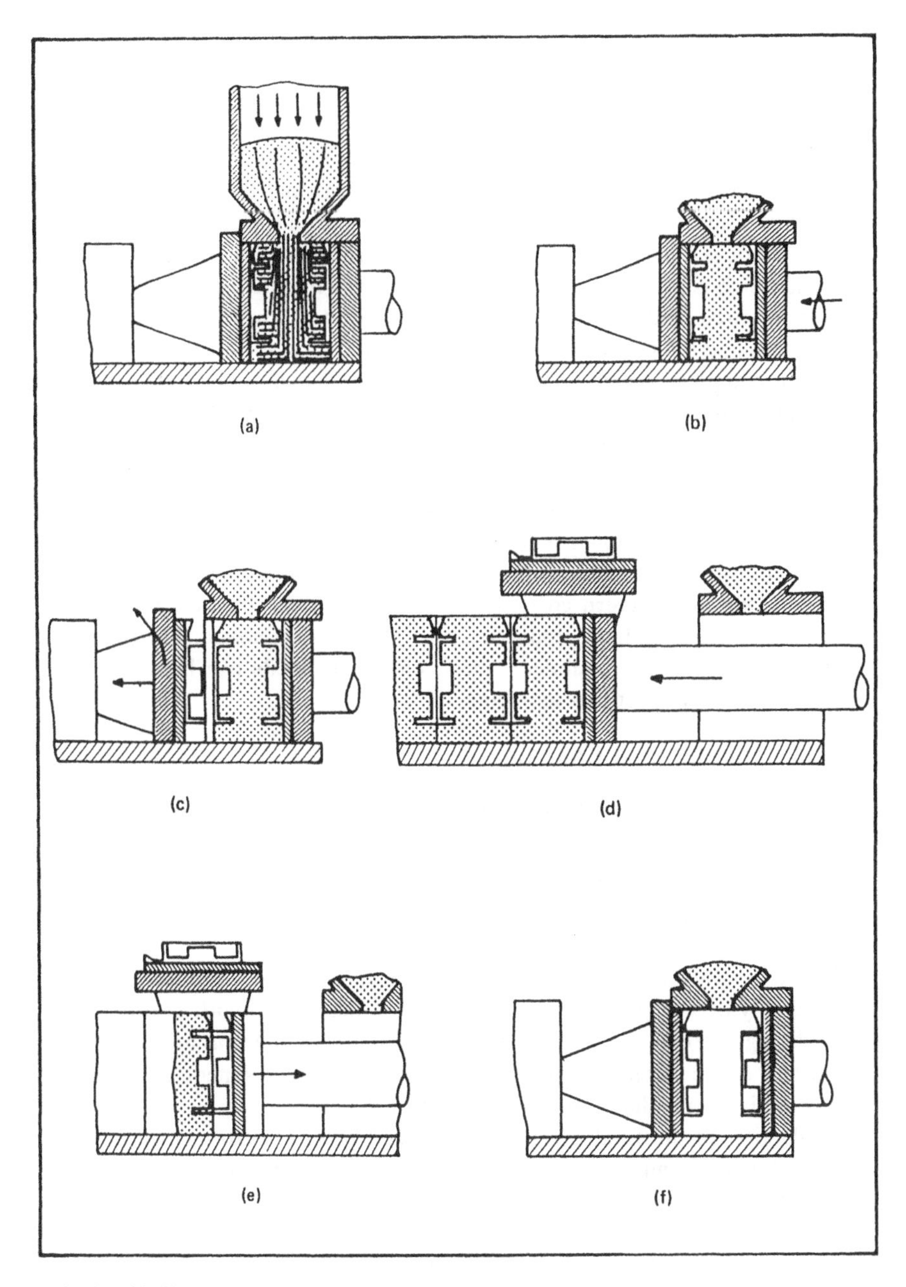

FIGURE 10.12 Disamatic flaskless automatic molding machine: (a) injection of sand; (b) compressing; (c) separation; (d) removal of mold; (e) separation; (f) ready [36].

baked green sand molds), core sand casting, and so on. An important variant is shell mold casting, which is described next.

10.5.2 Shell Mold Casting

In shell mold casting, which is a variant of sand casting, the mold is produced from dried silica sand mixed with a thermosetting resin (phenolic). An accurate metal pattern is heated to 150–250°C, and the sand mixture is dumped on the pattern, which is placed in a mold box. After a few minutes, a layer of the sand mixture is cured, and the excess mixture is removed by inverting the mold box. The pattern and the partially cured shell are baked in an oven for a few minutes to obtain complete curing. The pattern and shell are now separated, and the mold halves assembled with clamps, glue, or other devices. The shell mold is placed in a pouring jacket and backed up or supported by shot or sand. If it is sufficiently rigid, the shell mold can be supported by pins.

The advantages of the shell molding process are: high dimensional accuracy (0.02–0.05 per dimensional unit, for example cm/cm), smooth surfaces, high reproducibility, a minimum of cleaning, and easy automation. The main disadvantage is the high pattern costs, but even at moderate volumes, the process, which has a relatively high production rate, becomes economic. The process is used extensively to cast stainless steels, but most metals can be cast by this method. Maximum part mass is about 10–20 kg.

10.5.3 Investment Casting

In investment casting (also called "lost wax casting" or "precision casting") a pattern of wax is used. The main stages in the investment casting process are:

Production of a master pattern (normally in metal but wood or plastic is sometimes used), used to produce a master die (low-melting-point alloys or steel)

Production of wax patterns by pouring or injection of wax into the master die

Assembly of wax patterns and a common gating system with a sprue (a soldering iron can be used), called a cluster if several patterns are to be united

Coating of the pattern assembly with a thin layer of investment material (dipping in a thin slurry of fine-grained silica)

Production of the final investment by placing the coated pattern assembly into a flask and pouring investment material around (vibrated to remove entrapped air, etc.)

Drying and hardening for several hours

Melting the wax pattern assembly by warming the mold and inverting it to allow the wax to flow out

Heating the mold to higher temperatures (850–1000°C) to drive off moisture and volatile matter

Preheating the mold to 500–1000°C (facilitating flow of the molten metal to thin sections to give better dimensional control)
Pouring the metal (by gravity, pressure, or evacuation of the mold)
Removal of the casting from the mold after solidification

Polystyrene or frozen mercury can also be used as the pattern material. Rubber may be used for permanent patterns whenever it is possible to extract it after investment.

Fine silica (bonded by tetraethyl silicate or sodium ammonium phosphate), plaster (for low pouring temperatures, for example, with magnesium, aluminum, and some copper alloys) or ceramics can be used as mold material. The use of plaster molds and ceramic molds has increased rapidly during recent years. The molds may be reinforced by fibers. In many cases, shell molds are used (produced by dipping about five times), which reduces the quantity of investment material, eliminates flasks, reduces firing time, and simplifies removal of the casting from the mold.

The investment process can be used to produce castings in all ferrous and nonferrous alloys and is important in the casting of special metals such as unmachinable alloys and radioactive metals.

The main advantages of investment casting are: the production of very complicated shapes even in high melting temperature alloys [this includes thin sections (~0.4 mm), undercuts, etc.], very fine details, exceptionally good surface finish, and very high dimensional accuracy (0.003–0.005 per dimensional unit, cm/cm).

The labor costs in investment casting are high. The pattern costs are also high; consequently, the process is used mainly to produce components that require the special characteristics of the process (good surfaces, tolerances, high complexity, etc.). Examples are metals that are difficult to machine or to deform plastically.

10.5.4 Die Casting

As mentioned previously, die casting is characterized by a permanent metal mold and high pouring or injection pressures. The injection pressure, under which solidification also takes place, may vary from 2 to 300 MPa; the usual range is 10–50 MPa. Two different die casting methods are employed: (1) the hot-chamber method and (2) the cold-chamber method. The principal distinguishing feature is the location of the melting pot, which also reflects the final design of the equipment. In the hot-chamber method, the melting pot is included in the machine and the injection cylinder is immersed in the molten metal (see Fig. 10.13a and b). Figure 10.13a shows the metal being forced by air into the die (pressures ~0.5–5 MPa) and Fig. 10.13b shows the metal being forced into the die by a plunger (activated by air or hydraulic pressure), resulting in injection and solidification pressures in the range 10–40 MPa. Hot-chamber die casting is used mainly for the casting of alloys of zinc, tin, lead, and magnesium.

The cold-chamber process has a separate melting furnace, and the molten metal is transferred from the furnace to the cold-chamber machine by hand or mechanically (see Fig. 10.13c). In the machine, the metal is forced into the die by a hydraulically activated plunger. The injection and solidification pressure is in the range 30–150 MPa. Machines of 25 MN plunger force and more are available, allowing the casting of components of up to 50 kg. The cold-chamber method is used mainly for brass, bronze, aluminum, and magnesium castings.

The die casting process is rapid (production rates of up to 1000 castings per hour) and it gives smooth surfaces, good dimensional accuracy (0.002–0.003 per dimensional unit, cm/cm of length) and thin sections (0.6 mm in zinc, 0.90 mm in aluminum, and 1.25 mm in magnesium, brass, and bronze). The draft necessary varies between 0.125 and 0.35 per dimensional unit (cm/cm), depending on the material.

Die casting requires, in general, no machining except for the drilling of holes and threading. Flash and fins must be removed.

The dies for die casting are made from heat-resistant steel and are water-cooled through internal channels. For large castings, a single-cavity die is used and for small castings, multiple cavity dies are used, often built up from inserts.

The die life varies from 10,000 for brass to several million castings for zinc. The optimum numbers of castings are approximately:

Alloys	Optimum number of castings
Zinc/zinc alloys	~200,000 (thickness > 1 mm)
Aluminum/aluminum alloys	~50,000 (thickness 1.5–2.5 mm)
Magnesium/magnesium alloys	~60,000 (thickness > 1.25 mm)
Copper alloys	~5,000 (thickness > 2.0 mm)

The high number of castings necessary to give economical production is due to the high machinery and die costs.

10.5.5 Low-Pressure Die Casting

In this process, a mold—made from graphite or metal—is mounted above an induction furnace, and the metal is forced into the die cavity by inert gas. The gas pressure varies between 0.12 and 0.20 MPa. Castings made by this method are dense, free of inclusions, and have high dimensional accuracy. The process is used for railroad-car wheels, steel ingots, and other applications.

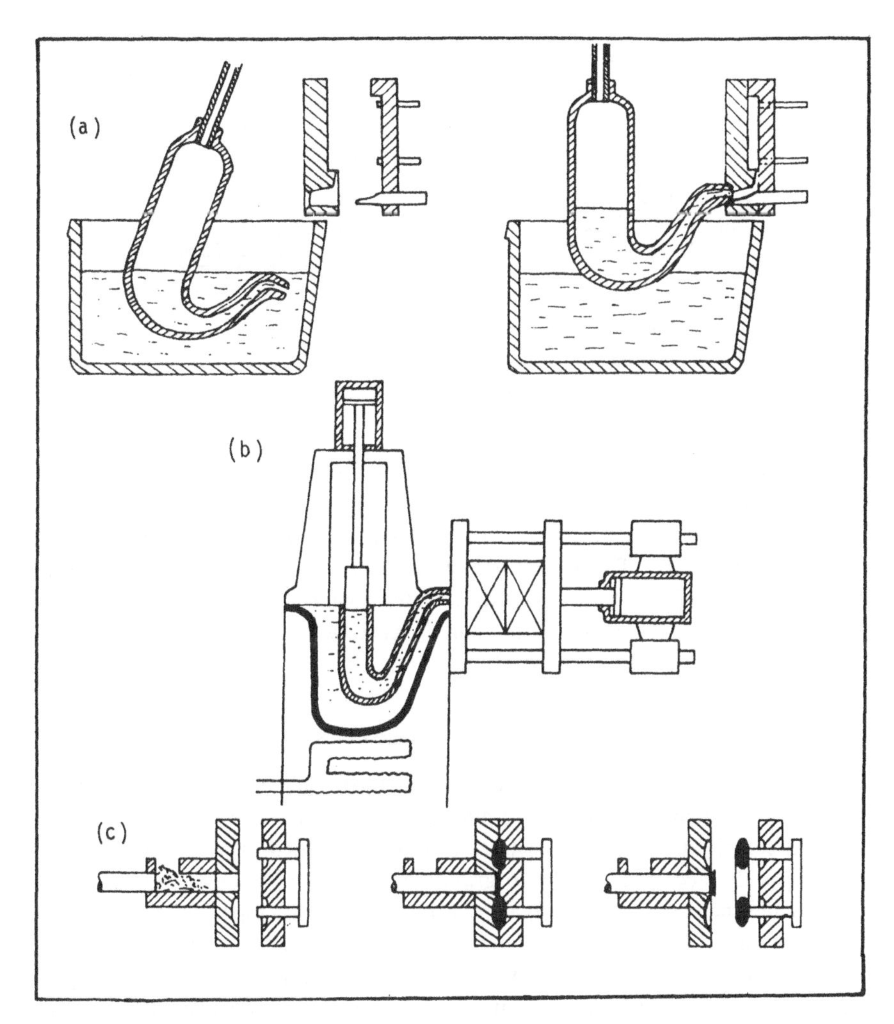

FIGURE 10.13 Die casting: (a) hot-chamber die casting, air activated; (b) plunger activated; (c) cold-chamber die casting.

10.5.6 Gravity or Permanent Mold Casting

In gravity or permanent mold casting, the mold is made from metal or graphite, perhaps coated with a refractory material. Pouring is by gravity, which means that the process is similar to sand casting, but the sand mold is replaced by a metal mold, often made from cast iron. The process, which gives good surface finish, good reproduction of details, and tolerances in the range ~0.02 (cm/cm), is used mainly for aluminum, aluminum alloys, zinc/zinc alloys, and sometimes

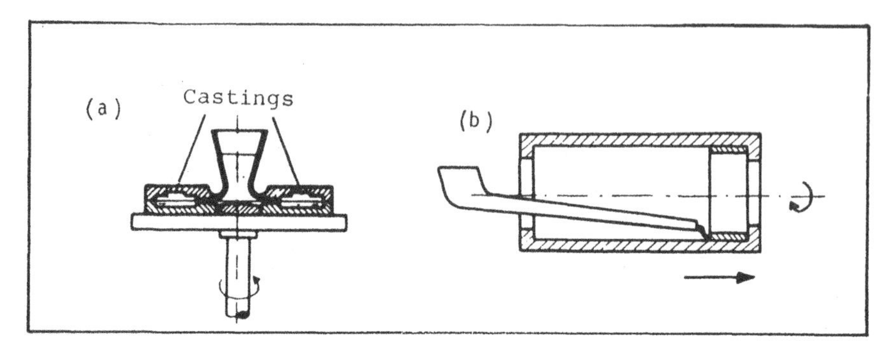

FIGURE 10.14 (a) Centrifuge casting and (b) centrifugal casting [32].

for brass and aluminum bronze. In recent years, the casting of gray cast iron has gained in interest.

Depending on the requirements, the permanent mold can be operated manually or mechanically. The economical number of castings can vary from 500 to 40,000, and in critical places the die is water cooled.

Compared to sand casting, permanent die casting gives better tolerances, surface finish, and mechanical properties.

10.5.7 Miscellaneous Casting Processes

The casting of pipes is often carried out by centrifugal casting, where a permanent or nonpermanent mold rotates during solidification (see Fig. 10.14b). The required centrifugal force necessitates a rotational speed of $n = 300 \sqrt{g}/\pi\sqrt{r}$, where g is the acceleration due to gravity and r is the radius of the pipe. Impurities and inclusions are segregated at the inside of the pipe, which is usually to be machined. Pipes or bushings in cast iron, stainless steel, copper alloys, and bearing metal are produced by this method. A variant of centrifugal casting is centrifuge casting, in which centrifugal force is used to create a suitable pouring pressure (see Fig. 10.14a). This process is limited to symmetrical components.

If a stock of molds with a common sprue and riser is situated along the rotational axis, the semicentrifugal casting process is obtained. The impurities along the axis are removed by machining.

In slush casting, which gives a shell-like component, liquid metal is poured into the mold and, after solidification of a thin layer, the still-molten metal is allowed to flow out by inverting the mold. This process is used for ornamental works, sculptures, toys, and so on. The common materials are tin, lead, zinc, and bronze.

Several other special-purpose casting processes exist, but they will not be described here.

10.6 GEOMETRICAL POSSIBILITIES

10.6.1 Introduction

One of the major advantages of casting results from the virtually unlimited geometrical possibilities (i.e., very complex shaped components can be produced). But the cost of the components depends directly on the complexity. Most foundries have cost estimation methods involving a complexity factor. To achieve the cheapest possible casting for a given complexity, certain design rules must be followed to obtain a good casting. When designing a component for casting, it is necessary to know the casting and molding method, so that the possibilities and limitations of the particular process concerning geometry, size, material, surfaces, tolerances, and so on, can be considered.

An important factor is the minimum wall thickness that can be cast. In normal sand casting, this is about 3 mm for aluminum and 1–6 mm for cast iron, depending on the actual size. These figures correspond to castings having a maximum dimension of about 400 mm.

Table 10.5 shows a rough comparison between some important casting methods [12]. The comparison is intended only as a rough guide. Table 10.6 shows another comparison, where the properties of the castings are graded from 1 to 7 (1 is best and 7 poorest) [16]. The comparisons in these figures are based on general information; in actual cases, detailed information about the particular process must be obtained.

10.6.2 General Design Rules

A few of the main design rules that must be followed to obtain sound castings are described briefly below. Figures 10.15 and 10.16 illustrate the most important rules.

When designing a component for casting, several main factors must be considered, since if forgotten they can each contribute to considerably increased cost. The most important factors are:

Pattern and pattern costs (plane parting surface, simple, draft, etc.)
Mold production (method and properties)
Core production (simplicity, methods, and properties)
Casting [gating system, risers as a function of geometry (Fig. 10.16)]
Cleaning (easy access to core cavities, easy removal of gating system)
Machining (where, how, and what allowances, Table 10.7)
Thermal stresses [reduced by uniform section thickness or when changes are necessary, gradual changes (Fig. 10.15), by permitting the contraction to take place without restraints]

TABLE 10.5 General Characteristics of Some Casting Processes

Casting method	Range of metals	Tolerances per dimensional unit (cm/cm)	Normal mass range	Minimum section thickness (mm)	Surface roughness R_a (μm)	Economical lot size
Sand casting	No limitations (examples: cast iron, steel, Al and Cu alloys)	~0.03 (~0.05 precision sand casting)	Few grams to Mg	2.5–5	5–25	Unit to mass production
Die casting	Al, Mg, Zn, and Cu alloys	~0.0015 (~0.003 for Cu)	Up to 50 kg	0.5–2	1–2	Minimum of 1000
Permanent mold casting	Al, Mg, and Cu alloys; cast iron	~0.01–0.025	Few grams to 50 kg	3	2–3	More than 1000
Shell mold casting	No limitations	~0.02–0.05	Up to 20 kg	1–3	2–5	More than 100
Investment casting	No limitations	~0.003–0.005	Few grams to 10 kg	0.5–0.8	1.5–2	100–5000
Centrifugal casting	No limitations	~0.03–0.10	Up to several Mg	6	5–25	A few hundred

Source: From Ref. 12.

TABLE 10.6 Comparison of Various Casting Processes

Property	Casting process					
	Sand casting	Die casting (hot chamber)	Die casting (cold chamber)	Permanent mold casting	Investment casting	Centrifugal casting
Porosity	7	5	4–3	3–4	2–1	1–2
Surface quality	6–7	2–3	3–2	4–5	1	5–4
Dimensional accuracy	6–7	2	1	5–4	3–2	5–4
Strength	7–5	2–1	1–2	4	7–5	3
Minimum section thickness	5–7	3–2	2–3	4	1	6
Mold cost	1–2	6–7	7–6	5	4	3
Production rate						
Small lot sizes	2	6–7	7–6	4	3	5
Large lot sizes	3	1–2	2–1	3	5	4

Source: From Ref. 16.

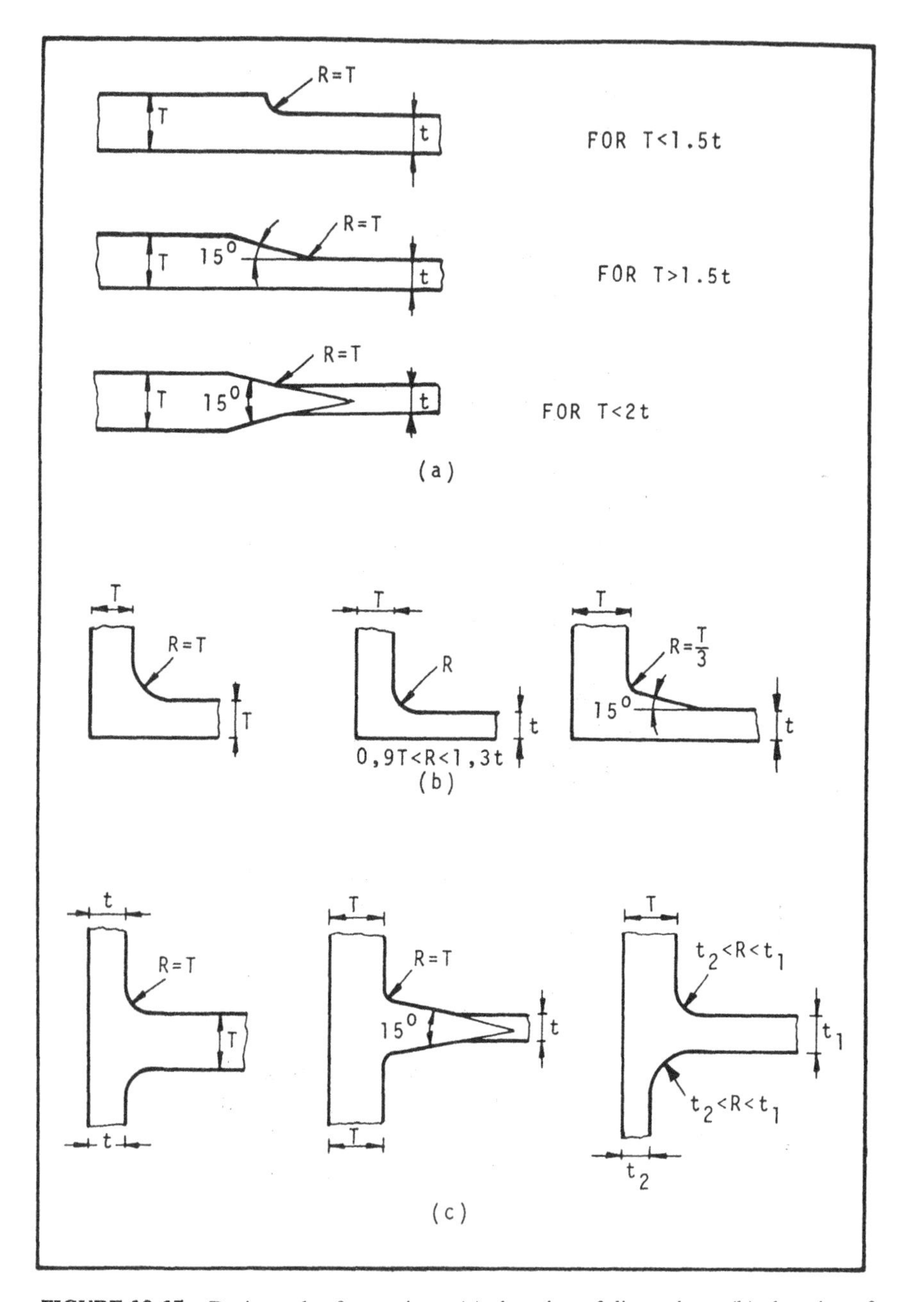

FIGURE 10.15 Design rules for castings: (a) changing of dimensions; (b) changing of directions; (c) T-joints.

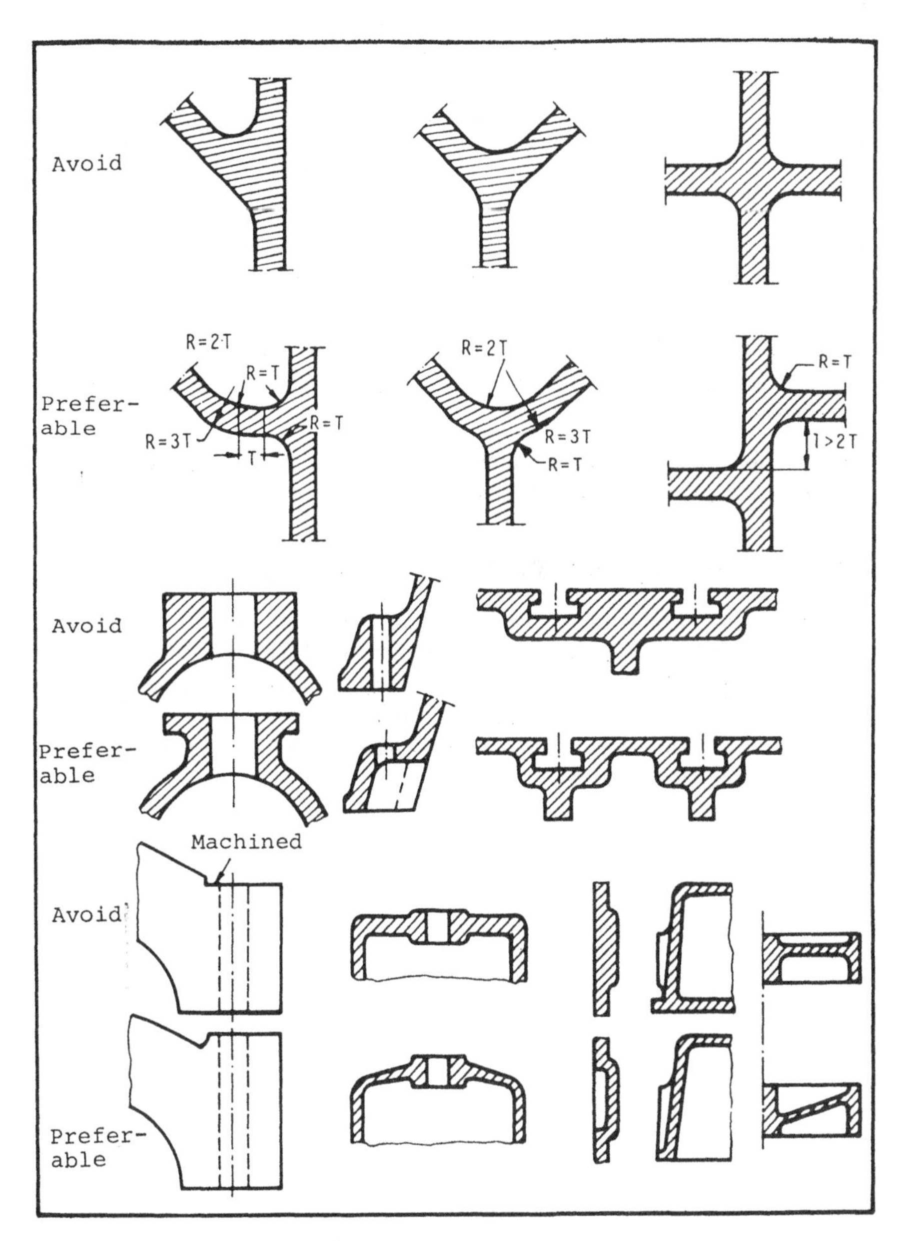

FIGURE 10.16 Design rules for castings.

TABLE 10.7 Machining Allowances for Sand Castings

	Main dimension (mm)			
Material	0–300	300–600	600–1000	1000–1500
Nonferrous metals	1.5	2.5	9.0	—
Cast iron	2.5	3.5	5.0	6.5
Steel	3.5	5	8.0	10.0

Mechanical loadings (in the shaping of the component, proper force transfer must be considered along with the characteristics of the casting process)
Appearance

More detailed information concerning the design of castings, casting processes, and so on, can be found in the literature [32–35,37,38].

10.7 EXAMPLES OF TYPICAL CASTING PROCESSES

In this section are found short descriptions of some typical casting processes, classified according to the fundamental elements and the predominant state of stress in a way similar to that used in Chapters 6, 7, and 8. Only those processes most frequently used in industry are described.

Figures are provided by courtesy of The Manufacturing Consortium, Brigham Young University, Provo, Utah.

PROCESS 1: Sand Casting (M, Me, Ri, TF, Co)

Description: In sand casting the molten metal is poured into a prepared sand mold, dimensioned, and contoured to match the desired casting. Internal shapes in castings are obtained by placing baked cores consisting of silica sand and a binder in the mold cavity. The melt is poured into the pouring basin and flows to the mold cavity through a gate or a system of gates. After filling the mold cavity, the melt enters the risers, which act as a reservoir of excess metal to compensate for shrinkage during solidification of the casting. A new mold must be made for each casting.

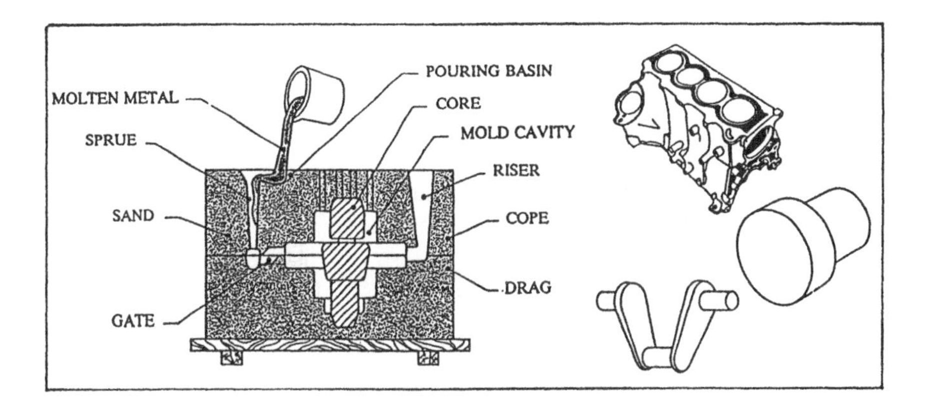

Applications: Sand casting of cast iron, steel, and metals is extensively used in industry. It is, in general, a relatively cheap method of obtaining very complicated components in one stage. It is a low-volume production process when not automated/mechanized and a high-volume process when automated. Typical component weights range from 500 g to 50 kg, but the method is feasible for weights of several metric tons. Wall thicknesses are generally 5–50 mm. Examples are engine blocks, crankshafts, connecting rods, bearing pedestals, machine tool beds, turbine housings, etc.

Tolerances/Surfaces: Sand castings are rougher surfaced and less accurate dimensionally than any other type of casting. Tolerance and surface finish depend upon accuracy of pattern, cores and inserts, pattern smoothness and stability, sand compaction, gating and rising system, etc., necessitating machining allowances of 3–10 mm for the castings.

Machinery: Equipment for shaping the molds consists of patterns commonly made of wood or metal and flasks with handles and guide pins for manual applications. For automated applications flaskless molds are used in a continuous string. Many different machines have been developed to facilitate compressing the sand to a suitable strength at a high rate.

PROCESS 2: Shell Mold Casting (M, Me, Ri, TF, Co)

Description: This method is very much the same as sand casting, but with a molding mixture of fine, sharp sand mixed with thermosetting resin instead of plain sand. As in sand casting, the mold is nonreusable. A metal pattern of the casting complete with sprue and gate is heated to 150–250°C and covered with a layer of sand mixture. After a few minutes, a shell about 5 mm thick is partially cured and the excess mixture removed. The shell is then baked in an oven and becomes a rigid, strong, and heat-resisting mold that must be destroyed to remove the cast component after its solidification.

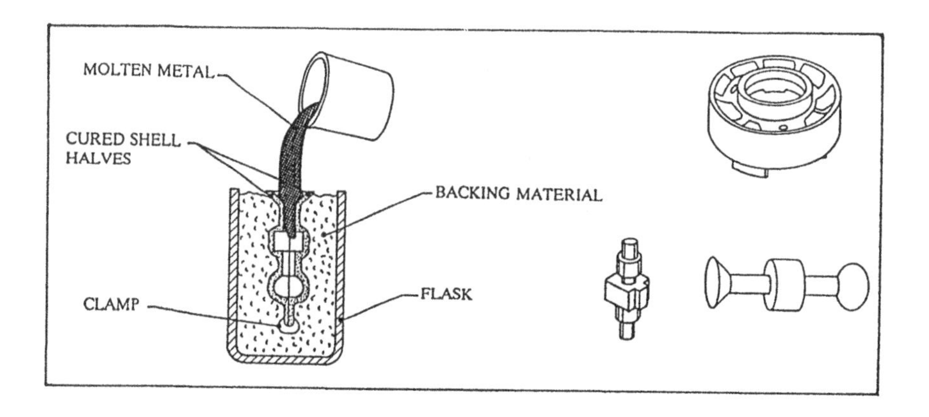

Applications: Shell mold casting offers greater dimensional accuracy and better surface finish than sand casting. It also offers better capabilities for sharp corners, intricate contours, small holes, etc. But the need for metal patterns makes it too expensive if only a few identical workpieces are needed. Components needing as-cast surfaces smooth enough to avoid subsequent machining and components with cooling fins, etc., as encountered in air-cooled combustion engines are typically produced by shell mold casting. Almost any type of metal or alloy can be cast. Maximum part mass is about 10–20 kg.

Tolerances/Surfaces: Shell mold castings need very little cleaning after separation from the mold and the reproducibility of the process is high. Dimensional accuracies from ±0.2% to ±0.5% and smooth surfaces, R_a-values of about 50–100 μm, are generally obtainable.

Machinery: Equipment for producing the mold shells consists of a metal pattern, pattern plate, and a dump box for the shell mix. Further, a burner and an oven for curing the shell mold and a furnace for melting the casting material. The process may also be mechanized.

PROCESS 3: Investment Casting (M, Me, Ri, TF, Co)

Description: A process in which molten metal is poured into a preheated mold made by means of a disposable pattern of wax or plastic coated with a thin layer of silica, plaster, or ceramic. The use of shell molds reduces the quantity of investment material, eliminates flasks, and simplifies removal of the casting from the mold. Before casting, the pattern is melted by warming the mold and inverting it, allowing the wax to flow out. No draft is necessary and very complex shapes, which cannot be made by other casting processes are possible.

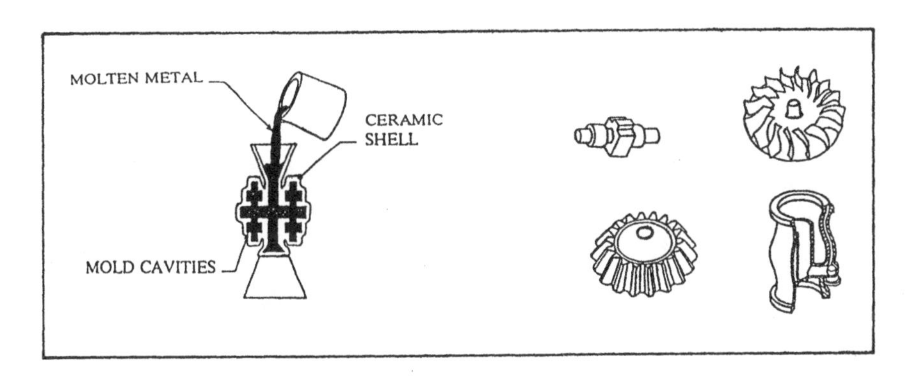

Applications: Geometrical possibilities of investment casting are almost limitless since mold geometry and draft are not a consideration because pattern and workpiece do not need to be removed from the mold. Workpieces with wall thicknesses as thin as 0.4 mm can be cast by this method. The castings typically range in size from 100 g to 20 kg. Turbine wheels, gears, and other products calling for exact dimensions in metals with a high melting point are examples of applications. The investment process can be used to produce castings in all ferrous and nonferrous alloys and is important in the casting of special metals such as unmachinable alloys and radioactive metals.

Tolerances/Surfaces: Tolerances of ±0.075% are obtainable for components up to 15 mm. For large components ±0.7% may be expected. The surface roughness of investment castings range from 1.5 to 3 μm R_a. It should be remembered, however, that the precision attainable involves a lot of preliminary work and skill, making investment casting too expensive for routine foundry work.

Machinery: Equipment for producing the ceramic mold consists of slurry tank and baking oven. Equipment for making the wax or plastic pattern is also needed as well as furnaces for melting the patterns from the mold and for melting the casting metal. The facilities may be more or less automated.

PROCESS 4: Permanent Mold Casting (M, Me, Ri, TF, Co)

Description: In permanent mold casting the molten metal is poured into a permanent (reusable) mold under gravity or low pressure and held until solidification begins. This means that the process is similar to sand casting except for the mold material, which in the case of permanent molding typically is fine-grain cast iron or steel. It is possible to heat or cool the mold in critical places, giving more favorable cooling conditions for the workpiece, resulting in better mechanical properties.

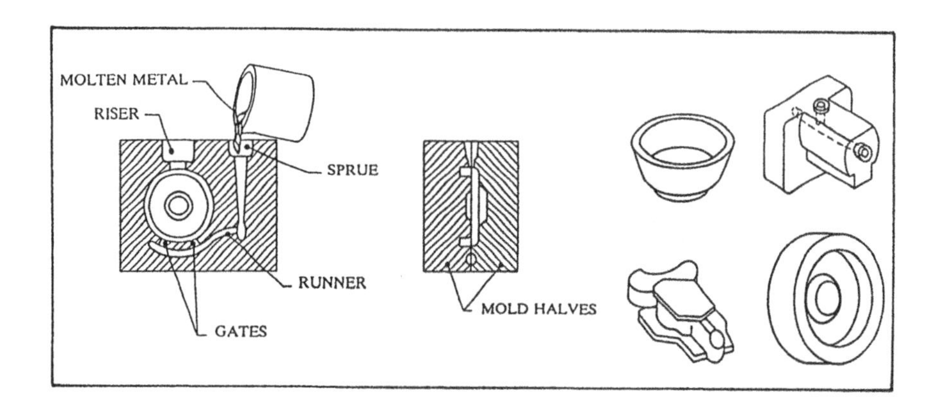

Applications: The permanent mold casting process can produce a wide variety of parts, but complex shapes require considerable expense in mold design and fabrication. It is used primarily for aluminum, zinc, magnesium, copper and their alloys, and recently also for gray cast iron. Component weights typically range between 30 g and 10 kg. The initial costs of the molds are high.

Tolerances/Surfaces: Dimensional accuracy is affected by the quality of the mold and by changes in position of moving parts in it. Typical tolerances are ±0.25 mm to ±4 mm across parting lines. Surface finish depends on the finish of the mold wall, mold coatings used, venting, casting temperature, gating design, etc. Typical ranges for surface finish are 25–100 μm R_a.

Machinery: The setup normally consists of two mold halves containing a gate and runner system and a pouring basin, alignment pins, an ejection system, and clamps to hold the mold halves together. Simple cores made of metal or sand are sometimes used. Automation is possible.

PROCESS 5: Die Casting (M, Me, Ri, TF, Co)

Description: Die casting is a process where molten metal is forced by a ram under high injection pressure (10–100 MPa) into a reusable mold and held under pressure until solidification occurs. Cooling of the workpiece is hastened by circulating water in the die, which also increases the life of the very expensive die. Die casting is economical for large production runs only. Two different methods are employed: the cold-chamber process, which has a separate melting furnace from which the molten metal is transferred to the machine by hand or mechanically, and the hot-chamber method, which has a melting furnace as an integral part of the machine.

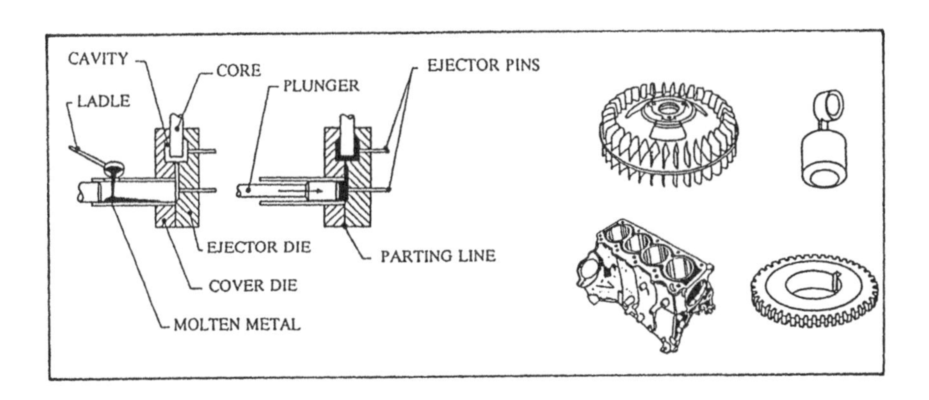

Applications: The die casting process is rapid—production rates of up to 1000 castings per hour—and is used for components of a wide range of sizes and geometries, such as engine parts, gears, fly wheels, rotors, frames, covers, etc. The cold-chamber method is used mainly for brass, bronze, aluminum, and magnesium castings while the hot-chamber method takes care of alloys with lower melting points (e.g., zinc, tin, lead, and magnesium). Wall thickness ranges from 1 to 40 mm, weights range from 10 g to about 50 kg.

Tolerances/Surfaces: Tolerances for die cast components vary according to the method and material used. Values of ±0.1 to ±0.005 mm/cm of length may be obtained. Surface finish values range from 5 to 25 μm R_a. In general, no machining is required, except drilling and threading of holes. Small amounts of flash around the casting edges may be present.

Machinery: A die casting machine consists of pressure and power cylinders, plunger, and mold or die cavity. Typically, production rates range from 50 to 500 shots/hour, although high production outputs of 2000 to 5000 shots/hour are feasible for small components.

11

Plastics and Plastic Processing

11.1 INTRODUCTION

In this chapter a short description is given of the manufacturing properties of plastics. Then some of the more important methods of processing plastics are discussed. It should be emphasized that the manufacturing methods for plastics are as diverse as the plastic materials themselves; consequently, this chapter presents only a selection of the most important methods. To acquire a more complete picture, the literature must be studied and contact with the manufacturers must be established to obtain detailed information about particular materials.

Plastics have become important engineering materials in recent years, and the number of applications is increasing steadily. Therefore, it is necessary to have a fundamental understanding of plastic materials and methods of processing.

11.2 MANUFACTURING PROPERTIES OF PLASTICS

In Chapter 3 the most important characteristics of plastic materials were discussed, based on their molecular structure. Depending on the structure, plastics can be classified as:

1. Thermoplastics (having a linear/chain structure)
2. Thermosetting plastics (having a cross-linked structure)

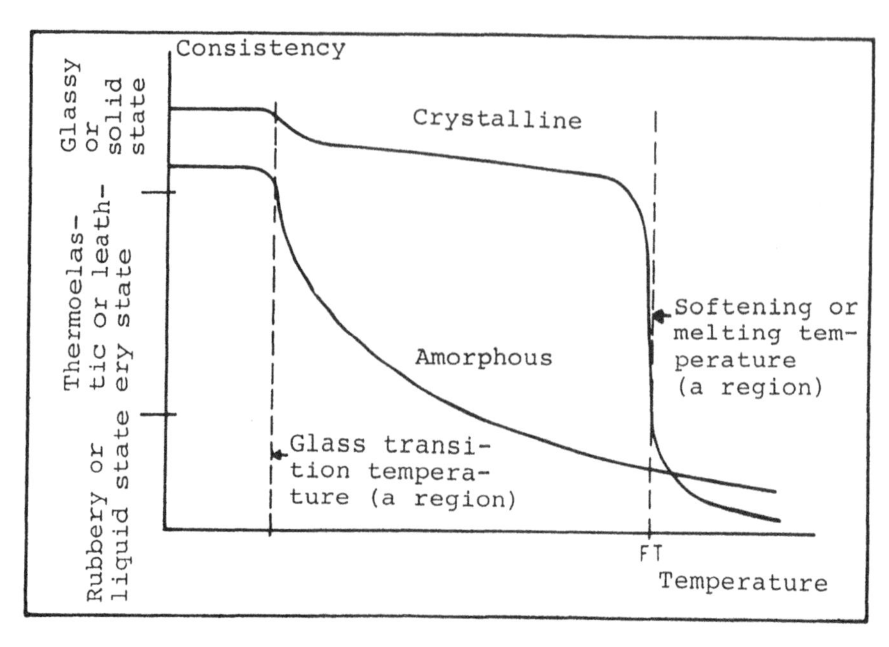

FIGURE 11.1 The consistency (e.g., elastic modules) of thermoplastics as a function of temperature.

The secondary bonding between the chains in thermoplastics is weakened by heating, allowing the chain molecules to move relative to each other. This change is reversible; increasing temperature softens thermoplastics and decreasing temperature makes them harder and stronger. The strong primary bondings (cross-links) between the chains in thermosetting plastics are, in general, not influenced by temperature; and if influenced, only to a very limited extent.

Classification of plastic materials into thermoplastics and thermosetting plastics thus gives valuable information about their behavior in processing and service.

11.2.1 Thermoplastics

The consistency (e.g., measured by the elastic modules) of thermoplastics as a class varies considerably with temperature. This is shown schematically in Fig. 11.1 for both amorphous and crystalline thermoplastics.

Below the glass transition temperature, the materials are hard, brittle, or ductile, depending on the actual material. It must be emphasized that the regions and levels, and even the existence of certain portions of the curve indicated in Fig. 11.1, can be quite different for different materials.

Above the glass transition temperature, amorphous and crystalline thermoplastics behave differently. For amorphous materials, the structural changes allow short segments of the structure to rotate and slide (reversibly) relative to each other. The state changes gradually from solid to thermoelastic (leathery) to rubbery or liquid (i.e., decreasing viscosity with increasing temperature). For the crystalline materials, the behavior depends on the degree of crystallinity (i.e., the relative sizes of crystalline and amorphous regions). Figure 11.1 shows that the consistency changes gradually with increasing temperature until the crystalline melting point is reached (i.e., the crystalline materials remain solid to higher temperatures). This means that the rubbery region for crystalline material is very small or is not present. Most thermoplastics contain both crystalline and amorphous regions, resulting in consistency-temperature curves lying between the limits shown in Fig. 11.1. In the rubbery state (rubbery flow is utilized in shaping crystalline materials) the temperature must be controlled very carefully.

The different states of the thermoplastic materials enable the application of different processing methods. In the glassy or solid state, the materials can be processed by machining (turning, drilling, milling, etc.). The existence of the thermoelastic or leathery state has allowed the development of many different forming processes. The material can, in this state, undergo large reversible deformations accompanied by orientation of the long-chain molecules. By cooling, this orientation will cause high internal stresses, which tend to establish the previous shape when heated again (i.e., components formed in this state have a low thermal stability). Examples of processes are blow molding and vacuum forming.

In the rubbery flow or liquid state, permanent molecular sliding takes place (i.e., after forming and cooling, a stable product results). It is in this state that most thermoplastic materials are shaped using processes such as extrusion and injection molding.

11.2.2 Thermosetting Plastics

Thermosetting plastics do not exhibit any significant change in consistency or properties with increasing temperature because of their cross-linked molecular structure. At a sufficiently high temperature, a chemical decomposition (i.e., destruction) takes place. In general, thermosetting plastics have a higher thermal or heat resistivity than do thermoplastics.

In the solid state (hardened or cured) thermosetting plastics can be processed only by machining. However, forming of thermosetting plastics, in general, is carried out in the uncured or partially cured state. If forming is carried out in the partially cured state, the final network structure (curing) then develops during and/or after forming.

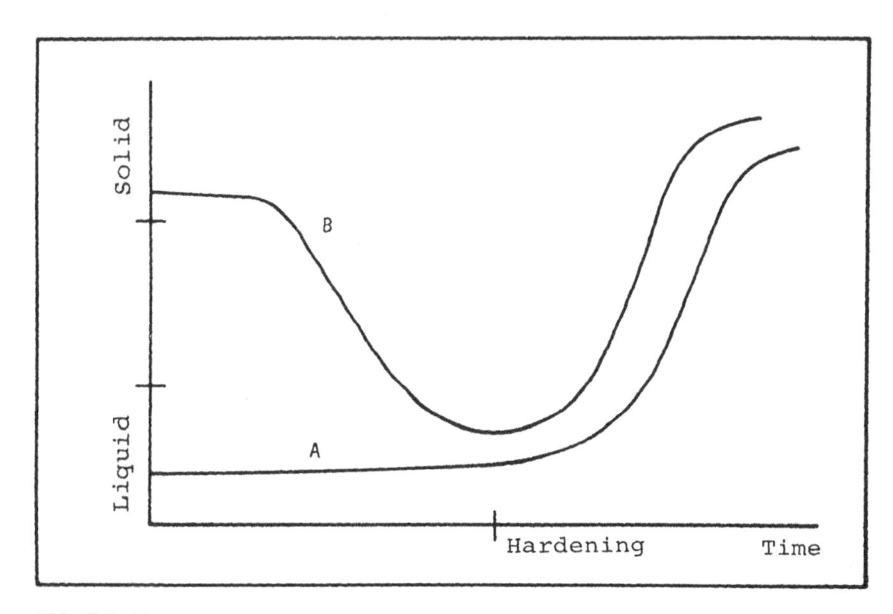

FIGURE 11.2 The change of consistency for thermosetting plastics during forming (time): A, liquid; B, granular (solid) raw materials.

The raw materials for thermosetting plastics are available in granular or liquid form. Figure 11.2 shows how the consistency changes with time (during forming) for a liquid (A) and granular (solid) (B) material. Curve A may illustrate, for example, a polyester region with the necessary initiation (using agitators) to cause hardening. The heat developed during curing will accelerate the process until the final solid state has been reached.

Curve B represents thermosetting plastics, which are available in the powder or granular state. Here forming is combined with the heating of the material. The heating has two different effects: the material becomes softer and attains a plastic state, so that processes can be used to shape it; and the hardening or curing is initiated. After some time, the curing process dominates and changes the state from thermoplastic to solid. In the forming of thermosetting plastics, the importance of control of the process, so that the final curing starts after the termination of forming, must be emphasized.

When the curing process has reached a certain level (i.e., sufficient strength is obtained), the component can be removed from the mold.

The main parameters in forming both thermoplastic and thermosetting materials are: the actual plastic material (with the necessary chemical agents), the temperature, and the pressure. The influence of temperature has been discussed briefly. The pressure is generally used to acquire sufficient geometrical accuracy, i.e., to force the soft or liquid plastic material to flow into all portions of the mold or die.

Based on a knowledge of the particular material and the influence of temperature and pressure, the morphological procedure described in Chapter 1 can be used to analyze and describe plastic forming processes. It is important to have detailed information about the particular material (from the manufacturer) before process parameters and surface creation principles can be selected. More detailed information can be found in Chapter 3 and in the literature.

In the following, some of the most important processes used to manufacture plastic components are discussed briefly.

11.3 PLASTIC PROCESSING METHODS

As mentioned, a great variety of methods to process plastic materials are available. Based on a broad knowledge of plastic materials and their behavior, the morphological approach (Chapter 1) will reveal the common factors in the processing of many plastics. Therefore, the important examples described below should be considered as specific cases derived from the morphological structure.

11.3.1 Casting

The casting of plastics involves, in general, the same main phases as the casting of metals: the production of a suitable die or mold, including cores; the melting of the material or mixing of liquid agents; pouring by gravitation; and "solidification." The "solidification" process depends on the actual material and may involve oven curing.

The necessary molds and cores are produced as simply and cheaply as possible. Often molds of lead are used, produced by dipping a steel or metal model in molten lead and stripping the shell of lead off the model after solidification. Cores can be produced in plaster, lead, rubber, and so on. Molds can also be produced from plaster, glass, wood, synthetic rubber (to allow undercuts), or other suitable materials.

Casting is used primarily for thermosetting materials such as phenolics, polyesters, epoxies, and allyl resins, but also thermoplastic materials such as ethyl cellulose, cellulose acetate butyrate, acrylics, and vinyls can be cast.

The casting process has many industrial applications, including short rods and tubes, toys, jewelry, clock and instrument cases, handles, knobs, drilling jigs, and punches and dies for sheet metal forming, including drop-hammer and stretch dies. The case components can be reinforced in various manners (steel, glass fibers, etc.).

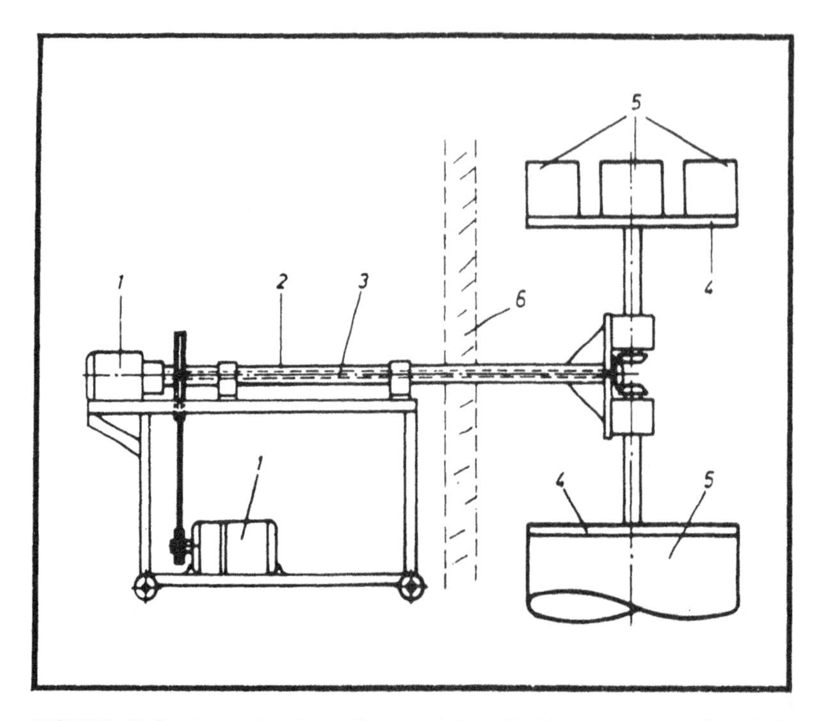

FIGURE 11.3 A rotational molding machine. (1) Motor and gear, (2) tubular shaft, (3) solid shaft, (4) mold tables, (5) molds, (6) oven wall.

11.3.2 Rotational Molding

In rotational molding, a simultaneous rotation of thin-walled models (of metal sheets) about two axes (perpendicular to each other) distributes the heated and molten material in the mold in a layer covering the inside of the mold. The thickness of the layer is determined by the amount of material fed into the mold. After cooling, the component is removed from the mold. A rotational molding machine is shown in Fig. 11.3. Heating and melting is carried out in a large oven using hot air and, after forming, cooling can be carried out by cold air, water, and so on, in the heating oven or in a special cooling chamber. Oil within the mold wall can be used for both heating and cooling, allowing for close temperature control.

For many years, only small products were produced by rotational molding (doll heads, toys, ornaments, bulbs, etc.), but in the last 10 to 15 years the process has been used primarily in the production of larger components such as garbage containers, chairs, tanks of various sorts, and drums for various purposes. The wall thicknesses obtainable vary between 2 and 12 mm, and components with main dimensions of 2–3 m are not unusual.

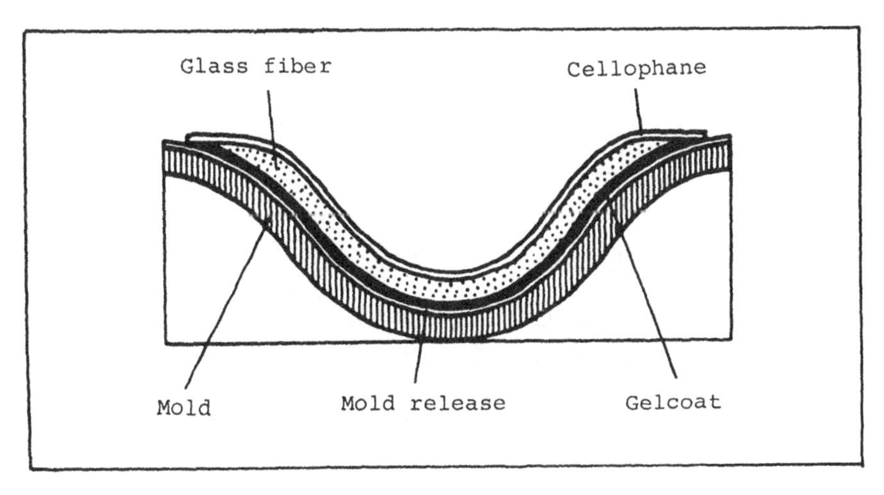

FIGURE 11.4 Open-mold forming of reinforced plastics (manual).

The materials used for the rotational molding of thermoplastics are, for example, powdered (100–300 μm) polyethylene resins and polyvinyl chloride paste, which are the most important, but most of the casting plastics may be used.

Advantages of the rotational molding process are: flexibility (various components can be produced on the same equipment, even in the same cycle), low tooling costs, closed- and open-ended components, good surface finish, and low cost. The molds can be produced from sheet metal, cast aluminum, and other materials.

11.3.3 Reinforced and Laminated Plastics

The reinforcement and lamination of plastics includes a wide variety of methods to produce composite materials. These methods can be carried out under temperature action alone or temperature action combined with a pressure action. The main feature is that a fiber material is impregnated with a liquid plastic (resin) which, after forming, is allowed to "solidify." The most commonly used materials are primarily polyester and epoxy resins, but many other materials can be used. The fiber materials can be glass (predominantly) in random or woven form, asbestos, cotton, and synthetic fibers. Fiberglass and other composite materials can be made by a variety of processes.

In open-mold forming, a male or female mold is used (see Fig. 11.4). The mold can be made from wood, metal, or other materials. Glass fibers and resins are placed in the mold and the entrapped air removed. In most cases, no pressure is applied, but vacuum or pressure bags can be used to give a low pressure. Cur-

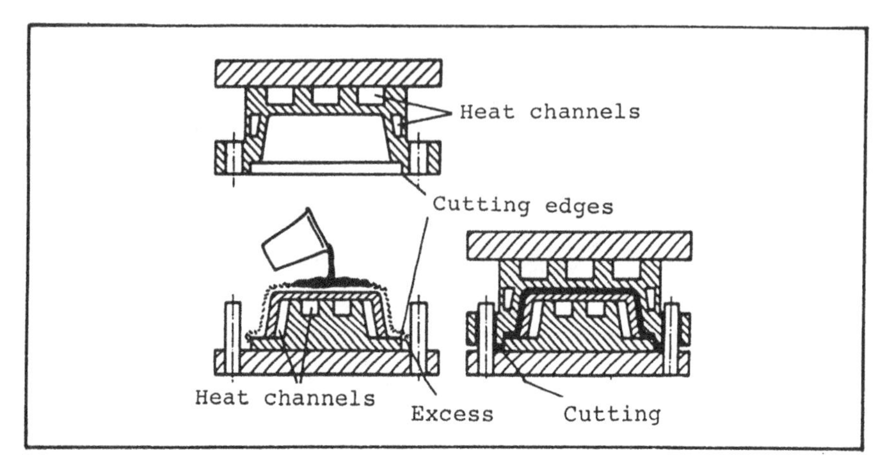

FIGURE 11.5 Closed-mold forming of reinforced plastics.

ing normally takes place in air. Examples of open-mold processing are fiberglass boats, aircraft parts, luggage, truck and bus components, and containers.

Closed-mold forming or pressing takes place in two-part molds (see Fig. 11.5). The molds are generally made from metal, and the process gives a good surface finish (both sides), good tolerances, and high production rates. Examples of closed-mold pressing include trays, helmets, machinery housings, and luggage. Closed-mold forming requires relatively expensive tooling and a hydraulic press.

Open-mold forming generally requires a minimum of equipment, but if larger number of components are desired, special machinery can be developed. Figure 11.4 shows the simple manual method, but in many cases, for example, boats and other large components, the spray-up method shown in Fig. 11.6 is used. Here glass fiber (cut into small lengths) and resin are deposited simultaneously.

The production of tubes, pressure cases, circular cylindrical bodies, and other products is carried out by filament winding. Single strands of fiber yarn are used as reinforcing material, which is passed through a bath of resin and wound onto a mandrel. This process has become important in recent years.

Laminated plastics consist of sheets of paper, fabric/cloth, asbestos, wood, and so on, which are impregnated or coated with resin and formed under heat and pressure to the desired shape. The resulting "sandwich" is cured under pressure at elevated temperatures. These products, which are commercially available as stock materials in sheet form, rod, tubes, and many special shapes, have excellent strength properties, high impact resistance, and good electrical insulation. Common applications include gears, electrical insulating parts, handles, and furniture. Laminated plastics have good machining properties. Many

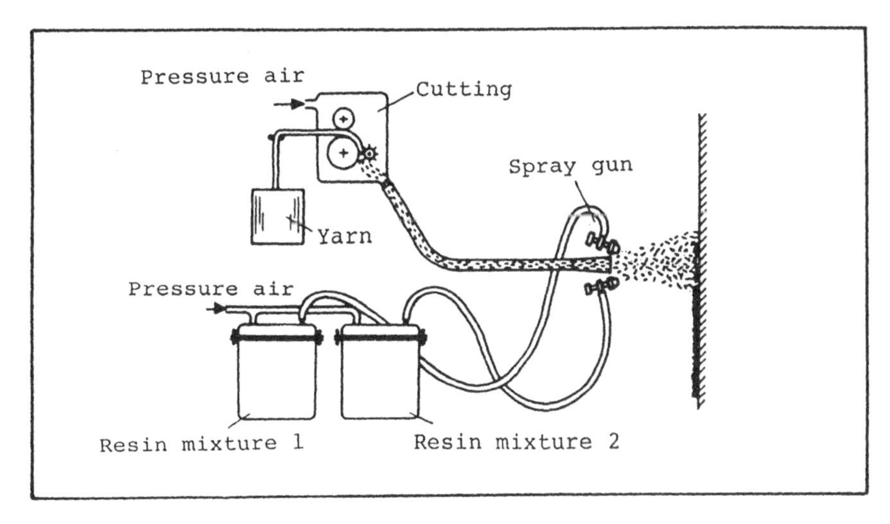

FIGURE 11.6 Spray-up of fiber-reinforced plastics.

special processes have been developed and various laminated products produced; for further details, see the literature.

11.3.4 Extrusion

The extrusion of plastics is extensively used to produce various geometries, such as long profiles, rods, tubes, sheets, and foils, in different lengths. A special but important application is the extrusion coating of the electrical insulators on wires and cables.

The extrusion process requires that the material is in a rubbery flow state, so that shaping can take place. Most extruded products are thermoplastic materials, but if special precautions are taken, thermosetting materials can also be extruded. Figure 11.7 shows a typical extrusion press. The plastic material is fed from the hopper into the screw chamber, where it is heated, compressed, and forced through a heated die. On leaving the extrusion die, the product is cooled by air or water, giving sufficient strength through hardening for further handling. Extrusion is a cheap and rapid method of molding.

Products requiring close tolerances can be calibrated (i.e., finish formed) just after the extruded shape has left the extrusion die.

11.3.5 Blow Molding

The blow molding process is used to produce thin-walled hollow components in thermoplastic materials. A cylinder of plastic material (a parison) is extruded and positioned in the opened two-part mold (see Fig. 11.8).

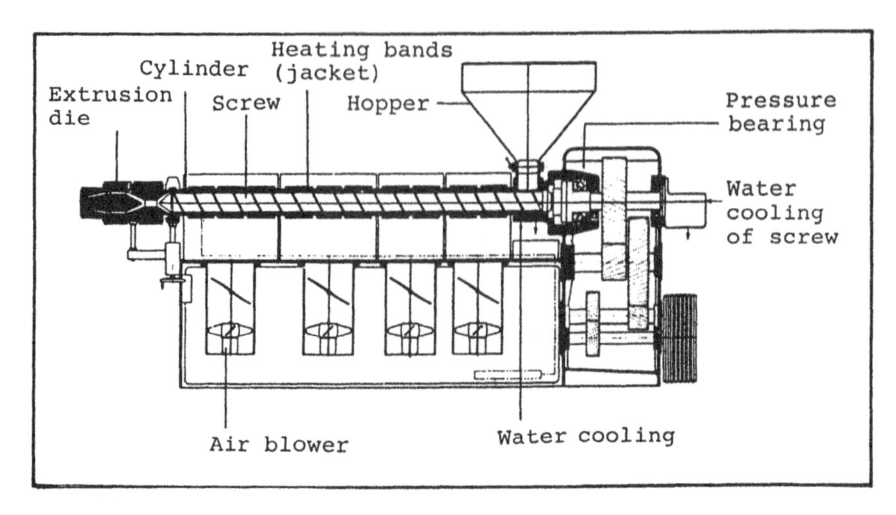

FIGURE 11.7 Single-screw extrusion press.

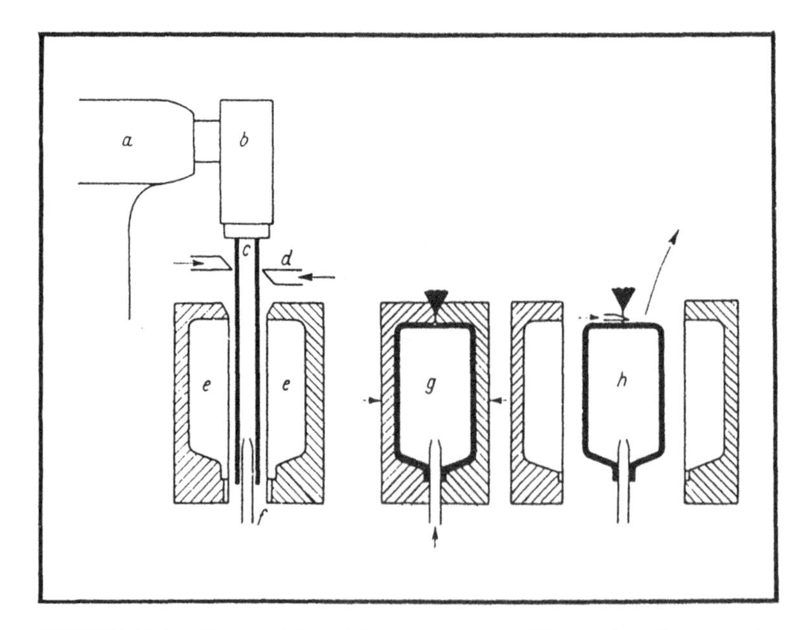

FIGURE 11.8 Blow molding: (a) extrusion press; (b) extrusion die; (c) parison; (d) cutting; (e) split mold; (f) blow mandrel or core tube; (g) closed mold; (h) opened mold and removal of the component.

The material is cut off, the mold closed, and air pressure fed into the hollow tube, expanding it toward the walls of the mold. After cooling, the mold is opened and the component removed.

Examples of blow-molded components include bottles, containers, floats, heater ducts, and cosmetic packaging. Suitable materials are polyethylene, polypropylene, and cellulose acetate, for example.

11.3.6 Injection Molding

Injection molding is the most widely used process for the manufacturing of components of thermoplastic materials. The process is similar to the die casting of metals. The granular material is fed from a hopper into a screw chamber, where it is heated and melted and then injected under high pressure into the mold or die and allowed to solidify (see Fig. 11.9). The solidification takes place when the material is still under pressure, giving rather close tolerances. The die is often water cooled. The injection pressure is normally in the range 50–200 MPa. Many different injection molding machines are available, but they will not be described here.

Examples of the applications of this important process include toys, cans, boxes, fittings, pumps, propellers, gears, bearings, guiding elements, caps, and housings for appliances. Because of the high production rates obtainable, it is a cheap mass-production process. Tolerances in the range ±0.1 to ±0.5 mm, depending on the size, are readily obtainable.

The materials are primarily thermoplastics, but in recent years the injection molding of thermosetting plastics has become quite extensively used, and molding machines capable of processing thermoplastic and thermosetting materials—as well as rubbers—have been developed.

The molding machines are characterized by the injection capacity (typically 1–5 kg) and the clamping force (up to 30 MN).

It should be mentioned that the properties of the components depend to a high degree on the proper filling of the mold, gates, and runners and on the cooling conditions. Consequently, mold or die design plays an important role in injection molding. Since the mold system is rather expensive, a large number of components must be manufactured to make the process economical.

11.3.7 Compression and Transfer Molding

Compression and transfer molding are closely related processes since both produce components using a two-part mold.

In compression molding, the plastic material is placed in the heated lower part of the mold and the mold is closed under the application of a high pressure (up to 60 MPa), causing the softened material to fill the cavity (see Fig. 11.10).

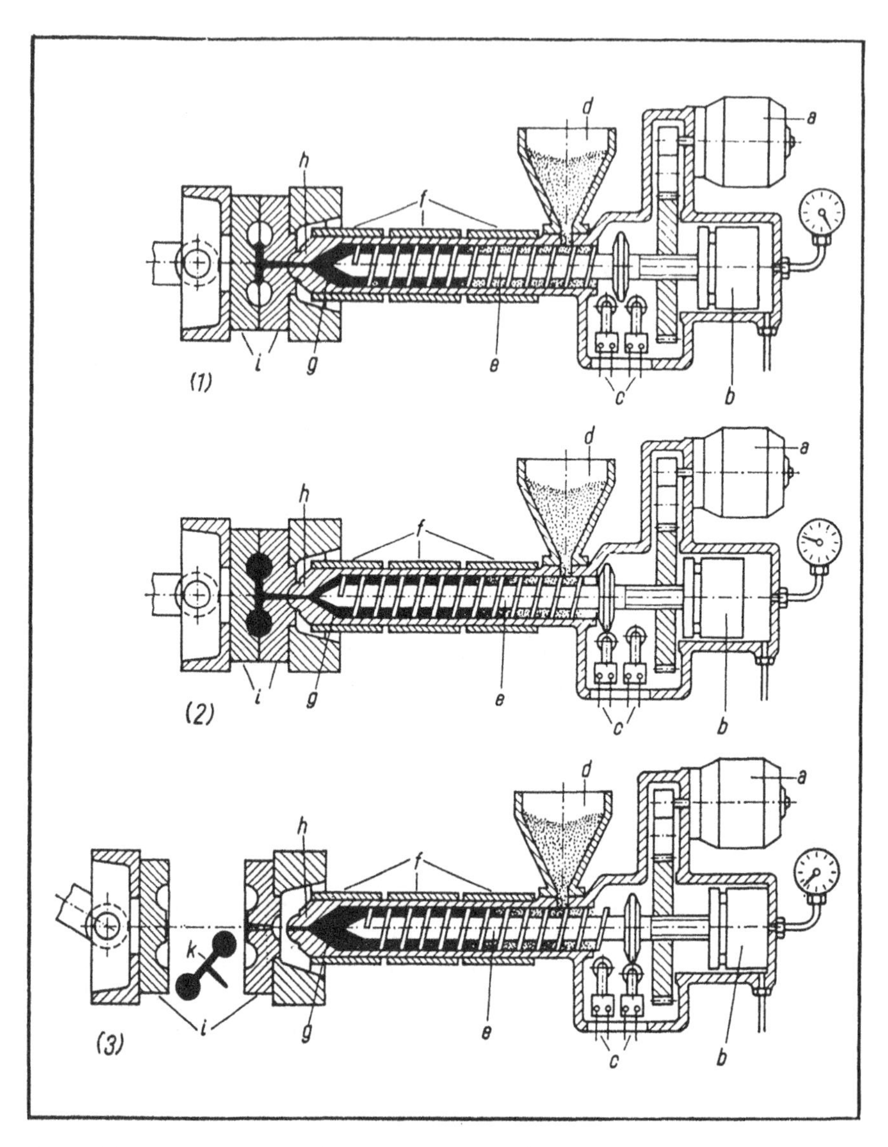

FIGURE 11.9 Injection molding screw machine. (1) Injection: the screw is forced forward, (2) solidification under pressure, (3) plasticizing of next shot terminated: the mold is opened and the component ejected. (a, electric motor; b, hydraulic cylinder; c, stroke adjustment; d, feed hopper; e, screw; f, heating elements for the plasticizing cylinder; g, plasticized material; h, injection nozzle; i, mold; k, component.)

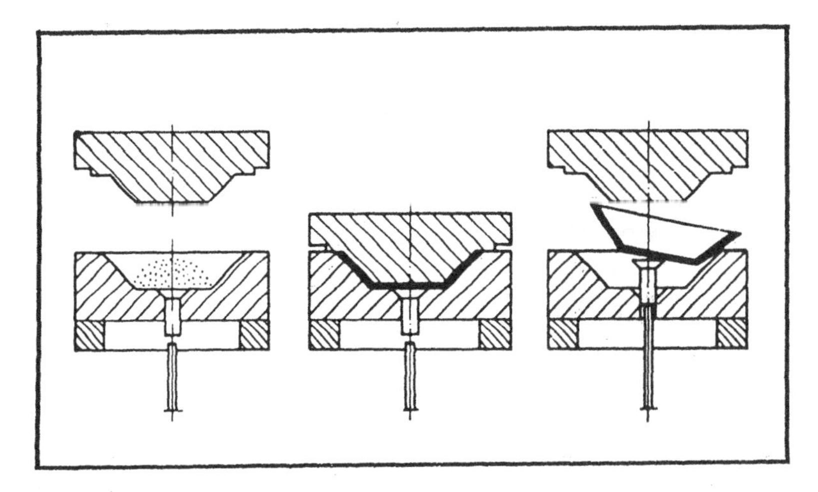

FIGURE 11.10 Compression molding.

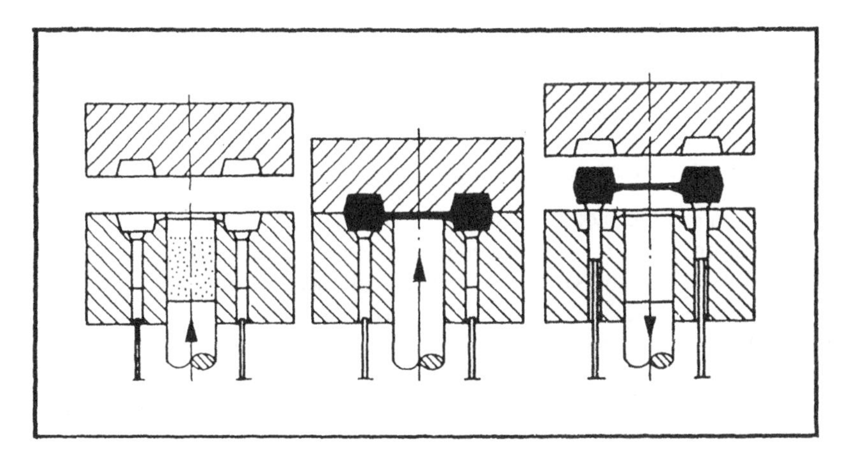

FIGURE 11.11 Transfer molding.

In transfer molding, the plastic material is placed in a separate well or chamber, where it is heated and forced under pressure into the cavity (see Fig. 11.11).

Both processes are used primarily to process thermosetting plastics. The molds are heated to about 160°C. The material can be supplied in granular or tablet form.

Transfer molding provides better geometrical possibilities than compression molding (more complicated shapes can be obtained), but for simple geometries, compression molding is usually preferable.

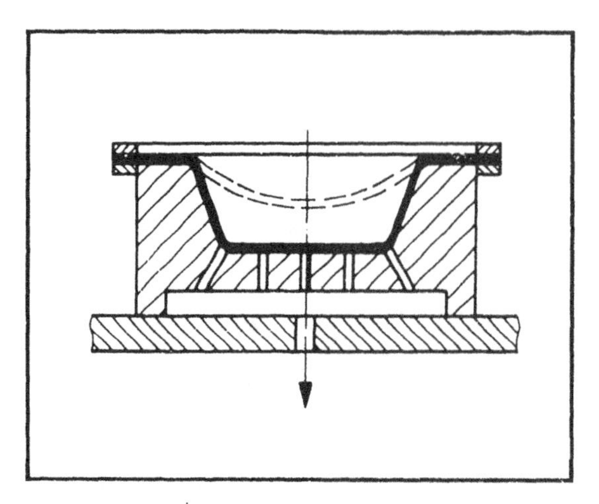

FIGURE 11.12 Thermoforming (by vacuum).

Examples of compression and transfer molded parts include electrical installation materials, insulating parts, appliances, knobs, handles, pulleys, and bearings. Many components are reinforced with glass fibers, for example, and different filler materials may be added depending on the requirements.

11.3.8 Thermoforming

In thermoforming, thermoplastic sheets are heated until they soften (the thermoelastic region, Fig. 11.1) and forced into the mold by vacuum, pressurized air (i.e., differential pressures), or by mechanical means (see Figs. 11.12 and 11.13). The shape obtained is stabilized by cooling.

Figure 11.12 shows the vacuum forming process, where the material, after heating by infrared radiation, is sucked into the mold by evacuating the mold cavity (i.e., the atmospheric pressure forms the sheet). Cooling of the component takes place partly by the sheet contacting the mold and partly by blowing cold air over it.

Figure 11.13 shows a more complicated process. Here, the sheet is heated and drawn onto the mold by pressurized air as shown by the dashed lines. The male mold is then introduced into the formed sheet and the vacuum gradually increased through the mold. These stages can be balanced so that reasonably uniform wall thicknesses can be obtained.

The thermoforming process is used primarily to produce shell-type components of a size up to about 3 m × 7 m. Examples includes boxes, car body components, skylight covers, panels (aircraft, automobile industry), shields, small boats, appliances, housings, and panels. The molds are made of wood, metal, plaster, or plastic.

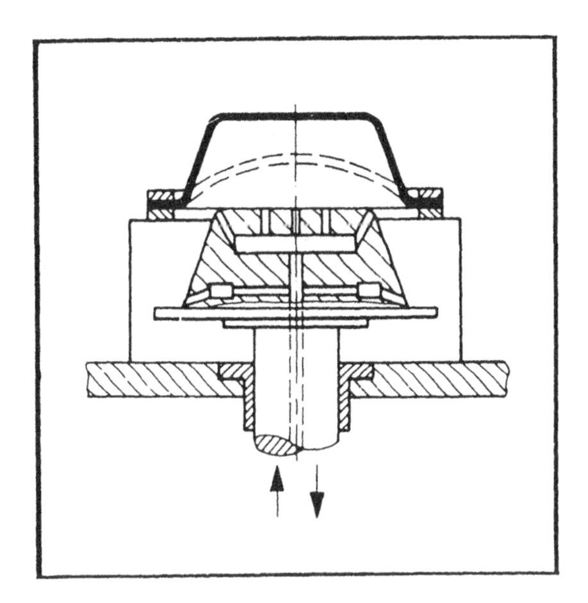

FIGURE 11.13 Thermoforming.

11.4 EXAMPLES OF TYPICAL PLASTIC PROCESSING METHODS

A few typical plastic processing methods are described in the following. The abbreviations used are the same as in chapters 6, 7, 8, and 10.

Figures are provided by courtesy of The Manufacturing Consortium, Brigham Young University, Provo, Utah.

PROCESS 1: Blow Molding (M, Me, Ri, TF, Te)

Description: A forming process in which hollow products are formed by extruding a heated thermoplastic cylinder—a parison—into an open two-part mold where pressurized air expands the parison to match the inner contours of the now closed mold. After cooling, the mold is opened and the component removed. Parts are usually quite uniform in thickness and the cycle time is short. Raw materials for blow molding are generally either pellets or granular compounds.

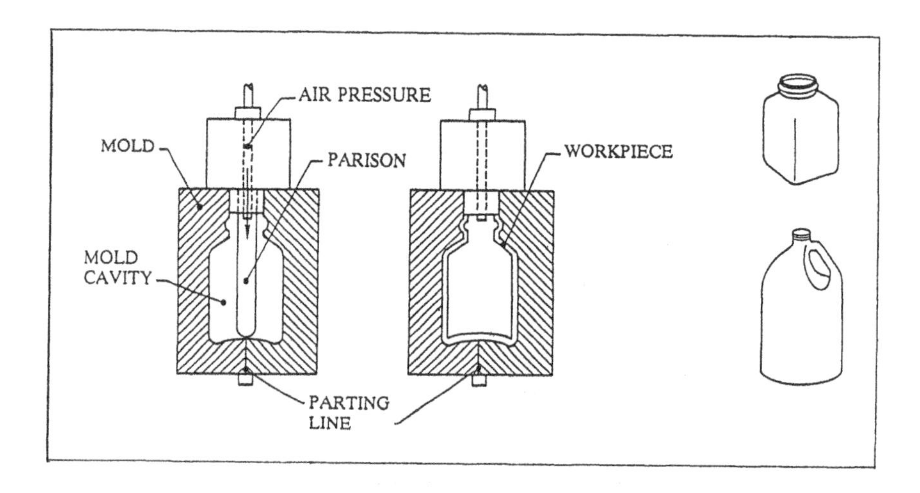

Applications: Blow molding is used to form a wide variety of thin-walled components such as bottles, containers, floats, ducts, etc. Wall thicknesses are typically 0.5–1 mm in bottle-sized components and up to 5 mm in larger containers (cans). Suitable materials are polyethylene, polypropylene, and cellulose acetate.

Tolerances/Surfaces: Surface quality of blow-molded components closely matches the quality of the mold surfaces. Parting lines from the mold parts are present as well as a minimal amount of flash. Tolerances of $\pm 1\%$ are obtainable for external dimensions.

Machinery: The blow-molding machine consists of extrusion press, extrusion die for producing the parison, a two-part mold, and a pressurized air supply.

PROCESS 2: Extrusion Molding (M, Me, Ri, ODF, Te)

Description: In extrusion molding, thermoplastic materials are fed from a hopper into the heated barrel of an extruder. A rotating helical screw inside the barrel pushes the material through the barrel toward a die located at the end of the machine. A heating jacket carefully controls the temperature of the plastic. The molten material is then continuously forced through the die opening and around a mandrel to produce a hollow workpiece. To maintain the desired shape, cooling fixtures are used. When the component has cooled enough to have regained some strength, it is cut to desired length.

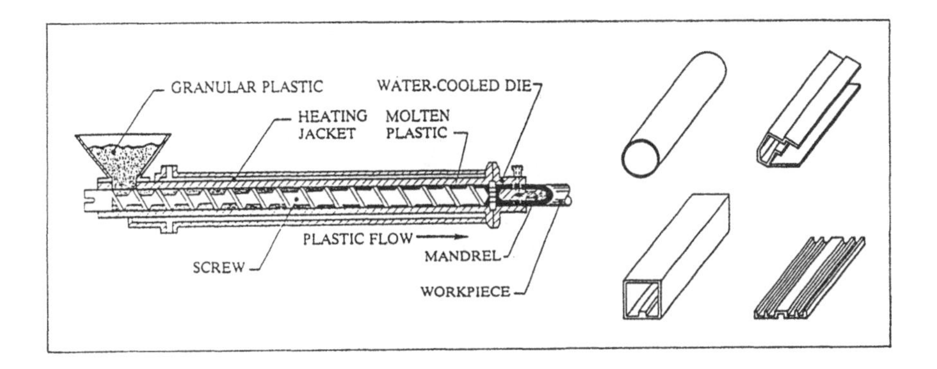

Applications: Granular compounds and pellets are typical material forms used in extrusion molding. Scrap materials may be chopped and mixed with virgin materials. Generally, thermoplastic materials are the only ones extruded. Typical profile extrusions are pipe, film or sheet, rain gutter, and window components. Extrusion molding is a high volume process with accurately controlled material thicknesses, ranging from 0.1 to 10 mm. Intricate profiles can be produced, lengths typically between 3 and 30 m.

Tolerances/Surfaces: Depending on the condition and geometry of the dies, type of workpiece material, working pressure, temperatures of process etc, dimensional tolerances typically range from 8 to 10% on thickness and around 4 to 5 degrees on angles. The surface finish varies between 0.2 and 2 μm (R_a). Extruded parts thus require very little or no finishing work.

Machinery: Specialized extrusion machine with hopper, extrusion screw, heating/cooling systems, die, mandrel, conveyor with cooling fixtures, and a cutting system.

PROCESS 3: Injection Molding (M, Me, Ri, TF, Te)

Description: In injection molding granular thermoplastic material is fed from a hopper into a heated barrel. As the granules slowly move forward by the screw-type plunger, the plastic is forced into a heating chamber where it is molten. The plunger then advances and the melted plastic is forced through a nozzle, allowing it to enter the mold cavity through a gate and runner system. The mold remains cold so the plastic solidifies to its configuration almost as soon as the mold is filled. In some circumstances thermoset plastics can also be used with this process.

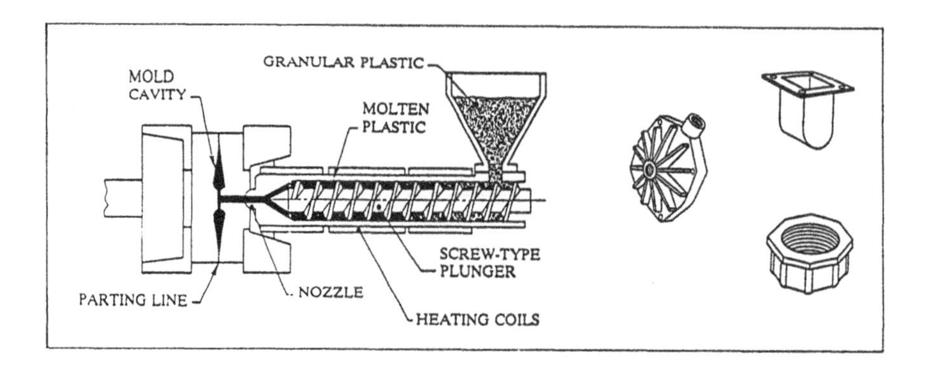

Applications: Injection molding is used to produce more thermoplastic products than any other process. Applications include toys, boxes, fittings, propeller gears, bearings, housing, etc. Molds can be designed to produce only a single part or many parts in one cycle. A mold made for one plastic material usually cannot be used with another material because of different shrinkage for various plastics. One of the major advantages of the process is its ability to produce very complex geometric shapes. The typical size ranges are between 50 and 500 mm in length or width and up to about 400 mm in depth.

Tolerances/Surfaces: To prevent undesired stresses and weak spots, injection molded parts should have a uniform wall thickness and streamline shape. Tolerances of ±0.2 to ±0.05 mm can be obtained. Surface finishes of 1 to 2 μm (R_a) or better are easily obtained as are rough textured or pebbled surfaces.

Machinery: Injection molding machines normally consist of a material hopper, injection ram or screw type plunger, and a heating system. Practically all injection molding machines may be operated on an automatic cycle. They are rated according to the number of ounces of material displaced by one forward stroke of the injection plunger or by the closing force of the dies. CNC-controlled machines are available.

PROCESS 4: Thermoform Molding (M, Me, Ri, TF, Co)

Description: Thermoforming is a forming process in which a sheet of thermoplastic material is heated to its softening point by infrared radiation and then pressed against the contours of a mold by a plug or pressurized air. A vacuum is applied to pull the plastic tightly against the contours of the mold where it is allowed to cool. An excess of material is needed for proper holding, forming, and sealing of the sheet to the mold. This excess is trimmed off in a secondary operation.

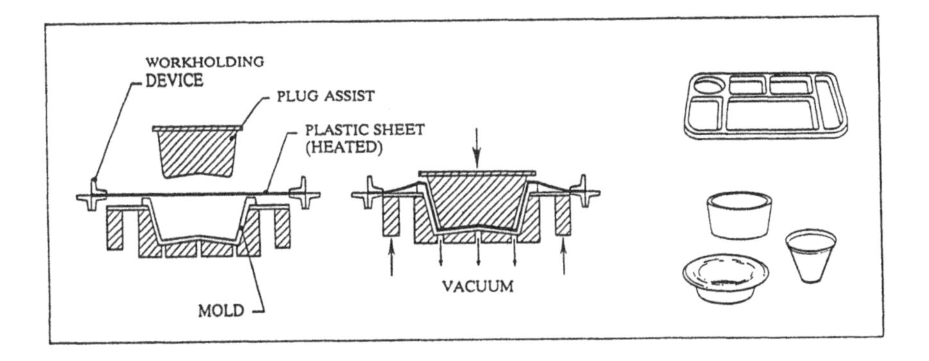

Applications: Production of simple geometries with large radii and no undercuts. Varying thicknesses on the finished product are to be expected. Trays, bowls, cups, skylights, car accessories, dinghies, etc., are typical examples formed from plastic sheet or film. Sizes can be as large as 3 m × 10 m.

Tolerances/Surfaces: Tolerances are often quite rough, and dimensional stability of the finished parts suffer from residual stresses, particularly at elevated working temperatures. Uneven distribution of material is a major problem. High surface quality may be obtained, especially in free forming.

Machinery: Thermoforming machines normally consist of an oven, a mold, a plug assist, and a workpiece holding device. They are made in a wide range of sizes.

12

Nontraditional Manufacturing Processes

12.1 INTRODUCTION

In the previous chapters several of the most commonly used manufacturing processes have been described. However, over the years, increasingly sophisticated designs and new materials have imposed demands on these processes. They have been able to meet these demands only with great difficulty or, sometimes, not at all. Since the 1940s, a revolution in manufacturing has therefore taken place in order to remedy the shortcomings of the traditional processes.

As already described, the conventional manufacturing processes in use today for material removal primarily rely on electric motors and carbide tool materials to perform tasks such as sawing, drilling, milling, turning, broaching, and so on. Conventional forming operations are performed with the energy from electric motors or hydraulics, and material joining is accomplished with thermal energy sources such as burning gases and electric arcs.

In contrast, nontraditional manufacturing processes utilize traditional energy sources in new ways or sources considered unconventional not many years ago. Material removal can now be accomplished with beams of light, sparks, electrochemical reactions, high-temperature plasmas, and high-velocity jets of liquids and abrasives. Materials that in the past have been extremely difficult to form are now formed with magnetic fields, explosives, and shock waves from powerful electric sparks. Material-joining capabilities have been expanded with the use of high-frequency sound waves and beams of electrons and coherent light.

As discussed in the following sections, the nontraditional manufacturing processes may be applied to:

- Increase productivity either by reducing the number of overall manufacturing operations needed to produce a product or by performing operations faster than the previously used method
- Reduce the number of rejects experienced by the old manufacturing method by increasing repeatability, reducing in-process breakage of fragile workpieces, or minimizing detrimental effects on workpiece properties
- Provide a capability that cannot be met with conventional techniques, either because of the hardness of the workpiece or because complicated geometries are required

These attributes have secured a steady growth for the nontraditional processes since their introduction, and many of them today are commonly used alongside their traditional counterparts.

The purpose of this chapter is to give a more detailed description of some of the nontraditional processes, partly to balance the traditional processes described previously, and partly to show that there is still a large potential in developing new principles of materials processing.

12.2 PROCESSES

The nontraditional processes will be grouped into four sections according to the type of energy source they utilize: electrical, mechanical, thermal, and chemical/electrochemical. Only those processes most commonly used in industry are discussed.

12.2.1 Electrical Discharge Machining (EDM)

In electrical discharge machining or ''spark erosion'' the removal of metal from the workpiece is obtained by means of energy released by repetitive spark discharges arranged to take place between two conductors referred to, respectively, as the electrode or tool, and the workpiece in which the eroded cavity is required.

As seen in Fig. 12.1, the electrically conductive workpiece is positioned in the EDM machine and connected to one pole of a pulsed power supply. An electrically conductive electrode, shaped to match the dimensions of the desired cavity or hole, is connected to the other pole of the supply. The two parts are separated by a small gap flooded by an insulating (dielectric) fluid to provide a controlled amount of electrical resistance in the gap. An increasing voltage is

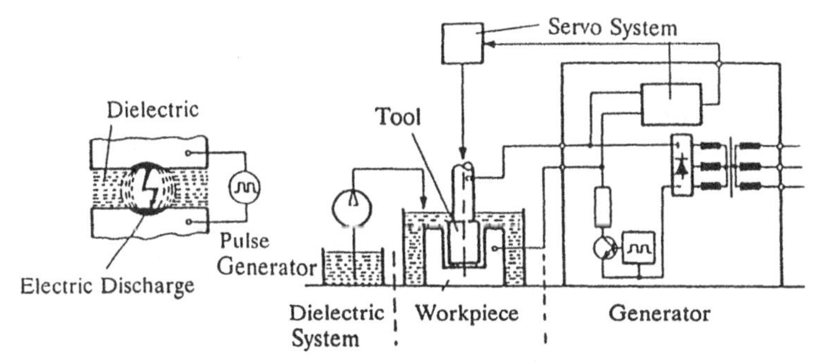

FIGURE 12.1 Schematic of an EDM system.(From Berger, VDI, Düsseldorf, 1987.)

then applied to the electrodes, resulting in an increasing stress being created on the fluid between them until it is eventually ionized, and the gap, normally nonconductive, suddenly becomes conductive, allowing current to flow from one electrode to the other in the form of a spark discharge. The spark channel in the first few microseconds has a very small cross-sectional area resulting in a correspondingly high current density, calculated to be on the order of 10^4–10^6 A/cm^2. Because of these extreme densities, the temperature in the channel is also considerable, between 5,000 and 10,000°C, resulting in the melting and vaporization of a small amount of material from the surfaces of both the electrode and the workpiece at the points of spark contact. Fed by the gaseous byproducts of vaporization, a rapidly expanding bubble is created in the dielectric fluid around the spark channel.

When the electrical pulse is terminated, the spark and heating action are stopped instantly. This causes both the spark channel and the vapor bubble to collapse. The violent inrush of cool dielectric fluid results in an explosive expulsion of molten metal from both the electrode and workpiece surfaces, resulting in the formation of a small crater in the surfaces of the two conductors and small, rapidly solidifying hollow balls of material, which are removed from the gap by the fluid.

The sequence just described is repeated anywhere from a few hundred to several hundred thousands of times per second. Each spark discharge occurs at a point that is the shortest distance between the electrode and the workpiece. An erosion of the workpiece, and to a smaller extent also of the tool, takes place locally, increasing the distance. The repetitive discharges wander across the electrode surface, always seeking the shortest distances, and erode the material from the workpiece in a form that matches the contour of the electrode. As the process progresses and the electrode is advanced automatically toward the workpiece to maintain a constant gap distance, a hole or cavity is generated in a reverse image of the electrode.

On modern machinery the operating parameters may be selected within a wide spectrum of settings. This, combined with a suitable choice of polarity and electrode material, makes it possible to minimize the material removal or wear on the tool electrode to a small fraction of that experienced by the workpiece.

Since the cross section of the resulting workpiece cavity corresponds precisely to the shape of the electrode, it follows that very complex geometries, impossible to machine by conventional means, can be accurately formed by EDM. Another advantage stems from the high operating temperatures, well above the melting point of any known material. Metal removal therefore proceeds independently of the hardness of the workpiece material. Heat treatment for hardening metals can thus be applied at an early stage, eliminating the risk of distortion of the work after machining. Also, working with difficult-to-machine materials such as tungsten and cobalt constitutes no problem.

During the cutting operation, the spark alone is in contact with the workpiece, which is therefore not subjected to the mechanical stresses of a cutting tool. This results in the ability of the EDM process to machine extremely fragile workpieces without damage. It is also worth mentioning that the surface finish produced by the process is nondirectional, with none of the lifts and scores that are characteristic of chip-forming machines. This is an asset where polishing is required as a final finish since the surface uniformity reduces the time necessary for this operation.

Equipment

Although there are several types of spark erosion machines, each with its own refinements and equipped to meet a user's particular specification, the typical EDM system is comprised of four major subassemblies that include a power supply, dielectric system, electrode, and servosystem. Figure 12.1 shows a schematic EDM system.

Power Source The power source converts the input ac power to dc power by conventional solid-state rectification. A crystal-controlled oscillator governed by a part of the dc power generates a square-wave signal that triggers a bank of power transistors, which act as high-speed switches to control the flow of the remaining dc power. In addition, the circuit makes it possible to obtain desired variations in power and pulse repetition frequency since the technique in spark erosion machining is to remove the bulk of the workpiece material at high power and low frequency (the roughing operation) and then to remove remaining small amounts of material at low power/high frequency (the finishing operation). Figure 12.2 is a schematic arrangement of a transistor circuit providing square voltage pulses. This type of circuit can be modified to control heavier currents by introducing transistors arranged in parallel, or to vary the on/off duty ratio of the transistor by introducing different values of circuit components and thus varying the pulse width and repetition frequency.

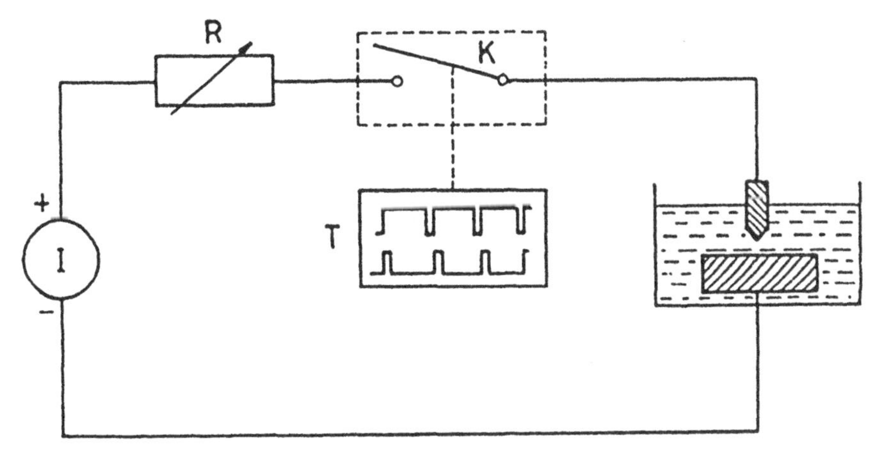

FIGURE 12.2 Transistor circuit providing square voltage pulses.

Sensing the voltage between the electrode and the workpiece is an additional function of the power supply. Because a direct relationship exists between this voltage and the electrode gap, the voltage is used to control the servosystem, enabling it to maintain a constant gap distance throughout the process. This is achieved by means of a velocity servo which "senses" the error between the required feed rate and the actual cutting rate and makes the appropriate correction (see below, under Servosystem).

Dielectric System The standing requirements of a dielectric fluid are:

- That it should serve as an insulation in the spark gap until the breakdown voltage is attained.
- That it should rapidly quench the spark once a discharge has occurred. This deionizes the dielectric in the spark, which is essential if a continuous arc is to be prevented and the cycle leading up to another spark discharge is to be established.
- That ionization should occur as quickly as possible at a consistent value of breakdown voltage.
- That it should serve as a flushing agent.

Of the four functions, flushing is by far the most critical for optimum process efficiency. Poor flushing results in stagnation of the fluid and a buildup of tiny machining residue particles in the gap, resulting in low removal rates or short circuits. As shown in Fig. 12.3, several methods are available for flushing the dielectric fluid through the cutting zone.

The dielectrics used for spark machining are almost invariably commercially available paraffins and light oils, which are reasonably cheap, have a suitably low viscosity, and have a flash point high enough to make them safe to work

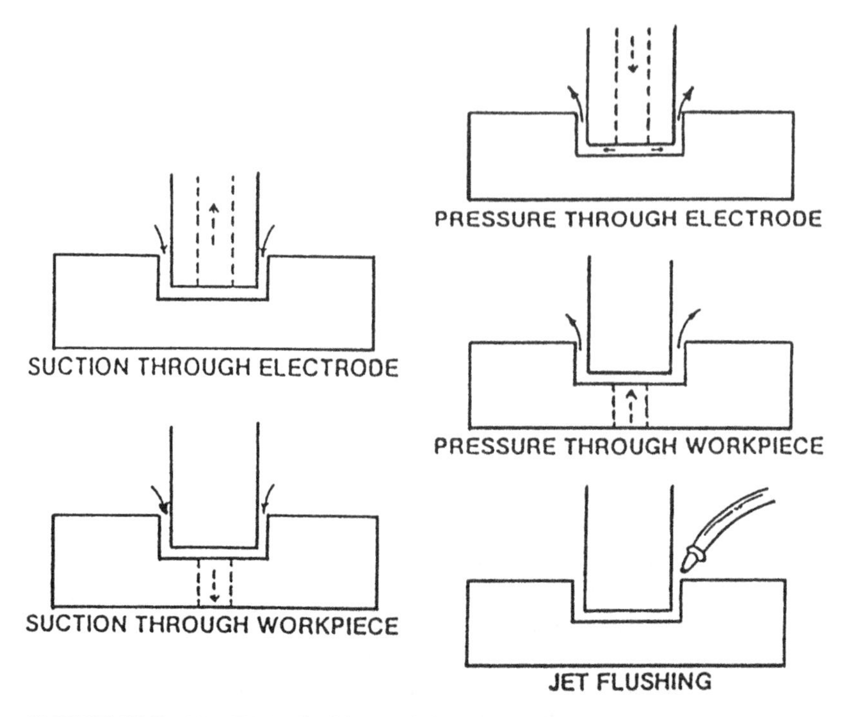

FIGURE 12.3 The EDM flushing techniques. (Courtesy, Hanstvedt EDM Division, Urbana, Illinois).

with. The fluid is continually cleaned, recycled, and returned to the cutting gap by means of pumps and filters.

Electrodes The main limitation to the accuracy of the shape produced by EDM is the unavoidable electrode wear. This wear depends on the material of the workpiece and of the electrode itself, the type of dielectric, flushing conditions, rate of cutting, and type of power source. Many materials have been successfully used for EDM applications. Common requirements are that electrode materials must be easily machinable, exhibit low wear, be electrically conductive, provide good surface finishes on the workpiece, and be readily available. It is important to bear in mind that the performance of any one type of electrode may vary considerably depending on the type of machining operation. For example, the wear of a given electrode might be low during a roughing operation but high during a finishing operation.

Copper and brass have high thermal conductivities and give good cutting rates; the former has a low wear factor. Both materials are reasonably cheap and are easily formed. Copper tungsten provides much lower wear rates than copper

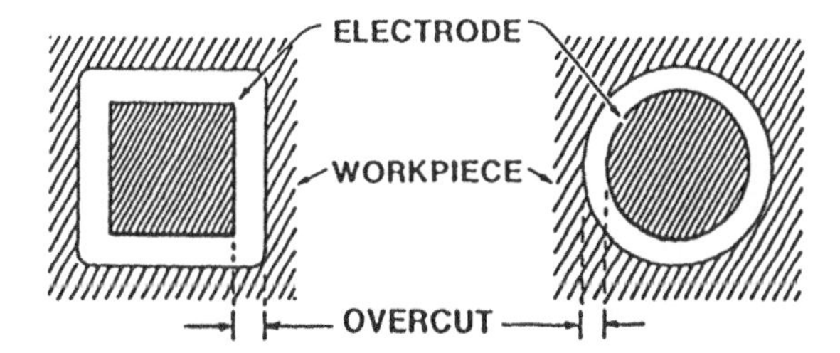

FIGURE 12.4 EDM electrodes and resulting overcut in the workpiece.

but is a difficult material to machine. Techniques (expensive) are available for molding electrodes from this material. Besides machining of copper electrodes, electroforming is used to produce a copper shell corresponding to the component; the shell is then filled with a suitable material to allow fastening.

Graphite and copper graphite are very popular electrode materials. Both are easy to machine and are available in a number of grades for application to all workpiece materials. The wear rate of graphite is low because of the very high temperature at which it vaporizes.

All electrode materials and configurations produce an overcut in the workpiece (see Fig. 12.4). The amount of overcut and hence the necessary compensation are predictable once the workpiece and electrode materials as well as the EDM operating parameters are known.

Allowances must also be made for electrode wear.

Servosystems The efficiency of the EDM process is closely related to the gap distance between electrode and workpiece, and the servosystem which continuously feeds the electrode to the workpiece during the cutting operation must maintain the optimum gap distance by controlling the infeed of the electrode to precisely match the rate of material removal.

Figure 12.5 shows the principle for controlling the electrode feed: the required gap distance is set corresponding to a certain gap voltage V_g. During the process the actual gap voltage V_a is constantly monitored. The difference between V_g and V_a, known as the *error voltage*, is fed to a servo amplifier, which in turn controls a servo valve and therefore the flow of hydraulic fluid to a cylinder with a piston rod connected to the work head. If the gap is as set, there is no error voltage, and the work head remains stationary. Subsequent increase of the gap distance causes a corresponding increase in V_a and the resultant error voltage causes the valve to lower the work head until V_a and V_g are equal again. Movement in the opposite direction is, of course, accomplished if a piece of electrically conductive material has bridged the gap between the electrode and workpiece. In this case V_a is less than V_g and the valve will react by increasing the gap distance until the dielectric fluid flushes the gap clear.

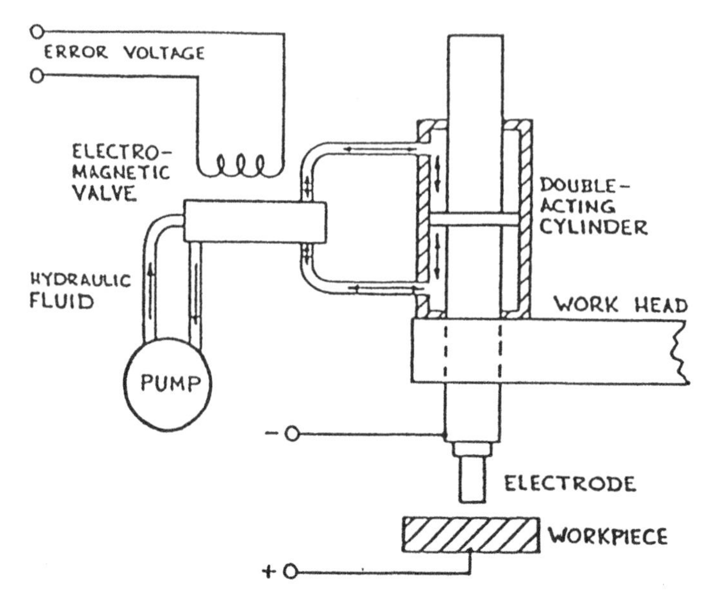

FIGURE 12.5 Schematic of EDM velocity servosystem.

Process Capabilities and Applications

EDM is capable of machining all electrically conductive materials regardless of hardness. The process is particularly well suited for producing irregularly shaped holes, slots, and cavities with accuracies of ±0.025 to ±0.1 mm. With special care, accuracies as fine as ±0.005 mm are obtainable.

Taper is affected by electrode wear and by the method of flushing chosen. When taper is not acceptable, it can be eliminated by using separate electrodes for roughing and finishing passes.

Volumetric metal removal rates are strongly dependent on the parameters chosen for current, voltage, and frequency, (i.e., the power of each discharge). Further, electrode material, dielectric, and method of flushing influence the working speed. High power settings are commonly used for roughing passes with a removal rate of 25–50 cm^3/h as typical values, whereas the rate for finishing may be as low as 0.01 cm^3/h.

The same parameters also strongly influence the surface finish of the workpiece. Values commonly range from 1 μm RMS for low power, high-frequency work to 10 μm RMS for high-power, low-frequency machining. Newer machines, combined with special care in operation and parameter selection, are now able to maintain surface finishes as fine as 0.2–0.3 μm RMS.

A limitation of the process is that because material is removed from the workpiece by thermal action, a layer of melted and solidified material, known as

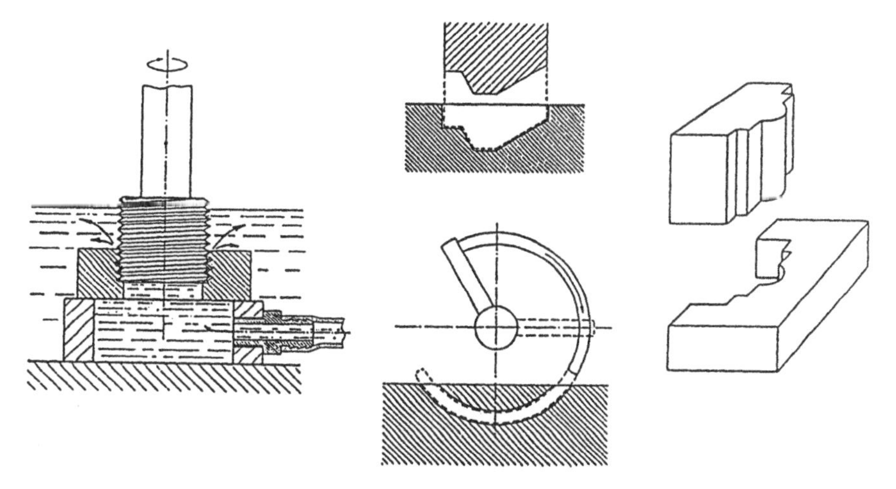

FIGURE 12.6 Some typical spark erosion machining techniques [48].

recast, remains on the machined surface. The recast is typically between 0.002 and 0.05 mm thick and is very hard, in excess of HRC 65, and brittle. Because of the poor physical properties of such a surface, recast must be removed from surfaces of products that require high levels of fatigue resistance. For many purposes, however, subsequent finishing is unnecessary and the matte finish from the overlapping minute, saucer-shaped craters can be an asset for the retention of lubricants.

The EDM process is used extensively in industry, from the production or modification of tools for the automobile industry—for instance, making of stamping or forging punches and dies weighing from a few kilograms to several tons—up to the manufacture of tools for the watchmaking, electronics, and high-precision industries, or even series production of parts such as injector nozzle holes, where its ability to produce highly complex geometries in otherwise difficult-to-machine materials is turned to good account. Figure 12.6 shows some typical operations.

Processes Derived from the EDM Process

From the conventional spark erosion process just described, a number of subprocesses have been derived, among which electrical discharge wire cutting and electrical discharge grinding have found widespread industrial use. These processes are dealt with briefly in the following:

Electrical Discharge Wire Cutting (EDWM) EDWM differs from conventional EDM in that a thin wire is used as an electrode instead of a formed electrode. The wire unwinds from a spool, feeds through the workpiece, and is taken

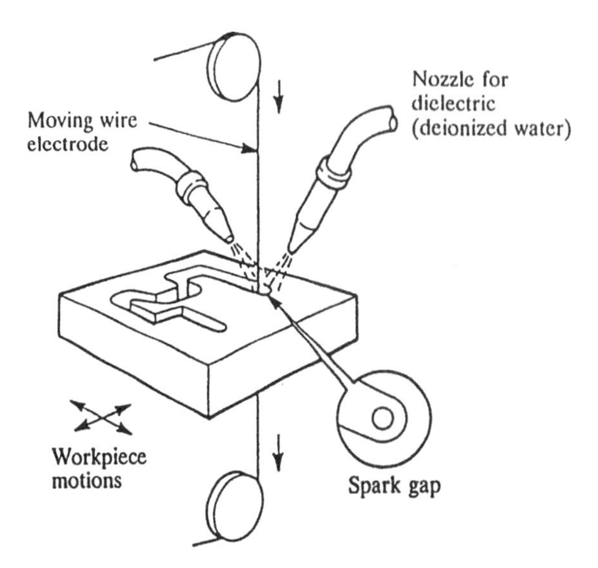

FIGURE 12.7 A schematic of an EDWC machine. (From *SME Tool & Manufacturing Engineers Handbook*, 4th ed.)

up on a second spool. A dc power supply delivers high-frequency pulses of electricity to the wire and to the workpiece, and material is eroded ahead of the wire by spark discharges. Either the workpiece or the wire is moved, causing the wire to cut like a bandsaw. The spark gap is flooded with a dielectric, usually deionized water. Figure 12.7 shows the fundamentals of a wire-EDM machine.

In this way, complicated cutouts can be made without the need to use high-cost conventional cutting processes or expensive formed EDM electrodes. High accuracy and fine surface finishes are obtainable and make wire-EDM particularly valuable in the manufacture of stamping dies, extrusion dies, dies for powder metal compaction, prototype parts (if the final stamping die configuration has not been fully determined), and even for the fabrication of conventional EDM electrodes.

Electrical Discharge Grinding (EDG) As shown in Fig. 12.8, in an EDG operation a rotating, electrically conductive wheel is used as the electrode. Both the wheel and the workpiece are submerged in a tank containing a dielectric. Pulsed electrical energy is delivered to the wheel and workpiece at rates of up to 250,000 pulses/s, resulting in a flow of sparks, each removing a small quantity of material from the workpiece. The rotating motion of the wheel, often made from graphite, ensures a constant flow of dielectric to the gap, thus eliminating any flushing problems.

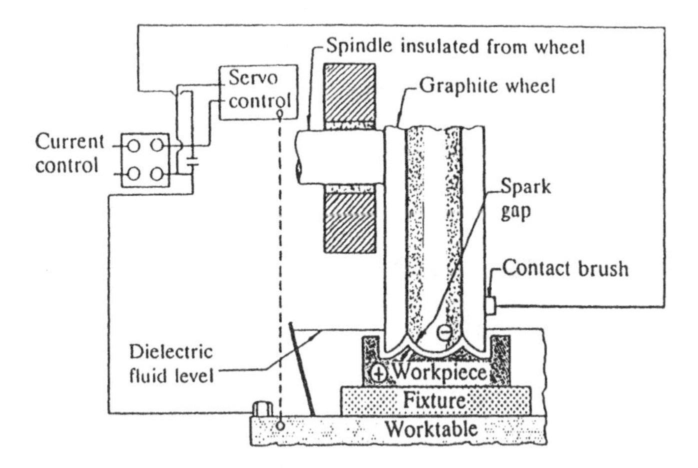

FIGURE 12.8 An EDG system, depicting wheel and interaction zone. (Courtesy, American Society for Metals, Ohio.)

Wheels can be dressed to produce complex shapes in a single pass, and since there is no mechanical contact between the wheel and the workpiece, EDG is often used to perform operations on very fragile parts or to produce thin sections without damage or distortions.

Electrical discharge grinding is popular for form-grinding carbide thread chasers, turbine blade fir-tree root forms, and similar geometries with thin cross sections. Conventional grinding would tend to distort the thin sections thermally, rendering the workpiece unacceptable.

12.2.2 Electron Beam Processes

Electron beams are used in many types of industrial equipment today because the electrons can be accelerated and formed into a narrow beam by an electric field. The beam thus formed can be focused and bent by electrostatic and electromagnetic fields, much as light rays can be focused by glass lenses.

In electron beam machining and welding equipment, relatively high-power beams are used with electron velocities from 30 to 75% of the speed of light. This high-speed stream of electrons is focused on a very small spot where it impinges upon the material to be treated, converting the kinetic energy of the electrons into thermal energy with an efficiency of nearly 100%, vaporizing or melting the material locally, depending on whether cutting or welding is desired. The process is usually carried out in a vacuum to prevent collisions between electrons and gas molecules, which would scatter the electron beam.

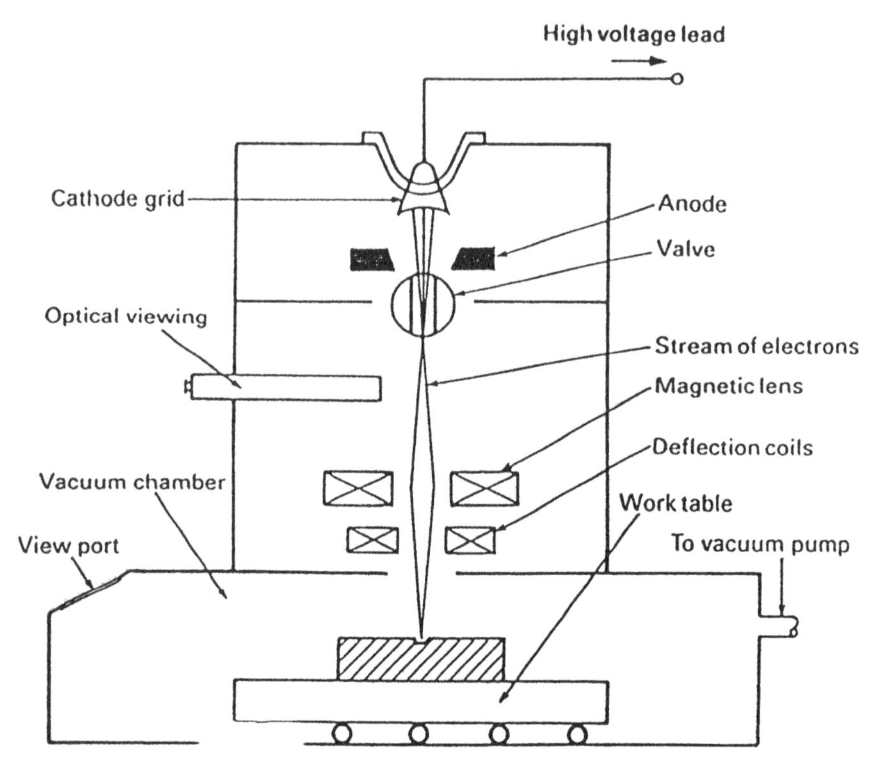

FIGURE 12.9 Components of an electron beam machine [45].

Electron beams are used for a wide variety of applications including welding, cutting, heat-treating, drilling, and glazing. Different applications require different types of electron beams since specific power density or the amount of power focused on a particular area varies with the application. However, the principles for creating, focusing, guiding, and so on, the beam to its desired point of attack are the same, regardless of the final use of it, which means that all the mentioned applications may be achieved by the same equipment simply by varying different parameters.

Equipment

The three major subsystems that make up an electronic beam system are the electron beam gun, power supply, and vacuum system. These systems work together to generate the electron beam and to provide the optimal environment for the process. A typical electron beam machine is shown schematically in Fig. 12.9.

Electron Beam Gun The electron beam is formed inside an electron gun, which is basically a triode consisting of:

1. A cathode, which is a tungsten filament heated inductively to approximately 2500°C, thus emitting high-negative-potential electrons
2. A grid cup, negatively biased with respect to the filament
3. An anode at ground potential through which the accelerated ions pass

A stream of electrons is emitted from the surface of the hot cathode and accelerated toward the anode by a high potential between the anode and the cathode. The degree of negative bias applied to the grid cup controls the flow of electrons and is also used to turn the beam on and off. Because of the shape of the electrostatic field formed by the grid cup, the electrons are electrostatically focused and pass as a converging beam through the hole in the anode without colliding with the anode itself. Final focusing is provided by an electromagnetic field produced by a focusing coil. As soon as the electrons have passed through the anode, they have reached their maximum velocity for a given accelerating voltage and will maintain this velocity since the process takes place in a collision-free environment until they impinge on the workpiece.

Most EBW guns incorporate a final set of electromagnetic coils after the focusing system. This set of coils is known as the beam deflection system and is used to provide a small amount of programmable beam motion.

The electron beam is a heat source which, with its power density, precision, and mobility, exceeds any known commercial heat source. Light rays such as those emitted by a laser produce electromagnetic wave radiation whose energy content depends on the temperature of the light source. Light rays cannot be accelerated to increase the energy content.

Electron emission is different. The beam consists of negatively charged particles whose energy content is determined by the mass and velocity of the individual particles. During the acceleration process, due to the potential between anode and cathode, the energy content in the beam can reach intensities far in excess of those obtainable from light. By refocusing the beam through a variable strength electromagnetic lens before the electrons colide with the workpiece over a well defined area (typically 0.01–0.02 mm in diameter), power densities of 10^7–10^8 W/cm^2 are reached at the point of impact, immediately vaporizing any type of workpiece material. Table 12.1 provides a comparison of power density of some thermal energy sources used for production-welding applications.

Power Supply The power supply provides energy for heating the filament, accelerating the electrons, controlling the bias grid, focusing the beam, and operating the deflection coil. The most important of these are the accelerating voltage and current, since these determine what kind of applications the particular system will be capable of performing.

TABLE 12.1 Power Densities of Various Heat Sources

Gas flame	$O_2 + C_2H_2$	10 W/cm^2
Hydrogen flame	$O_2 + H_2$	30 W/cm^2
TIG torch	200 A	150 W/cm^2
Electron beam	continuous	10^7 W/cm^2
Laser beam	continuous	10^7 W/cm^2

Source: Adapted from Terai, 1978.

Electron beam systems are available that generate from 30 to 300 kV of accelerating voltage and between 0.0005 and 1.5 A of beam current, resulting in a total beam power generally ranging from 15 W to 200 kW. The electrical conversion efficiency is high, approximately 60–70% of the power supplied to the machine being delivered to the workpiece. For comparison, laser systems utilizing high-power carbon dioxide lasers exhibit an efficiency of 10–20%.

Vacuum System The vacuum system comprises two subsystems: the pumping system and the vacuum chamber. All electron beam systems require some sort of pumping system to generate a high vacuum in the electron gun. The first systems available were of the high-vacuum type and required the workpiece as well as the electron gun to be enclosed in the vacuum chamber, which of course limited the productivity because of the time necessary to evacuate the chamber after placing a new workpiece in it. In order to increase productivity, medium-vacuum and even nonvacuum systems have been developed. These systems still require a high vacuum in the electron gun but deliver the beam to a workpiece at low vacuum/atmospheric pressure, thus reducing or avoiding nonproductive pumpdown cycles. Current penetration capabilities for electron beam welding using a high-vacuum system are about 300 mm, while welding with a medium vacuum is limited to approximately 50 mm and nonvacuum welding to about 10 mm penetration.

A vacuum chamber is required since the generation and transmission of the electron beam takes place in a vacuum of 10^{-4}–10^{-6} torr. The impingement of high-velocity electrons results in X-ray emission, making it necessary to shield the vacuum chamber with suitable materials—thick stainless steel or lead—to absorb this radiation. Appropriate workholding and positioning mechanisms are installed in the vacuum chamber.

Process Capabilities and Applications

As mentioned previously, electron beams are used for a number of different applications, usually subdivided into two categories:

Electron beam machining (EBM)
Electron beam welding (EBW) and heat treatment

Electron Beam Machining The EBM process is used to cut or drill a wide range of materials, metallic as well as nonmetallic, such as ceramics, leather, and plastics. Since it is a thermal process, some thermal effects remain on the machined edge after processing, but because of the extremely high beam power and the short duration of the beam/workpiece interaction time, thermal effects are usually limited to a recast layer and a heat-affected zone, which seldom exceeds 0.025 mm. Typically, no burr is generated on the exit side of the hole. Table 12.2 gives data on drilling holes in various materials.

Another capability of the EBM process due to the high-power density is its ability to drill deep, high-aspect ratio holes. In most materials, ratios as high as 15:1 can be achieved with hole diameters from 0.1 to 1.5 mm in thicknesses up to 10 to 12 mm. The tolerance on the hole diameter is typically ±5% of the diameter. Since the beam does not apply any force to the workpiece, brittle or fragile materials can be processed without danger of fracturing. Minimum permissible distance between holes is generally twice the hole diameter. In practice, this limitation poses no difficulty, and workpieces can be perforated with small holes at up to 1000 holes/cm^2, which is one of the really strong features of the process.

EBM is also adaptable to cutting narrow slots in thin-gage materials. To minimize heating and melting adjacent to the cut, extremely short beam pulses are used with considerably longer periods between pulses to permit dissipation by thermal conductivity of any incidental heating adjacent to the cut. The cutting efficiency is therefore much lower than the actual efficiency of the equipment since the power is off a large percentage of time. Figure 12.10 shows metal removal rates versus power.

Only relatively small cuts are economically feasible with EBM techniques since the material removal rate is approximately 0.2–0.5 mg/s. However, EBM makes possible the production of very precise and fine cuts of any desired contour in any material.

Most current applications of EBM are for aerospace, insulation, food and chemical, and clothing industries. The drilling of a turbine engine combustor dome made of a CrNiCoMoW steel is a good example. The part has a wall thickness of 1.1 mm and is perforated with 3748 holes that are 0.9 ± 0.05 mm in diameter. Each part is drilled in 60 min, that is, a drilling rate of approximately one hole per second [47]. Filters and screens used in the food processing industry also require thousands of holes to be drilled through relatively thin, formed sheet metal.

A rather new use of electron beam perforation involves the shoe manufacturing industry. A percentage of shoes made today are fabricated from artificial leather consisting of a plastic-coated textile substrate. This artificial leather is not permeable to moisture and air, which makes its level of comfort poor. Partial perforating the material by EDM makes it acceptable for use in shoes. A similar

TABLE 12.2 Holes Drilled by EBM in Various Materials

Work Material	Workpiece Thickness		Hole Diameter		Drilling Speed	Accelerating Voltage	Average Beam Current	Pulse Width	Pulse Frequency
	in	mm	in	mm	s	kV	μA	μs	Hz
400 Series stainless steel	0.010	0.25	0.0005	0.013	<1	130	60	4	3,000
Alumina Al_2O_3	0.030	0.76	0.012	0.30	30	125	60	80	50
Tungsten	0.010	0.25	0.001	0.025	<1	140	50	20	50
90-10 Tantalum-tungsten	0.040	1.0	0.005	0.13	<1	140	100	80	50
90-10 Tantalum-tungsten	0.080	2.0	0.005	0.13	10	140	100	80	50
90-10 Tantalum-tungsten	0.100	2.5	0.005	0.13	10	140	100	80	50
Stainless steel	0.040	1.0	0.005	0.13	<1	140	100	80	50
Stainless steel	0.080	0.2	0.005	0.13	10	140	100	80	50
Stainless steel	0.100	2.5	0.005	1.3	10	140	100	80	50
Aluminum	0.100	2.5	0.005	0.13	10	140	100	80	50
Tungsten	0.016	0.41	0.003	0.076	<1	130	100	80	50
Quartz	0.125	3.18	0.001	0.025	<1	140	10	12	50

NOTE: The main control parameters for shaping the hole are the pulse width for the depth of the hole, the beam current for the diameter of the hole and the power distribution within the beam as well as the position of the focus with respect to the workpiece.
(Source: E. J. Weller et al, Society of Manufacturing Engineers, Dearborn Mich. 1984).

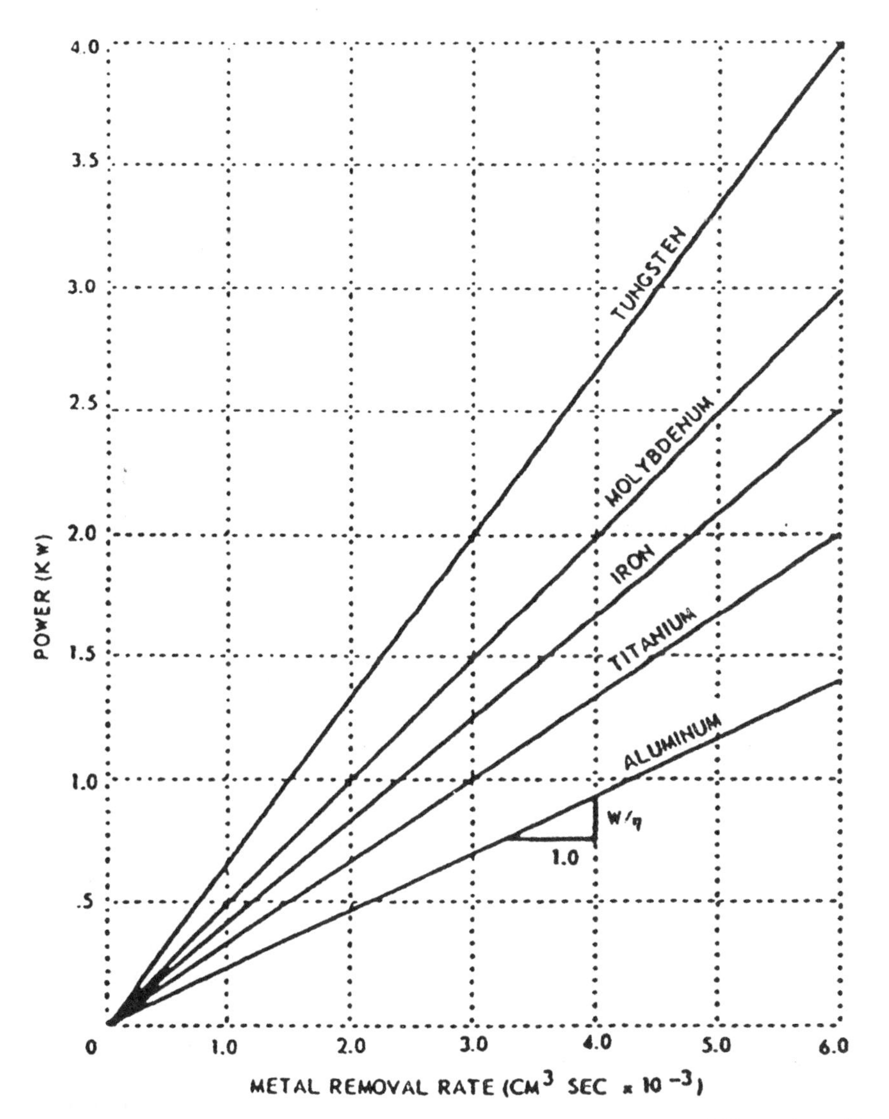

FIGURE 12.10 Metal removal rates versus power, assuming a 15% cutting efficiency [46].

method is used for material for rainwear, drilling these materials with 0.05-mm-diameter holes at a rate of 5000 holes/s.

Electron Beam Welding The high intensity of the electron beam as a heat source results not only in deep, narrow penetration but also in the associated

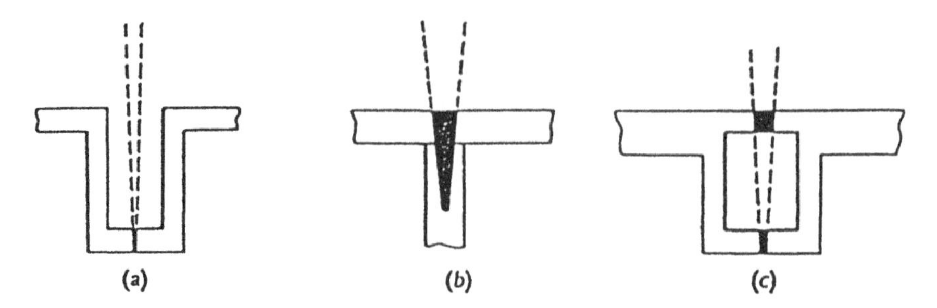

FIGURE 12.11 Types of joint specifically applicable to electron beam welding. (Houldcroft, Cambridge, England 1973.)

effect of narrow heat-affected zones. Both effects together are of considerable importance, as they result in reduced distortion and improved mechanical properties in joints when compared with other welding processes. The necessity to use a vacuum is at once both an advantage and a disadvantage. It enables fusion welds of unequaled quality to be made, but the size and shape of the workpiece is limited by the vacuum chamber.

A variety of joint types can be welded by the EBW process. Because of the high speeds attainable, however, and the difficulty of feeding filler wire through the vacuum chamber, it is customary to use close butt edge preparations which do not require filler. Joints must be machined so that edge faces are square; surface finish must be equal to or better than 3 μm RMS, and a gap between parts no larger that 0.15 mm must be provided. Cleaning is also a very important step for successfully joining parts with EBW in order to avoid porosity in the weld and possible contamination of the pumps and vacuum chamber. As shown in Fig. 12.11, three types of joints are unique to electron beam welding: (a) with beams of narrow convergence it is possible to weld in inaccessible positions (e.g., in the bottom of a deep narrow cavity); (b) the deep penetration technique permits the making of a fillet joint from the outside using what is called the spike technique, and (c) deep penetration with high-density fine beams also permits two or more welds to be made at once, one above the other.

A wide range of metals and alloys can be welded by EBW. Materials exhibiting very good weldability are stainless steels, low-carbon steels, copper, nickel, aluminum and their alloys, and refractories such as zirconium, tantalum, titanium, and niobium. Tungsten, tungsten carbide, and cast iron are more difficult to manage; and brass, cadmium, and zinc are examples of materials that are impossible to weld.

Electron beam welding is applied by a wide spectrum of industries, ranging from aeronautics and space over energy and heavy industry to mass production in the automobile industry. Because of the restrictions imposed by the vacuum chamber, most EBW parts are of moderate size, although giant chambers with a volume of more than 260 m^3 exist for welding components in the nuclear industry. Components of jet engines—compressor guide vanes, rotors, nozzles, etc.—and welding of automotive catalytic converters are practical examples.

Heat Treatment Use of an electron beam for heat treatment is finding increasing popularity and is performed on systems operating at high to medium levels of vacuum. The method is used to selectively surface-harden carbon-bearing steels capable of being transformation-hardened.

A moderately powerful beam, with power density of about 1.5×10^4 W/cm^2, is scanned over the surface to be hardened. The beam density causes a thin portion of the part surface to rise almost immediately to the materials austenitizing temperature. As the beam is either turned off or moved to a new location, the cool core of the material rapidly draws the heat from the surface to provide a self-quenching action, thus transforming a thin layer of the surface material to hard martensite.

Because the process is so rapid, and because so little energy is used, distortion is negligible, allowing the part to be fabricated to final dimensions and then electron-beam-hardened without requiring any further finishing operations.

12.2.3 Laser Processing (LP)

Laser processing is based on principles discovered only recently. The word *laser* is an acronym of *l*ight *a*mplification by *s*timulated *e*mission of *r*adiation. The process depends on the interaction of an intense, highly directional, coherent, and monochromatic beam of light with a workpiece to remove or melt material from it or even thermally modify the material. Lasers are able to perform a wide variety of tasks, ranging from cutting, drilling, and welding to marking, heat-treatment, and selective cladding.

The light energy emitted by the laser has several characteristics that distinguish it from other light sources:

Spectral purity: The light emitted by lasers is monocromatic. The beam can be focused using simple optics.

Directivity: The beam is highly collimated with typical divergence angles of 10^{-2} to 10^{-4} radians.

High power density: Because of its small beam divergence, all of the laser beam energy can be collected and focused onto a very small area, producing power densities as high as 10^7 W/mm^2.

Before discussing types of laser systems and their applications to machining, a brief explanation of the fundamental principles involved will aid understanding. These principles must be explained at the atomic level.

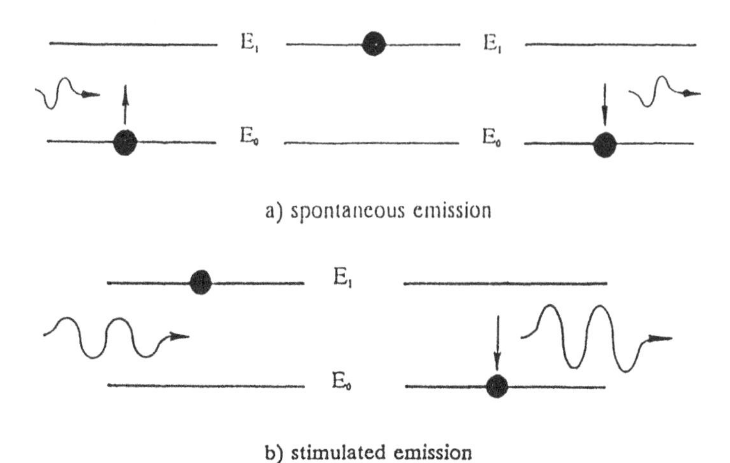

FIGURE 12.12 Interaction between light and matter [52].

An atom's orbital electrons can jump to higher energy levels (orbits further away from the nucleus) by absorbing quanta of stimulating energy, e.g., heat, light, chemical reaction, When this occurs, the atom is said to be in the "excited" state and may then spontaneously emit or radiate the absorbed energy. Simultaneously, the electron drops back to its original orbit or to an intermediate level. If another quantum of energy is absorbed by the electron while the atom is in the excited state, two quanta of energy or photons are radiated, and the electron drops to its original level. The radiated energy has precisely the same wavelength as that of the stimulating energy. As a result, the stimulating energy is amplified as shown in Fig. 12.12. This principle is the basis of laser operation.

Many materials can be made to undergo stimulated emission. However, to build a working laser, additional conditions must be met. First, the energy source that provides the initial stimulation must be powerful enough to ensure that there are more electrons in the upper energy level than in the lower. This condition is known as *population inversion* and is a condition necessary for laser operation.

The second condition required to produce a laser is to provide a feedback mechanism. This captures and redirects a portion of the coherent photons back into the active medium to stimulate the emission of still more photons of the same frequency and phase. The feedback mechanism is designed to allow a small percentage of the photons to escape the system in the form of laser light, but the majority of them will still be available to maintain the amplification process.

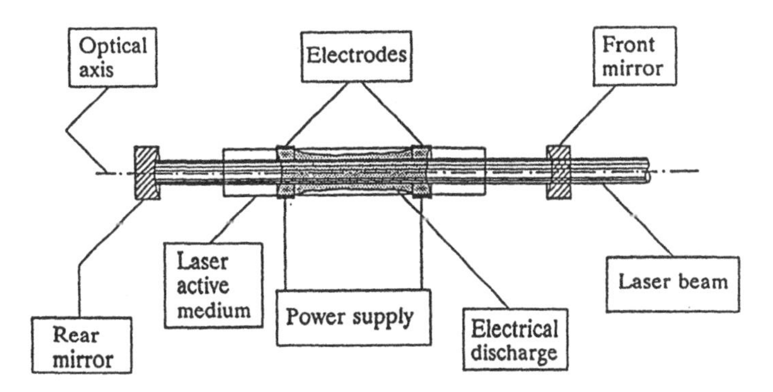

FIGURE 12.13 Schematic of a gas laser [52].

Figure 12.13 shows the working principle of a gas laser. It consists of a laser cavity, which is a glass tube containing the laser active medium, in this case a gas at low pressure. Energy to stimulate the atoms or molecules of the gas is delivered by electric discharges supplied from a high voltage power supply. By placing parallel mirrors at both ends of the tube, the necessary feedback mechanism is established. One mirror is made to be as close to 100% reflective as possible, while the other is partially transparent to provide the laser output.

When electricity is applied to the gas, it forms a plasma, and photons are emitted in all directions. The small percentage of them that were emitted along the optical axis of the tube or resonator are reflected by the mirrors to provide amplification, while the photons that were not emitted along the axis are lost from the system and removed as waste heat. The result of this is a standing light wave between the mirrors (Fig. 12.14). Some of the light is emitted through the partially reflecting mirror, from where its power density is increased by focusing.

Equipment

Many mediums have lasing capability, each medium resulting in a laser having its own characteristic output parameters, but only a few types are powerful and reliable enough to be practical for machining operations. Lasers can be classified by their lasing medium. They are solid state, liquid, or gas. The two types most used in machining are the CO_2 gas laser and the solid state laser.

As already mentioned, gas lasers usually consist of an optically transparent tube filled with either a single gas or a gas mixture as the lasing medium. A typical commercial CO_2 gas laser contains CO_2, He, and N_2, where the CO_2 supplies required energy levels for laser operation, He provides intracavity cooling, and N_2 keeps the upper energy levels populated through collisions. The CO_2 laser operates at a wavelength of 10.6 μm either pulsed or in a continuous wave.

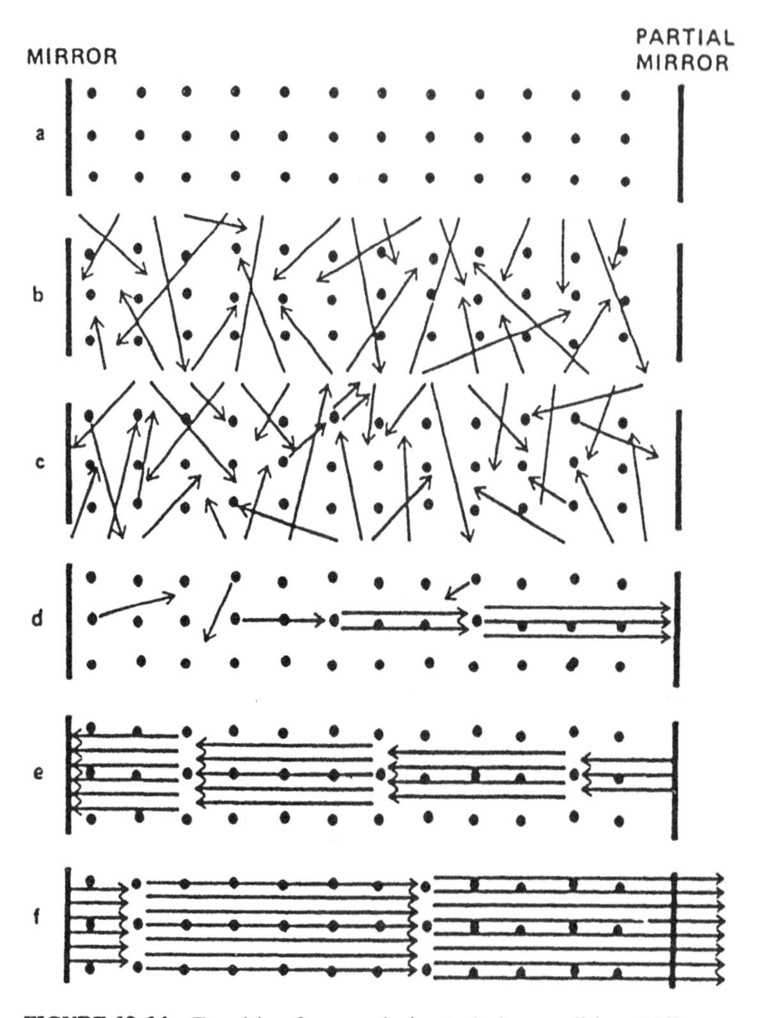

FIGURE 12.14 Transition from nonlasing to lasing conditions [46].

In high-power gas lasers, the available output is limited by the ability to cool the gas and to stabilize the gas discharge properly.

Solid-state lasers consist of a crystalline or glass host material and a doping additive to provide the reservoir of active ions needed for the lasing action. The original solid-state lasers used ruby (Al_2O_2) doped with approximately 0.05% of Cr_2O_3 as the lasing medium. Another common solid-state laser is the Nd:YAG laser, which uses a single crystal of yttrium aluminum garnet (YAG) shaped as a rod with parallel, flat ends, which are optically ground and polished and doped

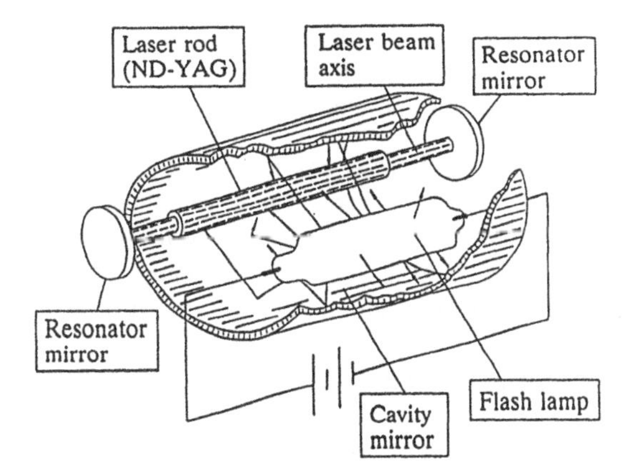

FIGURE 12.15 The Nd:YAG solid-state laser [52].

with neodymium (Nd) as the lasing medium. This laser is relatively efficient, allows for high pulse rates and can be operated with a simple cooling system. The Nd:YAG laser operates at a wavelength of 1.06 μm. In contrast to most other solid-state lasers, which operate only in the pulsed mode, the Nd:YAG laser may be operated either pulsed or in a continuous-wave (CW) mode.

Since the Nd:YAG laser material is electrically insulating, it cannot be powered by electrical excitation. Instead, xenon-flash and high-pressure mercury-discharge lamps are used to generate an intensive light flux. The light is absorbed by the medium and collimated into a laser beam. Figure 12.15 is a diagram of a solid-state laser system.

In solid-state lasers, removal of waste heat (i.e., the excess energy not usefully converted into laser radiation), is a fundamental problem. The radius of the rod is limited by the need to conduct surplus heat to its cooled periphery. This requirement sets a practical upper limit to the power that can be extracted per unit length of rod, and thus per system. Typical cooling systems are water, water to air, water to water, and refrigerated, recirculating water.

Because of the high processing speeds inherent with lasers, essentially all laser manufacturing systems are computer controlled. Machine designs vary, depending upon the requirements of the particular application, but they can be as simple as a two-axis driller or as complex as a seven-axis cutter. The processing motions can be accomplished by moving the workpiece, by moving the laser beam, or through a combination of the two.

Process Capabilities and Applications

Lasers are used to drill, cut, mark, weld, and heat-treat materials, but laser processing is not usually employed as a mass material removal or heating process. Successful application requires good coupling of laser energy to the part to be machined since laser processing is essentially a controlled heating process. The most important material properties and characteristics are:

Those affecting the manner in which the light is absorbed by the material. These are reflectivity of the surface at the particular wavelengths being used and the absorption coefficient of the bulk material. Materials with good electrical conductivity such as gold, copper, and aluminum are poor light energy absorbers, while plastics and wood are almost perfect absorbers.

Those governing the flow of heat in a material: thermal conductivity and diffusivity.

Those relating to the amount of energy required to cause a desired phase change: density, heat capacity, heat of fusion, and heat of vaporization.

Drilling Laser drilling is a process for producing small holes by using either single or multiple pulses from a stationary laser beam to penetrate the material. This technique is often referred to as laser percussion hole drilling. It is accomplished by placing the workpiece at or near the focal point of the beam. A short pulse from the laser causes a small volume of the illuminated workpiece material to both partially melt and partially vaporize. The explosive escape of the vaporizing material causes most of the molten volume to be removed as a spray of droplets.

Holes with diameters ranging from 0.1 to 1 mm can be produced with length/diameter ratios of about 50:1. They are characterized by a tapered, rough shape lacking a high degree of roundness. The walls of the hole exhibit a recast layer and a heat-affected zone that can vary in thickness from 0.002 to 0.10 mm depending upon the material type, thickness, and drilling parameters. Almost any material can be drilled, and hole entry angles can be as shallow as 15° from the surface. Diametral repeatability is about ±10% of the diameter. Application of the ability to drill small holes may be found in industries producing fuel filters, carburetor nozzles, and cooling holes in jet engine blades.

Cutting Holes with diameters larger than 1 to 1.5 mm cannot be produced by percussion drilling because power density is lacking in the defocused beam. In common practice, when holes are larger in diameter than approximately 0.5 mm they are more often laser-cut than drilled. Systems are available for rotating a focusing lens in a horizontal plane on an axis coincident with the incoming stationary beam. The focused spot will always be on the focal axis of the lens and will be rotated in a circle with the lens. The effective radius of operation is limited by the size of the lens used. Other systems use an X–Y table controlled by a CNC system to move the workpiece relative to a stationary laser beam, enabling cutting of any desired shape.

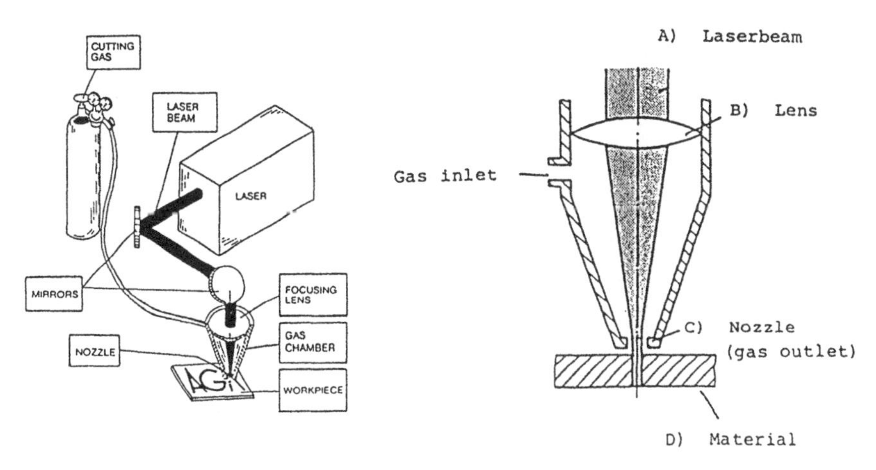

FIGURE 12.16 Principle of laser cutting system [52].

Laser cutting and drilling is usually assisted by a flow of gas. Figure 12.16 shows a typical gas-assist system. The flow of gas performs several functions, such as cooling the area around the cut and blowing away swarf and slag and also helps to keep debris from contaminating the focusing lens. Reactive gasses increase cutting speed, but self-burning of the material can cause poor kerf quality. It is important that the proper gap between the gas jet and the workpiece be maintained. This may be accomplished with self-adjusting, height-sensing units that control the gap automatically, regardless of surface unevenness.

When cutting, the laser beam can be operated either CW or rapidly pulsed. The CW beam results in the highest cutting speeds, while a pulsed beam results in the lowest thermal effects and the least distortion of the workpiece. The laser is able to produce a very narrow kerf, a narrow heat affected zone and minimal slag. Kerf widths of 0.1–0.4 mm are typically achieved with the focal point positioned about one-third of the material thickness below the surface, using a 1.25-mm-diameter gas jet nozzle positioned about 0.5 mm above the surface.

Focusing becomes progressively more critical for a given material as its thickness is increased. It is also critical for materials with very high melting points. Since a short wavelength gives better focusing ability than a long wavelength, it follows that the Nd:YAG laser for any given power output is preferable to a CO_2 laser for these types of work.

Examples of applications may be found in almost any industry with a need to cut metallic or other types of material in thicknesses between 0.5 and 10 mm with a relatively high degree of precision and complex cut contours, preferably in small batches. Figure 12.17 is an example of a layout for a cutting sequence

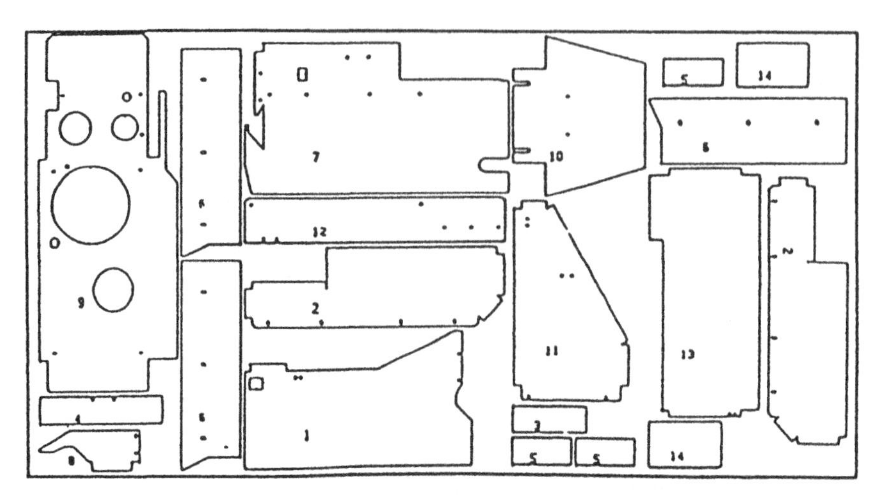

FIGURE 12.17 Layout for laser-cut items in 22-mm mild steel for just-in-time production [52].

in 2-mm mild steel. The items are used in a just-in-time production of photocopy machines. Notice that waste material from the metal sheet has been reduced to a minimum.

A more advanced example of three-dimensional cutting of items for the automobile industry is seen in Fig. 12.18. The item is punch-pressed into shape, and a seven-axis gantry cutting machine is then used for trimming and producing holes and openings.

Welding Lasers are used in place of more conventional joining methods when low-distortion, high-speed welds are desired. Both spot and seam welding can be performed, although only pulsed lasers capable of high repetition rates or CW lasers can be used effectively for seam welds. Parts to be welded must have a tight fit with a gap less than 5% of the thickness of the material. Various shielding gases, preferably helium, nitrogen, or argon are used locally to protect the weld puddle from oxidation contamination.

Depending on the parameters being used, laser welds are accomplished primarily as either surface welds or penetration welds, of which the former occurs with low beam power densities or high welding speeds. The penetration depth is in this case limited by thermal conduction of the energy, restricting penetration to a maximum of 2.5 mm.

Deep penetration welds are accomplished by high-power lasers producing power densities high enough (5.10^5 W/cm^2 or more) to actually vaporize and "drill" a small molten channel through the workpiece. This hole, often referred to as a keyhole, is held open by vapor pressure as the beam is traversed across

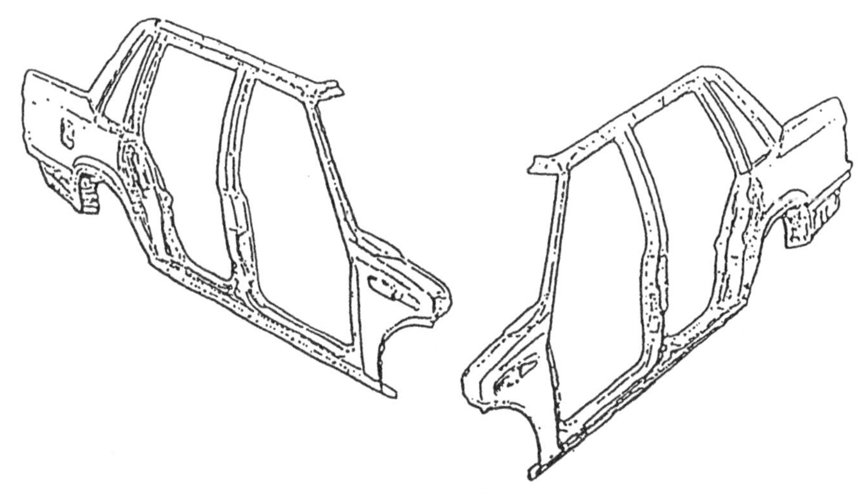

FIGURE 12.18 Three-dimensional laser cutting of item for automobile. (Courtesy, Volvo, Gothenburg, Sweden.)

the workpiece. This allows the beam energy to be deposited deep within the material instead of only on the surface. As the beam is traversed across the workpiece, the front surface of the keyhole is exposed more directly to the beam than the rear surface. The major part of the vaporization therefore takes place at the front, and the ablation pressure forces the liquid metal to flow from the front of the molten pool around the keyhole cavity to solidify at the trailing edge of the pool in a characteristic ripple pattern. As shown in Fig. 12.19, the final penetration depth depends on workpiece material, welding speed, and beam power.

In laser welding the same types of joints as those encountered in traditional welding are found. Figure 12.20 illustrates some examples. Although lasers have been used for welding since the mid-1970s, it is only in the last decade or so that they have gained real industrial acceptance through the development of high-power lasers with high beam quality and reasonable prices. Table 12.3 shows some typical industrial laser welding applications.

Other Laser Processes While drilling, cutting, and welding, are the most important industrial implementations of the laser, it is also used successfully for marking, surface treatment, and cladding.

Laser marking utilizes short laser pulses—duration 1 μs or less—to ensure removal of material from the top layer of the material only. At such short pulse lengths, material a few tenths of a millimeter below the surface is not thermally affected by the radiation, although temperatures well above the boiling temperature have been reached at the surface. As the beam is scanned across the surface, it vaporizes a series of overlapping blind holes to produce smooth-

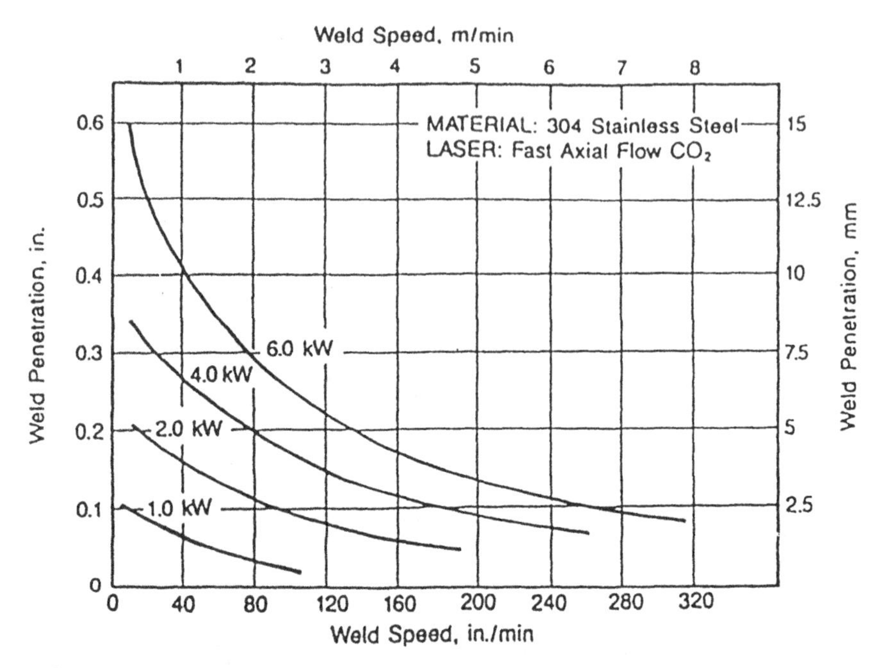

FIGURE 12.19 Possible weld penetration depths for a CO_2 laser [53].

bottomed grooves that make up the identification letters or symbols. Often a computer is used to control the laser beam positioning and the timing of the energy pulses. The laser marking process is very fast and offers significant improvements in legibility compared to conventional marking methods.

Laser surface treatment is a transformation hardening process used for nearly distortionless, localized surface hardening caused by rapid heating of a localized area by irradiance from a laser beam, followed by a rapid quenching. As in conventional hardening, the material must contain sufficient carbon to produce the martensitic phase that is the sole source of the hardening mechanism. The laser beam is defocused to produce a power density of only a few hundreds to a few thousands watts per square centimeter at the part surface and is traversed across the work at a rate fast enough to avoid melting. Heat is conducted from the surface into a thin volume of metal beneath the beam. The volume is rapidly heated beyond its upper critical temperature, transforming it to austenite. As the beam moves on, self-quenching of the heated layer occurs instantly because of the rapid flow of heat into the cold substrate. Laser hardening is most often performed by CO_2 lasers.

In laser cladding, a defocused beam and a local shield gas are used to melt and selectively deposit special alloys onto part surfaces. Parts suitable for cladding are those that require small areas to be protected from corrosion or wear,

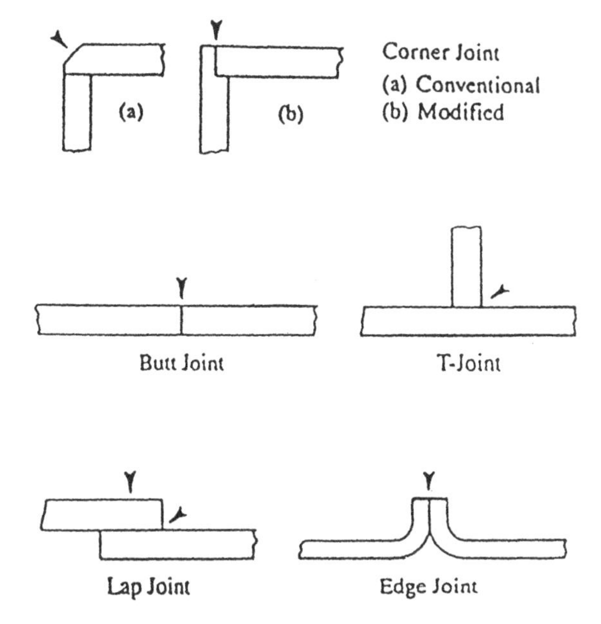

FIGURE 12.20 Commonly used joint types in laser welding [52].

but do not require that expensive materials be used for the entire part. The applications include valve seats, piston rings, turbine blades, and rock drills.

12.2.4 Abrasive Jet Machining

Abrasive jet machining (AJM) removes material by the impingement on the workpiece of fine, abrasive particles entrained in a high-velocity gas stream. Abrasive jet machining differs from conventional sand blasting in that the abrasive is much finer, and process parameters and cutting action are carefully controlled. Material removal occurs through a chipping action, which is especially effective on hard, brittle materials such as glass, silicon, tungsten, and ceramics. Soft, resilient materials, such as rubber and some plastics, resist the chipping action and thus are not effectively processed by AJM. The process is inherently free from chatter and vibration problems because the tool is not in contact with the workpiece, enabling AJM to produce fine, intricate detail in extremely sensitive objects. In addition, the cutting action is cool, since the carrier gas serves as a coolant, and workpieces experience no thermal damage. Figure 12.21 schematically depicts the main elements of an AJM system.

Process Parameters

Major process variables that affect the removal rate are nozzle tip distance, abrasive type, abrasive flow rate, and gas pressure. Each of these variables is treated brieflv in the following discussion.

TABLE 12.3 Typical Industrial Laser Welding Applications

Nd:YAG laser welding	CO_2 laser welding
Electrical and electronics	Precision engineering
Terminal ports	Membranes
Gas sealed batteries.	Sheet metal plates
Cathode picture tubes	Tubes
Precision engineering	Mechanical engineering
Pressure sensors	Mixer parts
Spectacle frames	Air condition compressor pulley
Dental instruments	
Sealing containers	
Membranes	Metalworking
Thin foils	Coal mine idler rolls
Mechanical Engineering	Automotive industry
Springs	Gear parts
Thermostats	Motor components
	Fuel filters
Etc.	Body parts (doors, roofs, windows, etc.)
	Diaphragms
	Caburetors
	Nuclear power plants
	Tubes
	Pipes
	Etc.

Source: Ref. 52.

Various nozzle tip distances (NTD) are used depending upon the application. When exacting definition is required, as in cutting, the nozzle is positioned very close to the workpiece, typically 0.8 mm. At this close distance, cutting rates are sacrificed for the sake of increased accuracy. As the NTD is increased, the particles are accelerated to much higher speeds, increasing the cutting speed until an optimum is reached. At even larger distances—about 7–13 mm—the expanding gas accelerates radially as well as axially and energy is lost, resulting in decreasing cutting speeds as shown in Fig. 12.22 for a glass workpiece. In addition, as the nozzle is moved away from the work, the diameter of the hole or width of the cut increases. At the same time, the walls of the cut assume a tapered shape (see Fig. 12.23).

The various abrasives used in AJM are selected by application. Aluminum oxide or silicon carbide is commonly used for cleaning, cutting, and deburring, while polishing or peening often is accomplished with glass beads. For very soft materials, sodium bicarbonate may be used. A summary of abrasives for AJM and their applications is given in Table 12.4. Particle size is important. Best cut-

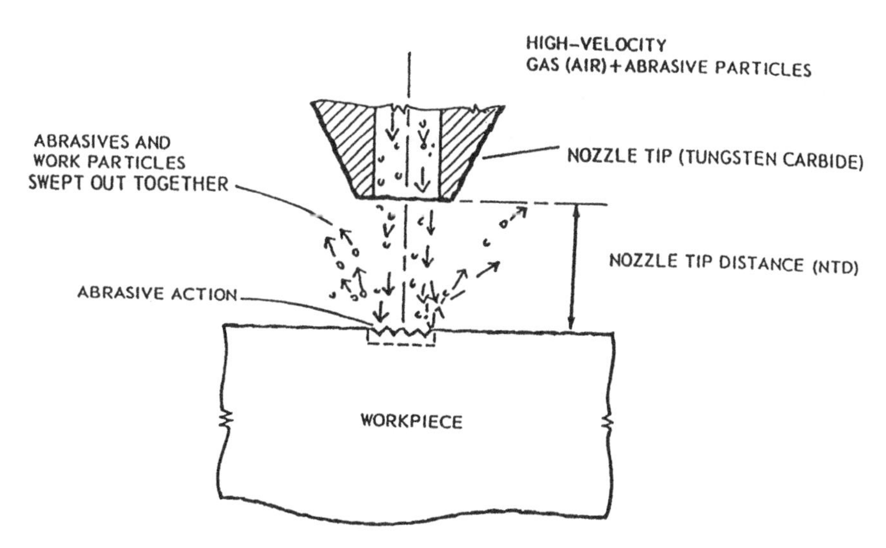

FIGURE 12.21 Abrasive jet machining (AJM) [46].

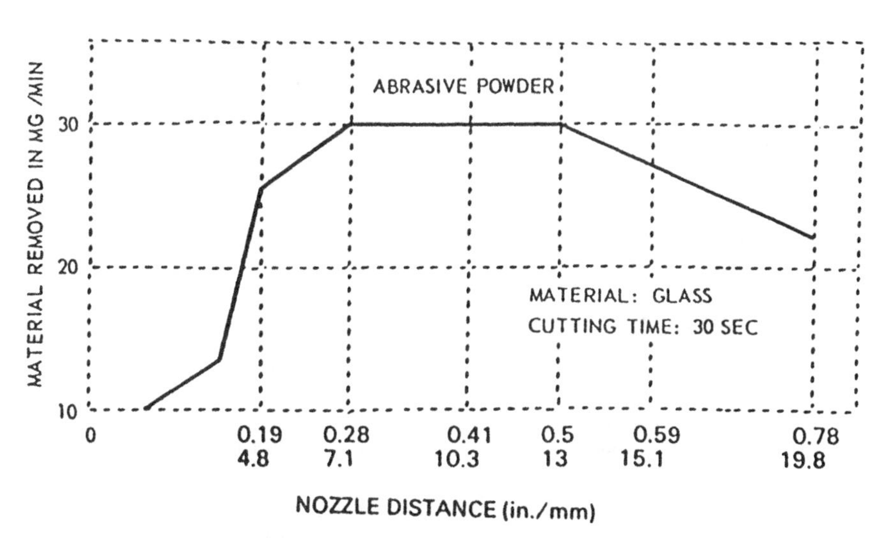

FIGURE 12.22 Influence of nozzle tip distance on cutting speed in glass. (Courtesy, S. S. White Company.)

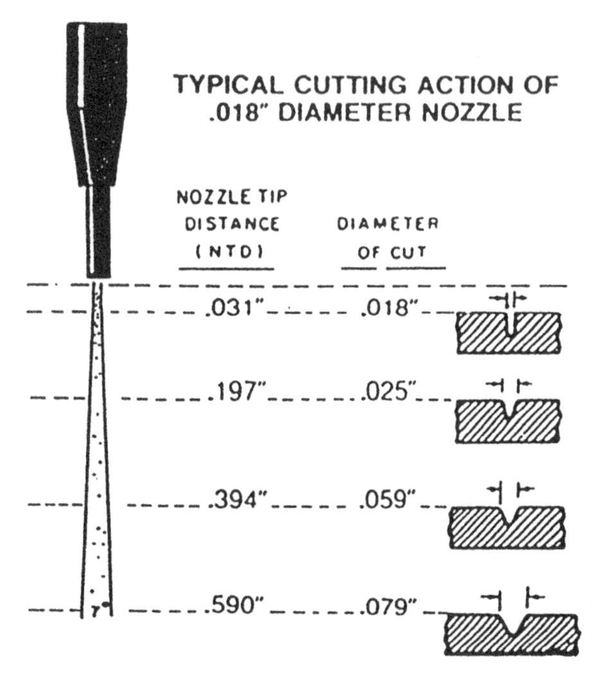

FIGURE 12.23 Kerfwidth is a function of nozzle tip distance (Courtesy, S. S. White Company.)

ting results are obtained when the bulk of particles vary between 15 and 45 μm. Abrasive powder should not be reused because its cutting and abrading action decreases and because it becomes contaminated with foreign material.

The flow rate of the abrasive particles is directly related to the metal removal rate, as shown in Fig. 12.24. The curve shows a maximum because, in the beginning, increasing the flow rate means more abrasive particles available for cutting. However, as the powder flow is further increased, the abrasive velocity

TABLE 12.4 The AJM Abrasives and Their Application

Abrasives	Applications
Aluminum oxide	Cleaning, cutting, deburring
Silicon Carbide	As above, but for harder materials
Glass beads	Matte polishing, cleaning
Crushed glass	Peening, cleaning
Sodium bicarbonate	Cleaning, cutting-soft materials

Source: Ref. 47.

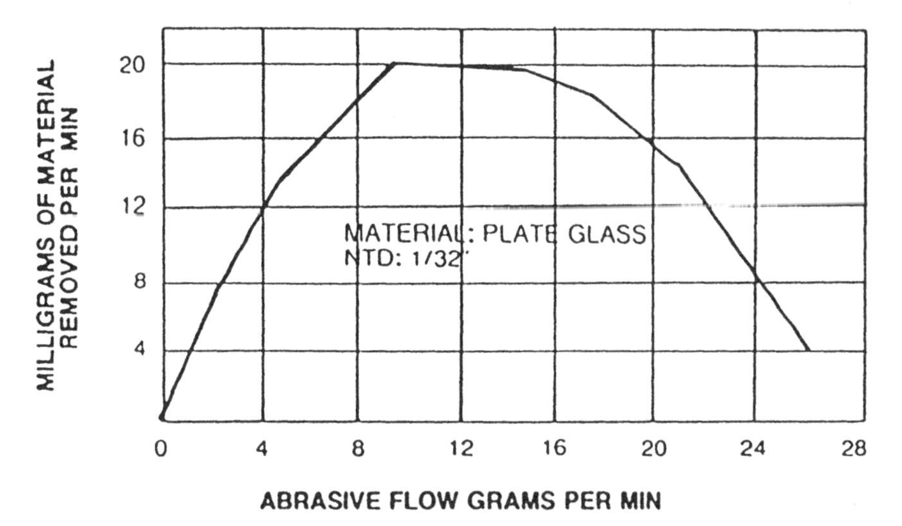

FIGURE 12.24 Effect of powder flow rate on material removal. (Courtesy, S. S. White Company.)

decreases, reducing the removal rate. This effect becomes apparent with flow rates greater than about 15 g/min, and consequently most operations are performed at 10 g/min, which also helps to conserve nozzle life.

Increasing the nozzle pressure results in a small increase in removal rate, but the effect is modest compared with the other process variables. However, these small increases are offset by decreased nozzle life, and pressures higher than 20–100 N/cm^2 are, therefore, seldomly used.

Process Capabilities and Applications

The mass rate of removal is low, usually about 50–100 mg/min, but this is more than compensated for by the ability to produce intricate detail in very hard materials. Slots as narrow as 0.15 mm can be produced when stray cutting is minimized with rectangular nozzles. Tolerances better than ±0.1 mm are easily obtained while surface finishes range from 0.3 to 1.5 μm, with the finer abrasives achieving the best finishes. Steel as thick as 1.5 mm and glass 6.3 mm thick have been cut, but at very slow rates and with large amounts of taper.

AJM has been successfully employed in the electronics industry to shape ceramic elements and for resistor adjustment through accurate and controlled removal of conductive material. Semiconductor materials such as germanium, silicon, and gallium are cut, cleaned, drilled, beveled, and so on.

It is possible to make small adjustments in steel molds and dies after they have been given a final hardening treatment. Precision deburring is another area

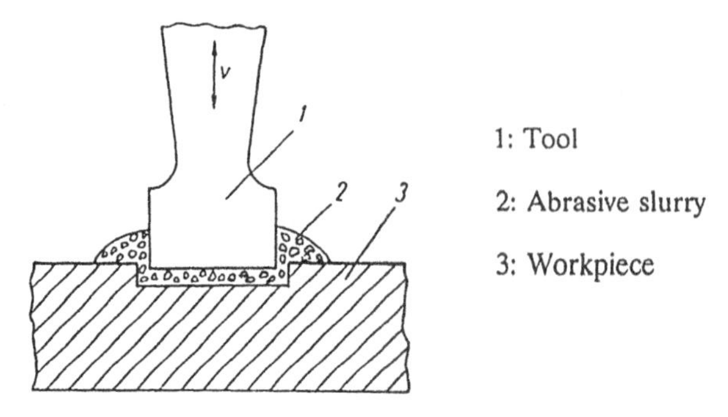

FIGURE 12.25 Ultrasonic machining.

for AJM applications since high quality standards are required in such technologies as aerospace, medical equipment, and computers.

It is also worth mentioning that abrasive jet machining can be used for safe removal of metallic smears on ceramics, oxides on metals, resistive coatings, etc., especially from parts too delicate to withstand manual scraping or conventional grinding. The process is not practical for removing heavy burrs or large amounts of material. Also, it should not be used for large parts or surfaces or low value components.

12.2.5 Ultrasonic Machining

Ultrasonic machining (USM) is a mechanical material removal process used to erode holes and cavities in hard or brittle workpieces by using shaped tools, high-frequency mechanical motion, and an abrasive slurry. USM is able to effectively machine all hard materials whether they are electrically conductive or not.

The process is performed by a cutting tool, which oscillates at high frequency, typically 20–40 kHz, in an abrasive slurry. The shape of the tool corresponds to the shape to be produced in the workpiece. The high-speed reciprocations of the tool drive the abrasive grains across a small gap against the workpiece (see Fig. 12.25). The tool is gradually fed with a uniform force. The impact of the abrasive is the energy principally responsible for material removal in the form of small wear particles that are carried away by the abrasive slurry. The tool material, being tough and ductile, wears out at a much slower rate.

The tool is oscillated by exploitation of an effect known as longitudinal magnetostriction. With this phenomenon, a magnetic field undergoing variation at ultrasonic frequencies causes corresponding changes in the length of a ferromagnetic object placed within its region of influence. A magnetostriction transducer, such as that illustrated in Fig. 12.26, or (more common today) a piezoelectric

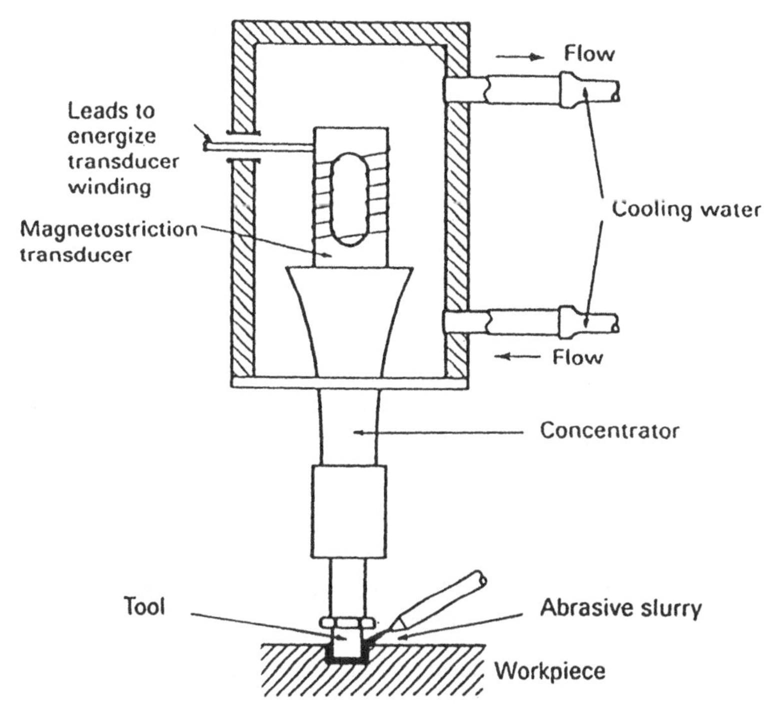

FIGURE 12.26 Elements of ultrasonic machining [45].

transducer is used. The tool vibrates with a stroke of only a few hundredths of a millimeter in a direction parallel to the axis of the feed. For efficient material removal to take place, the tool and toolholder must be designed so that resonance can be achieved within the frequency range of the machine.

Material removal occurs when the abrasive particles, suspended in the slurry between the tool and workpiece, are struck by the downstroke of the vibrating tool. The impact propels the particles across the cutting gap, hammering them into the surface of both tool and workpiece. Since the tool is made of a ductile material, the abrasive grits only give rise to plastic deformation here, whereas at the workpiece actual disintegration occurs by the chipping out of small pockets at the surface.

Besides direct impact, researchers also report that cavitation erosion contributes to disintegration. Collapse of the cavitation bubbles in the abrasive suspension results in very high local pressures. Under the action of the associated shock waves on the abrasive particles, microcracks are generated at the interface of the workpiece. The effects of successive shock waves lead to chipping of particles from the workpiece.

The relative contribution made to material removal by the two effects have been found to vary with the operational conditions, but it seems that the cavitation effect in general accounts for less than 5% of the total volumetric removal rate.

Equipment

The machines for USM range from small, tabletop-sized units to large-capacity machine tools, supplied as cutting heads for installation in other machine tools, as bench units, and as self-contained machine tools. Power is rated in watts and can range from about 40 W to 2.5 kW. The power rating determines the area of the tool that can be accommodated and thereby strongly influences the material removal rate.

All USM machines are made up of common subsystems regardless of their physical size or power. The most important of these subsystems are the power supply, transducer, toolholder, tools, and abrasive.

Power Supply The power supply is a sine-wave generator that offers the user control over both the frequency and power of the generated signal. It converts low-frequency (50/60 Hz) power to high-frequency (~10–25 kHz) power, which is supplied to the transducer for conversion into mechanical motion.

Transducer Two types of transducers are used in USM to convert the supplied energy to mechanical motion. They are based on two different principles of operation: magnetostriction and piezoelectricity.

Magnetostrictive transducers are usually constructed from a laminated stack of nickel or nickel alloy sheets. Magnetostriction is explained in terms of domain theory. Domains are very small regions, of the order of 10^{-8}–10^{-9} cm^3, in which there are forces that cause the magnetic moments of the atoms to be oriented in a single direction. In each domain the atomic magnetic moments are oriented in one of the directions of easy magnetization, which coincide with the directions of the crystallographic axes of the given crystallite. In the cubic-lattice crystals of iron and nickel there are six directions of easy magnetization. In unmagnetized material all these directions are present in equal numbers and, therefore, the magnetic moments of the orderless, unorientated domains compensate one another. When the material is placed in a sufficiently strong magnetic field, the magnetic moments of the domains rotate into the direction of the applied magnetic field and become parallel to it. During this process the material expands or contracts until all the domains have become parallel to one another.

Among practical materials, iron-cobalt, iron-aluminum, and nickel have the highest magnetostriction. As the temperature is raised due to hysteresis and eddy current losses, the amount of magnetostrictive strain diminishes, and magnetostrictive transducers therefore require cooling by fans or water. They have power ratings up to 2.5 kW.

Piezoelectric transducers generate mechanical motion through the piezoelectric effect by which certain materials such as quartz or lead zirconate titanate generate a small electric current when compressed. Conversely, when an electric current is applied, the material increases minutely in size. When the current is removed, the material instantly returns to its original shape. Piezoelectric materials are composed of small particles bound together by sintering. The material undergoes polarization by heating it above the Curie point; when it is placed in an electric field on cooling, the orientation is preserved. Such transducers exhibit a high electromechanical conversion efficiency that eliminates the need for cooling. They are available with power capabilities up to 1 kW.

The magnitude of the length change that can be achieved by the two types of transducers is limited by the strength of the particular transducer material. The limit is approximately 0.025 mm.

Toolholder The toolholder provides the link between the transducer and the tool. Its function is to increase the tool vibration amplitude and to match the vibrator to the acoustic load. In this capacity, the toolholder must be detachable and is therefore often threaded. It must be constructed of a material with good acoustic properties and be highly resistant to fatigue cracking. Monel and titanium have good acoustic properties and are often used together with stainless steel, which is cheaper. However, stainless steel has acoustical and fatigue properties that are inferior to those of monel and titanium, limiting it to low-amplitude applications.

Toolholders are available in two configurations: nonamplifying and amplifying. Nonamplifying holders are cylindrical and result in the same stroke amplitude at the output end as at the input end. Amplifying toolholders have a cross section that diminishes toward the tool, often following an exponential function. They are designed to increase the tool stroke through stretching and relaxation of the toolholder material. An amplifying toolholder is also called a concentrator. Figure 12.27 shows how sonic energy is propagated in an exponential concentrator.

Because of the gain in tool stroke, amplifying holders are able to remove material up to 10 times faster than the nonamplifying type. The disadvantages of using amplifying toolholders include increased cost to fabricate, a reduction in surface finish quality, and the requirement of much more frequent tuning to maintain resonance.

Tools Tools should be constructed from relatively ductile materials such as stainless steels, brass, and mild steels. The harder the tool material, the faster its wear rate will be. It is important to realize that finishing or polishing operations on the tools are sometimes necessary because their surface finish will be reproduced in the workpiece.

The geometry of the tool generally corresponds to the geometry of the cut to be made, but because of the unavoidable overcut, allowances must be made to

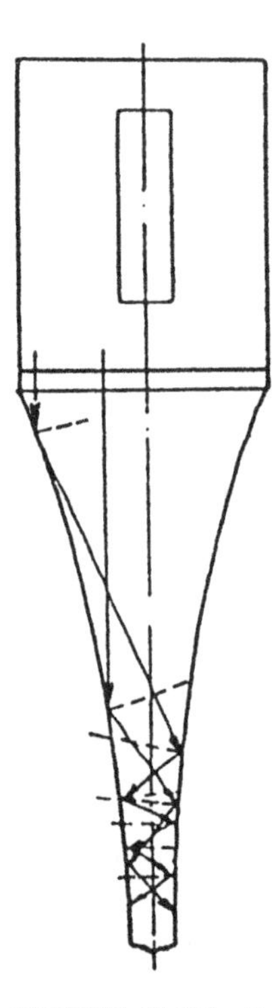

FIGURE 12.27 Energy propagation in an exponential toolholder [51].

use tools that are slightly smaller than the desired hole or cavity. Tool and toolholder are often attached by silver brazing.

Abrasives Slurries of synthetic abrasive powders in a liquid, usually water, are generally used in ultrasonic machining. The criteria for selection of an abrasive for a particular application include hardness,usable life, cost, and particle size. Diamond is by far the fastest abrasive, but is not practical because of its cost. Boron carbide is economical and yields good machining rates. It is therefore one of the most commonly used abrasives for USM, but silicon carbide and aluminum oxide are also widely used.

Grain size has a strong influence on removal rate and surface finish. Coarse grits exhibit the highest removal rates, but when the grain size becomes comparable with the tool amplitude, a maximum is reached and larger grains cut more slowly. As would be expected, the larger the grit size, the rougher the machined surface. Water is the liquid medium predominantly used, with an abrasive concentration of about 50% by weight, but thinner mixtures are used to promote efficient flow when drilling deep holes or when forming complex cavities.

TABLE 12.5 Penetration and Tool Wear Rates in Ultrasonic Machining at 700 W Input

Material	Ratio Stock Removed To Tool Wear*	Maximum Practical Machining Area		Average Penetrating Rate**	
		in.^{2}	cm^2	in./min	mm/min
Glass	100:1	4.0	25.8	0.150	3.81
Ceramic	75:1	3.0	19.4	0.060	1.52
Germanium	100:1	3.5	22.6	0.085	2.16
Tungsten carbide	1.5:1	1.2	7.7	0.010	0.25
Tool steel	1:1	0.875	5.6	0.005	0.13
Mother of pearl	100:1	4.0	25.8	0.150	3.81
Synthetic ruby	2:1	0.875	5.6	0.020	0.51
Carbon-graphite	100:1	3.0	19.4	0.080	2.00
Ferrite	100:1	3.5	22.6	0.125	3.18
Quartz	50:1	3.0	19.4	0.065	1.65
Boron carbide	2:1	0.875	5.6	0.008	0.20
Glass-bonded mica	100:1	3.5	22.6	0.125	3.18

Source: Data from Raytheon Company, *Impact Grinders for Ultrasonic Machining*, 1961.
*Tool material; cold rolled steel in all cases; #320 mesh boron carbide abrasive.
**½" (12.7 mm) diam. tool; ½" (12.7 mm) deep.

Process Capabilities and Applications

Ultrasonic machining does not compete with conventional material removal operations on the basis of stock removal rates. The productivity of the process depends to a marked extent on the hardness and brittleness of the workpiece. The best machining rates are obtained on materials harder than HRc 60, with carbides, ferrites, germanium, ceramics, glasses, and tungsten representing groups that are difficult or impossible to process conventionally or by spark erosion, but are well suited for ultrasonic machining. A representative ranking of process performance for various hard and brittle materials is shown in table 12.5. Note that slow material rates are associated with high tool wear rates.

The USM process is particularly well suited to:

Making holes of any shape for which a master can be made, including multihole screens. The range of obtainable shapes can be increased by moving both tool and workpiece relative to each other during cutting, permitting operations such as threading to be carried out.

Coining and engraving operations on glass, hardened steel, and sintered carbides.

Parting and machining of diamonds and other precious stones.

The smallest holes that can be cut by USM are approximately 0.08 mm in diameter, hole size being limited by the strength of the tool and the clearance required for the abrasive. The largest diameter solid tool used so far has a 115-mm diameter, but larger holes can, of course, be cut by trepanning. Hole depth is limited by tool wear and by difficulties in feeding fresh cutting slurry to the end of the tool.

Surface finish is governed primarily by the abrasive particle size. The best finishes are on the order of 0.25 μm RMS (800-grit). For accurate holes, rough and finish cuts are advisable. The surface of an ultrasonically machined workpiece usually exhibits a nondirectional surface texture and is therefore easily polished.

Most successful applications of ultrasonic machining involve drilling holes or machining cavities in nonconductive ceramic materials, making the process a valuable supplement to spark erosion. USM is often used simultaneously to produce a multitude of holes in precise patterns in germanium or aluminum oxide substrates in the electronics industry and for the production of spinning nozzles in ceramics, metals, and minerals. Lately, the process has found use in a multistep production of silicon nitride (Si_3N_4) turbine blades.

12.2.6 Electrochemical Machining

In electrochemical machining, (ECM) metal is dissolved atom by atom according to the principles of electrolysis. The process is very simple in its principle of operation, but its dependence on electrical, chemical, and fluid flow phenomena gives rise to a number of technical problems that have proved themselves difficult to overcome, particularly in the field of electrode design where a great deal of trial-and-error work is still required before the exact shape of a new electrode for a given job is fully determined.

Like spark erosion, ECM can operate on particularly hard and tough electrically conducting materials, but with much higher material removal rates. Because of this, and bearing in mind the electrode design problems, the main field of applications so far is found in the aerospace industry, where the complexity of component shapes and toughness of the high-strength, heat-resistant alloys used in gas turbine engines can preclude the use of conventional methods of manufacture. The process can, however, also be economic for the machining of very complex workpieces in relatively soft materials since with electrochemical machining, in contrast to conventional machining, the whole surface can be machined simultaneously and the machining time can be very much less than by conventional machining.

As determined by Faraday (1791–1867), when two conductive poles are placed in a conductive electrolyte bath and energized by a direct current, metal

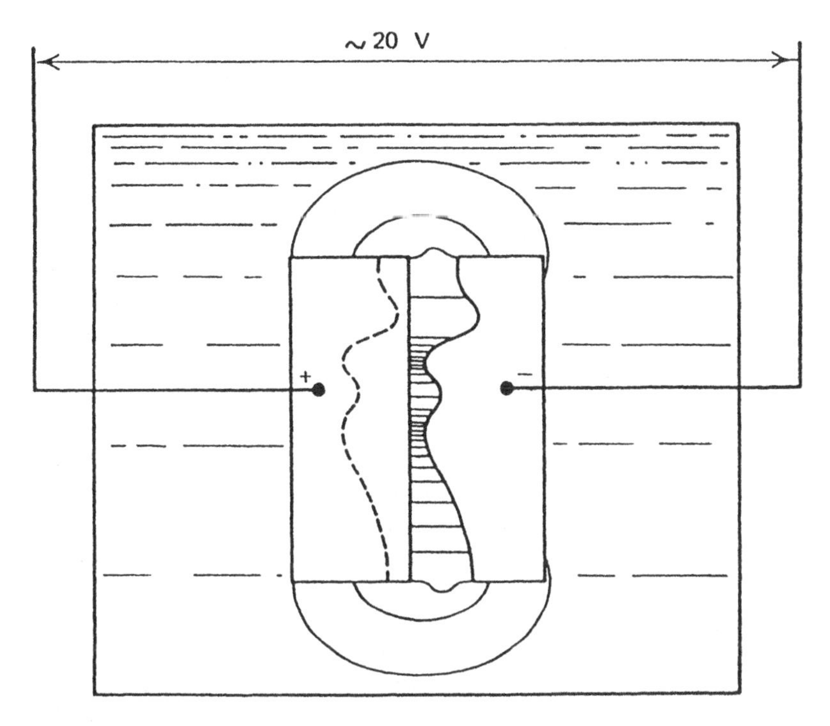

FIGURE 12.28 Principles of electrochemical machining. Distribution of current flow between closely spaced electrodes.

may be deplated from the positive pole, or anode, and plated onto the negative pole, or cathode. The plating mechanism has been used for many years for adding metals to the surface of parts, but it is the reverse action of metal deplating in a controlled manner from the anode that is used in modern electrochemical machining.

In order to obtain competitive machining rates, high current densities, about 800–1000 A/cm^2, must be used with a narrow gap between the workpiece and a suitably shaped tool (see Figure 12.28). The density and direction of current flow is indicated by lines joining the electrodes. The smaller the gap, the greater will be the current flow and rate of metal removal from the anode. The dotted line shows the shape of the anode after a period of time. It can be seen that even with stationary electrodes, the shape produced resembles that of the cathode.

The high current density promotes rapid generation of hydroxide solids and gas bubbles in the electrode gap. These become a barrier to the electrolyzing current after a few seconds and must be continuously removed by circulating the electrolyte at a high velocity through the gap. This forced circulation permits continuous fast metal removal from the anode, and the original gap increases in size, rapidly at first, but then at a progressively decreasing rate because the current falls as the gap, and hence the electrical resistance, increases. To maintain the current density at its initial high value, the tool is therefore moved toward the anode at the same rate at which the metal is being dissolved. This ensures a constant gap spacing, and current and metal removal rate will remain high.

Theoretically, the electrode (cathode) reproduces its shape accurately in the workpiece (anode) since current is conducted from all surfaces of the cathode to all surfaces of the anode. The smaller the gap spacing between confronting surfaces, the higher will be the current density and the greater the rate of removal. With progressive movement of the cathode towards the anode, the two surfaces will ultimately correspond closely. In fact, the final shape of the workpiece is not exactly the inverse of that of the tool, requiring adjustment of the cathode shape (usually by trial and error) since one of the outstanding problems of ECM is the calculation and design of tools to produce a given workpiece shape.

The theoretical rate of metal removal from the anode is calculated by Faraday's laws of electrolysis and constitutes approximately 1.5–2 cm^3/min for each 1000 A of current. Machines have been built that are capable of removing metal at a rate of up to 50 cm^3/min. ECM machines are available in a wide variety of sizes and configurations. Usually they have dc power outputs between 10,000 A and 40,000 A. Potentials of 5 to 25 V applied across the tool and work electrodes are required to circulate these currents through the resistive machining gap. Voltage must be closely controlled since it is a factor in determining the size of the machining gap, that is, the accuracy of the work contour.

Electrochemical machine tools are expensive to buy and operate. They are often characterized by their enormous size because a very solid construction is necessary to withstand deflection while containing the forces generated by the high electrolyte flow. Although the electrolyte pressures themselves are not high, the resulting forces can be tremendous when the pressures are applied to a large surface.

The corrosive nature of the ECM electrolytes promotes special considerations for ECM machines. Any portion of the machine or tooling that comes in contact with the electrolyte must be made of stainless steel, plastic, or other corrosion-resistant materials. Drive units such as ball screws and motors must be protected by sealing and may even require positive pressurization with dry, clean air.

Equipment

In broad categories, an ECM system consists of an electrolyte system that provides high-velocity electrolyte flow between the electrodes; an electrical

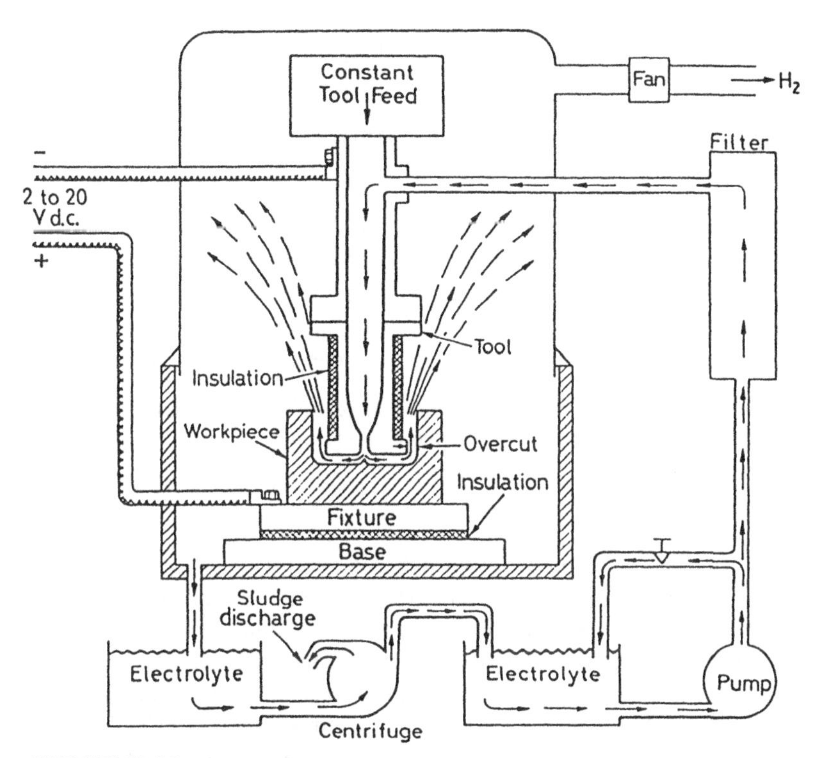

FIGURE 12.29 Schematic diagram of electrochemical machine [50].

power system that supplies the electrolyzing current to the electrodes, and a mechanical structure that locates and provides movement of the electrodes. These elements are shown diagramatically in Fig. 12.29.

Electrolytes

The electrolyte completes the circuit between the tool and the workpiece and permits the desired machining reactions to occur. It also carried heat and reaction products away from the machining zone. An effective electrolyte should therefore have good electrical conductivity and be inexpensive, nontoxic, safe, and as noncorrosive as possible. The most widely used electrolyte at present is sodium chloride in water, which has the desirable characteristics outlined above, although its corrosiveness presents a problem. A wide range of metals have been machined with sodium chloride. Sodium nitrate is another commonly used electrolyte, less corrosive than sodium chloride. Other chemicals that have been used as electrolytes include potassium chloride, sodium hydroxide, sodium

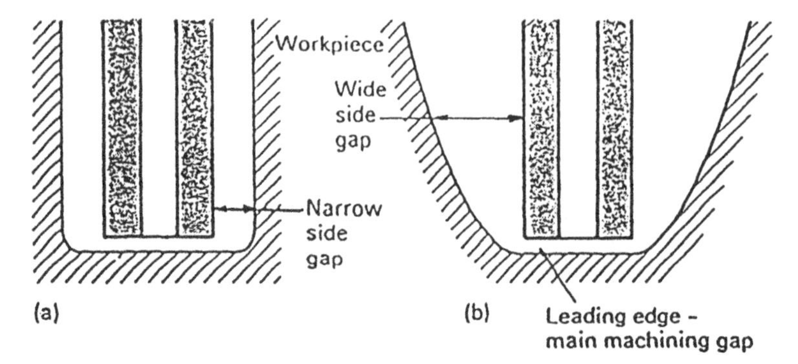

FIGURE 12.30 Schematic representation of effects of different electrolytes on overcut in ECM drilling: (a) sodium nitrate, (b) sodium chloride. [45].

chlorate, and sulfuric acid. Proper electrolyte selection is important. As seen in Fig. 12.30, sodium chloride yields much less accurate components than does sodium nitrate, the latter electrolyte having far better dimensional control because of its current efficiency and current density characteristics.

All electrolytes must be cooled to maintain a constant temperature since the electrical conductivity changes as a function of temperature, which results in variations in the machining gap. Further, higher electrolyte temperature may cause changes in the thermal expansion of tooling, fixtures, machine, etc. Both factors affect the dimensional repeatability of the machined components.

Most electrolytes need filtering and purification because during work they produce insoluble byproducts of hydroxides or hydrated oxides at a rate of 100–150 cm^3 per cubic centimeter of material removed. A properly functioning electrolyte should have no more than 2% sludge content. The sludge generally takes the form of tiny particles as small as 1 μm. Purification systems incorporate filtration, settling, and/or centrifuging.

Electrode Tools

Although almost any conducting material can be used as a cathode tool, there are many advantages in using either copper or stainless steel. Copper is normally the preferred material since it is corrosion resistant, easy to machine, and its high conductivity ensures distribution of the electrolyzing current to all parts of its operating surfaces without overheating or power loss. Stainless steel does not have the qualities of copper with regard to thermal and electrical conductivity or thermal capacity, but its resistance to chemical corrosion is good.

Determination of the shape of the tool necessary to produce the required workpiece constitutes a major problem in the application of ECM. Ideally, it should be possible to perform a theoretical calculation of the shape at the drawing board stage, but this is not yet practicable. Available methods of tool design allow calculation of only a first approximation to the final shape, to be followed by an often considerable amount of adjustment of tool shape on an empirical basis.

Process Capabilities and Applications

The ECM process is usually used to machine hard, intractable materials that would be less economical to work in other ways. Any electrically conductive material can be processed, although for some metals the method has proven difficult or uneconomical. High-silicon aluminum alloys, for example, cannot be machined to an acceptable surface finish. Basically, ECM and EDM hold the same possibilities regarding geometry and material of the workpiece, but with much higher removal rates for ECM.

Tolerances of approximately ±0.05 mm may be produced once the tool shape is completely determined. The typical overcut at the side of tools is about 0.15 mm, and taper can be limited to 0.001 mm/mm, depending on the specific tool design and process parameters being used. Surface finish is dependent upon the workpiece material, the type of tool, the electrolyte flow, and the current density. Typically, RMS-values range between 0.2 μm and 1 μm at the tip of the tool.

When compared with mechanical milling, a part produced by ECM will have slightly less fatigue resistance because the ECM process removes material without inducing any work hardening in the part surface, whereas mechanical machining often leaves the part surface in a favorable state of compressive residual stress. If necessary, shot peening can be used to improve the fatigue properties of ECM-processed workpieces.

A very common application is deburring, where a cathode tool, complementary in shape to the workpiece, is placed opposite it. The current densities at the peaks of the surface irregularities are higher than those elsewhere, and the irregularities are therefore removed preferentially so that the workpiece becomes smoothed. ECM deburring is a fast process, typical process times being 5–25 s, and can often be performed with a fixed, stationary tool. The process is particularly attractive for deburring the intersectional region of cross-drilled holes, as shown in Fig. 12.31.

Electrochemical drilling is commonly used for drilling the cooling holes in gas-turbine blades (see Fig. 12.32) and for many other jobs in the manufacture of gas-turbine engines. It is possible to drill holes of a diameter of 1 mm to depths of more than 300 mm.

Multiple-hole drilling is another task for which the ECM process is well suited. An example of this is a stainless steel burner plate with 198 holes in a pattern of 9 × 22 holes, each with a diameter of 1.25 mm. Because of the close spacing of the holes, this part was previously made by drilling the holes, one at

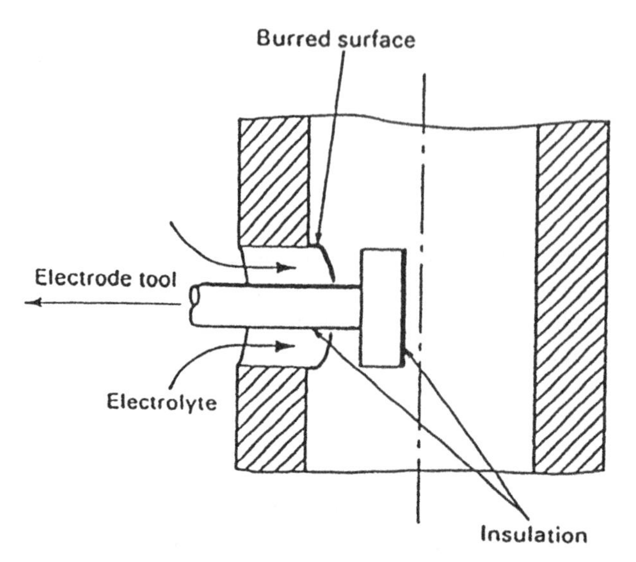

FIGURE 12.31 ECM deburring of cross-drilled hole.

a time, but converting to the use of ECM makes it possible to manufacture all the holes at the same time, substantially reducing machining time and cost. Also, ECM eliminated the need for subsequent deburring operations on the bottom of the burner plate.

Drilling by ECM is not restricted to round holes. Since the shape of the workpiece is determined by that of the tool-electrode, a cathode "drill" with any cross section will produce a corresponding shape on the workpiece. This is extensively utilized in die-sinking as well as in trepanning.

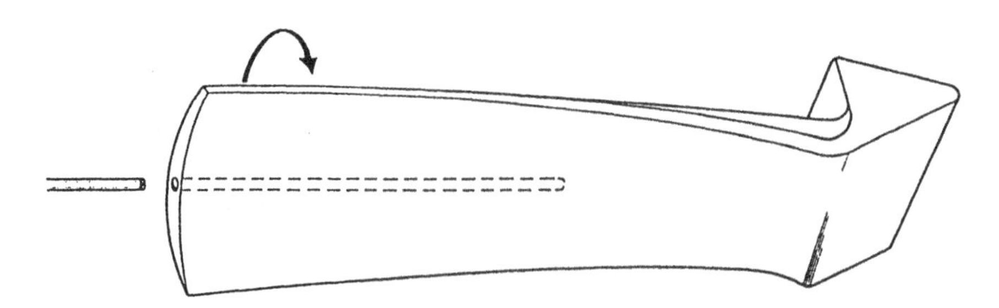

FIGURE 12.32 Deep-hole drilling by ECM [50].

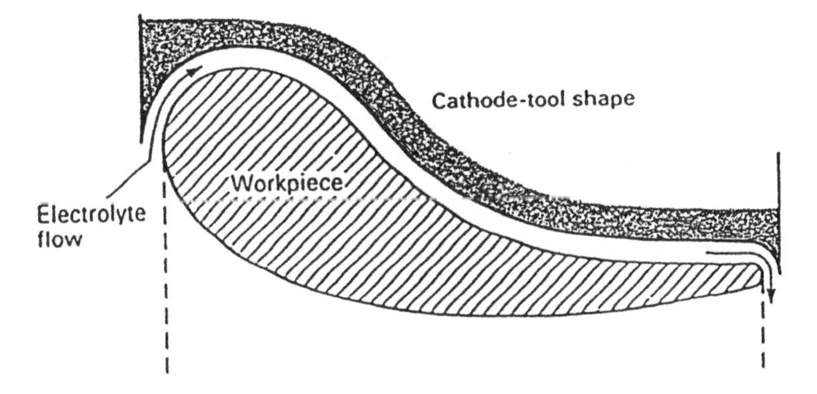

FIGURE 12.33 Design of cathode tool shape to obtain profile of turbine blade [45].

Full-form shaping utilizes a constant gap across the entire workpiece and a constant feed-rate in order to produce the type of shape illustrated in Fig. 12.33. Full-form shaping is well known for the production of compressor and turbine blades.

12.3 LAYER MANUFACTURING TECHNOLOGY

The processes mentioned in Section 12.2 are some of the most commonly used members of the family of nontraditional manufacturing processes, characterized by their capability of combining a high degree of geometrical flexibility with a capacity for economical machining of intractable materials. The processes described can all be classified as belonging to the material removal processes since basically they start from a larger amount of bulk material and remove all excess material.

Today a new group of manufacturing techniques, generally denoted layer manufacturing technology (LMT), based on the principle of adding, rather than removing or deforming material, is beginning to appear. So far, only materials with low melting points can be used, rendering the new processes useful for such purposes as producing prototype models without need to develop special tools, molds or dies, or perhaps even complicated three-dimensional shapes of the design displayed on the designers CAD station. The mechanical properties of the components are generally not sufficient to enable their use as functional parts in larger assemblies, but bearing in mind how swiftly shortcomings in otherwise promising processes tend to be overcome, it seems reasonable to assume that this will also be the case here. The rather esoteric processes commented on next may thus very well be among the important production processes of the future.

One major reason to believe in the future success of processes based on the principle of material increase is to be found in their basic CIM nature. All the processes mentioned in the following discussion are developed for and fully dependent on CAD and CIM. This contrasts drastically with CIM as applied to material removal and forming techniques. These techniques were developed long before computers entered the manufacturing environment and it has been a hard business to adapt them to CIM, CAD, and CAM.

The fact that material is deposited only where needed and without any tooling makes the LMT very well suited for CIM and eliminates most of the problems encountered with other methods. The result is a considerable reduction in lead time from product development to delivery and improved competitiveness.

Several names are used to describe the new techniques:

Rapid prototyping
CAD-oriented manufacturing
3D printing
Desktop manufacturing
Instant manufacturing
Layer manufacturing
Laminated object manufacturing
Solid freeform fabrication
Material deposit manufacturing
Material addition manufacturing
Material increase manufacturing

But no matter which name is preferred, they all function in basically the same way: by the application of a kind of selective solidification or binding of liquid or solid particles through glueing, welding, polymerization, or chemical reaction to build the part in successive layers or points created on top of each other.

A few typical and commercially available processes will be described.

12.3.1 Stereolithography

Stereolithography is an example of a layer-by-layer manufacturing process based on polymerization. It was the first system commercially available and is still the most popular one with more than 250 machines sold by the beginning of 1991.

The part to be produced is built on a horizontal platform placed in a vessel containing a liquid plastic monomer. Solidification is brought about by photopolymerization resulting from the impact of light from a laser on the upper surface of the liquid. Light absorption in the monomer limits the polymerization to a few tenths of a millimeter below the surface, roughly corresponding to the

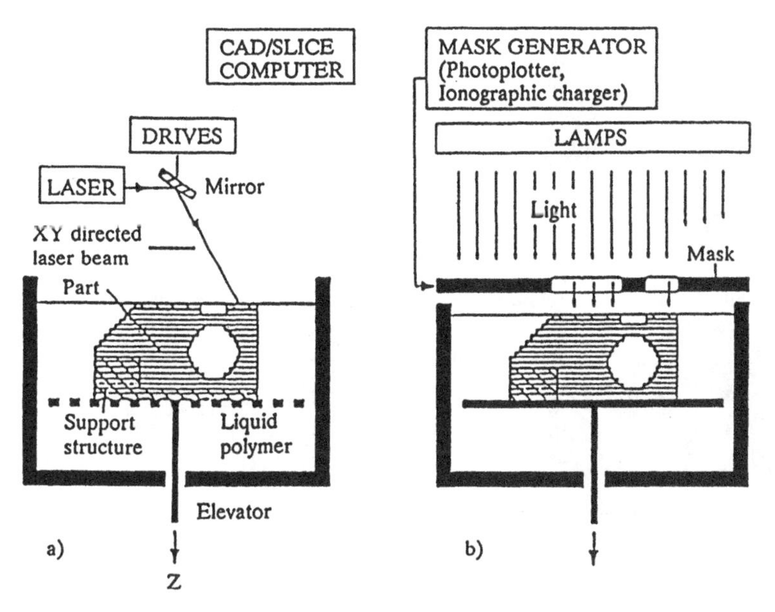

FIGURE 12.34 Principle of stereolithography: (a) point-by-point scanning, (b) layer-by-layer scanning [56].

layer thickness. Illumination of the liquid surface is restricted to a pattern that corresponds to the cross section of the part (see Fig. 12.34). Once a layer is solidified, the part is flooded with a new thin layer of liquid monomer by lowering the elevator platform, and a new cross section of the part is solidified. The whole cycle is repeated until the part is totally formed.

Most stereolithography machines apply a point-by-point solidification (Fig. 12.34a). A laser beam scans the liquid surface in order to solidify a series of voxels, each large enough—typically 0.1–1 mm—to ensure connection with neighboring voxels and with the underlaying layer. Scanning speeds from 0.5–2.5 m/sec are used. The size of voxel overlap is controlled by adjusting the distance between voxels, the layer thickness, the laser power, and the scanning speed. Close control of these parameters is important for the accuracy of the finished part, which may be expected to be within ±0.5% for workpiece dimensions up to 500 mm cubed.

To save time, the workpiece's cross sections are often only partially scanned and solidified, i.e., the laser only scans the outer and inner contour of the cross section together with some cross-hatching pattern giving the part a sufficient initial stiffness as shown in Fig. 12.35. Solidification of the liquid polymer still contained within the cross-hatching pattern is done after all layers have been generated. This happens by further exposing the "green part" to light.

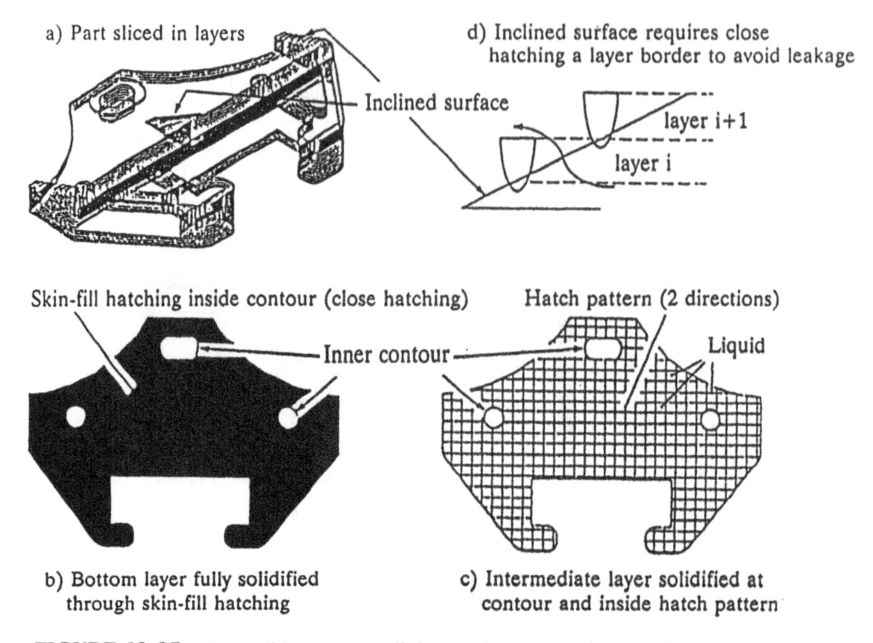

FIGURE 12.35 Stereolithography: slicing and scanning layers [56].

Some more recent equipment solidifies a whole layer at once instead of using point-by-point solidification. Illumination of a whole layer is often accomplished through a mask representing the cross section of the workpiece as shown in Fig. 12.34b. These masks are made from translucent photosensitive plastic foils by a photoplotter by charging electrostatically a glass plate with a toner as done in a photocopy machine. This allows the glass plate to be reused for successive masks since a new mask must be made and deposited for each different layer to be solidified. Recent innovations describe systems that eliminate the need for creating large numbers of foil masks by using programmable masks or lighting arrays that can be put in direct contact with the liquid monomer.

12.3.2 Solid Foil Polymerization

This process applies solid-to-solid polymerization rather than the liquid-to-solid used in stereolithography. Raw material consists of semipolymerized plastic foils progressively stacked on top of each other. Each separate foil is illuminated locally, which causes the illuminated parts to polymerize further and adhere to the foil underneath. Illuminated parts also become indissoluble, allowing the parts not exposed to be dissolved afterward, leaving the desired product.

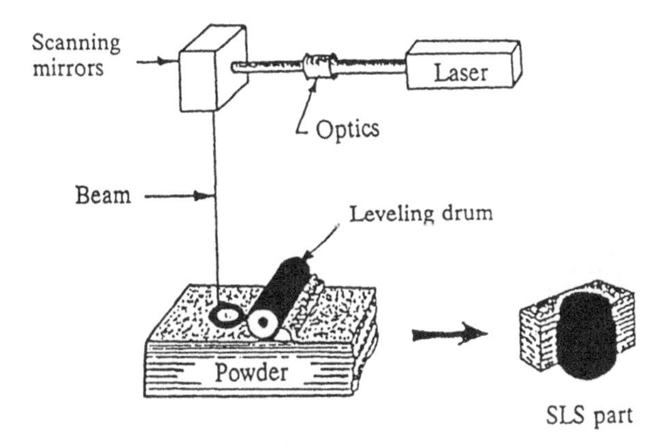

FIGURE 12.36 Principle of selective laser sintering [56].

12.3.3 Selective Laser Sintering

The layout of a selective laser sintering machine (SLS) (see Fig. 12.36) is similar to that of the point-by-point stereolithography machine: a laser beam scans and solidifies successive layers of the product. But in SLS processing, the liquid polymer is replaced by bulk powder material, preheated to a temperature slightly below its melting point. Selective solidification occurs by further heating to the sintering temperature by means of the XY controlled, pulsed laser beam. The powder that is not scanned by the laser is unaffected and remains in place to support the next layer of powder and possible overhangs of the product.

No binder material is needed and no postcuring is required except for ceramics, the application of which requires polymer-coated ceramics and post-sintering in an oven. Today's industrial applications involve thermoplastics and investment casting wax, and successful laboratory tests have been carried out with brass, copper, steel, and phosphate coated ceramics.

Overall accuracy of about 1%, layer thickness of 0.1 mm, and scanning speeds up to 1 m/s are typical values for the SLS process, which is used for a working envelope of up to 350 mm cubed.

12.3.4 Ballistic Particle Manufacturing (BPM)

Ballistic particle manufacturing produces parts by shooting droplets of molten material on top of each other (see Fig. 12.37). The droplets are produced by piezoelectric ink-jet printing nozzles generating droplets of about 50-μm diameter. The technique is primarily applied for creating wax models for investment casting without need for dies but could easily be extended to other materials with low melting and solidification points. A prototype machine able to deposit up to 1 kg of aluminum droplets per hour has been built.

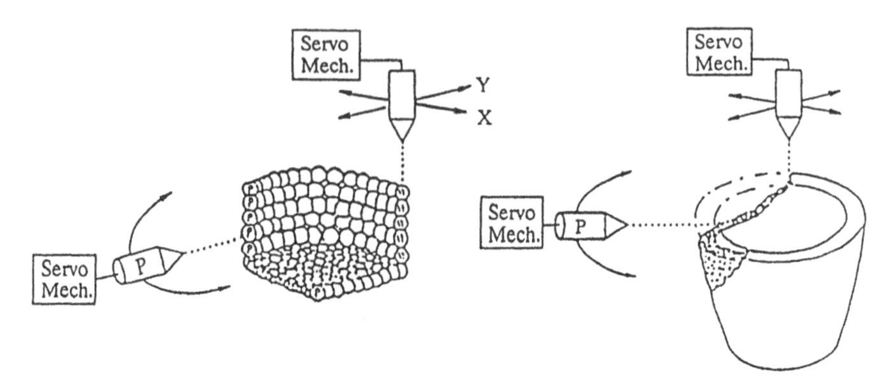

FIGURE 12.37 Ballistic Particle Manufacturing (BMP) [56].

A layer of material is created by moving the droplet gun in *X* and *Y* directions. Each time a layer is completed, the elevator moves downward (*Z*) and a new layer is deposited on top of the previous one. Printing heads with up to 32 parallel nozzles operating at 10 kHz each bring about high deposition rates, although certain problems exist in regard to the control of individual drops so that after collision, fusing, and curing, they are deposited exactly within the wanted geometry of the object.

A major advantage of BPM is that it permits application of different materials or colors in a single part. Working speed, accuracy, and so on, correspond to the values given for stereolithography.

12.3.5 Summary: Advantages and Disadvantages

Compared with the traditional or nontraditional processes based on removal or deforming of material, the processes based on the layer manufacturing technique possess a number of advantages and disadvantages.

Advantages

Geometrical Possibilities There are almost no limitations to the complexity of the parts: no problems of accessibility, no need for cores, and no problem of demolding cores and dies. The processes are well suited to tiny details, thin-walled products, or sculptured surfaces.

Multi-Material and Anisotropic Parts Most processes allow a change of material during the building process and thus permit creation of parts combining different materials, colors, and mechanical or thermal properties.

Fast Prototyping There is no need to develop special tools, molds, or dies.

Small Series of Parts This mainly applies to plastic products where the breakeven point justifying a mold still requires series of several thousands of components. Series in the order of tens or hundreds can be produced economically by layer manufacturing, especially when different parts are produced simultaneously, since it is then possible to divide the relatively long layer formation time over several products.

Disadvantages

Accuracy Since the object is formed by joining cross-sectional slices, only an approximation of the real object is obtained. The description is correct only at the slices, not between them. Obviously, the thinner the layers, the closer the approximation to the real object.

Materials Only materials with low melting points are applicable.

13

Manufacturing Systems

Up to this point in the book we have focused on capabilities of manufacturing processes as well as methods to optimize the individual processes. Normally, many manufacturing processes are involved in the fabrication of products. The productivity of the manufacturing process seen from a holistic point of view is then very much determined by how the production system is designed and how we plan and control the production activities.

This chapter presents some of the fundamental concepts and technologies in modern manufacturing systems within the mechanical industry, including discussion of state-of-the-art production equipment as well as developments in production planning and management philosophies.

13.1 THE FUNDAMENTALS OF MANUFACTURING SYSTEMS

As an introduction to the subsequent discussions of production systems and advanced manufacturing technologies it is useful to present a definition of the term *manufacturing system*. A manufacturing system can be defined as a series of value-adding manufacturing processes converting the raw materials into more useful forms and eventually finished products.

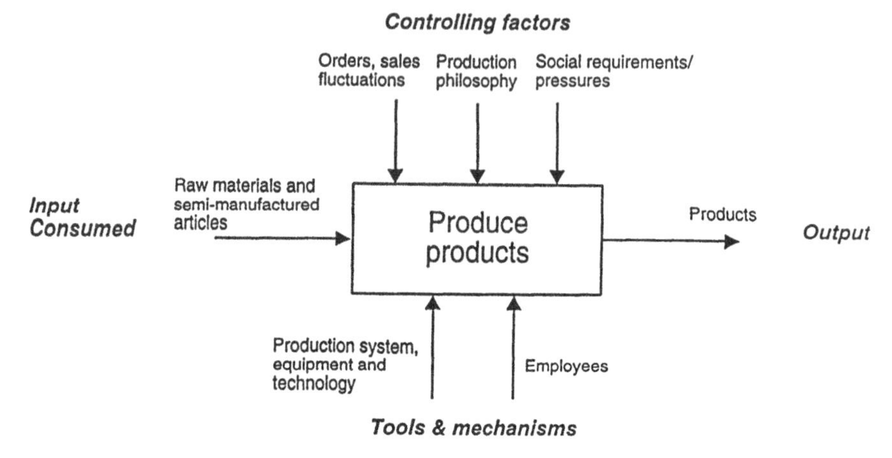

FIGURE 13.1 A model of manufacturing systems as a value-adding process.

Figure 13.1 illustrates basic means of manufacturing. The process transforms raw materials and semimanufacturing articles (input) to finished goods of a higher value (output). The basic mechanisms and tools used in carrying out this transformation are:

1. The production systems and manufacturing equipment involved in the physical flow and transformation of materials
2. The employees, directly involved in either manufacturing operations or management of production activities

The manufacturing process is mainly controlled by three factors (controlling factors in Figure 13.1):

1. The production orders determine the desired output from the production system, and fluctuations of the market often require adjustment of production plans. Frequent changes of production plans to respond to market fluctuations impose an increased need for flexibility in the production system.
2. The applied production philosophy contains the control and management principles that guide the way in which production is carried out.
3. Social requirements and pressures and conditions in the surrounding society are factors that need to be considered when establishing and managing production facilities. At present, environmental protection and improvement of working conditions are considerations that have a major impact on the development of production systems and manufacturing technology.

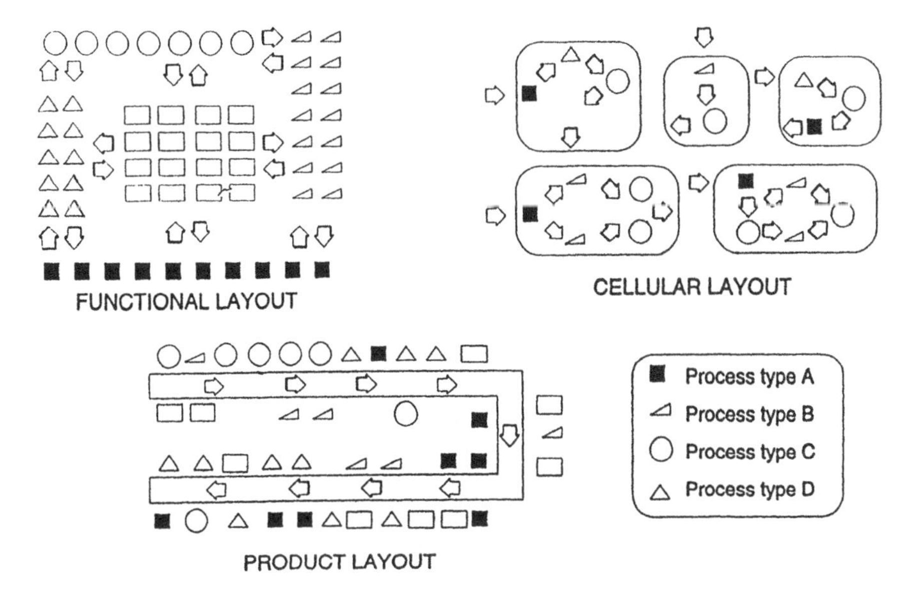

FIGURE 13.2 The basic production layouts.

In the rest of this chapter we will examine the basic manufacturing tools and mechanisms as well as the controlling aspects illustrated in Figure 13.1.

In the mechanical industry, production systems are categorized by the layout, that is, by how the manufacturing equipment is organized on the shop floor. Figure 13.2 shows the three basic layouts in their conceptual form:

Functional layout
Product layout
Cellular layout

In the *functional layout* production equipment is located in functional departments according to the manufacturing process. Typically, one sees a milling department, turning department, and so on. In production systems organized according by this principle, parts (orders) are transported from department to department depending on the actual process plan. The result is normally a large amount of work in progress and long production lead times. The advantage of and original reason for applying the functional layout are that knowledge and capacity are concentrated in a department where employees become specialists in carrying out a specific manufacturing process. Often, the functional layout is called the job-shop layout and production systems using the functional layout are called *job shops*.

In the *product layout* (or flow-line production) a dedicated production line is constructed to manufacture a specific product. This is the layout principle developed by Henry Ford in the automotive industry, which revolutionized industrial manufacturing. The basic principle is that production is divided into the smallest possible operations, machines and workers are positioned in lines, and the products move from workplace to workplace where a minor operation is done on the product. The advantage of the product line is the high productivity achieved with a small level of work in progress. The drawbacks of the product line include limited flexibility concerning changing to other types of products and the vulnerability of the production system to breakdowns and other stoppages in the line. Breakdowns will stop the whole production line because only minor buffers are placed between the workplaces.

The *cellular layout* was developed to combine some of the advantages of the functional layout and the product layout, that is, flexibility and productivity. When implementing the cellular layout the group technology concept is used.

Group technology (GT) is the concept of gathering parts into a number of groups based on similarities of the parts. Figure 13.3 shows a conceptual illustration of GT. The criteria applied for classifying parts into part families are usually similarities of shape, materials, and manufacturing processes. For such a family of parts a manufacturing cell can be established containing all the processes needed to finish production of parts in the specific family. The group technology cells can be divided into two categories depending on the flow of parts and materials within the cell [64]:

1. *GT manufacturing cells* are stand-alone machines of dissimilar functional types grouped together and dedicated to producing a family of parts. In the GT manufacturing cell there is no fixed flow of materials; in essence, each cell is a small dedicated job shop.
2. *GT flow-line cells* are configurated as small flow lines dedicated to producing a family of parts. The continuous flow of parts within the GT flow-line cells provides the efficiency of product lines in a low- or medium-volume manufacturing environment.

Compared to the functional layout, cellular manufacturing typically has major advantages:

1. A 70–90% reduction in production lead time and work in progress.
2. A 75–90% reduction in material handling.
3. A 65–80% reduction in setup times because of the similarities of the parts in a family.
4. A 50–80% reduction in quality-related problems.

Furthermore, the shop-floor control task is simplified and decentralized, which raises job satisfaction among employees on the shop floor. The disadvantages and limitations of cellular manufacturing are:

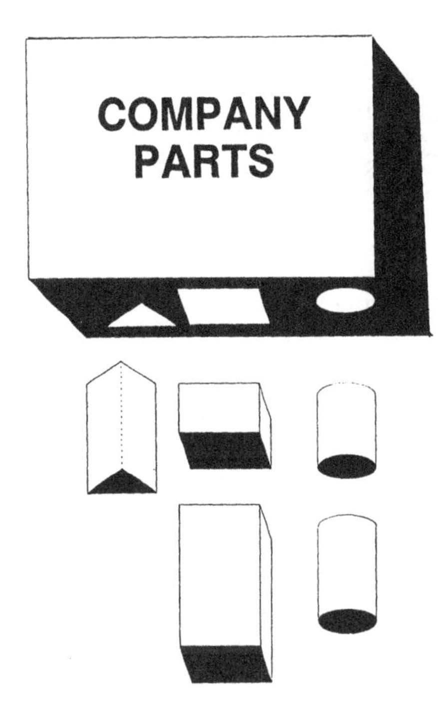

FIGURE 13.3 The concept of group technology.

1. Machine utilization decreases.
2. Flexibility is not as great as in the functional layout, and all parts have to be standardized and designed into a GT family, which is not always is possible.
3. The education, motivation, and engagement of shop-floor employees are essential for achieving the outlined benefits.

Cellular manufacturing has been one of the key technologies in the Japanese philosophy of just-in-time (JIT) manufacturing. In a later section further attention will be devoted to Japanese production philosophy as well as to how cellular manufacturing can improve production planning and control.

Figure 13.4 shows the relationship between part volume and flexibility of the production system for these basic layout configurations. Besides the discussed basic layouts, it also illustrates how FMS (flexible manufacturing systems) can combine the strengths of product lines and job-shop manufacturing better than the cellular layout. FMSs are highly automated computer-controlled manufacturing facilities, which because of automation can produce small batches with

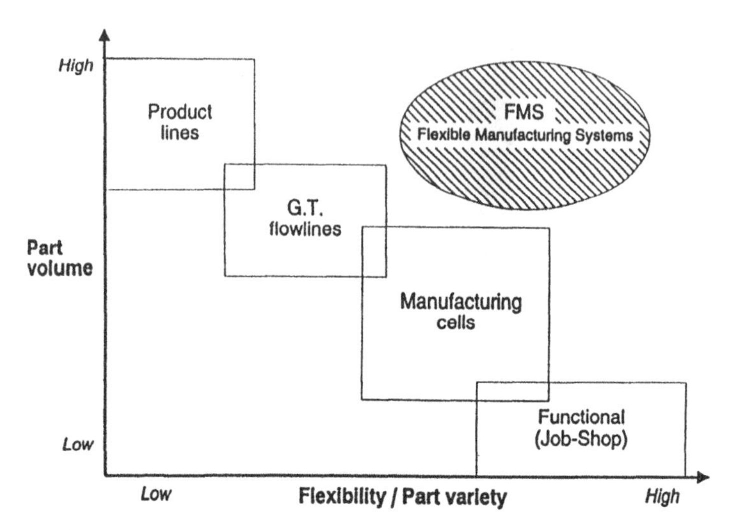

FIGURE 13.4 Part volume and flexibility relationship for the different types of layout configurations.

the efficiency of mass production. FMS technology will be discussed in a later section of this chapter.

13.2 ADVANCED PRODUCTION EQUIPMENT

The introduction of low-cost microcomputers has in the last decades totally changed the shop floor. However, the application of computers to manufacturing has been somewhat slow, compared to other application areas. This section will discuss the facilities and capabilities of modern production equipment. The discussion will focus on two areas, machine tools and industrial robots, because it is in these areas that the major developments have been made and new manufacturing opportunities are opening up for the mechanical industry.

13.2.1 Advanced Machine Tools

Conventional manual machine tools have problems in manufacturing complex and accurate parts. This started the development of NC (numerical controlled) machine tools. By 1952 the first NC machine, with three controlled axes, was developed at the Massachusetts Institute of Technology (MIT). In the 1960s NC machines reached a stage where they became reasonably reliable and productive. These first NC systems used electronic hardware, based upon digital circuit technology. The next improvement was the CNC (computer numerical controlled) systems, which were introduced in the early 1970s. CNC systems employ a microcomputer and eliminate, as far as possible, additional hardware circuits in the machine controller. This development from hardware-based NC to software-based CNC increased flexibility and provided the possibility of introducing new features in the machine control unit. Some of the advantages and new features are:

Ease of updating the controller software, which securing the investment in CNC machines
Correction of the part programs directly on the CNC
Built-in machining cycles, for example, for milling of pockets
Graphical simulation of the tool path on the CNC display
Sophisticated monitoring of the machine tool; generation of alarm and status messages to the operator
Control of auxiliary equipment such as manipulators and pallet exchangers
Management of tool data and tool life

To improve productivity, most modern CNC machining centers have a tool magazine typically containing from 20 to more than a hundred tools which prolong the time for unmanned operation. A built-in tool manipulator can automatically change the tool during production. To gain the greatest advantage from having large tool magazines, advanced facilities are required in the CNC controller to manage all the information related to each tool. Modern CNCs can therefore handle a lot of tool information, such as:

Tool offset
Tool nose radius
Tool placement and tool groups
Tool life
Remaining tool life

The management of tool data is a comprehensive task because it is not unusual for there to be more tools than parts on the shop floor and each job on a machine tool requires a specific setup of tools. Often, the tool data has to be put in manually at the CNC controller and is stored in the memory of the CNC system. The disadvantage of this is that tool data has to be loaded into the CNC each time the tooling is changed. A much nicer solution is to integrate a microchip into the tool base, which can store all the specific tool information. Then the microchip can be programmed with all the tool information when the tool is assembled and measured, activities which typically take place in a tool preparation room isolated from the shop floor. When the tool is placed in the tool holder the CNC can read the tool data directly from the microchip as well as updating the microchip with information such as the remaining lifetime of the tool. Thus all relevant tool information always follows the tool, and tool management is made less complicated. Many controllers can operate with "sister" tools, identical tools that can replace each other. When the lifetime has expired, the controller automatically exchanges the tool with a "sister" tool, which can continue the process.

Automation of the loading and unloading of parts is another feature often used to improve the productivity of machine tools. Most vendors of machining centers offer a pallet magazine as an extension to the machine. The operator can then prepare the fixturing of raw parts on an empty pallet outside the machine while manufacturing another part. These pallet magazines typically consist of two to ten pallets that contain work for several hours of unmanned manufacturing. CNC turning centers can be enhanced with a loading robot taking out the finished parts and loading the machine with new raw parts. The CNC controller of such machine tools can be programmed to operate according to a cyclic schedule, loading parts, executing programs, unloading parts, and so on. Figure 13.5 shows a small two-pallet changer system allowing workpiece setup during machining.

Today the programming of machine tools is most often done off-line using CAD/CAM systems. CAD (computer aided design) systems are computer-based drawing systems for product modeling. CAM (computer aided manufacturing) systems are computer-based systems on which the part programs for CNC machine tools can be generated on the basis of a CAD model of the part to be manufactured. The application of CAD/CAM for off-line programming of CNCs has two important advantages:

1. It improves the productivity of the machine tool by not taking up the machine while developing the part programs.
2. It enables development of programs for complex three-dimensional operations that were "impossible" to program directly on the machine controller.

The transfer of NC programs can either be done on tapes, read by a punch reader at the CNC, or transferred electronically through a DNC (direct numerical control) communication link between a computer and the CNC. It is convenient to distinguish between two categories of DNC communications:

"Simple" DNC
"Full" DNC protocol

A "*simple*" *DNC* is a communication interface based on the punch-reader interface available on almost all CNC controllers. Using this type of interface it is possible to upload and download NC programs between a computer and the CNC. The "simple" DNC normally uses serial communication, as used when connecting a plotter or a printer to a computer. This itself, has some disadvantages:

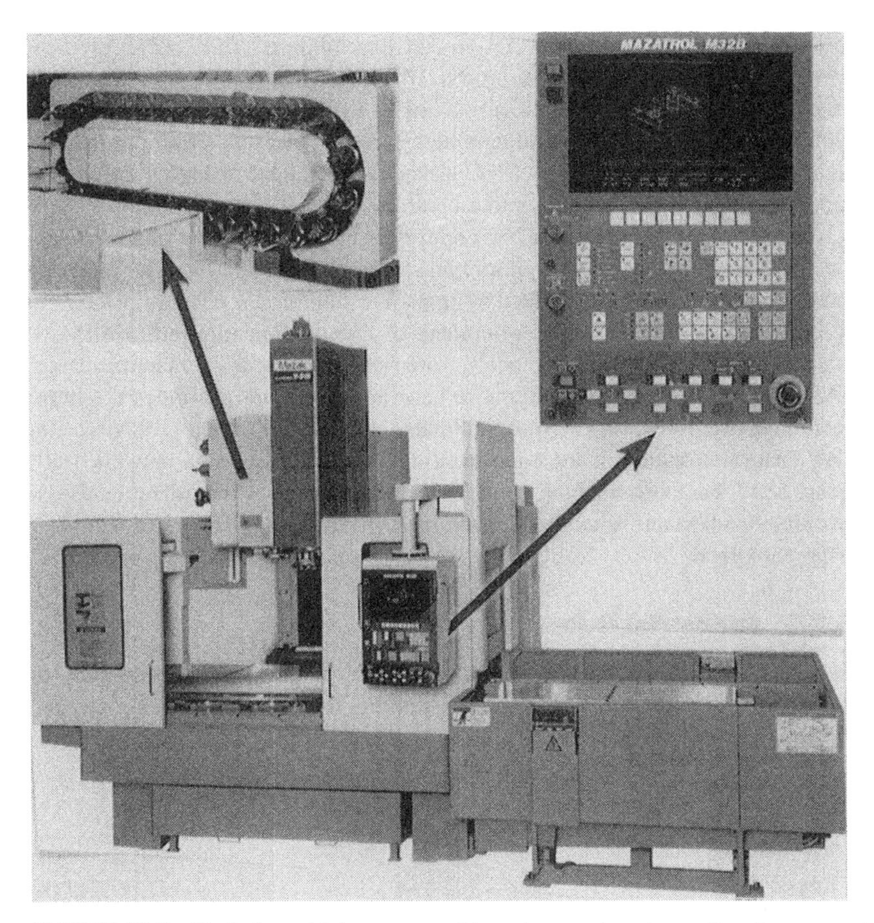

FIGURE 13.5 Typical machining center with tool magazine, two-pallet changer, and high-level CNC control. (Courtesy, SINDAL Machiine Tools A/S.)

1. The communication speed is limited; transmission of large programs can take several minutes.
2. The communication has a problem coping with the electromagnetic noise in a shop-floor environment. Furthermore the communication protocol has very limited error-correcting facilities.

A "*full*" DNC *protocol* is an advanced communication protocol enabling the computer not only to upload and download programs but also to perform full remote control of the CNC controller. Normally, this type of DNC protocol enables performance of all the functions that are available on the controller itself from a computer. Normally, an option card providing a "full" DNC protocol can be purchased as an extension for the controller. These DNC protocols either can be based on a serial communication link as the "simple" DNC or can apply a communication network providing a higher communication speed. At the moment the application of "full" DNC protocols is limited to highly automated manufacturing systems where remote control of the CNC is necessary.

At present most vendors of CNC controllers have a proprietary communication protocol. This causes many problems for companies that want to integrate machine tools into more integrated systems. For that reason, General Motors in the early 1980s initiated the development of a communication protocol MAP (manufacturing automation protocol), satisfying the needs of manufacturing. MAP specifies a set of instructions to be available on various types of equipment; machine tools, PLCs (programmable logical controllers), robots, etc. MAP enables coupling of the equipment directly to a local area network [62]. Today, MAP has been adapted as an ISO (International Standardization Organization) standard and is available from most of the major vendors of manufacturing equipment.

13.2.2 Industrial Robots

The technology of industrial robots has matured, and robots are beginning to revolutionize the industry. Robots are particularly useful in applications such as:

Material handling
Spray painting
Spot welding
Arc welding
Inspection
Assembly

It is difficult to define what is a robot and what is not. Many definitions are presented in the literature. The Robot Institute of America (RIA) defines an industrial robot as a "re-programmable multi-functional manipulator designed to move materials, parts, tools, or other specialized devices through variable programmed motions for the performance of a variety of tasks." The technology of robots is related to NC technology but differs somewhat, since robots include higher velocities and movement in more axes. Basically, the robot needs six axes of motion (or degrees of freedom) to reach a point with a specific orientation in space. Industrial robots can be classified into three generations. The first-generation robots follow relatively simple control strategies and are often referred to as "pick-and-place" robots. The feedback devices of these robots are simply pairs of limit switches and stoppers for each axes of motion. The second-generation robots are generally computerized devices controlled in closed loops

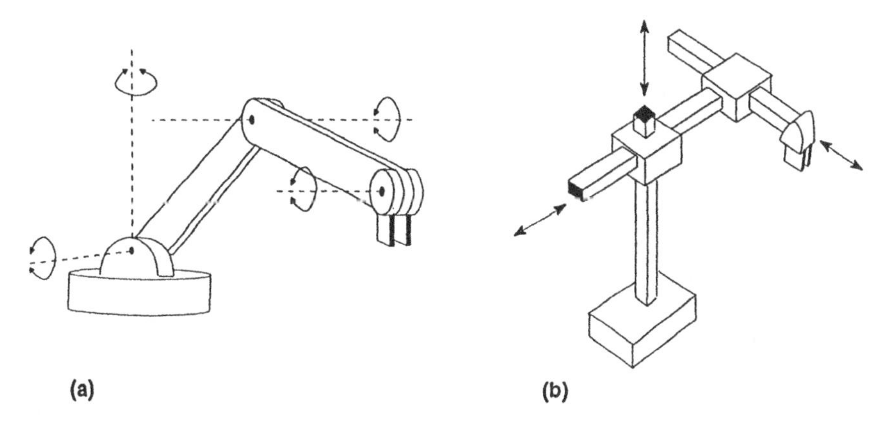

FIGURE 13.6 Two of the basic robot manipulators. (a) The articulated manipulator. (b) The Cartesian manipulator.

by servo-drivers. These robots have a high level of programming and control flexibility. The third-generation robots are more "intelligent" devices capable of making decisions and generating unprogrammed motions based upon information captured by sensors such as vision cameras, force and pressure transducers, and so on. It is second- and third-generation robots that are usually referred to as industrial robots.

The appearance of an industrial robot can differ very much from robot to robot. Structurally the robots can be classified according to the coordinate system of the main frame (the manipulator or arm), which has three axes of motion. Figure 13.6 illustrates two of the four basic coordinate systems in industrial robots:

- The cartesian manipulator consists of three orthogonal linear sliding axes, as shown in Figure 13.6.
- The cylindrical manipulator consists of two linear axes and one rotary axis. Typically a horizontal column is mounted on a vertical column, which in turn, is mounted on a rotary base. Thus the working volume is the annular space of a cylinder.
- The spherical coordinate robot consists of one linear and two rotary axes.
- The articulated manipulator consists of three rotary axes, as in the human arm (see Fig. 13.6). This type of robot has a relatively low resolution and accuracy. On the other hand, it has excellent mechanical flexibility and moves with high speed, which help make it the most popular medium-sized robot on the market.

Most industrial robots are used for automating repetitive processes such as spray painting, spot welding, deburring, and so on, and many robots are never reprogrammed. Therefore, only a few robot applications fully utilize the flexibility of the robots. The programming of robots can be done in several ways; at least four programming methods are commonly used in robotics:

Manual teaching
Lead-through teaching
Programming languages
Computer-aided off-line programming

Manual teaching is the simplest and most frequently used in point-to-point robotics systems. Teaching is done by moving each axis of the robot manually to the desired position. When the desired position is reached, the operator stores the coordinates on the point in the robot memory.

In *lead-through teaching* the end-effector (tool) is manually guided through the desired path at the required speed, while simultaneously recording the position of the axes. This teaching method is widely used in spray painting of auto parts or other products.

Programming languages are high-level descriptive languages enabling one to describe the movement of the end-effector in order to perform a specific task. These languages are very similar to the languages known in the NC programming.

Computer-aided off-line programming is the generation of robot programs by a computer, based on a geometric model of the part and the environment in which the specific task is going to be carried out. These off-line programming packages (e.g., ROBCAD) are very similar to the CAM systems for machine tool programming and they can also be integrated with the CAD systems, in which products and facility layouts are modeled.

Development in robotics is heavily focused on creating a "smart" robot that can hear, see, and touch and consequently make decisions based on the artificial sense organs. In principle the goal is to achieve the human's ability to sense and to react to sense impressions. In arc welding much research and development have been carried out to develop efficient sensor systems for searching/tracking the many types of joints as well as continuously securing a high-quality welding seam. These systems can apply techniques as different as vision systems for localization and measurement or dedicated laser-based tracking systems for arc welding.

An industrial case story showing many state-of-the-art applications of robotics is found at Odense Steel Shipyard Ltd. in Denmark. The company produces cargo carriers and oil tankers. At Odense Steel Shipyard automation of arc welding by industrial robots has been an important goal given considerable attention. This has positioned the shipyard as a world leader in application as well as research in the use of industrial robots for arc welding. Several circumstances complicate the use of robotics for arc welding in the shipbuilding industry:

- The part cannot be moved during the process; the robot must move relative to the ship section being manufactured.
- The variance in quality of joints provides a need for advanced sensor systems for on-line process adjustments.
- Because very few welding seams are identical, the number of robot programs becomes enormous and off-line programming is necessary if satisfactory utilization of the robots is to be achieved.

The ships are modeled on a three-dimensional CAD system, HICADEC, an advanced CAD system for shipbuilding on which complete three-dimensional models can be generated and checked for design errors. From the CAD model an IGES processor generates an IGES (Initial Graphics Exchange Specifications) file of the section to be welded. Using the off-line programming and robot simulation software, ROBCAD, it is tested whether or not a joint can be welded using the robots available in production. If it is possible, the programs are generated and verified using the simulation software, or the design of the ship section can be changed to enable robot welding.

Today, Odense Steel Shipyard has about 30 arc welding robots used in daily production. Typically, two or three of the robots are mounted on a gantry from which they can be lowered into the different sections and cavities to be welded. Figure 13.7 shows such an installation in action with two Hirobo robots mounted on a gantry—both robots are welding on large ship sections and are operated by one worker. Figure 13.8 shows a close-up of one of the Hirobo welding robots in action.

13.3 FLEXIBLE MANUFACTURING SYSTEMS

13.3.1 The Elements of Flexible Manufacturing Systems

The purpose of flexible manufacturing systems (FMS) is to achieve the efficiency of mass production (product lines) for batch production (see Figure 13.4). A FMS typically encompasses:

- Process equipment e.g., machine tools, assembly stations, and robots
- Material handling equipment e.g., robots, conveyors, and AGVs (automated guided vehicles)
- A communication system
- A computer control system

The name FMS does not directly say that the system should be fully automated and computer controlled since there are a number of other means for being flexible in manufacturing. But over the years FMS has become the acronym for the

FIGURE 13.7 Picture of robots mounted on a gantry, welding on a large ship section at Odense Steel Shipyard. (Courtesy, Odense Steel Shipyard.)

fully automated manufacturing system as well, as it represents a vision of the factory of the future. FMS is not really new. An installation established in England during 1968 is regarded as the first FMS and was developed by D. T. N. Williamson, who was working for the Molins Machine Tool Company. But developments in computer and advanced machine tools have vitalized the FMS concept and today it is easy to find examples of reliable "true" FMS in industry. In Fig. 13.9 the layout and components of a typical FMS are illustrated. At the *loading area* parts are mounted on fixtures and introduced to the system. The use of fixtures enables the transportation equipment to handle a large number of different parts within the system.

Transport of materials and palettes is essential to an unmanned and flexible production. In most systems it is carried out by an AGV, which is a computer controlled vehicle following predefined paths on the shop floor. The most common AGV is wire-guided, controlled by the magnetic field from electrical wires concreted into the shop floor. Some systems use conveyors for transporting palettes in the FMS. Conveyors are less flexible than the AGV but, on the other hand, they are more reliable than the more sophisticated AGVs, which are sensitive to chips and oil on the floor. Robots are normally not used as trans-

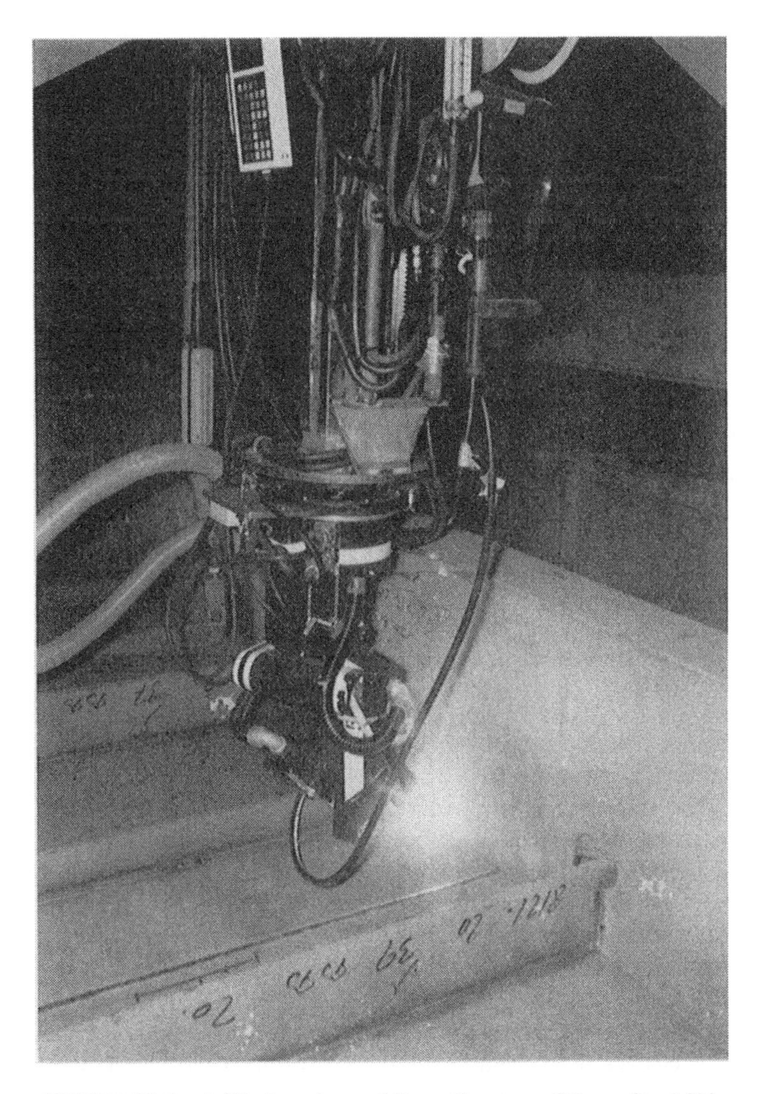

FIGURE 13.8 A Hirobo robot welding. (Courtesy, Odense Steel Shipyard.)

porters in FMS because of the limited lifting capacity and radius of action. But robots are used in FMSs for feeding machine tools, assembly, painting, or similar operations.

The *machining area* may contain from two to an "unlimited" number of CNC machines. The majority of CNCs in FMS are machining centers (milling and drilling) because it is easy to load/unload the machines with palettes directly

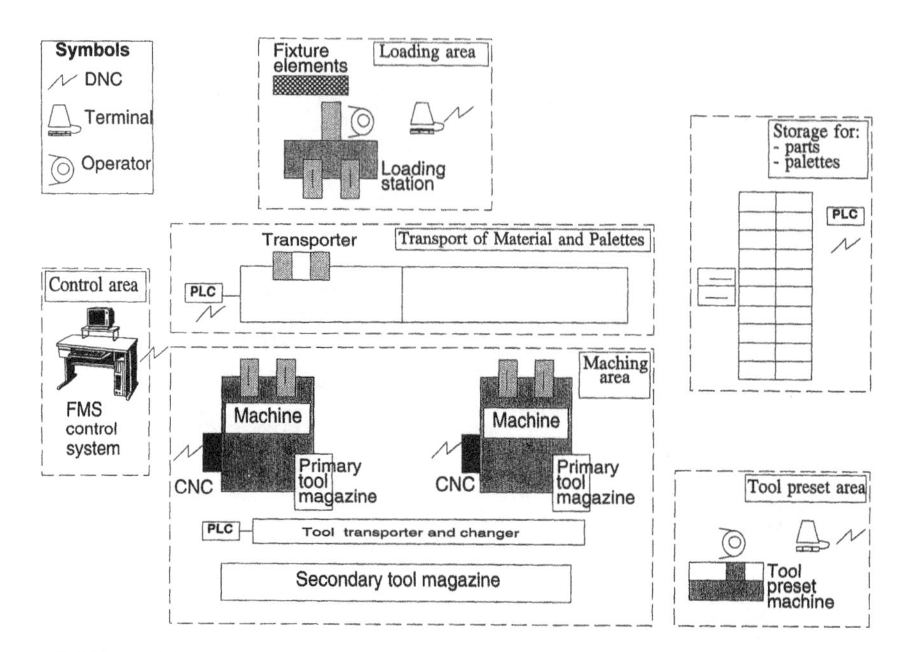

FIGURE 13.9 The components in a typical Flexible Manufacturing System.

from the transporter. Turning centers can in some cases also appear in a FMS. Here palettes of turning parts are moved to a fixed position near the machine where a loading and unloading robot can take the parts one by one and feed them into the turning center. Other types of equipment such as coordinate measuring machines, washing machines, and so on, can also be found. On the machine tool there is a primary tool magazine normally containing 40 to 100 tools. When this is not sufficient, a secondary tool magazine is installed with a tool transporter and changer enabling automatic tool exchange and substitution during production (see Figure 13.9).

An *automated storage and retrieval system* (AS/RS) is very often integrated in the FMS. This enables the system to run one or two unmanned shifts with reduced workforce. In the day shift raw parts are fixtured on palettes loaded into the system and stored in the AS/RS waiting for available capacity on the machines required. When the process is finished, the transporter can move the part back to the AS/RS where it waits for the next process or to be unloaded from the FMS when the workforce reports for work next morning. The fully unmanned operation of FMSs is very rare; in most cases one or more operators will always be present, not so much to intervene in critical situations (this is normally done automatically by the computer control system) but to correct minor faults and keep the system running with highest possible utilization.

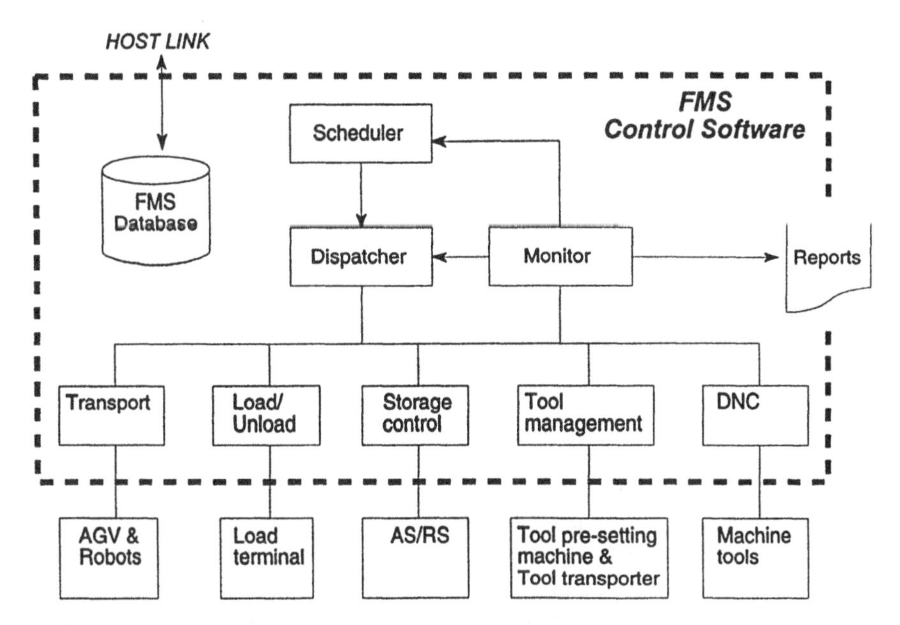

FIGURE 13.10 Diagram of functions in advanced FMS control software.

The *control area* with the computer running the FMS control system is the center from which all activities in the FMS are controlled and monitored. The FMS control software is rather complicated and sophisticated since it has to carry out many different task simultaneously. Despite the considerable research that has been carried out in this area, there is no general answer to designing the functions and architecture of FMS software.

Some typical functions of a FMS control system are illustrated in Fig. 13.10. The *scheduler* function involves planning how to produce the current volume of orders in the FMS, considering the current status of machine tools, work-in-progress, tooling, fixtures, and so on. The scheduling can be done automatically or can be assisted by an operator. Most FMS control systems combine automatic and manual scheduling; the system generates an initial schedule that then can be changed manually by the operator. The *dispatcher* function involves carrying out the schedule and coordinating the activities on the shop floor, that is, deciding when and where to transport a pallet, when to start a process on a machining center, and so on. The *monitor* function is concerned with monitoring work progress, machine status, alarm messages, and so on, and providing input to the scheduler and dispatcher as well as generating various production reports and alarm messages. A *transport* control module manages the transportation of parts and palettes within the system. Having an AGV system with multiple vehicles, the routing control logic can become rather sophisticated and become a critical part of the FMS control software. A *load/unload* module with a terminal at the loading area shows the operators which parts to introduce to the system and enables him or her to update the status of the control system when parts are ready for collection at the loading area. A *storage control* module keeps an account of which parts are stored in the AS/RS as well as their exact location. The *tool management* module keeps an account of all relevant tool data and the actual location of tools in the FMS. Tool management can be rather comprehensive since the number of tools normally exceeds the number of parts in the system, and furthermore the module must control the preparation and flow of tools. The *DNC* function provides interfaces between the FMS control program and machine tools and devices on the shop floor. The DNC capabilities of the shop-floor equipment are essential to a FMS; a "full" DNC communication protocol enabling remote control of the machines is required (see the discussion on DNC in the previous section).

The fact that most vendors of machine tools have developed proprietary communication protocols is complicating the development and integration of FMSs including multivendor equipment. Furthermore, the physical integration of multivendor equipment is difficult; for example, the differences in pallet load/unload mechanisms complicate the use of machine tools from different vendors. Therefore, the only advisable approach for implementing a FMS is to purchase a turn-key system from one of the main machine tool manufacturers. These systems are reliable and well tested and should the system not function satisfactorily a single vendor responsibility will facilitate remedy of malfunctions. Fig. 13.11 shows a picture of a turn-key Mazak FMS, which is built on a modular concept, enabling the system to be configured with the number of machines required by the customer.

13.3.2 FMS at Fanuc

Most of the large FMS installations in industry are found at the machine tool manufacturers, undoubtedly because they are mastering all the technologies involved and need the installations as working showrooms of the capabilities of their highly sophisticated automation products. Fanuc Ltd. is one of the leading companies in manufacturing of machine tools as well as factory automation products. The Fanuc company of Japan is also one of the automation success stories of our time. They have several highly automated factories applying FMS to manufacture their own products.

To illustrate the state-of-the-art in FMS, Fig. 13.12 shows one of the Fanuc manufacturing plants in Japan, the Machining Factory. This factory has a machining capacity of parts for some 1100 Fanuc products, such as robots and wire-cut EDMs, each month. The Machining Factory was started in 1980 as the first facility in Japan to adopt a FMS, which enabled unmanned operation during the

FIGURE 13.11 A Mazak turn-key FMS with tool management and an integrated transportation and storage system. (Courtesy, SINDAL Machine tools A/S.)

night. The system later on was enhanced with five of the latest "Fanuc Cell 60" controllers, which has enabled continuous unmanned operation for 72 hours, even through weekends.

Fanuc's motor factory is another highly automated manufacturing facility (see Fig. 13.13). It has a capacity of 25,000 motors for some 500 models per month and is located within a two-floor building. On the lower floor is the machining area, comprising about 60 machining cells with over 50 robots. On the second floor, assembly operations are carried out. The two floors are linked together with an automatic warehouse system that is used to store raw materials and machined components, as well as transporting components between the two floors. Materials stored in the automatic warehouse are supplied to each machining cell, by AGVs, and unmanned machining is carried out at night using handling robots.

Finally, a look is taken at the Fanuc Injection Modeling Machine Factory (Fig. 13.14). This factory is capable of manufacturing 100 plastic injection molding machines each month. This factory is also capable of carrying out unmanned operation, but it is interesting mainly because the machining department features the latest facilities with MAP, as the communication protocol on the shop floor. Machine operation monitoring, as well as information on machining output and NC data, and so on, are available through MAP interfaces at the machine controllers. This proves Fanuc's commitment to the international ISO MAP standard.

To illustrate the potential in automation, the operating revenue of Fanuc Ltd. in 1972 was $51 million, and it had risen to $1378 million in 1990. For the same period the number of employees only rose from 615 to 1810. This is an outstand-

FIGURE 13.12 Inside Fanuc Machining Factory producing parts for some 1100 Fanuc products each month. (Courtesy, GE Fanuc Automation Europe S.A.)

FIGURE 13.13 Inside Fanuc's motor factory. (Courtesy, GE Fanuc Automation Europe S.A.)

ing achievement and is largely due to the successful application of advanced technology within production and products.

13.3.3 The Hierarchical Nature of Production Control

When describing and discussing production control and factory automation in advanced manufacturing systems applying FMS and CIM (computer integrated manufacturing) technologies it is convenient to put the many planning and control activities in a hierarchical perspective, with the overall strategic planning at the top and the operational control of manufacturing processes at the bottom. There are several candidates for such a hierarchical model (see Fig. 13.15). The two most important are formulated by NBS (National Bureau of Standards) now called NIST (National Institute of Standards and Technology) in the United States (the NBS-model) and by ISO (International Standardizations Organization), which proposed an international standards for the definition of control levels in advanced manufacturing systems (the ISO-model) [59]. The NBS-model recognizes five levels, namely, facility, shop, cell, workstation, and equipment. The ISO-model has added a layer and includes six levels in their model, namely, enterprise, facility/plant, section/area, cell, workstation, equipment. These hierarchical models have mainly been used as frames of references in planning and implementation of computer integrated manufacturing systems, but they are also applicable for discussing production planning and control activities in general.

It is relevant to discuss the definition of manufacturing control levels because these include some of the terms often used in the literature and in descriptions of commercial products for factory automation. In Fig. 13.15 an illustration of the hierarchical structure is presented according to the ISO-model. It illustrates the fundamental characteristics, namely, that each control function (at level n) controls a minor set of functions at the lower level (level $n - 1$). The typical tasks and responsibilities at the different levels are as follows:

- *Enterprise control* includes the overall strategic planning for the enterprise. This is product portfolio planning, market strategy, and division of work between divisions in the enterprise. The manufacturing control carried out at the enterprise level is responsible for achievement of the mission of the enterprise; its planning horizon is measured in years and should not be changed too often.
- *Facility control* is responsible for implementing the enterprise strategy. It includes such functions as manufacturing and product engineering, information management, production management, and other long-term activities in controlling and running a manufacturing facility.
- *Area control* is responsible for the allocation of resources and the coordination of production on the shop floor. This control level typically operates within a horizon of several days or weeks. This level is also often

FIGURE 13.14 Inside the machining department of Fanuc Injection Molding Machine Factory. (Courtesy, GE Fanuc Automation Europe S.A.)

titled "shop level" and the term "shop-floor control" covers the area control and the levels below. In U.S. literature shop-floor control is often called "production activity control." In flexible manufacturing systems the FMS control software will carry out the area control.

- *Cell control* is responsible for the direction of jobs through the workstations within the manufacturing cell. This includes resource assignment, making decisions on job routings, dispatching jobs to individual workstations, and monitoring of jobs and the working condition of workstations.
- *Workstation control* is responsible for the coordination of tasks to be carried out on a workstation to perform a job assigned to the workstation. This function operates within a time horizon of milliseconds to hours.
- *Equipment level* realizes the physical execution of tasks on a machine. In relation to a machine tool, this level will typically be the local machine controller controlling movement of axes, spindle speed, coolant, and so on.

13.4 COMPUTER INTEGRATED MANUFACTURING

The concept of computer integrated manufacturing (CIM) was developed in parallel with the FMS concept, and often the two concepts are mixed up. CIM is not a product or a system that can be purchased, and it is therefore less well

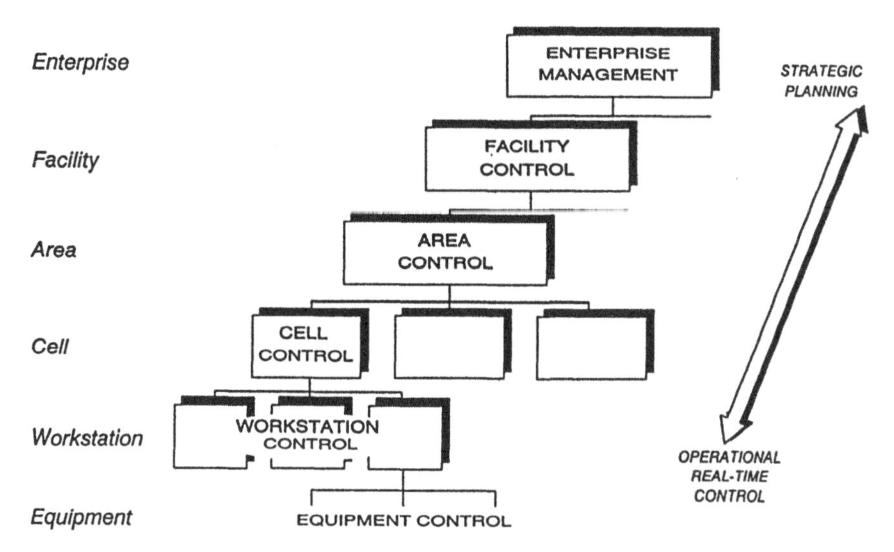

FIGURE 13.15 A hierarchical perspective of the manufacturing control model.

defined than FMS. Here, CIM is defined as *a strategy of using information technology in a strategic manner to integrate manufacturing functions and improve the flow of information*. CIM has been one of the most discussed topics in manufacturing spheres during the 1980s. Today, CIM is attracting less attention because it has become a matter of course to apply computers and information technology in almost all manufacturing functions. CIM has to a wide extent been succeeded by concepts such as concurrent engineering, which basically addresses the same problems: to improve and speed up the flow of information within the manufacturing organization. But concurrent engineering focuses more on the organizational and product design aspects of integration than the application and interfacing of computer systems.

There are many different points of views on what CIM is. In practice, implementation of CIM, a computer integrated manufacturing system, ends up with a company-specific system. Looking at manufacturing enterprises within the mechanical and electromechanical industry, there are some typical main areas in CIM implementation. Figure 13.16 illustrates the main areas which have to be integrated with each other to obtain the designation CIM.

The business information system covers all factory-level planning and control activities: production planning, inventory control, purchasing, financial management, and so on. These activities were the first to utilize information technology. Because business administration area tends to be conservative, many of the systems implemented were a direct computerization of old paper-based

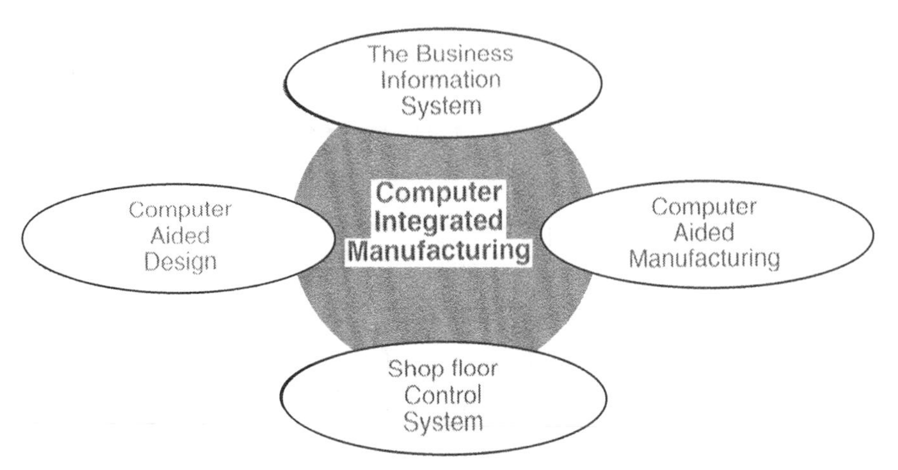

FIGURE 13.16 The main areas within the CIM concept [63].

management systems. A major enhancement came with the development of computer-based materials planning systems (MRP systems) introducing new possibilities for planning and controlling large complex manufacturing enterprises (see the next section).

Computer-aided design has been in a period of tremendous development during the last 15 years and with the appearance of low-cost workstations and personal computers, CAD technology has been made available for all sizes of companies. The big market for CAD systems has stimulated the development of software which today has reached a very powerful stage, offering advanced modules for, for example, finite element and kinematic analysis.

The area of *Computer-aided manufacturing* (CAM) has developed in close connection with CAD and today offers good modules for off-line NC and robot programming, shortening the running-in time for new products in production, as well as it has made it possible to generate part programs for geometries (e.g., three-dimensional surfaces) that could not have been programmed manually. The area of CAM also embraces computer-aided process planning (CAPP)—computer systems supporting the process planner in selecting the sequence of processes and process parameters needed to manufacture a product.

The area of *shop-floor control* systems embraces a wide range of systems varying from fully automated FMSs to manual shop-floor information systems using computer terminals and bar-code technology to integrate the shop floor with the overall production management system.

The most important computer technologies for integrating the manufacturing activities into a CIM system are computer networks and database technology. Communication networks enable us to distribute applications and exchange in-

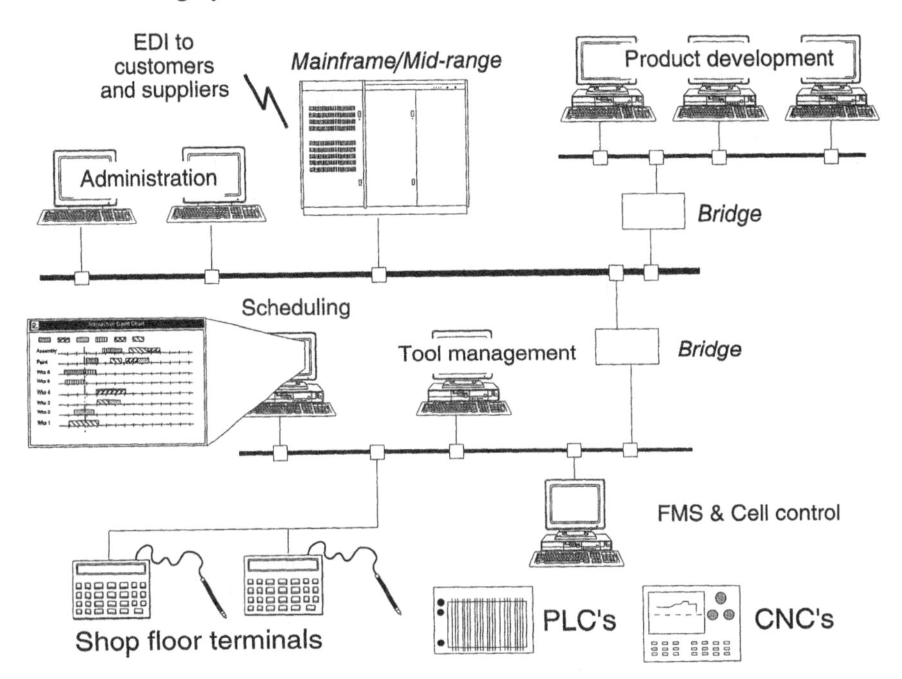

FIGURE 13.17 The computer architecture of a CIM enterprise.

formation throughout the enterprise. Figure 13.17 illustrates a typical computer architecture for a CIM enterprise where most of the manufacturing activities are carried out using various computer support tools. As illustrated, the enterprise is fully network integrated, enabling the different distributed computer systems to exchange data and access the common database. Interface problems and lack of communication standards have represented the main difficulties in realizing the CIM enterprise. The development of international ISO standards for open systems interconnections (OSI) will make it easier to integrate computer systems in the future. The aim of these standards is to ensure an unproblematic integration of computer hardware and software from different vendors by specifying standards for communication network interfaces and protocols.

The database technology is important, because databases store information in a structured and secure manner. Common databases make information available wherever it is needed and ensure that decisions are made on the newest available information. The design and implementation of the common database is a key issue in the realization of CIM.

Integration of manufacturing activities is not just a question of integrating the activities within the enterprise itself. All enterprises have extensive communication with external organizations (e.g., customer, vendors, forwarding agencies, and taxation authorities). This communication with external organizations can entail a considerable delay in business transactions as well as be a potential source of errors. Many companies see EDI (electronic data interchange) as a solution to these problems and a means of improving communication with customers and vendors. EDI contains standards for electronic exchange of most of the forms and information exchanged between business partners (e.g., requisitions, orders, inviting quotations, invoices). In most countries EDI service agencies take care of telecommunication and ensure that messages are distributed to the receiver. EDI is a natural extension of CIM, and the development of international EDI standards will provide new possibilities for international trade free from language and legal barriers.

13.5 EFFICIENT MANUFACTURING

Until now, this chapter has focused on the mechanisms (from Fig. 13.1) in manufacturing systems: facility layout, advanced CNC equipment, robotics, FMS, and CIM. The remaining sections will discuss the principles in production planning and control and how these influence the overall efficiency of production systems.

First, one can ask the question: What is efficient manufacturing? No single answer can be given to this question. Efficiency will depend on the enterprise in question. Generally speaking, efficiency can be defined as the ratio between the value of the produced output and the value of the organizational resources needed in the manufacturing activities. Being an efficient manufacturer is not just a question of how many parts you can produce in a given time. It is just as much a question of how to enhance the value of the products to the customer and meet the needs of the customer. The value qualities of a product will differ from product to product. Normally, they will be related to properties of the physical product as well as the services provided before, during, and after the sale. Some of the value qualities are:

Product price
Product reliability
Manufacturing quality
Product design
Product lifetime
Maintenance cost
Product customization
Speed of delivery
Delivery reliability
Volume flexibility

Looking only at the production activities transforming raw materials into value-added products, the efficiency of these activities can be labeled *productivity*, which can be defined as:

$$\text{Productivity} = \frac{\text{value of output}}{\text{value of input}}$$

The definition of productivity can be applied on different levels: a single manufacturing process or workstation (e.g., a machining center), a department within the production system, or the whole production system. In Western industry the attitude toward productivity has throughout this century been dominated by the management philosophy called *scientific management*. The father of scientific management was Frederick W. Taylor (1856–1915), who was an engineer at the Bethlehem Steel Company. His goal was to increase the productivity of workers by analyzing the work and finding the "*one best way*" to perform a task. The best illustration of Taylor's scientific method lies in a description of how he analyzed the jobs of employees whose sole responsibility was shoveling materials at the Bethlehem Steel Company [58]. He increased the workers' productivity by matching shovel size with such factors as materials, men, height, and distance to throw. In three years the total number of shovelers needed was reduced from 600 to 140.

These principles of scientific management and division of labor were developed further by Henry Ford by the introduction of the automobile assembly line. This innovation was the breakthrough for industrialism, compared to traditional tradesman-oriented production. New standards for production volume, productivity, and uniform product quality were established. The way of thinking founded by Frederick W. Taylor and Henry Ford has throughout the century been—and in many places still is—a dominating production management philosophy in the Western world. Today, the market as well as society have changed. The basic needs for material goods are fulfilled in the industrial world, and customers are demanding new product variants, customized products, quick and on-time delivery, zero defects, and so on. This requires a high level of flexibility in the manufacturing system which traditional mass production has difficulties in matching. The work force is now well educated and is not satisfied with the high decomposition and scientific specification of work tasks. They need job satisfaction and substance in the work.

The Japanese production philosophy, which also is dominant in other Asian countries, has been successful in fulfilling many of these new demands by implementing new ways of managing, organizing, and controlling manufacturing activities. Most of the old Western enterprises were taken by surprise in the 1970s and 1980s, and they have had substantial difficulty in changing and becoming competitive with the newer Asian manufacturing enterprises. This is most significant within the automotive and consumer electronic industries. A succeeding section will discuss the principles of the Japanese production philosophy.

13.6 PRODUCTION PLANNING AND CONTROL

13.6.1 Production Planning by MRP

Because there are so many different and complex products, the planning of purchasing and production becomes a huge task that no human can handle without assistance from production planning and control (PPC) systems. Take, for instance, an automobile. It consists of thousands of components—some manufactured in-house, other purchased from suppliers. Since the early 1970s the material requirements planning (MRP) type of large-scale production planning and control system has been the most widely used. The definitive textbook on the MRP technique is undoubtedly Joseph Orlicky's 1975 publication [60]. Orlicky realized a computer-based technique (MRP) for effective management of manufacturing inventories at a detailed level.

The MRP technique is built around a bill of materials (BOM). Figure 13.18 illustrates the product structure of a lamp as seen in a BOM. At the top level (level 0) we have the end product, and its immediate components and/or subassemblies are shown at level 1. Each of these levels is similarly divided into successively lower levels, ending at the fundamental components, which are raw materials and/or purchased parts. A BOM must also include the quantities needed of the fundamental as well as the intermediate components to produce one end product. In Fig. 13.18 the quantities are typed at each branch in the BOM.

The MRP calculation takes the master production schedule giving the requirements for finished products in the coming period of time (e.g., a total pe-

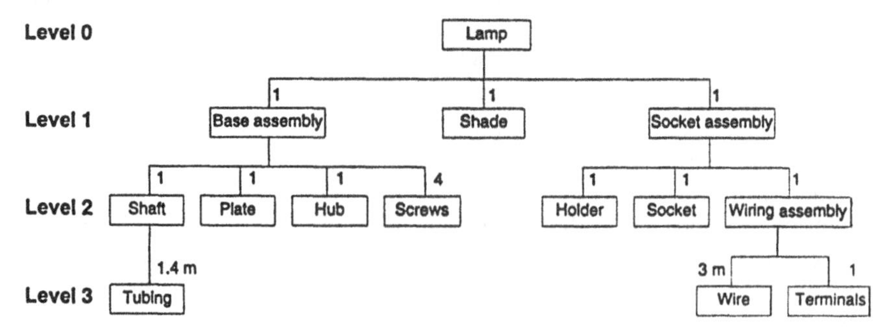

FIGURE 13.18 Simplified Bill of Materials structure of a lamp.

riod of 6 months) and calculates the gross requirements for subassemblies and components in the BOM. When this calculation is carried out on all products manufactured in the company, the gross requirements are added up, providing the total gross requirements for all subassemblies and components each week (or each day). Taking into account the amount materials in stock and the scheduled deliveries of materials from suppliers, the net requirements can be calculated. These net requirements give the amount of materials and components to purchase as well as the production orders to be produced in-house.

The MRP type of production planning follows the "push-principle." The incoming orders and/or forecasts tell the need for finished products. These requirements for finished products are then broken down, by the MRP calculation, to production orders and purchase needs, which are "pushed" down to the departments and suppliers handling the various production phases. The coordination of shop-floor activities is centralized in the production planning department and the periodic MRP calculation. This is in direct contrast to the principle often applied in JIT (just-in-time) production.

13.6.2 Production Planning in JIT

While almost all large manufacturing companies were applying and implementing push systems modeled on MRP, Toyota Motor Corporation developed its own production management system—a pull system. The so-called Toyota production system was developed in the years following World War II by Taiichi Ohne, then machining department manager at the Honsha plant. Figure 13.20 illustrates the difference between a pull system and a push system illustrated in Figure 13.19. In a pull system no detailed scheduling is done at any of the intermediate manufacturing stages. The assembly line receives a production schedule that corresponds to actual customer demands. The final assembly line

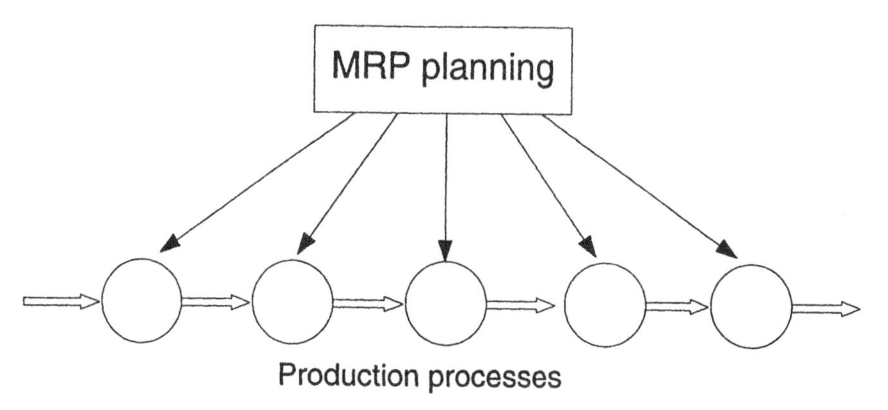

FIGURE 13.19 The push principle.

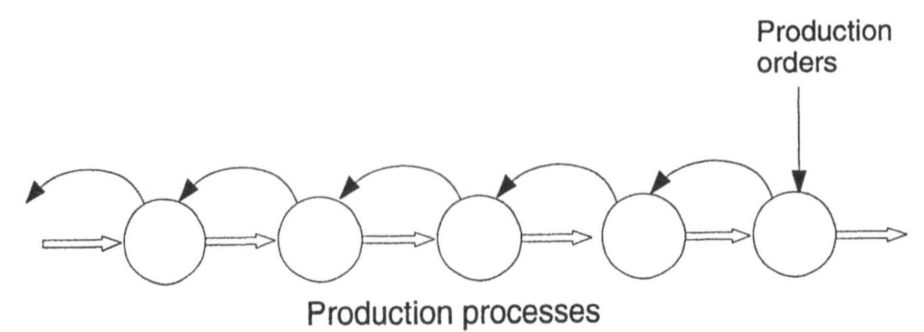

FIGURE 13.20 A pull system.

then pulls parts and subassemblies from the intermediate manufacturing stages based on actual needs. The pull system developed at Toyota Motor Corporation was a two-card Kanban system (Kanban is the Japanese word for card). The basic principle of Kanban is that each manufacturing stage at the assembly line has a minor buffer storage of the components used in the assembly process. These are divided into batches, typically represented by a container, and each batch of components is attached to a Kanban. When components are consumed at the assembly line, the Kanban is sent back to the manufacturing cell or the supplier producing the components. The employees in the manufacturing cell or the supplier must then produce and deliver the quantity of components represented by the incoming Kanbans. When this principle is applied at all manufacturing stages, no detailed centralized requirements planning, as MRP, is needed. The manufacturing system controls itself through the flow of Kanbans.

The Kanban system is very much related to the concept of manufacturing cells but can just as well be used for controlling the supply of components and materials from suppliers. Figure 13.21 illustrates the basic configuration of an assembly line pulling components from manufacturing cells dedicated to production of specific part families. The major advantages of the Kanban system are:

- It is an autonomous system; very little centralized planning is needed.
- It is easy to control the inventories. Reduction of inventory and work-in-progress can be done by withdrawing some of the Kanbans.

Many have investigated the size of the logistics advantages gained by applying JIT and a Kanban production control system. Some typical figures are [61]:

90% reduction of inventories
90% reduction of production lead time
10–30% reduction of the work force
75% reduction of setup times
50% reduction of plant floor needed

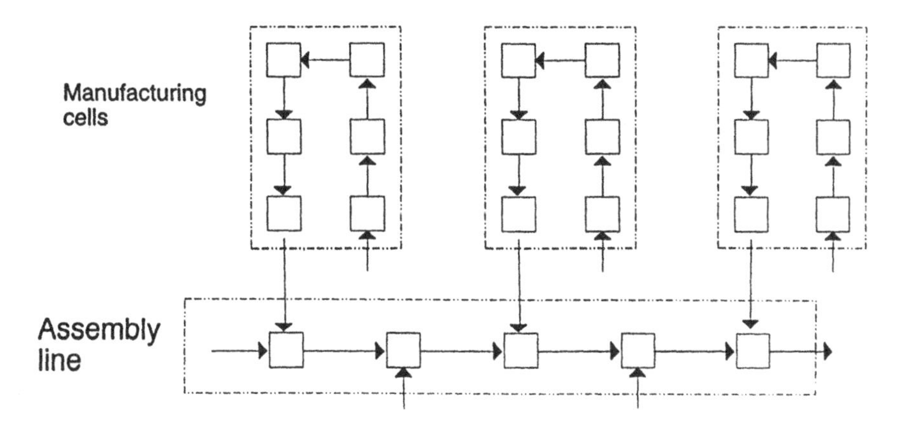

FIGURE 13.21 Cellular production system layout in a JIT manufacturing system.

50–60% reduction of indirect costs
75–90% improvement of quality

There are some preconditions for applying a Kanban system:

Repetitive production: A Kanban pull system requires a mass or repetitive manufacturing environment, as in the automobile industry.

Steady production volume: The high reduction of inventories can only be achieved if production volume is high and does not vary more than around 10%.

Small setup times: If the cells must be able to operate with small batches and react quickly to need at the assembly lines, setup times have to be small.

Uniformity in process and delivery times: Components with, for example, a delivery time of half a year cannot be controlled with Kanban. Such components have to be purchased using forecasts that may reduce the possibility of successful application of Kanban.

These preconditions reduce the number of manufacturing systems that can base their production planning on a Kanban system. Therefore, there will continue to be a need for push-oriented production planning systems such as MRP.

13.7 THE JAPANESE PRODUCTION PHILOSOPHY

The previous section discussed how a Kanban pull system operates in a JIT manufacturing system. Kanban has been a key element in Japanese industrial success and realization of JIT production but other factors have facilitated the success we have seen. Lack of awareness of these factors has been a reason for the failure of many cases of JIT implementation in Western enterprises.

The foundation of Japanese production philosophy is JIT. Basically, JIT is a manufacturing philosophy saying that we must eliminate all waste in all manufacturing activities. With focus on production this can be put down to a set of operational goals:

- Zero defects
- Zero setup time
- Zero inventories
- Zero material transportation
- Zero machine breakdown
- Zero lead time
- Batch sizes of one

Of cause, many of these objectives are unreachable and in direct opposition to each other. But this is one of the important characteristics of Japanese production philosophy—we are never good enough, improvements can always be made. The seven listed items all represent important improvement areas for eliminating waste in production systems. In Japan some tools and techniques for reducing waste in these areas have been developed. In the following discussion we will stress a few of the areas where Japanese enterprises have applied different approaches than Western enterprises.

The Kanban pull system and cellular manufacturing are important tools used to reduce the inventories (and especially work-in-progress), material transportation, lead time, and batch sizes to a minimum. But in purchasing and supplier management, the Japanese enterprises have also broken with paradigms prevailing in Western enterprises, which prefer to have a number of suppliers competing in getting an order. In general, this has been shown to result in big ordering quantities, pure quality, and so on. The Japanese philosophy is to reduce the number of suppliers to one or a few, which are regarded as partners and linked to the company by long-term agreements. Furthermore, suppliers are chosen from nearby, enabling them to deliver small batches just in time for the assembly line. Close cooperation with suppliers is also used in product development. When Toyota Motor Company is developing a new car, the suppliers are involved at a very early stage; for example, when specifications are finished for the brake system, the detailed design of the system is handed over to the supplier, which has the best knowledge regarding its production capabilities and therefore can design the subsystem for manufacturing under the conditions of production at the specific supplier plant.

Quality control is another area where new approaches have been applied. Quality of raw materials and manufacturing processes is essential in JIT manufacturing, with buffers and inventories reduced to a minimum. There is no time for inspection of incoming goods; they are supplied directly to the assembly line. Their quality is a responsibility of the supplier and based on the relationship of trust gained through the long-term collaboration between manufacturer and supplier. The production quality control responsibility is delegated down the organization to the lowest possible level to eliminate the production of huge batches of components with bad quality. The operators are responsible for the quality of parts produced at their workstation, and they are educated and provided the right tools for controlling the parts produced, planning the actions needed, and implementing these corrections without assistance from a supervisor. Furthermore, quality circles are a deeply rooted part of the culture in many Japanese manufacturing enterprises. A quality circle is a group of 5 to 10 employees who meet on a regular basis discussing how to improve manufacturing quality. Japanese companies educate the participants of quality circles in systematic techniques for problem solving that are essential to their achievements.

Many Western enterprises have tried to copy some of the Japanese manufacturing techniques and often the results have not lived up to the expectations. As a result of this many have concluded that the techniques cannot be transferred to Western companies because of differences between Western and Japanese culture. This, however, is no more than a poor excuse for bad implementation and lack of a full understanding of the substance of Japanese manufacturing philosophy. The roles of involvement and education of employees have especially been neglected or underestimated. It took decades for the Japanese to develop and implement their manufacturing philosophy; therefore, we cannot copy and implement it in a rush. Alteration of company culture and old ways of thinking take time.

13.8 CONCLUSIONS

This chapter has discussed many of the technologies and aspects affecting the efficiency of manufacturing systems. Examples of advanced modern production equipment and their application have been presented, including the flexible manufacturing system, which combine the characteristics of high flexibility of the job shop and the productivity of mass production.

Attention has also been drawn to the fact that the quality of production planning and control has a major impact on the efficiency of the manufacturing enterprise in general. The fundamental production planning approaches push and pull (illustrated by the production control techniques MRP and Kanban) have been described.

14
Cleaner Manufacturing

14.1 INTRODUCTION

During the past few years, the issues of environmental protection, safe working conditions, and resource utilization (materials and energy) have caused changes in the demand for industrial products. The implication is that manufacturing companies now have to cope with these concerns as well as those of cost, quality, and schedule that have already been established by global competition. A new approach called life cycle design can help companies meet these two set of demands at the same time.

Life cycle design actually takes concurrent engineering one step further by incorporating the production, distribution, usage, and disposal and recycling issues at an early stage of product design and development. The term *life cycle design* thus means that all the life cycle phases of a product are considered at the conceptual design stage, and the consequences for the environment, working conditions, and resource utilization are assessed in all phases. This means that a product is systematically designed for the life cycle phases so that damage to the environment and workers health is minimized and resource utilization is optimized [65]. In the production phase, materials, processes, systems, and so on, must be applied so that all environmental, occupational health, and resource requirements are fulfilled.

The question now is, "How do we make manufacturing cleaner?" In the following sections, some guidelines that can stimulate cleaner technologies will be discussed; further details on the life cycle design approach can be found in Alting [65].

14.2 CLEANER MANUFACTURING

The term *cleaner manufacturing* implies an attitude of creating production with minimal environmental and occupational health damages. The decision to create cleaner technologies necessitates application of some of the following guidelines.

A manufacturing process uses

Raw materials
Secondary materials (chemicals, water, etc.)
Energy
Information

to produce

Useful products
Waste of raw materials
Waste of secondary materials
Waste of energy
Emissions to air, water, and ground (pollution)

The products are the primary goal which leads to the other secondary "products."

Waste of raw materials means that not all incoming raw material is utilized in the product since—depending on the processes—chips, cutoffs, gates and runners, sheet and plate cutoffs, unacceptable products, and so on, are undesired byproducts. Since production of raw materials requires much energy, pollution of air, water, and ground is the result. Therefore it is very important to utilize raw materials efficiently.

Most processes use some secondary materials such as chemicals, lubricants, degreasing agents, cooling fluids, cooling water, process water (plating), and so on, which create more or less pollution (to air, water, ground) as well as dangerous working conditions.

Many processes have a low energy efficiency, leading to energy loss as heat. Often the process equipment has several support functions that also require energy and here also much energy is wasted.

Most processes produce various emissions: oil and oil mist, vapor from various chemicals, heavy metals in waste water, noise, vibrations, radiation and so on. To establish cleaner manufacturing, it is necessary to focus on the sources producing environmental pollution and unsafe working conditions. It is also necessary to focus on efficient utilization of the resources (materials and energy).

In the following discussion a few internal company programs aiming at creating cleaner manufacturing will be outlined.

14.3 SELECTION OF MANUFACTURING PROCESSES AND MATERIALS

At the design stage many decisions are made that determine what happens in the production phase. Materials are selected, and most often all the main manufacturing processes too. This means that the manufacturing engineers are not given many degrees of freedom to change materials or processes. It is absolutely necessary that the designers and the manufacturing engineers work together and evaluate the consequences of the design decisions. For example, if the designer chooses an epoxy that is dangerous to work with, he has to be aware that operating under safe working conditions will require extra investment. If he chooses a process with poor material utilization, he has to be aware of the costs and the problems in handling waste.

Therefore the manufacturing engineer must take part in material and process selection so that minimal environmental and no occupational health damages are produced and high utilization of the resources is obtained.

As mentioned previously, the life cycle approach at the design stage is of vital importance. When selecting a material, it is necessary to consult databases describing the material properties, for example, RTEC (U.S. National Institute for Occupational Safety and Health Registry of Toxic Effects of Chemical Substances), ELDRIN (Environmental Chemical Data Information Network).

When selecting a process it is important to look at:

Material utilization
Energy efficiency
Equipment and setup
Chemicals, lubricants, etc., to be used and their properties
Emissions (substances—solid/liquid, vapor, waste water, aerosols, etc.)
Exposure of the operator in setup, operation, repair/maintenance
Ergonomy
Risk of accidents
Environmental pollution

If a specific material or process involves occupational health risks and environmental pollution potentials, it is up to the manufacturing engineer to take the necessary measures to protect the operator and the environment. It is the responsibility of the manufacturing engineer to see that no laws and directives are violated.

14.4 WASTE REDUCTION PROGRAM

In manufacturing, much waste is produced and often not enough attention is paid to it. A careful analysis will show that much money can be saved by systematic waste management.

Waste from manufacturing may include

Cutoffs, unacceptable products or components
Chips
Worn-out toolings
Used chemicals, lubricants, coolants, etc.
Waste water
Cooling water
Packaging

In most manufacturing companies waste adds up to a large volume of material. It can be recommended that a waste management organization be formed that

- Makes detailed records of any waste
- Sets up programs jointly with the involved personnel to:
  - Use raw materials better
  - Save expensive chemicals, lubricants, etc.
  - Reduce use of water and minimize polluted waste by recycling, extraction of pollutants by filtration, reverse osmosis, electrolysis, etc.
  - Reduce cooling (utilize the heat) or recirculate cooling water (heat exchanger)
  - Keep waste clean—do not mix chemical wastes since this prevents later reworking of the chemicals

A systematically applied waste management program saves money, prevents pollution, and creates better working conditions.

14.5 ENERGY SAVING PROGRAM

Energy saving must be aimed at all levels where energy is used, such as:

The process equipment
Support functions (hydraulics, compressed air)
Heating
Lighting

Often the process equipment has built-in several functions using energy. Each of these has to be analyzed and evaluated. Idle energy may be too high. Idle energy in hydraulic systems is often spent in pressure reduction valves, excess compressed air, pumps for lubricants, coolants, and so on.

Support functions may involve compressed air, cranes, cooling systems, heating systems, degreasing systems, and so on. Energy analyses are only rarely performed. Heating in buildings can often be reduced by improving doors, thereby reducing cold air circulation, using waste heat, and so on. Light is necessary but it is often on when not needed.

The main message here is that an energy analysis often leads to 30–40% reduction in consumption.

14.6 POLLUTION MINIMIZATION

A program focusing on pollution prevention is of major importance. Every pollution source must be investigated and described. And for each source, an analysis of reduction measures must be performed. The best and most cost-effective reduction is carried out at the source.

Pollution includes emissions sent to air, water, and ground as well as emissions from the waste produced. It is necessary to investigate all activities in the manufacturing facility to account for emissions. The emissions are related to or produced by the raw materials, the processes, the equipment, transportations, storage, and so on. Pollution may be reduced or prevented by proper selection of materials, processes, or procedures.

14.7 BETTER WORKING CONDITIONS

A program aiming at better working conditions is of great importance. Good working conditions lead to higher efficiency. Working conditions include the social and the psychological climate as well as the physical environment. The physical environment must be investigated to identify exposures to various emissions (noise, vibrations, radiation, heat, vapors, aerosols, liquids, dust, material contact, etc.) and the physical loads (ergonomy).

In most countries laws require a company to form a department responsible for working conditions. There is a tendency to form an organization that has responsibility for both environmental and occupational health issues, since they are closely related.

14.8 CONCLUSION

The manufacturing engineer must be aware of the importance of environmental, occupational health, and resource issues. In the future these issues must become a part of company policy as customers, agencies, and authorities enforce this development. It will become a competitive necessity.

15

Notes on Industrial Safety

15.1 INTRODUCTION

In earlier chapters the manufacturing processes have been considered only from a technical point of view. But when it comes to industrial applications, a number of factors, such as economy, energy, and material utilization and safety, must be carefully analyzed before a selection can be made. Energy and material utilization are closely related to economy, whereas the safety aspects, which of course have economic consequences, are of a different nature. Safety aspects ought to be taken into consideration right from the beginning so that the establishment of a safe working place is integrated in the planning and design phases.

In this chapter the term *industrial safety* will be used in connection with the subject of the book and as a discipline in itself with which all engineers should be well acquainted. We will only give a brief introduction here to emphasize that the safety area is of utmost importance in manufacturing; details and solutions can be found elsewhere [39,40].

15.2 INDUSTRIAL SAFETY

The concept of industrial safety has become more and more important in the last few years because of rapid developments within industrial societies, where engineering technology plays a major role in all types of production and material developments. There are two reasons for the increased importance of industrial safety:

1. Society is dependent on industrial enterprise and on technological developments. Society now requires that this enterprise meet certain requirements concerning environment, resources, safety, and so on.
2. Society reacts, because of fear and uncertainty, to consequences of industrial and technological developments. The fear and uncertainty are directed mainly toward the risks faced by people and by the environment.

The safety concept can be considered by asking two questions: (1) safety for whom (what), and (2) safety against what?

It should be realized that complete safety will never exist but that safety must be considered by a degree of probability or frequency of events that may cause undesirable consequences. The consequences can be measured in terms of

Death
Accidents (workdays lost, disability)
Sickness and long-term disability
Damage to the environment
Loss of production (money)
Damage to equipment (money)

For technical purposes we can define risk as expected loss or expected consequence, which means that safety is high when risk is low.

15.3 RISKS IN INDUSTRY AND RISK ANALYSES

Most often, when talking about risk in an industrial context, what is meant is risk for the people involved in the industrial enterprise. This risk or safety is mainly influenced by the technological system itself and the objectives and the working routines for the operation of the system. These two factors determine to what extent safety is dependent on human errors, the understanding of safety requirements, and other factors. A reliable and safe technical system may become unsafe as a result of poor management and operation.

Typical risk factors in industrial systems include:

Overloading of the system (too-high production rates)
Damages due to long-term loading (design errors)
Poor reliability (design errors)
Poor maintenance
Human-machine errors (the machine is not satisfactory ergonomically or a person makes errors)

Insufficient personnel training
Poor management
Unsatisfactory environment

How can a safe industrial situation be established? To obtain the best result, the design and production engineer must carry out risk analyses. Many risk analysis methods exist, but only during recent years have they been applied to industrial production systems.

Two types of risk analysis are important: qualitative analysis and quantitative analysis. Qualitative analysis determines the kinds of failures that can occur, and the combinations of failures leading to larger accidents. Quantitative analysis determines the probability or frequency of failures and can be used to estimate expected losses and to optimize safety systems.

Qualitative analysis is a necessary first step before quantitative analysis. Often, a rough quantitative analysis is required to focus the analyst's attention on the most important part of the production system.

Therefore, it is strongly recommended that risk analysis methods be studied and applied to industrial production systems. This is the only way to cope with these problems systematically.

15.4 GOVERNMENTAL LAWS AND REGULATIONS

It is the responsibility of the engineer to have a fundamental knowledge of federal and state laws and regulations so that he or she can take proper action in planning, design, and management. In manufacturing he or she must have a knowledge of the safety problems in welding, casting, cutting (cutting fluids), and all similar processes. Consequently, it is strongly recommended that engineers, as a part of their education, take courses, study laws, and so on, so that they can operate on a safe basis.

References

1. S. Kalpakjian, *Mechanical Processing of Materials*, D. Van Nostrand, Princeton, N.J., 1967.
2. M. I. Begeman and B. H. Amstead, *Manufacturing Processes*, Wiley, New York, 1969.
3. Various standards and recommendations on materials and materials testing. Danish Standards DS 10010, 10110, 10230, 10410, 10411, 10412, 12012, etc., have been used. The local and international standards should be consulted: for example, ANSI standards, ASTM standards, SAE standards, AISI standards, ISO standards, etc.
4. E. P. DeGarmo, *Materials and Processes in Manufacturing*, Collier-Macmillan, London, 1969.
5. A. Almar-Naess, *Metalliske Materialer* [Metallic Materials], Tapir, Trondheim, Norway, 1969.
6. B. Chalmers, *Physical Metallurgy*, Wiley, New York, 1962.
7. J. B. Moss, *Properties of Engineering Materials*, Butterworths, London, 1971.
8. Brochures and catalogues from steel manufacturers. The major steel manufacturers normally supply data for their steels. Since European and American manufacturers use different steel types and designations, data from the local producers (American) should be obtained.
9. O. Hoff, *Mekanisk Teknologi* [Manufacturing Technology], Teknisk Forlag, Copenhagen, 1964.
10. B. Schifter-Holm, *Introduktion til Plastmaterialer* [Introduction to Plastic Materials], Akademisk Forlag, Copenhagen, 1963.

11. G. Schreyer, *Konstruieren mit Kunststoffen* [Design with Plastics], Carl Hanser Verlag, Munich, 1972.
12. J. M. Alexander and R. C. Brewer, *Manufacturing Properties of Materials*, D. Van Nostrand, London, 1963.
13. T. Wanheim, *Teknologisk Plasticitetslaere* [Technological Theory of Plasticity], I, II, and III, AMT-kompendium M7.13.66, DTH, Denmark, 1966.
14. J. T. Videm, *Sponskjaerende Bearbejdning* [Metal Cutting], Universitetsforlaget, Bergen, Norway, 1970.
15. N. H. Cook, *Manufacturing Analysis*, Addison-Wesley, Reading, Mass., 1966.
16. *Karlebo Handbok* [Mechanical Engineering Handbook], Maskinaktiebolaget Karlebo, Stockholm, 1960.
17. "IVF-Skärdata" [Metal Cutting Data], Institutet för Verkstadsteknisk Forskning, Göteborg, Sweden, 1968. (A continuous series of test results; in the United States, Metcut, Inc., Cincinnati, Ohio, produces the *Machining Data Handbook* based on test results.)
18. "Bearbetningsdata för Boforsstål" [Cutting Data for Bofors Steel], Aktiebolaget Boforsstål, Bofors, Sweden, 1971.
19. *Skärteknik* [Cutting Mechanics], Sveriges Mekanförbund, Stockholm, 1968.
20. W. Charchut, *Spanende Werkzeugmaschinen* [Machine Tools for Cutting], Carl Hanser Verlag, Munich, 1968.
21. P. Ettrup Petersen, *Svejsning* [Welding], Teknisk Forlag, Copenhagen, 1964.
22. P. Ettrup Petersen, *Håndbog i Lysbuesvejsning* [Handbook in Arc-Welding], Teknisk Forlag, Copenhagen, 1964.
23. F. Koenigsberger and J. R. Adair, *Welding Technology*, Macmillan, London, 1965.
24. *ASME Handbook—Metals Engineering: Processes*, McGraw-Hill, New York, 1958.
25. L. Alting, "Eksplosionssvejsning—en ny Metode" [Explosive Welding—A New Method], *Ingeniørens Ugeblad*, No. 11, Copenhagen, 1971.
26. Acker Pohl, *Klebverbindungen* [Adhesive Bonding], Blaue TR-Reihe Heft 44, Verlag Hallwag, Bern and Stuttgart, 1969.
27. Acker Pohl, Limning af Metaller [Adhesive Bonding of Metals], Sveriges Mekan Förbund, Stockholm, 1961.
28. L. Olsson and B. Anulf, *Pulversmidning* [Powder Forging], IVF-resultat 73627, Institutet för Verkstadsteknisk Forskning, Göteborg, Sweden, 1973.
29. H. Fischmeister, "Pulvermetallurgi" [Powder Metallurgy], Lecture notes from Chalmars Technical University, Göteborg, Sweden.
30. *Iron Powder Handbook*, Höganäs AB, Höganäs, Sweden, 1969.
31. *Pulvermetallurgisk Tilverkning* [Design Rules in Powder Metallurgy], Konstruktionsanvisninger, Mekanresultat No. 66004, Sveriges Mekanförbund, Stockholm.
32. V. DeLange Davies, *Stöperiteknik* [Casting Technology], Tapir, Trondheim, Norway, 1970.
33. R. A. Flinn, *Fundamentals of Metalcasting*, Addison-Wesley, Reading, Mass., 1963.
34. K. Karmark Hansen, *Støberiteknik* [Casting Technology], Polyteknisk Forlag, Copenhagen, 1968.
35. E. Storm, *Gjutning* [Casting], Läromedelsförlagen, Teknik och Ekonomi, Stockholm, 1968.

36. V. Aa. Jeppesen, ''En ny Automatisk Formeproces'' [A New Automatic Mold Production Process], *Ingeniøren*, No. 18, Copenhagen, 1962.
37. N. Helner, *Konstruktionsregler for Støbegods* [Design Rules for Castings], Teknisk Forlag, Copenhagen, 1960.
38. R. W. Bolz, *Production Processes*, Industrial Press, New York, 1963.
39. *Industriel Sikkerhet—Risikoanalyse* [Industrial Safety—Risk Analysis], VF 77/9, Verkstedsindustriens Forskningskomité, Oslo, 1977.
40. R. Østvik, *Risikoanalyse—Personsikkerhet* [Risk Analysis—Personal Safety], Statusrapport 1976, SINTEF, STF 18 A 77069, Trondheim, Norway.
41. F. E. Gorczyca, *Application of Metal Cutting*, Industrial Press, New York, 1987.
42. E. M. Trent, *Metal Cutting*, Butterworth-Heinemann Ltd., Oxford, 1991.
43. E. Paul DeGarmo, J. Temple Black, and R. A. Kohser, *Materials and Processes in Manufacturing*, Macmillan, 1988.
44. S. Kalpakjian, *Manufacturing Engineering and Technology*, Addison-Wesley, New York, 1989.
45. J. A. McGeough, *Advanced Methods of Machining*, Chapman & Hall, London, 1988.
46. E. J. Weller (ed.), *Nontraditional Machining Processes*, Society of Manufacturing Engineers, Dearborn, 1984.
47. G. F. Benedict, *Nontraditional Manufacturing Processes*, Marcel Dekker, New York, 1987.
48. N. Mironoff, *Introduction into the Study of Spark-Erosion*, Microtecnic-Scriptar S.A., Lausanne, 1967.
49. N. Mironoff, *La Décharge Électrique dans un Milieu Diélectrique Liquide*, Microtecnic-Scriptar S.A., Lausanne, 1974.
50. A. E. deBarr and D. A. Oliver, *Electrochemical Machining*, MacDonald, London, 1968.
51. A. I. Markov, *Ultrasonic Machining of Intractable Materials*, Iliffe Ltd., London, 1966.
52. F. O. Olsen (Ed.), *Laser Materials Processing*, Technical University of Denmark, 1992.
53. D. Belforte and M. Levitt, *Industrial Laser Annual Handbook*, Penn Well Books, Tulsa, 1989.
54. T. R. Dombrowski, The how and why of abrasive jet machining, *Modern Machine Shop*, February 1983, pp. 76–79.
55. E. Gajdusek, Advances in non-vacuum electron beam technology, *Welding Journal*, July 1980, pp. 51–57.
56. J. P. Kruth (Ed.), Material incress manufacturing by rapid prototyping techniques, *Annals of the CIRP*, Vol. 40/2, 1991, pp. 603–614.
57. Ø. Bjørke (Ed.), *Layer Manufacturing: A Challenge of the Future*, Tapir Publishers, Trondheim, Norway, 1992.
58. F. W. Taylor: *The Principles of Scientific Management*. Harper & Row, New York, 1947.
59. Bauer, A., Browden, Browne, Duggan, and Lyons: *Shop Floor Control Systems: From Design to Implementation*. Chapman & Hall, New York, 1991.

60. J. Orlicky: *Material Requirements Planning: The New Way of Life in Production and Inventory Management.* McGraw-Hill, New York, 1975.
61. W. Gattermeyer (Arthur Andersen & Co., Frankfurt am Main): The "Softside" of JIT/CIM. Proceedings from the 3rd International Conference on Just-In-Time Manufacturing, IFS Publications, 1988.
62. V. C. Jones: *MAP/TOP Networking*. McGraw-Hill, New York, 1988.
63. P. G. Ranky: *Computer Integrated Manufacturing*. Prentice-Hall, Englewood Cliffs, NJ, 1986.
64. H. J. Steudel and P. Desruelle: *Manufacturing in the Nineties*. Van Nostrand Reinhold, New York, 1992.
65. L. Alting, Life Cycle design, *Concurrent Engineering: Issues, Technology and Practice*, Vol. 1, No. 6 (1991).

Problems

CHAPTER 1

1.1 What are the basic processes, primary and secondary, in:
   a. Forging (hot and cold)
   b. Rolling
   c. Powder compaction
   d. Casting
   e. Turning
   f. Electrochemical machining
   g. Electrical discharge machining
   h. Flame cutting

1.2 Describe the energy flow systems for the processes listed in Problem 1.1. The answer should distinguish between (1) energy delivered to the basic process from the tool/die system, and (2) energy delivered to the tool/die system from the process equipment.

1.3 Discuss the energy flow systems for:
   a. Electromagnetic forming (Fig. 1.12a)
   b. Ultrasonic machining (Fig. 1.13)

1.4 Describe the information impressing systems for the processes listed in Problem 1.1 and discuss how the information (impressing) and energy supplies are integrated (Table 1.10).

1.5 Discuss, in general, the possibilities and limitations of the four types of mass-reducing processes (Fig. 1.26) with respect to:
   a. Materials
   b. Tolerances
   c. Geometries
1.6 A sheet metal dome is to be produced in a total forming process (see Fig. P1.6). The die material is solid. Discuss and describe in a systematic manner the possible energy supply systems (Table 1.4).

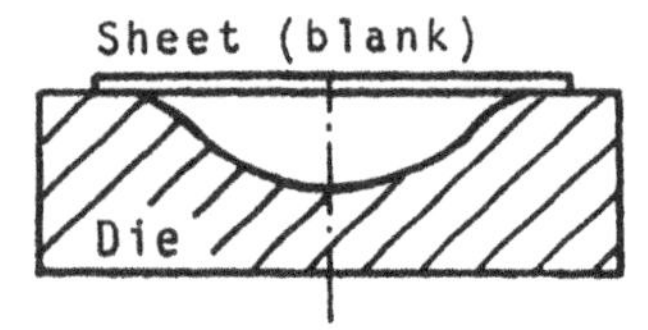

FIGURE P1.6

1.7 Produce illustrations showing each of the principles of converting electrical energy into heat shown in Table 1.5.
1.8 Classify the following processes as either total forming, one-dimensional forming, two-dimensional forming, or free forming.
   a. Drop forging
   b. Rolling
   c. Powder compaction
   d. Sand casting
   e. Turning
   f. Drilling
   g. Electrical discharge machining
   h. Electrochemical machining
   i. Flame cutting
1.9 Discuss the possibilities of using the morphological process model in other manufacturing areas, such as the glass industry or the building industry.

CHAPTER 2

2.1 Based on Fig. 2.4:
   a. Define the engineering stress.
   b. Define the engineering strain.
   c. What happens physically at points A, B, K, C, F, G, and H? Label the stresses and strains at points A, G, and H.
   d. When does permanent straining start?
   e. In which area is Hooke's law valid?

2.2 In a tensile test of stainless steel, the following data were obtained:

Load (N)	Elongation (mm)
7,200	0
11,250	0.508
13,500	2.03
16,200	5.08
18,900	10.16
20,200	15.24
20,700 (max)	21.80
13,200 (fracture)	24.90

The dimensions of the cylindrical specimen were initial length $l_1 = 50.8$ mm and initial area $A_1 = 35.8\ mm^2$. The area at fracture was $A_{min} = 10.0\ mm^2$. Draw the engineering stress–strain diagram.

2.3 Describe the difference between Vickers hardness, Brinell hardness, and Rockwell C hardness. Can you expect the same hardness values from these three methods for a given material?

CHAPTER 3

3.1 Discuss the basic material requirements when forming takes place from the liquid state.

3.2 Discuss the basic material requirements when forming takes place from the solid state for:

a. Mass-conserving processes
b. Mass-reducing processes
c. Joining processes

3.3 Discuss briefly how the processes may change the material properties.

3.4 How can the material properties be changed before or during the process in mass-conserving forming?

3.5 Write a brief essay on the strength-increasing mechanisms for metals.

3.6 Why are steels alloyed?

3.7 Why are nickel and chromium often used together as alloying elements?

3.9 Discuss briefly the difference in properties of amorphous and crystalline plastic materials.

3.10 Write a brief essay on the application of ceramic materials as cutting tools, with particular reference to their properties.

3.11 Describe in general terms those properties of nonferrous metals and alloys which account for their extensive use.

3.12 Describe briefly some typical applications for aluminum/aluminum alloys, copper/copper alloys, magnesium/magnesium alloys, and zinc/zinc alloys.

CHAPTER 4

4.1 Figure P4.1 shows a cylindrical tensile specimen. Consider a cross section inclined at the angle θ to the longitudinal axis:

FIGURE P4.1

a. Show on the figure the directions of the normal force P_n and the shear force P_t.

b. Determine the normal stress σ_θ and the shear stress τ_θ when $D = 10$ mm, $\theta = 60°$, and $P = 15{,}000$ N.

c. Draw a diagram showing σ_θ and τ_θ as a function of θ ($0° < \theta < 90°$).

d. What are the maxima for σ_θ and τ_θ, and what are their directions?

4.2 The component shown in Fig. P4.2 has the main dimensions height h, top width b_1, bottom width b_2, and thickness l. It is subjected to a force P that is uniformly distributed over the end surfaces. Determine the stresses at section A and section B when $P = 28{,}000$ N, $h = 50$ mm, $b_1 = 10$ mm, $b_2 = 30$ mm, $l = 10$ mm, $h_1 = 10$ mm, and $h_2 = 30$ mm.

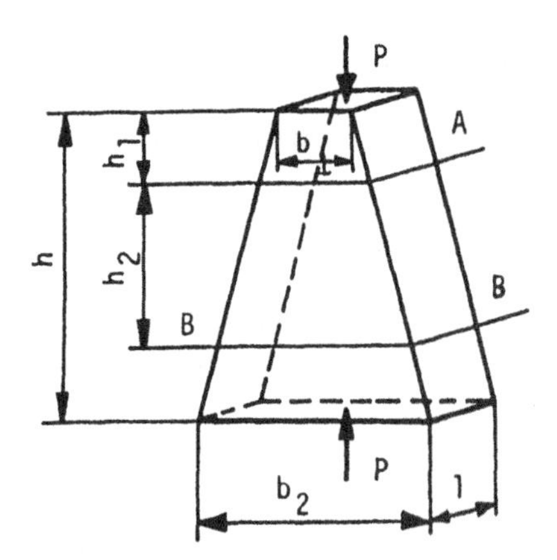

FIGURE P4.2

4.3 A plane stress situation is given by $\sigma_x = -210\ \mathrm{N/mm^2}$, $\sigma_y = -70\ \mathrm{N/mm^2}$, and $\tau_{xy} = 70\ \mathrm{N/mm^2}$. Determine:

a. The magnitudes and directions of the principal stresses

b. The magnitudes and directions of the maximum shear stresses

4.4 A thin-walled tube (outer diameter D and wall thickness t) (see Fig. P4.4) is subjected to a torque T and an axial force P. Determine the magnitudes and directions of:

a. The circumferential stress σ_θ
b. The axial stress σ_a
c. The radial stress σ_r
d. The shear stresses $\tau_{a\theta}$, $\tau_{\theta r}$, and τ_{ra}
e. The principal stresses

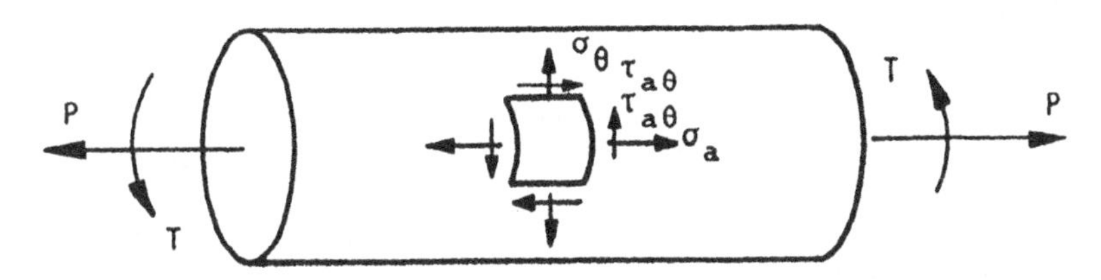

FIGURE P4.4

4.5 How are the nominal and true stresses defined? Derive the expression that relates σ_{nom} and σ.

4.6 How are the nominal and natural (true) strains defined? Derive the expression that relates e and ϵ.

4.7 In Problem 2.2, the results from a tensile test were given. Based on these data, draw a true stress-natural strain diagram (σ, ϵ) and compare this to the nominal stress-strain diagram found in Problem 2.2.

4.8 It is assumed that the σ-ϵ curve in Problem 4.7 can be described by the analytical expression $\bar{\sigma} = c\bar{\epsilon}^n$. Determine c and n.

4.9 A 50-mm cube is subjected to the following loads on the three pairs of faces: $P_1 = -375$ kN, $P_2 = 250$ kN, and $P_3 = x$ kN. The material has a yield stress of 250 N/mm^2. Determine the load P_3 for yielding.

4.10 For the cube in Problem 4.9, determine the uniaxial compression force necessary to start yielding. Determine the force necessary to cause yielding when the two other pairs of faces are subjected to compressive stresses of 150 and 200 N/mm^2.

4.11 A plane stress system is defined by $\sigma_x = 780$ N/mm^2, $\sigma_y = 160$ Nmm^2, and $\tau_{xy} = 160$ N/mm^2. Determine the yield stress based on the yield criteria (a) Tresca, and (b) von Mises, using the assumption that the stress system can initiate yielding.

4.12 A thin-walled closed-end cylinder of thickness t and diameter D is subjected to an internal pressure p and an axial force P. The yield stress of the material is σ_0. Determine, according to both Tresca and von Mises, the expression for the pressure p required to cause yielding for $P = 0$ and $P \neq 0$.

4.13 Estimate the plastic work necessary to stretch a tensile specimen (initial area $A = 35.8$ mm^2 and initial length $l_1 = 50.8$ mm) to instability when it is assumed that the material can be described by $\bar{\sigma} = 1200\bar{\epsilon}^{0.35}$ N/mm^2.

4.14 A thin-walled sphere ($D_1 = 750$ mm, $t = 1$ mm) is expanded by internal pressure to a diameter of $D_2 = 850$ mm. Determine the plastic work necessary. The material can be described by the expression $\bar{\sigma} = 1200\bar{\epsilon}^{0.35}$ N/mm^2.

4.15 A cylindrical specimen $D_1 = 30$ mm, $h_1 = 20$ mm is plastically deformed to $h_2 = \frac{1}{3}h_1$. The material can be described by $\bar{\sigma} = 160\bar{\epsilon}^{0.25}$ N/mm^2. Determine:

a. The true stress at the end of the deformation
b. The nominal stress at the end of the deformation
c. The plastic work necessary

CHAPTER 6

6.1 Write a short essay on the characteristics of hot and cold working.

6.2 Why are the tolerances not as good in hot working as in cold working?

6.3 Why can hot working be carried out to very large compressive deformations (strains) without fracture?

6.4 List the main parameters that influence the deformation process and describe briefly the nature of this influence.

6.5 Discuss the principles in partial and total deformations and illustrate how these can be used to reduce the necessary forces.

6.6 Determine the rolling force necessary to reduce the thickness of an 1800-mm-wide aluminum sheet from 2.5 mm to 2.0 mm. The material follows $\bar{\sigma} = 150\bar{\epsilon}^{0.25}$ N/mm^2 and the radius of the rolls is $R = 250$ mm.

a. What is the change in rolling forces if the reduction is increased from 0.5 mm to 0.75 mm?
b. What is the power necessary per roll (both are driven) when the rotational speed is $n = 50$ rpm?

6.7 The component shown in Fig. P6.7 is produced in a direct or back-extrusion process. Determine the extrusion force necessary for a mild steel material described by $\bar{\sigma} = 650\bar{\epsilon}^{0.22}$ N/mm^2.

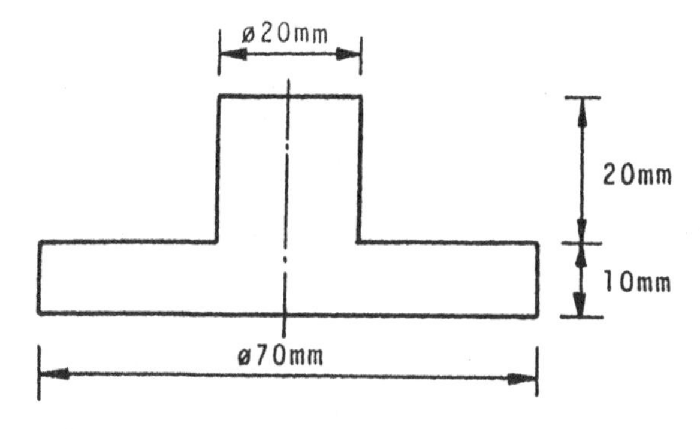

FIGURE P6.7

6.8 A tube with an outside diameter of $D_1 = 60$ mm and a thickness of $t_1 = 10$ mm is reduced to an outside diameter of $D_2 = 50$ mm in a simple drawing process (through a cylindrical die with an internal mandrel; Fig. 6.3). The material is copper with $\bar{\sigma} = 440\bar{\epsilon}^{0.3}$ N/mm^2.

a. What is the drawing force when the final wall thickness is $t_2 = 7$ mm?

b. Determine the maximum possible reduction in thickness.

6.9 The component shown in Fig. P6.9 is produced in stainless steel by expansion over the length L_3 from a tube with an outside diameter $D_1 = 50$ mm. $L_1 = 40$ mm, $L_2 = 15$ mm, $L_3 = 80$ mm, $t = 2$ mm. In a tensile test with a cylindrical specimen of the same material having initial dimensions $l_1 = 56.8$ mm and $A_1 = 37.0$ mm^2, the maximum force was found to be $P_{max} = 21{,}400$ N and the corresponding elongation was $\Delta l = 24.4$ mm.

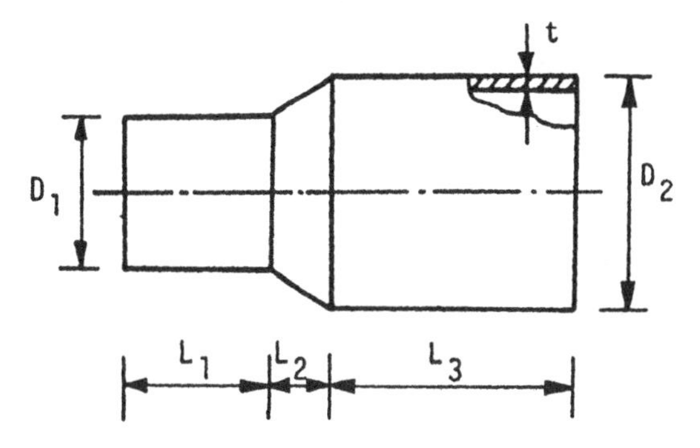

FIGURE P6.9

a. Can the tube be expanded to a diameter of $D_2 = 75$ mm without instability when the axial strain is zero?

b. The expansion can for example be carried out by explosive forming. For $D_2 = 60$ mm determine the plastic work necessary. An approximation must be made for the transition length L_2. The material can be described by $\bar{\sigma} = 1200\bar{\epsilon}^n$ N/mm^2.

6.10 The component shown in Fig. P6.10 is produced in a stretch-forming process ($\Delta b = 0$) from a sheet having the dimensions $2a \times b \times t_1$. The forming limit of the process is instability. Determine the instability strain necessary for the material.

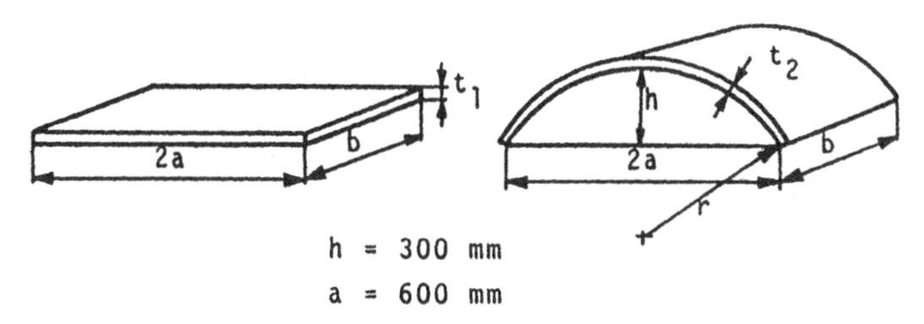

FIGURE P6.10

6.11 The component shown in Fig. P6.11 is produced by extrusion from a cylindrical rod with a diameter of $D = 60$ mm. The material has the stress–strain curve $\bar{\sigma} = 150\bar{\varepsilon}^{0.22}$ N/mm^2. Make a sketch of the required die and determine the extrusion force and the plastic work.

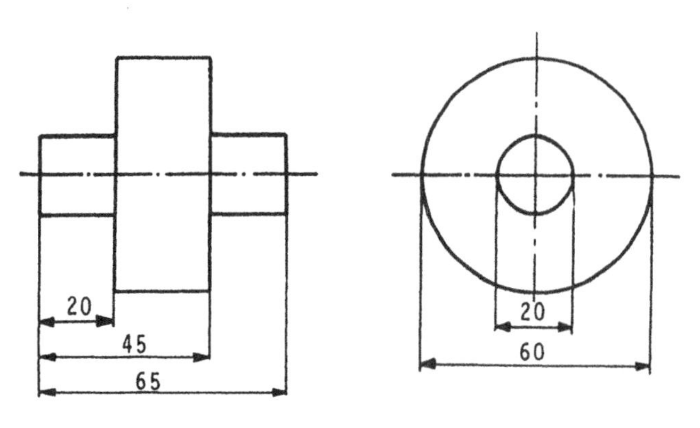

FIGURE P6.11

CHAPTER 7

7.1 Define: cutting speed, feed, chip thickness, chip width, area of cut, and removal rate for turning, drilling, and milling.

7.2 Define the tool terminology: major cutting edge, minor cutting edge, corner radius, tool face, and major and minor flanks. Also, define the angles: major cutting edge angle, minor cutting edge angle, cutting-edge inclination, normal rake, normal wedge angle, normal clearance angle, and included angle.

7.3 What are the three main factors in metal cutting?

7.4 Why is the deformed chip thickness always greater than the undeformed chip thickness ($h_2 > h_1$)?

7.5 Describe the three types of chips and their characteristics. Discuss the problems with a built-up edge on the tool face.

7.6 What are the most important cutting tool materials?

7.7 Define the term *tool life* and discuss the parameters affecting it.

7.8 Discuss how cutting data can be selected.

7.9 Write a brief essay on the parameters affecting the machinability of the work material.

7.10 In turning a cylindrical shaft of steel (0.5%C) the diameter is to be reduced from 100 mm to 88 mm in one pass. Sintered carbide is used as the tool material with $\gamma = +6°$ and the lathe is running at a number of revolutions of $n = 255$ rpm. The power of the lathe is 10 kW and the efficiency in $\eta = 0.75$. Determine the feed so that the power is fully utilized.

7.11 A hole of 25-mm diameter is to be drilled in a steel material (0.5%C). A HSS twist drill with $\gamma = 30°$ is used. The cutting conditions are $f = 0.3$ mm/rev, $v =$ 35 m/min, and $\eta = 0.70$. Determine:
   a. The torque on the drill
   b. The feed force
   c. The power necessary

7.12 Determine the largest possible working engagement (a_e) in milling with a HSS side milling cutter (diameter 320 mm, width 22 mm, and number of teeth $z = 26$). The material is carbon steel ($\sigma_{uts} = 1200$ N/mm^2), the power of the milling machine 6.5 kW, the efficiency 0.75, the cutting speed $v = 45$ m/min, and the feed per tooth $f_z = 0.12$ mm.

7.13 For a plain milling cutter the following data are available: diameter $D = 80$ mm, length $L = 50$ mm, number of teeth $z = 14$, and material HSS. In the milling of a certain material it is found that the average tangential force per tooth is $P_{tl}, =$ 4600 N. The cutting data were $a_e = 12$ mm, $f_z = 0.1$ mm/tooth, $v = 35$ m/min, and $\eta = 0.80$. Determine the specific removal rate (V_s) for the material.

7.14 In an orthogonal planing process for mild steel (0.15%C) at a speed of $v =$ 20 m/min, a power of $N = 7$ kW is used. The cutting tool has a width of $b =$ 12.5 mm, the clearance angle is 5°, and the wedge angle is 75°. The thickness of the undeformed chip is 2.2 mm. Determine the shear angle ø and discuss what this indicates regarding the process efficiency.

7.15 In a turning process a sintered carbide tool is used. The following data apply: clearance angle $\alpha = +6°$, edge angle $\beta = 84°$, major cutting-edge angle $\kappa =$ 60°, minor cutting-edge angle $\kappa' = 15°$, and corner radius $r = 0.25$ mm.
   a. Using the assumption that the feed is larger than the corner radius ($f > r$), find an analytical expression for the roughness height $R_{max} = F(f, r, \kappa, \kappa')$.
   b. In the turning of a steel with $k_s = 2150$ N/mm^2 (at $f = 0.4$ mm/rev, $\gamma = -6°$) the following interrelation between v and T is found: $v = 120$ m/min at $T =$ 30 min and $v = 165$ m/min at $T = 10$ min. Determine the rotational speed n (rpm) for which $T = 45$ min for turning a shaft with a diameter of $D = 90$ mm.
   c. What final diameter can be obtained for this material when the power of the machine of 10 kW is fully utilized ($\eta = 0.80$) and the resulting roughness height R_{max} is 40 μm?

CHAPTER 8

8.1 Describe the three basic principles of joining.

8.2 What are the basic requirements to produce a satisfactory permanent joint using cohesion and/or adhesion?

8.3 Discuss the characteristics of fusion welding, pressure welding and brazing, soldering, and adhesive bonding.

8.4 Write a brief essay on the process characteristics of metal-electrode arc welding, shielded metal-arc welding, submerged-arc welding, gas metal-arc welding (MIG), and gas tungsten-arc welding (TIG).

8.5 Discuss the application characteristics of the welding processes in Problem 8.4.
8.6 Describe the advantages and limitations of electron beam welding and laser welding.
8.7 What are the basic characteristics of resistance welding? Discuss the applicational aspects of spot welding, seam welding, and projection welding.
8.8 What are the typical applications of friction welding?
8.9 Define brazing and soldering, and discuss the joint requirements for these processes.
8.10 Write a brief essay on the applicational advantages and limitations of soldering and brazing.
8.11 Why is joining by adhesive bonding becoming increasingly common?

CHAPTER 9

9.1 Define and describe the four steps normally involved in the production of parts by powder metallurgy.
9.2 What are the characteristics of reduced and atomized powders?
9.3 Give the main factors determining the strength of a sintered component.
9.4 Discuss the four main principles in die design.
9.5 Discuss the main parameters of sintering.
9.6 What is meant by impregnation and infiltration?
9.7 Why has the use of powder metallurgy increased so rapidly in recent years?
9.8 The component shown in Fig. P9.8 is to be produced by powder metallurgy (reduced iron powder).
   a. How can the geometry be changed in minor ways so that it is more suited to the process?

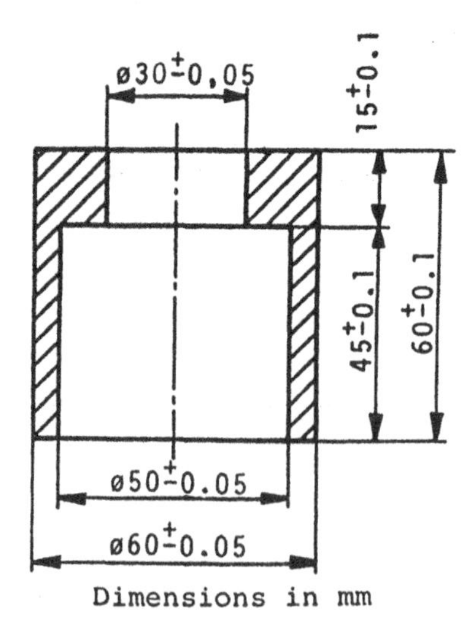

FIGURE P9.8

b. To which treatments must the green compact be subjected?
c. Make a sketch of the die system and sketch the steps from filling to ejection of the finished component.
d. What number of components must be produced to make the process economical?

CHAPTER 10

10.1 Describe the main stages in a casting process.
10.2 What are the most frequently used types of furnaces for melting? Give a brief description of their characteristics.
10.3 Describe the three basic mold production methods.
10.4 Write a brief essay on patterns (categories, types, and materials). Define the allowances necessary.
10.5 Make a few sketches showing typical gating systems for sand casting.
10.6 Based on Tables 10.1 and 10.5, write a brief essay on the application characteristics of the various casting processes.
10.7 Describe hot and cold chamber die casting.
10.8 Why is the industrial application of casting increasing in recent years?
10.9 Discuss how the pattern in casting differs from the part to be cast.
10.10 The clutch part shown in Fig. P10.10 is to be made by sand casting in a batch of 100.
  a. Give a description using sketches of the manufacture of the necessary sand mold. A cross section of the mold and the pattern must be shown.
  b. Suggest design changes which, without changing main dimensions and the functional requirements of the product, make the mold production simpler. Describe the new mold production procedure and the pattern.

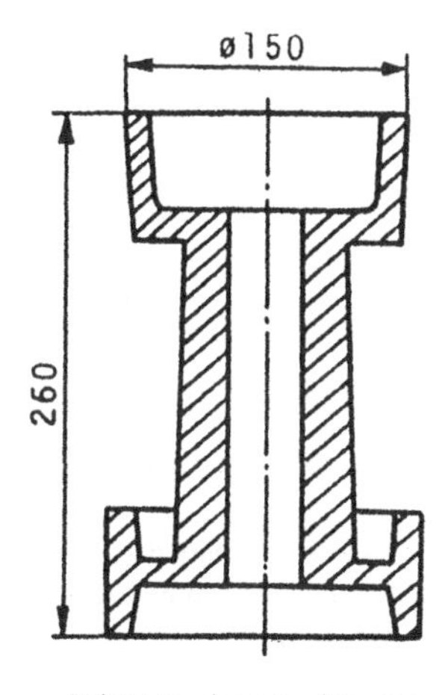

Dimensions in mm

FIGURE P10.10

10.11 The part shown in Fig. P10.11 is to be made by sand casting in a limited number.
   a. Give a description of how the mold can be produced, using sketches.
   b. Describe the gating system and discuss the pattern necessary, including the parting line and allowances.
   c. Which operations are necessary after casting to fulfill the specifications on the drawing?

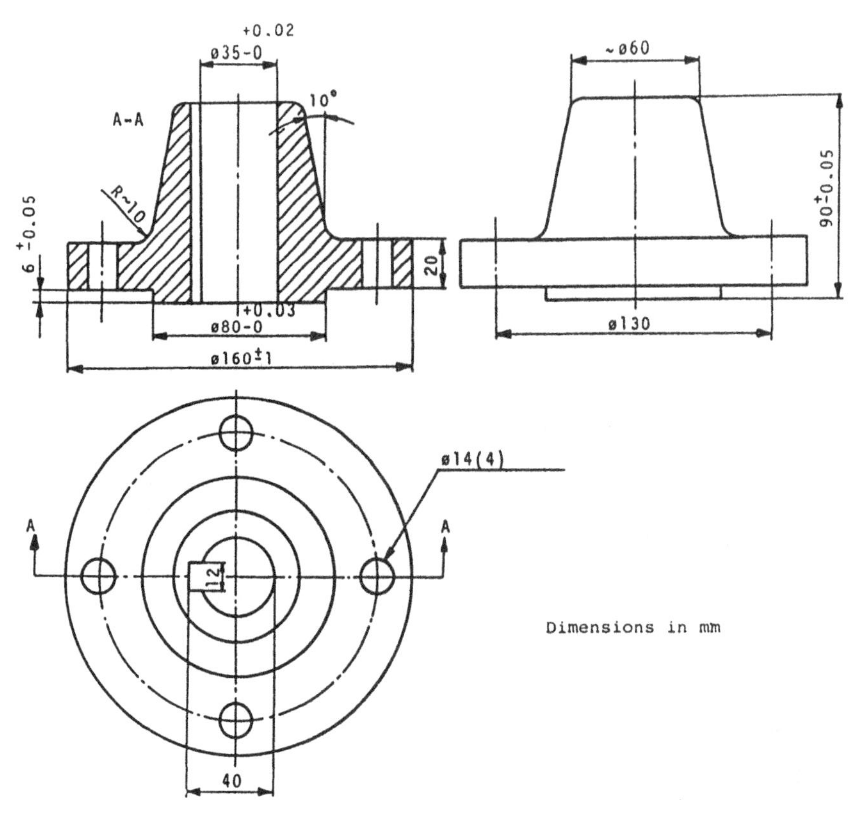

FIGURE P10.11

CHAPTER 11

11.1 Write a brief essay on the manufacturing properties of thermoplastics and thermosetting plastics.

11.2 Discuss the differences and similarities in the casting of metals and plastics.

11.3 Describe typical applications of rotational molding. Which materials can be used, and what are the characteristics of the process?

11.4 Discuss the principles of the closed-mold forming of plastics and list typical applications. Which materials can be used for reinforcement?

11.5 Make a sketch showing the principles of tube and sheet extrusion.
11.6 Describe the characteristics of injection molding. What is the injection pressure?
11.7 Which features make injection molding suitable for mass production?
11.8 What is transfer molding? List a few typical applications.
11.9 Describe typical applications of thermoforming. What are the mold materials, and for which plastic materials can the process be used?

ADDITIONAL PROBLEMS

1. The part shown Fig. 1 is to be produced in numbers of 50,000 per year for 5 years. The material is to be an aluminum ($\bar{\sigma} = 150\,\bar{\epsilon}^{0.18}$ N/mm^2) or a thermoplastic. The processes can be mass conserving or mass reducing.

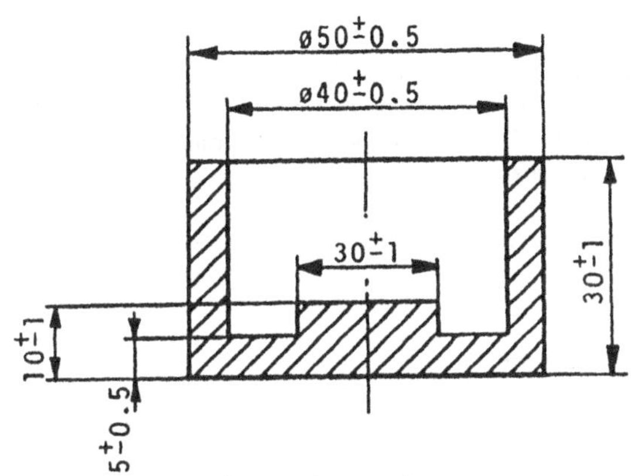

FIGURE 1

   a. Make a systematic survey of the relevant production methods for the two materials. For the actual processes, sketches of tools, dies, patterns, and so on, should be shown. If design changes are necessary, discuss the advantages obtained. (See Chapter 1.)
   b. Which process would you select, and why?
2. The part shown in Fig. 2 is made of steel ($\bar{\sigma} = 650\,\bar{\epsilon}^{0.22}$ N/mm^2).
   a. Make a systematic survey of the relevant production methods (Chapter 1).
   b. Which process will you select for the following production rates:
      (1) 1000 per year?
      (2) 50,000 per year?
      (3) 500,000 per year?
   c. Determine the maximum forces when the part is:
      (1) Extruded
      (2) Cold forged (headed)
   d. Make a sketch of both the extrusion and the forging dies.

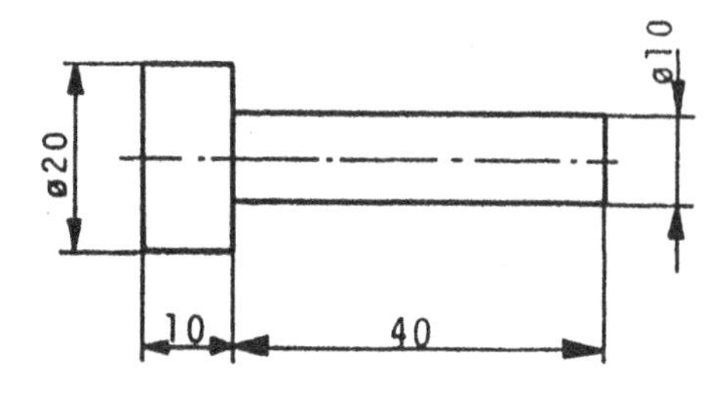

Figure 2

3. The coffee pot shown Fig. 3 is made from stainless steel except for the handle, which is made from a plastic material. The pot may be produced by joining various numbers of components. Make a systematic survey of the relevant production methods for the components and the joining methods.

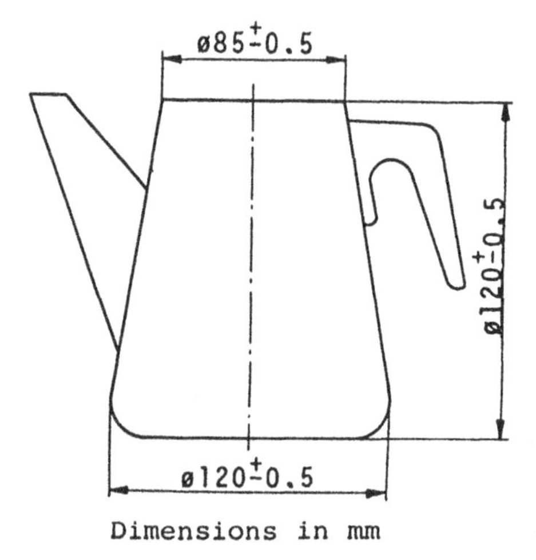

Figure 3

4. The connecting rod shown Fig. 4 is to be produced in large numbers. Make a systematic survey of the relevant production methods.
5. The part shown in Fig. 5 is produced from steel or gray cast iron. The production number is 5000 per year. Discuss the relevant production methods, using sketches. If design changes are necessary, describe the advantages obtained.

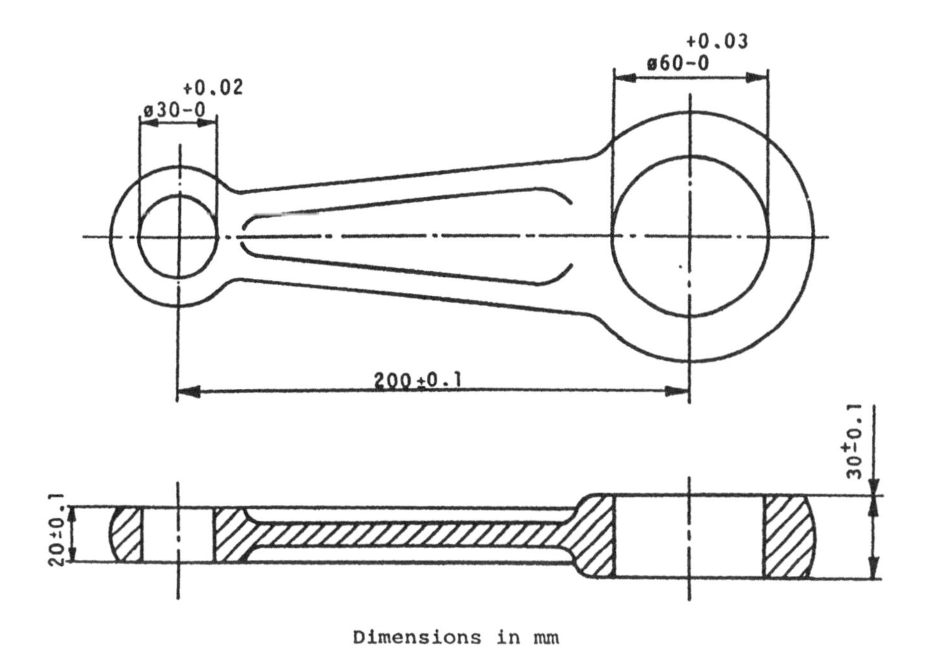

Dimensions in mm

Figure 4

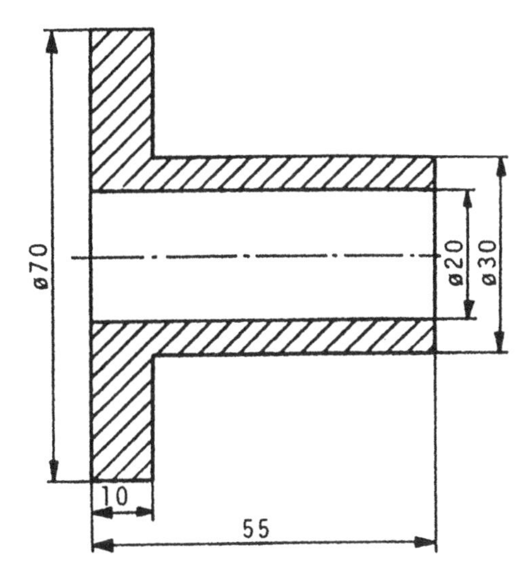

Dimensions in mm

Figure 5

Answers to Selected Problems

Answers are given only for problems having numerical results.

4.1 $\sigma_\theta = 143\ \text{N/mm}^2$, $\tau_\theta = 83\ \text{N/mm}^2$
σ_θ = max for $\theta = 90°$ (and min for $\theta = 0°$)
τ_θ = max for $\theta = 45°$ (and min for $\theta = 0°$ and $90°$)

4.2 $\sigma_{\text{nom},AA} = 200\ \text{N/mm}^2$
$\sigma_{\text{nom},BB} = 108\ \text{N/mm}^2$

4.3 $\sigma_1 = -41\ \text{N/mm}^2$, $\sigma_2 = -239\ \text{N/mm}^2$
$\theta = 3\pi/8,\ 7\pi/8$
$\tau_{\max} = 99\ \text{N/mm}^2$, $\theta_{\tau\max} = 5\pi/8$

4.8 $c = 1200\ \text{N/mm}^2$, $n = 0.36$

4.9 $P_3 = 2.5\ 10^5\ \text{N}$

4.10 $P_{\text{yield},1} = 6.25\ 10^4$ N (compression)
$P_{\text{yield},2} = 1.05\ 10^6$ N (compression)

4.11 Tresca: $\sigma_0 = 820\ \text{N/mm}^2$
von Mises: $\sigma_0 = 770\ \text{N/mm}^2$

4.13 $W_p = 392$ J

4.14 $W_p = 240$ kJ

4.15 $\sigma = -164\ \text{N/mm}^2$ (compression)
$\sigma_{\text{nom}} = 492\ \text{N/mm}^2$ (compression)
$W_\text{p} = 2040$ J

6.6 $P = 2340$ kN
$P = 3200$ kN
$N = 69$ kW
6.7 $P = 6300$ kN (without friction)
6.8 $P = 140$ kN
$t_2 = 4.7$ mm $\Rightarrow$ reduction ratio $r = 57\%$
6.9 $\epsilon_{inst} = n = 0.36$; $\bar{\epsilon} = 0.47 > \epsilon_{inst}$ (which means that the desired expansion cannot be accomplished)
$W_p \simeq 3400$ J
6.10 $\epsilon_{inst} \geq 0.17$
6.11 $P = 3800$ kN
$W_p = 17$ kJ
7.10 $f_{max} = 0.50$ mm/rev
7.11 $M_v = 40$ Nm
$P_A = 5.6$ kN
$N = 2.7$ kW
7.12 $a_e = 20.8$ mm
7.13 $V_s = 20 \text{ cm}^3 / \text{kW·min}$
7.14 $\phi \sim 35°$
7.15

$$R_{max} = \frac{f}{\cot \kappa + \cot \kappa'} + r\left(1 - \frac{\sin \kappa' + \sin \kappa}{\sin (\kappa' + \kappa)}\right)$$

$n = 375$ rpm
$D = 83$ mm
A2 Extruded, $P = 250$ kN. Cold forged, $P = 220$ kN.

Appendix: Unit Conversions

	SI units	ISO-accepted units
Length	m, mm	
Surface	m^2, mm^2	
Volume	m^3	dm^3 = liter (l)
Velocity	m/s	km/h, m/min[a]
Mass	kg, g	
Density	kg/m^3	g/cm^3
Force	N (1 kg m/s^2)	
Moment of force	Nm	
Power	W (1 J/s)	
Pressure, stress	N/m^2 = Pa, N/mm^2 = MPa	bar
Energy, work	J (1 N·m)	kWh

[a]Within material cutting, m/min is accepted by ISO because of the common use of this unit.

m = meter
mm = millimeter
kg = kilogram
g = gram
N = newton
W = watt
Pa = pascal (N/m^2)
J = joule
l = liter
h = hour
s = second

Of the units mentioned, m, kg, and s are basic SI units.

Multiples used are: M = mega = 10^6, k = kilo = 10^3, d = deca = 10^{-1}, c = centi = 10^{-2}, and m = milli = 10^{-3}.

To convert American units to SI units or ISO-accepted units:

	Multiply by:		Multiply by:
Length		*Force*	
in. to m	25.4×10^{-3}	lbf to N	4.448
in. to mm	25.4		
ft to m	0.3048	*Moment of force*	
ft to mm	304.8	lfb·in. to Nm	0.113
		lfb·ft to Nm	1.356
Surface			
in.^{2} to m^2	0.645×10^{-3}	*Power*	
in.^{2} to mm^2	645.16	ft·lbf/s to W	1.356
ft^2 to m^2	92.9×10^{-3}	hp to W	745.7
ft^2 to mm^2	92.9×10^{3}	Btu/h to W	0.2931
		kcal/h to W	1.163
Volume			
in.^{3} to m^3	16.387×10^{-6}	*Pressure, stress*	
in.^{3} to l	16.387×10^{-3}	lbf/in.^{2} to N/m^2	
ft^3 to m^3	28.317×10^{-3}	(Pa)	6.895×10^{3}
ft^3 to l	28.317	lbf/in.^{2} to N/mm^2	
		(MPa)	6.895×10^{-3}
Velocity		atm to N/m^2	
in./s to m/s	25.4×10^{-3}	(Pa)	101.3×10^{3}
ft/s to m/s	0.3048		
in./min to m/min	25.4×10^{-3}	*Energy, work*	
in./min to mm/min	25.4	ft·lbf to J	1.356
		Btu to J	1.055×10^{3}
Mass		ft·lbf to kWh	376.6×10^{-9}
lb to kg	0.4536	Btu to kWh	0.293×10^{-3}
lb to g	453.59	kcal to J	4.1868×10^{3}
slug to kg	14.594	kcal to kWh	1.163×10^{-3}
slug to g	14.5939×10^{3}		
short ton to kg	907.185		
Density			
lb/in.^{3} to kg/m^3	27.6799×10^{3}		
lb/in.^{3} to g/cm^3	27.68		
lb/ft^3 to kg/m^3	16.0185		
lb/ft^3 to g/cm^3	16.02×10^{-3}		

To convert SI units or ISO-accepted units to American units:

	Multiply by:
Length	
m to in.	39.3701
mm to in.	39.37×10^{-3}
m to ft	3.2808
mm to ft	3.281×10^{-3}
Surface	
m^2 to $in.^2$	1.55×10^{3}
mm^2 to $in.^2$	1.55×10^{-3}
m^2 to ft^2	10.76
mm^2 to ft^2	10.76×10^{-6}
Volume	
m^3 to $in.^3$	61.0237×10^{3}
l to $in.^3$	61.02
m^3 to ft^3	35.3147
l to ft^3	35.31×10^{-3}
Velocity	
m/s to in./s	39.3701
m/s to ft/s	3.2808
m/min to in./min	39.3701
mm/min to in./min	39.37×10^{-3}
Mass	
kg to lb	2.2046
g to lb	2.2×10^{-3}
kg to oz	35.274
g to oz	35.274×10^{-3}
kg to slug	68.25×10^{-3}
Desity	
kg/m^3 to $lb/in.^3$	36.13×10^{-6}
kg/m^3 to lb/ft^3	62.43×10^{-3}
g/cm^3 to $lb/in.^3$	36.13×10^{-3}
g/cm^3 to lb/ft^3	62.428

	Multiply by:
Force	
N to lbf	0.2248
Moment of force	
Nm to lbf·in	8.851
Nm to lbt·ft	0.7376
Power	
W to ft·lbf/s	0.7376
W to hp	1.341×10^{-3}
W to Btu/h	3.4118
W to kcal/h	0.8598
Pressure, stress	
N/m^2 (Pa) to $lbf/in.^2$	0.145×10^{-3}
N/mm^2 (MPa) to $lbf/in.^2$	145
N/m^2 (Pa) to atm	9.868×10^{-6}
Energy, work	
J to ft·lbf	0.7376
J to Btu	0.9479×10^{-3}
kWh to ft·lbf	2.655×10^{6}
kWh to Btu	3.4130×10^{-3}
kWh to kcal	859.8

Index

For Product Safety Concerns and Information please contact our EU
representative GPSR@taylorandfrancis.com Taylor & Francis Verlag GmbH,
Kaufingerstraße 24, 80331 München, Germany

Printed and bound by CPI Group (UK) Ltd, Croydon, CR0 4YY
01/05/2025
01858462-0001